Denira Deuling

P9-DNY-242

ELEMENTS OF PHOTOGRAMMETRY

With Air Photo Interpretation and Remote Sensing

Second Edition

Paul R. Wolf, Ph.D.

Professor of Civil and Environmental Engineering
The University of Wisconsin, Madison

McGraw-Hill, Inc.
New York   St. Louis   San Francisco   Auckland   Bogotá
Caracas   Lisbon   London   Madrid   Mexico City   Milan
Montreal   New Delhi   San Juan   Singapore
Sydney   Tokyo   Toronto

This book was set in Times Roman by The Total Book (WG).
The editors were Julienne V. Brown and Kiran Verma;
the production supervisor was Leroy A. Young.
The cover photograph was taken by the Freelance Photographers Guild.

ELEMENTS OF PHOTOGRAMMETRY

Copyright © 1983, 1974 by McGraw-Hill, Inc. All rights reserved. Printed in the United
States of America. Except as permitted under the United States Copyright Act of 1976, no
part of this publication may be reproduced or distributed in any form or by any means, or
stored in a data base or retrieval system, without the prior written permission of the publisher.

10 11 12 13 14 BKMBKM 9 9 8 7 6 5 4

ISBN 0-07-071345-6

Library of Congress Cataloging in Publication Data

Wolf, Paul R.
    Elements of photogrammetry, with air photo
interpretation and remote sensing.

    Includes bibliographies and index.
    1. Photogrammetry. 2. Aerial photogrammetry.
3. Photographic interpretation. 4. Remote sensing.
I. Title.
TR693.W64 1983   526.9'82   82-4700
ISBN 0-07-071345-6

CONTENTS

PREFACE

Since the first edition of this book was published, new technological developments have significantly influenced the practice of photogrammetry. One of the principal objectives in the preparation of this revision has been to update the book by incorporating these new developments. Another equally important objective, however, was to include the many valuable improvements that have been suggested to the author by professors and students who have used the first edition in their classes.

The level, scope of coverage, and style of presentation of this second edition have been maintained equivalent to those of the first edition. The author's intent continues to be the production of a book that will be suitable for an introductory course in photogrammetry at the college level. This includes courses given at universities, junior colleges, colleges of applied arts and technology, and military schools. The book has also been written so that it will be suitable for self-study and reference. Thus, it should continue to be a valuable addition to the libraries of practicing photogrammetrists, cartographers, engineers, foresters, geologists, geographers, landscape architects, and others who use maps and photographs in their work.

Besides the numerous small improvements that have been included in the second edition, many major changes and additions have been introduced. The order of several chapters has been revised so that the subjects of ground control and project planning are presented after stereoplotters, orthophotography, and aerotriangulation. Since the latter three subjects must be considered in ground control and project planning, this arrangement improves the sequence of presentation. The chapter on project planning has been expanded to include a separate new part on cost estimating and scheduling, and an example problem is presented. The chapter on aerotriangulation has been expanded and rearranged into three parts: Part I—analogue methods, Part II—semianalytical methods and Part III—analytical methods. In Part III, a new section has been included on aerotriangulation using image density measurements. The coverage on rectification of tilted photographs has been expanded and includes discussions of graphical, analytical, optical-mechanical, and electro-optical methods. The chapters on stereoplotters and orthophotography have been updated to include the latest equipment, and the discussion of analytical plotters has been significantly expanded in response to the rapid growth in the use of these instruments in practice. The chapter on remote sensing has been thoroughly revised to present the latest developments in this rapidly changing field.

As in the first edition, the material in the second edition has been written using elementary terms as much as possible, and extensive use has been made of illustrations and diagrams. Example problems have frequently been included to clarify computational procedures. Metric and English units are interspersed approximately 50 percent each throughout the book. In order that the book be suitable for students of varying levels of mathematical competence, the body of the text has been presented so that knowledge of only elementary algebra, geometry, and trigonometry is necessary. More challenging mathematical developments have been placed in the appendixes, where they are available either for reference or for students having the necessary mathematics background.

The order of presentation has been arranged so that early chapters establish fundamental principles, while later chapters discuss the more specialized aspects of photogrammetry with emphasis on practical applications. In general, each chapter is arranged so that the more important material is presented first. This makes it convenient to cover only the first parts of certain chapters if course-time limitations will not allow covering the entire book. While the material has been kept as elementary as possible, the depth of coverage is sufficient to make the book suitable for an advanced course in photogrammetry as well. The coverage of the later chapters and the appendixes would be particularly useful in an advanced course. Appendix A deals with random errors and least squares adjustment; Appendix B covers the subject of coordinate transformations; and Appendix C gives the development of the collinearity equations. This material forms the basis of modern computational photogrammetry.

A selected list of references is given at the end of each chapter, and from the materials cited, students may expand their knowledge on particular subjects of interest. To conserve space, only a limited number of key references have been given, but many of these also have excellent bibliographies that will lead the student to numerous additional articles. The number of after-chapter homework problems has been significantly increased, and a solutions manual is available to instructors from the publisher. In addition to providing answers to all numerical problems, this solutions manual contains computer program listings together with examples of their use in solving several of the more time-consuming computational photogrammetry problems.

The author wishes to again acknowledge his sincere appreciation to the many individuals who contributed to the first edition of this book which formed the basis for this edition. For the second edition, acknowledgments of thanks are extended specifically to Professors David Tyler of the University of Maine and Robert Turpin of Texas A & M University who reviewed the entire manuscript; to Professor James Scherz of the University of Wisconsin-Madison who revised Chap. 3 on the Principles of Photography, and who provided other valuable suggestions throughout the book; to Professor Alan Vonderohe of the University of Wisconsin-Madison who reviewed substantial portions of the manuscript; to Professor Ralph Kiefer of the University of Wisconsin-Madison who revised Chap. 19 on Photographic Interpretation; to Professor

Thomas Lillesand of the University of Minnesota who completely revised and updated Chap. 20 on Remote Sensing; to Professor Donald Graff of the University of Wisconsin-Madison who made substantial contributions throughout the book, but specifically in Chap. 14 on Aerotriangulation; to Professors Joseph Ulliman of the University of Idaho, Robert Schultz of Oregon State University, Steven Johnson of Virginia Polytechnic and State University, and Joseph Colcord of the University of Washington who made many valuable suggestions; to Professor Terrence Keating of the University of Maine who provided figures for and reviewed the new sections on Analytic Aerotriangulation Using Digitized Image Densities; to Mr. Randall Olson of the U.S. Geological Survey, Menlo Park, California who reviewed the new sections on Rectification; to Mr. David Smith of David Smith and Associates, Portland, Oregon who provided many valuable suggestions for the new material on Project Planning and Cost Estimating; to Mr. Bon Dewitt, graduate student at the University of Wisconsin-Madison who solved many of the after-chapter problems to ensure that they were workable and reasonable, and who also prepared computer programs for the Solutions Manual; and to the many others, including graduate and undergraduate students who have made numerous valuable contributions to this new edition.

The author also wishes to thank the instrument manufacturers, government mapping agencies and private photogrammetric firms who provided many of the figures and diagrams used in this book. Express thanks are due to Robert Holdridge and others of the Engineering Services Section of the Wisconsin Department of Transportation for their many contributions. Finally, special recognition is expressed to Louise Shafer who labored many hours to transform the author's scroll into a beautifully typed manuscript.

Paul R. Wolf

(Courtesy Gene Bosben)

ONE

INTRODUCTION

1-1 DEFINITION OF PHOTOGRAMMETRY

Photogrammetry is defined by the American Society of Photogrammetry as the art, science, and technology of obtaining reliable information about physical objects and the environment through processes of recording, measuring, and interpreting photographic images and patterns of recorded radiant electromagnetic energy and other phenomena. As implied by its name, the science originally consisted of analyzing photographs. Although photogrammetry has now expanded to include analysis of other records, such as radiated acoustical energy patterns and magnetic phenomena, photographs are still the principal source of information. In this text "photographic" photogrammetry is emphasized, but other sources of information are also discussed.

Included within the definition of photogrammetry are two distinct areas: (1) *metric* photogrammetry and (2) *interpretative* photogrammetry. Metric photogrammetry consists of making precise measurements from photos and other information sources to determine, in general, the relative locations of points. This enables finding distances, angles, areas, volumes, elevations, and the sizes and shapes of objects. The most common application of metric photogrammetry is the preparation of planimetric and topographic maps from photographs. The photographs are most often *aerial* (taken from an airborne vehicle), but *terrestrial* photos (taken from earth-based cameras) are also used.

Interpretative photogrammetry deals principally in recognizing and identifying objects and judging their significance through careful and systematic analysis. It includes branches of *photographic interpretation* and *remote* sensing. Photographic interpretation involves the study of photographic images, while remote sensing, which is a newer branch of interpretative photogrammetry, includes not only the analysis of photography but also the use of data gathered from a wide variety of sensing instruments, including multispectral cameras, infrared sensors, thermal scanners, and side-looking airborne radar. Remote sensing instruments, which are often carried in vehicles as remote as orbiting satellites, are capable of providing quantitative as well as qualitative information about objects. At the present time, with our recognition of the importance of preserving our environment and natural resources, photographic interpretation and remote sensing are both being employed extensively as tools in management and planning.

1-2 HISTORY OF PHOTOGRAMMETRY

Developments leading to the present-day science of photogrammetry occurred long before the invention of photography. As early as 350 B.C. Aristotle had referred to the process of projecting images optically. In the early eighteenth century Dr. Brook Taylor published his treatise on linear perspective, and soon thereafter, J. H. Lambert suggested that the principles of perspective could be used in preparing maps.

The actual practice of photogrammetry could not occur, of course, until a practical photographic process was developed. This occurred in 1839, when Louis Daguerre of Paris announced his direct photographic process. In his process the exposure was made on metal plates which had been light-sensitized with a coating of silver iodide. This is essentially the photographic process in use today.

A year after Daguerre's invention, Arago, a geodesist with the French Academy of Science, demonstrated the use of photographs in topographic surveying. The first actual experiments in using photogrammetry for topographic mapping occurred in 1849 under the direction of Colonel Aimé Laussedat of the French Army Corps of Engineers. Among Colonel Laussedat's experiments were the use of kites and balloons for taking aerial photographs. Due to difficulties encountered in obtaining aerial photographs, he curtailed this area of research and concentrated his efforts on mapping with terrestrial photographs. In 1859 Colonel Laussedat presented an accounting of his successes in mapping using photographs. His pioneering work and dedication to this subject earned him the title "Father of Photogrammetry."

Topographic mapping using photogrammetry was introduced to North America in 1886 by Captain Deville, the Surveyor General of Canada. He found Laussedat's principles extremely convenient for mapping the rugged mountains of western Canada. The U.S. Coast and Geodetic Survey (now the National Geodetic Survey), adopted photogrammetry in 1894 for mapping along the border between Canada and the Alaska Territory.

Meanwhile new developments in instrumentation, including improvements in cameras and films, continued to nurture the growth of photogrammetry. In 1861 a three-color photographic process was developed, and roll film was perfected in 1891. In 1909 Dr. Carl Pulfrich of Germany began to experiment with stereo pairs of photographs. His work formed much of the foundation for the development of many instrumental photogrammetric mapping techniques in use today.

The invention of the airplane by the Wright Brothers in 1902 provided the great impetus for the emergence of modern aerial photogrammetry. Until that time almost all photogrammetric work was, for the lack of a practical means of obtaining aerial photos, limited to terrestrial photography. The airplane was first used in 1913 for obtaining photographs for mapping purposes. Aerial photos were used extensively during World War I, primarily in reconnaissance. In the period between the two World Wars, aerial photogrammetry for topographic mapping progressed to the point of mass production of maps. During this period many private firms and governmental agencies in North America and in Europe became engaged in photogrammetric work.

During World War II photogrammetric techniques were used extensively to meet the great new demand for maps. Air photo interpretation was also employed more widely than ever before in reconnaissance and intelligence. Out of this war-accelerated mapping program came many new developments in instruments and techniques.

Contributions in instrumentation and techniques during the past 35 years have been too numerous to itemize. All the contributions taken collectively, however, have made photogrammetry so accurate, efficient, and advantageous that at the present time, except for mapping small parcels, very little topographic mapping is done by other means. Although topographic mapping constitutes the major use of photogrammetry, it is also being employed in many other fields. Many of these special applications are discussed in Sec. 1-6 and elsewhere throughout this book.

1-3 TYPES OF PHOTOGRAPHS

Two basic classifications of photography used in the science of photogrammetry are *terrestrial* and *aerial*. Terrestrial photographs are taken with ground-based cameras, the position and orientation of which are often measured directly at the time of exposure. A great variety of cameras are used for taking terrestrial photographs and may include anything from simple hobby cameras, which are hand-held, to precise specially designed cameras mounted on tripods. A *phototheodolite,* as shown in Fig. 1-1, is a combination camera and theodolite mounted on a tripod used for taking terrestrial

Figure 1-1 Phototheodolite used for terrestrial photographs. (*Courtesy Wild Heerbrugg Instruments, Inc.*)

photographs. The theodolite facilitates alignment of the camera in a desired or known azimuth and measurement of its position and elevation. Figure 1-2 shows a terrestrial photograph taken with a camera of the type shown in Fig. 1-1.

Another special type of terrestrial camera is the *ballistic camera*, an example of which is shown in Fig. 1-3. These large cameras were mounted at selected ground stations and used to obtain photographs of orbiting artificial satellites against a star background. The photographs were analyzed to calculate satellite trajectories, the size, shape, and gravity of the earth, and the precise positions of the camera stations. This procedure utilized precisely known camera constants, together with the known positions of the background stars at the instant of exposure. Ballistic cameras played an important role in the establishment of a worldwide network of control points and the accurate determination of the relative positions of the continents, remote ocean islands, etc.

Aerial photography is commonly classified as either vertical or oblique. *Vertical photos* are taken with the camera axis directed as nearly vertically as possible. If the camera axis were perfectly vertical when an exposure was made, the photo plane would be parallel to the datum plane and the resulting photograph would be termed *truly vertical*. In practice the camera axis is rarely held perfectly vertical due to unavoidable aircraft tilts. When the camera axis is unintentionally tilted slightly from vertical, the resulting photograph is called a *tilted photograph*. These unintentional tilts are usually less than 1° and seldom more than 3°. For many practical applications, the simple procedures suitable for analyzing truly vertical photos may also be used

Figure 1-2 Terrestrial photograph. (*Courtesy Wild Heerbrugg Instruments, Inc.*)

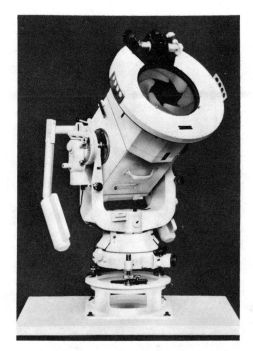

Figure 1-3 Wild BC-4 ballistic camera. (*Courtesy Wild Heerbrugg Instruments, Inc.*)

for tilted photos without serious consequence. Precise photogrammetric instruments and procedures have been developed, however, that make it possible to rigorously account for tilt with no loss of accuracy at all. Figure 1-4 shows an aerial camera with its electrical control mechanism and the mounting framework for placing it in an aircraft. The vertical photograph illustrated in Fig. 1-5 was taken with a camera of the type illustrated in Fig. 1-4 from an altitude of 1,500 ft above the terrain.

Figure 1-4 Aerial camera, model Zeiss RMK 15/23, with electrical controls and aircraft mountings. (*Courtesy Carl Zeiss Oberkochen.*)

Figure 1-5 Vertical aerial photograph. (*Courtesy Carl Zeiss, Oberkochen.*)

Oblique aerial photographs are exposed with the camera axis intentionally tilted away from vertical. A *high oblique* photograph includes the horizon; a *low oblique* does not. Figure 1-6 illustrates the orientation of the camera for vertical, low oblique, and high oblique photography and also shows how a grid of ground lines would appear in each of these types of photographs. Figures 1-7 and 1-8 are examples of low oblique and high oblique photographs, respectively, taken of the same area.

A rather spectacular classification of photographs called *extraterrestrial* has recently emerged as a result of space exploration. These are photos taken from high-altitude spacecraft, and many spectacular ones of the moon and near planets have

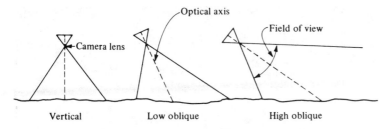

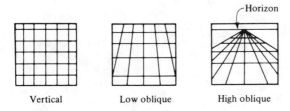

Figure 1-6. Camera orientation for various types of aerial photographs.

been obtained in recent years. Figure 1-9 is a high oblique extraterrestrial photograph taken during the Apollo space program. It shows the moon with the earth rising in the background.

1-4 TAKING VERTICAL AERIAL PHOTOGRAPHS

When an area is covered by vertical aerial photography, the photographs are usually taken along a series of parallel passes called *flight strips*. As illustrated in Fig. 1-10, the photographs are normally exposed in such a way that the area covered by each successive photograph along a flight strip duplicates or overlaps part of the coverage of the previous photo. This lapping along the flight strip is called *end lap,* and the area of coverage common to an adjacent pair of photographs in a flight strip is called the *stereoscopic overlap area.* The pair of photos is called a *stereopair.* For reasons which will be given in subsequent chapters, the amount of end lap is normally between 55 and 65 percent. The positions of the camera at each exposure, e.g., positions 1, 2, 3, etc., of Fig. 1-10, are called the *exposure stations,* and the altitude of the camera at exposure time is called *flying height.*

Adjacent flight strips are photographed so that there is also a lateral overlapping of adjacent strips. This condition, as illustrated in Fig. 1-11, is called *side lap,* and it is normally held at approximately 30 percent. The photographs of two or more side-lapping strips used to cover an area is referred to as a *block* of photos.

Figure 1-7 Low oblique photograph (note that the horizon is not shown). (*Courtesy State of Wisconsin, Department of Transportation.*)

1-5 EXISTING AERIAL PHOTOGRAPHY

Photogrammetrists and photo interpreters can obtain aerial photography in one of two ways: (1) they can purchase photographs from existing coverage, or (2) they can obtain new coverage. It is seldom economical to utilize existing coverage for mapping, because it rarely meets the needs of the user, but it may prove suitable for reconnaissance or for photo interpretation purposes. If existing photography is not satisfactory due to age, scale, camera, etc., it will be necessary to obtain new coverage. Of course, before the decision can be made whether to use existing photography or obtain new, it is necessary to ascertain exactly what coverage exists in a particular area.

Existing aerial photography is available for nearly all of the United States and Canada. Some areas have been covered several times, so that various scales and

Figure 1-8 High oblique photograph of the same area shown in Fig. 1-7 (note that the horizon shows on the photo). (*Courtesy State of Wisconsin, Department of Transportation.*)

qualities of photography are available. Most of the existing coverage is single-lens vertical photography.

The National Cartographic Information Center (NCIC) maintains an Aerial Photographic Summary Records System (APSRS), which provides the most comprehensive information available on aerial photographic coverage for the United States.† This computerized system identifies the agency holding the negatives and gives area of coverage, camera data, scale, date of photography, type of emulsion, and other information for over 100 branches of the federal and state governments and for numerous private firms. Requests for further information and orders for photos must be sent directly to the appropriate organization.

†Requests for APSRS information should be directed to the National Cartographic Information Center, United States Geological Survey, 507 National Center, Reston, Va. 22092.

Figure 1-9 High oblique extraterrestrial photograph of the moon with the earth in the background. (*Courtesy National Space Science Data Center.*)

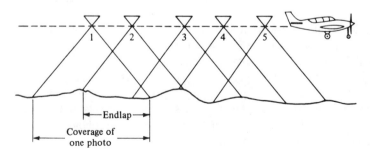

Figure 1-10 End lap of photographs in a flight strip.

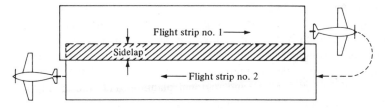

Figure 1-11 Side lap of adjacent flight strips.

1-6 USES OF PHOTOGRAMMETRY

The earliest applications of photogrammetry were in topographic mapping, and today that use is still the most common of photogrammetric activities. At the present time, the U.S. Geological Survey, the federal agency charged with mapping the United States, performs nearly 100 percent of its map compilation photogrammetrically. Besides map compilation, much of the needed ground control is extended from sparse networks of control monuments by photogrammetric means. Fieldwork cannot be eliminated entirely, because basic control surveying and field checking of photogrammetrically compiled maps is necessary. In addition to topographic mapping, many other special-purpose maps are also prepared photogrammetrically. These maps vary in scale from large to small and are used in planning and designing highways, railroads, rapid transit systems, bridges, pipelines, aqueducts, transmission lines, hydroelectric dams, flood-control structures, river and harbor improvements, urban renewal projects, etc.

Photogrammetry has become an exceptionally valuable tool in land surveying. To mention just a few uses in this field, aerial photos can be used as rough base maps for relocating existing property boundaries. If the point of beginning or any corners can be located with respect to ground features that can be identified on the photo, the entire parcel can be plotted on the photo from the property description. All corners can then be located on the photo in relation to identifiable ground features which, when located in the field, greatly assist in finding the actual property corners. Aerial photos can also be used in planning ground surveys. Through stereoscopic viewing, the area can be studied in three dimensions. Access routes to remote areas can be identified, and surveying lines of least resistance through difficult terrain or forests can be found. The photogrammetrist can prepare a map of an area without actually setting foot on the ground—an advantage which circumvents problems of gaining access to private land for ground surveys.

The field of highway planning and design provides an excellent example of how important photogrammetry has become in engineering. In this field, aerial mosaics are prepared to assist in area and corridor studies and to select the best route; small-scale topographic maps are prepared for use in preliminary planning; large-scale topographic maps are compiled for use in final design; and earthwork cross sections are taken to obtain contract quantities. In many cases the plan portion of the "plan-profile" sheets of highway plans are prepared from aerial photographs. Partial payments and even final pay quantities are often calculated from photogrammetric measurements. The use of photogrammetry in highway engineering not only has reduced costs but also has enabled better overall highway designs to be achieved.

Many areas outside of engineering have also benefited from photogrammetry. Nonengineering applications include the preparation of tax maps, soil maps, forest maps, geologic maps, and maps for city and regional planning and zoning. Photogrammetry is used in the fields of astronomy, architecture, archaeology, geomorphology, oceanography, hydrology and water resources, conservation, ecology, and mineralogy. Stereoscopic photography literally enables the outdoors to be brought into the comfortable confines of the laboratory or office for viewing and study in three dimensions.

Photogrammetry has been used successfully in traffic management and in traffic accident investigations. Its use in the latter area has the advantage that photographs overlook nothing that may be needed later to reconstruct the accident, and it is possible to restore normal traffic flow quickly. Even in the fields of medicine and dentistry, measurements from x-ray and other photographs have been useful in diagnosis and treatment. Of course, one of the oldest and still most important uses of aerial photography is in military intelligence. Space exploration is one of the new and exciting areas where photogrammetry is being utilized.

Photogrammetry has become a powerful research tool because it affords the unique advantage of permitting instantaneous recordings of dynamic occurrences on film. Measurements from photographs of quantities such as beam or pavement deflections under impact loads may easily be obtained photographically where such measurements would otherwise be nearly impossible.

It would be difficult to cover all the many situations in which photogrammetric principles and methods could be or have been used in solving measurement problems. Photogrammetry, although still a relatively new science, has already contributed substantially to engineering and nonengineering fields alike. New applications appear to be bounded only by our imagination, and the science should continue to grow in the future.

1-7 PROFESSIONAL PHOTOGRAMMETRY ORGANIZATIONS

There are several professional organizations in the United States serving the interests of photogrammetry. Generally these organizations have as their objectives advancement of knowledge in the field, encouragement of communication among photogrammetrists, and upgrading of standards and ethics in the practice of photogrammetry.

The American Society of Photogrammetry (ASP), founded in 1934, is the foremost professional photogrammetric organization in the United States. One of this society's most valuable contributions has been its publication of the "Manual of Photogrammetry," the "Manual of Photographic Interpretation," the "Manual of Color Aerial Photography," the "Manual of Remote Sensing," and the "Handbook of Non-Topographic Photogrammetry." In preparing these volumes, leading photogrammetrists from government agencies as well as private and commercial firms and educational institutions have authored and coauthored chapters in their various special areas of expertise. The American Society of Photogrammetry also publishes *Photogrammetric Engineering and Remote Sensing,*[1] a monthly journal which brings new developments and applications to the attention of its readers. The society regularly sponsors technical meetings, at various locations throughout the United States, which bring together large numbers of photogrammetrists for the presentation of papers, discussion of new ideas and problems, and first-hand viewing of the latest photogrammetric equipment.

The fields of photogrammetry and surveying are so closely knit that it is difficult to separate them. Both are measurement sciences dealing with the production of maps. The American Congress on Surveying and Mapping (ACSM), although primarily concerned with traditional ground surveying, is also vitally interested in photogrammetry. Founded in 1941, ACSM is regularly cosponsor with ASP of technical meetings. The quarterly journal of ACSM, *Surveying and Mapping,* frequently carries photogrammetry-related articles.

The Surveying and Mapping Division of the American Society of Civil Engineers (ASCE) is also dedicated to surveying and photogrammetry. Articles on photogrammetry are frequently published in its journal, the *Journal of the Surveying and Mapping Division.*

The Canadian Institute of Surveying (CIS) is the foremost professional organization of Canada concerned with photogrammetry. The CIS regularly sponsors technical meetings, and its journal, *The Canadian Surveyor,* carries photogrammetry articles. The *Australian Surveyor* and *Photogrammetric Record* are similar journals with wide circulation, published in English, by professional organizations in Australia and Great Britain, respectively.

The International Society for Photogrammetry and Remote Sensing (ISPRS), founded in 1910, fosters the exchange of ideas and information among photogrammetrists all over the world. Nearly a hundred foreign countries having professional organizations similar to the American Society of Photogrammetry form the membership of ISPRS. This society sponsors international conferences at four-year intervals. Its organization consists of seven technical commissions, each concerned with a specialized area in photogrammetry and remote sensing. Each commission holds periodic symposiums where photogrammetrists gather to hear presented papers on subjects of international interest. The official journal of ISPRS is *Photogrammetria,* which is published in English.

1-8 UNITS OF PHOTOGRAMMETRIC MEASUREMENTS

The solution of photogrammetric problems generally requires some type of length, angle, or area measurement. Length measurements may be in either the English system of inches and feet or the metric system of meters, centimeters, millimeters, and

---

[1]The title of this journal was changed from *Photogrammetric Engineering* to *Photogrammetric Engineering and Remote Sensing* in 1975.

micrometers (microns). Conversion from the English system to the metric system, or vice versa, is frequently necessary. In the United States steps are being taken toward the adoption of the metric system for linear measure. Textbooks should not lag in this regard, and therefore the metric system is used extensively in this book.

Angle measurements in the United States are usually given in the *sexigesimal* system of degrees, minutes, and seconds. Instruments of European manufacture are very common in photogrammetry, however, and these are graduated in the *grad* angle system. It is often necessary to convert back and forth between these two systems. Electronic computers commonly use radian measure for angles, and therefore conversion from degrees or grads to radians, and vice versa, is also frequently necessary.

The following list of length, angle, and area units should be helpful to the student of photogrammetry:

1. Length equivalents
   1 foot (ft) = 12 inches (in)
   1 yard (yd) = 3 ft
   1 rod = $16\frac{1}{2}$ ft
   1 mile (mi) = 5,280 ft
   1 millimeter (mm) = 1,000 micrometers (microns or μm)
   1 centimeter (cm) = 10 mm
   1 meter (m) = 100 cm
   1 kilometer (km) = 1,000 m
   1 in = 2.54 cm†
   1 ft = 304.80 mm
   1 m = 3.2808 ft
   1 m = 39.370 in
   1 km = 0.62137 mi
   1 Gunter's chain = 66 ft
   1 Gunter's chain = 100 links
   1 link = 0.66 ft
2. Angle equivalents
   pi (π) = 3.14159265
   180° = π rad
   1 circle = 360°
   1 circle = 2π rad
   1° = 60'

†In 1959, the United States officially adopted the inch as being equal to exactly 2.54 centimeters, or 1 foot equals 0.3048000 meter. This is now called the *U.S. Standard Foot*. Prior to that time, 1 meter had been accepted as exactly 39.37 inches, or 1 foot equals 0.3048006 meter. The difference between these two standards is only about 1 part in 500,000. Because all surveying done before 1959 was on the earlier standard, it would have been difficult and confusing to change. Thus, this earlier standard, now called the *U.S. Survey Foot* has been officially retained for all surveying measurements. In general, photogrammetric operations are not affected by these two standards because their difference is too minute. Students of surveying should be aware of the two standards, however, because it does take on some significance in precise control surveys. In this book, however, unless otherwise indicated, the U.S. Standard Foot will be assumed.

$1' = 60''$
$1 \text{ rad} = 180°/\pi = 57°17'44.8''$
$1 \text{ rad} = 57.295778°$
$1 \text{ rad} = 206,264.8''$
$1° = 0.01745329 \text{ rad}$
$1 \text{ circle} = 400 \text{ grads}$
$1 \text{ grad} = 100 \text{ centigrads}$
$1 \text{ grad} = 1,000 \text{ milligrads}$
$1 \text{ grad} = 0.9°$
$1 \text{ grad} = 0.01570796 \text{ rad}$
$1 \text{ centigrad} = 0.54'$
$1 \text{ milligrad} = 3.24''$

3. Area equivalents
$1 \text{ acre} = 43,560 \text{ ft}^2$
$1 \text{ acre} = 4,046.9 \text{ m}^2$
$1 \text{ acre} = 10 \text{ square Gunter's chains}$
$640 \text{ acres} = 1 \text{ mi}^2$
$247.1 \text{ acres} = 1 \text{ km}^2$
$1 \text{ hectare (ha)} = 2.471 \text{ acres}$
$1 \text{ acre} = 0.4047 \text{ ha}$

1-9 SIGNIFICANT FIGURES

By definition, measured values have a number of significant figures equal to the number of digits that are *certain*, plus one *estimated* digit. As an example, suppose a distance of 24.37 mm is measured with a scale whose smallest graduations are $\frac{1}{10}$ mm. The first three digits are certain and the fourth (7) is estimated on the scale between the 0.3- and 0.4-mm graduations; thus it is also significant. The value 24.37 therefore contains four significant figures. It is important to record *all* significant figures in measurement. Rounding off or failing to record the last estimated digit is a waste of the extra time spent in obtaining that added accuracy. Conversely, if more than the acutal number of significant figures is recorded, an accuracy is implied which, in fact, does not exist.

Zeros in a recorded value may or may not be significant, depending upon the circumstances. The following rules apply:

1. Zeros to the left of the first nonzero digit serve only to position the decimal point and are not significant.

Examples:

0.003	one significant figure
0.057	two significant figures
0.00281	three significant figures

2. Zeros to the right of nonzero digits which are also to the right of a decimal point are significant.

Examples:

0.10	two significant figures
7.50	three significant figures
483.000	six significant figures

3. Zeros to the right of a nonzero digit but to the left of the decimal point are not significant unless specified by placing a bar over them or by moving the decimal point to the left and expressing the number in powers of ten.

Examples:

380	$= 3.8 \times 10^2$	two significant figures
$38\bar{0}$	$= 3.80 \times 10^2$	three significant figures
$1,60\bar{0}$	$= 1.600 \times 10^3$	four significant figures

In computations using measured values, it is important that answers be given to a number of significant figures consistent with the number of significant figures in the data used to compute them. Specifying an answer to less than the proper number of significant figures does not take advantage of the accuracy achieved in measuring the quantities, and giving the answer to more significant figures than are justified is misleading because it implies accuracy that does not exist.

In adding and subtracting, the calculations are performed without regard to significant figures, but the rightmost significant figure in the final rounded-off answer occurs in the rightmost column filled with significant figures.

Examples:

```
 4.735
 2.05
24.
──────
30.785      answer = 31 (two significant figures governed by 24)
```

```
 1,130
− 83.073
─────────
1,046.927   answer = 1,050 (three significant figures governed by 1,130)
```

In multiplication and division, the number of significant figures in answers is equal to the least number of significant figures of any of the data used in the calculation.

Examples:

$1,738 \times 24 = 41,712$ answer $= 42,000$ (two significant figures governed by 24)
$648.1 \times 0.0523 = 33.89563$ answer $= 33.9$ (three significant figures governed by 0.0523)
$23.985/13 = 1.845$ answer $= 1.8$ (two significant figures governed by 13)

In multiplying or dividing by exact constants, the constants do not govern the number of significant figures in the answer.

Example:

Convert 15.73 ft to inches

Solution:

15.73×12 in/ft $= 188.76$ answer $= 188.8$ (four significant figures governed by 15.73, not 12)

In this text the term "nominal" is frequently used to imply nearness to a given value, e.g., a nominal 6-in-focal-length camera. "Nominal" in this context may be assumed to imply two additional significant figures (a nominal 6-in-focal-length camera lens therefore means a focal length of 6.00 in).

In intermediate computations it is customary to carry one more than the required number of significant figures and then round the final answer to the correct number of significant figures.

REFERENCES

American Society of Photogrammetry: "Handbook of Non-Topographic Photogrammetry," Falls Church, Va., 1979.

————: "Manual of Color Aerial Photography," Falls Church, Va., 1968.

————: "Manual of Photogrammetry," 4th ed., Falls Church, Va., 1980.

————: "Manual of Photographic Interpretation," Falls Church, Va., 1960.

————: "Manual of Remote Sensing," Falls Church, Va., 1975.

Avery, T. E., and D. M. Richter: An Airphoto Index to Physical and Cultural Features in Eastern U.S., *Photogrammetric Engineering,* vol. 31, no. 5, p. 896, 1965.

Brandenberger, A. J.: Surveying and Mapping in the Soviet Union, *Surveying and Mapping*, vol. 35, no. 2, p. 137, 1975.

————: World-Wide Mapping Survey, *Photogrammetric Engineering,* vol. 36, no. 4, p. 355, 1970.

Dix, W. S.: Surveying and Mapping—50 Years of Progress—1928–1978, *Surveying and Mapping*, vol. 38, no. 4, p. 301, 1978.

Doyle, F. J.: Photogrammetry: The Next Two Hundred Years, *Photogrammetric Engineering and Remote Sensing,* vol. 43, no. 5, p. 575, 1977.

Eldrige, W. H.: Photogrammetry for Property Surveying, *Surveying and Mapping,* vol. 27, no. 1, p. 63, 1967.

Eliel, L. T.: One Hundred Years of Photogrammetry, *Photogrammetric Engineering,* vol. 25, no. 3, p. 359, 1959.

Gruner, H.: Photogrammetry 1776–1976, *Photogrammetric Engineering and Remote Sensing,* vol. 43, no. 5, p. 569, 1977.

Landen, D.: History of Photogrammetry in the United States, *Photogrammetric Engineering,* vol. 18, no. 5, p. 854, 1952.

Latham, J. P.: Perspective on Education in Photogrammetry and Remote Sensing, *Photogrammetric Engineering and Remote Sensing,* vol. 43, no. 3, p. 257, 1977.

McCulloch, T.: The CIS and the International World of Surveying and Mapping, *Canadian Surveyor,* vol. 31, no. 4, p. 293, 1977.

Nealey, L. D.: Remote Sensing/Photogrammetry Education in the United States and Canada, *Photogrammetric Engineering and Remote Sensing,* vol. 43, no. 3, p. 259, 1977.

Parker, L.: Highway Plans from Photogrammetrically Compiled Maps, *ASCE Journal of the Surveying and Mapping Division,* vol. 90, no. SU1, p. 31, 1964.

Radlinski, W. A.: Surveying, Mapping, Photogrammetry, and Remote Sensing in Support of National Energy Programs, *Surveying and Mapping,* vol. 37, no. 4, p. 305, 1977.

Richter, D. M.: An Airphoto Index to Physical and Cultural Features in Western U.S., *Photogrammetric Engineering,* vol. 33, no. 12, p. 1402, 1967.

————: Urban Photo Index for Eastern U.S., *Photogrammetric Engineering,* vol. 32, no. 1, p. 54, 1971.

Southard, R. B.: The Changing Scene in Surveying and Mapping, *Photogrammetric Engineering and Remote Sensing,* vol. 46, no. 11, p. 1415, 1980.

Stanton, B. T.: Education in Photogrammetry, *Photogrammetric Engineering,* vol. 32, no. 3, p. 293, 1971.

Stone, K.: World Air Photo Coverage, *Photogrammetric Engineering,* vol. 27, no. 2, p. 214, 1961.

Thompson, M. M.: Surveying and Mapping Research 1978–1988: An Overview, *ASCE Journal of the Surveying and Mapping Division,* vol. 105, no. SU1, p. 43, 1979.

————: USGS Mapping: The Last Three Decades, *Photogrammetric Engineering and Remote Sensing,* vol. 45, no. 12, p. 1607, 1979.

Whitten, C. A.: Metrication for Surveying and Mapping, *ASCE Journal of the Surveying and Mapping Division,* vol. 104, no. SU1, p. 7, 1978.

Williams, O. W.: Outlook on Future Mapping, Charting, and Geodesy Systems, *Photogrammetric Engineering and Remote Sensing,* vol. 46, no. 4, p. 487, 1980.

Wolf, P. R.: Surveying—Current Status and Future Challenges, *Surveying and Mapping,* vol. 36, no. 2, p. 155, 1976.

PROBLEMS

1-1 Explain the differences between metric and interpretive photogrammetry.

1-2 Describe the different classifications of aerial photographs.

1-3 What is the basic difference between high and low oblique aerial photographs?

1-4 What are extraterrestrial photographs?

1-5 What is a phototheodolite?

1-6 Define the following photogrammetric terms: end lap, side lap, stereopair, exposure station, and flying height.

1-7 What is a ballistic camera and what is it used for?

1-8 Discuss some of the principal uses of aerial photogrammetry.

1-9 Discuss some of the principal uses of terrestrial photogrammetry.

1-10 Describe how you would go about obtaining existing aerial photographic coverage of an area.

1-11 To what extent is photogrammetry being used in highway planning and design in your state?

1-12 Convert the following lengths to inches:
  (*a*) 75.28 mm
  (*b*) 152.44 mm
  (*c*) 0.93 m
  (*d*) 37.21 cm

1-13 Convert the following lengths to millimeters:
  (*a*) 5.73 in
  (*b*) 0.85 in
  (*c*) 4.80 ft
  (*d*) 31.29 yards

1-14 Make the following length conversions:
  (*a*) Express 620 m in feet
  (*b*) Express 92 mi in kilometers

(c) Express 8,749 ft in meters

(d) Express 3.751 km in yards

1-15 Convert the following lengths to feet:

(a) 10.52 Gunter's chains

(b) 325.0 m

(c) 160 yards

(d) 7846.2 mm

1-16 Express the following in degrees, minutes, and seconds:

(a) 65 grads

(b) 921.48 grads

(c) 1.023 rad

(d) 0.904 rad

1-17 Convert the following to grads, centigrads, and milligrads:

(a) 42°08′

(b) 39°12′16″

(c) 1.301 rad

(d) 2.426 rad

1-18 Express the following angles in radians:

(a) 147°

(b) 26°53′30″

(c) 297.425 grads

(d) 39.48 grads

1-19 Make the following area conversions:

(a) Express 17.275 acres in square feet.

(b) Express 497.5 acres in hectares.

(c) Express 1.325 mi^2 in acres.

(d) Express 1.000 ft^2 in square centimeters.

1-20 How many significant figures are there in the following numbers?
17.05; 21; 420; 2.07503; 8.00; 19,300; 7,382.080

1-21 Express the following numbers using powers of 10 and retain the indicated number of significant figures: 28,074.2; 32.050; 26,00$\bar{0}$; 0.007130; 19,381,420; 12

1-22 Express the answers to the following problems to the correct number of significant figures:

(a) 38.121 + 170 + 0.0748 + 20.091

(b) 5,982 − 62.083

(c) 382.00 − 0.013 + 24

(d) 48 × 29,075

(e) $(42.5)^2$

(f) 17.593 ÷ 39.40

TWO

OPTICS FOR PHOTOGRAMMETRY

2-1 INTRODUCTION

Practically all photogrammetric instruments depend in some manner upon optical elements for their function. The number and kind of optical elements used varies widely with the type of equipment. Small pocket stereoscopes, for example, use only simple thin lenses; aerial cameras contain highly corrected and expensive compound lenses; and complex stereoscopic plotters often utilize many lenses, mirrors, and prisms.

The science of optics consists of two principal branches: *physical optics* and *geometrical optics*. In physical optics, light is considered to travel through a transmitting medium such as air in a series of electromagnetic waves emanating from a point source. Conceptually this can be visualized as a group of concentric circles expanding or *radiating* away from a light source, as illustrated in Fig. 2-1. In nature a good resemblance of this manner in which light waves propagate can be created by dropping a small pebble into a pool of still water to create waves radiating from the point where the pebble was dropped. As with water, each light wave has its own *frequency, amplitude,* and *wavelength.* Frequency is the number of wavelengths that pass a given point in a unit of time; amplitude is the measure of the height of the crests and depths of the troughs; and wavelength is the distance between any wave and the next succeeding one. The speed with which a wave moves from a light source is called its *velocity.* Velocity is related to frequency and wavelength according to the equation

$$V = f\lambda \qquad (2\text{-}1)$$

In Eq. (2-1), V is velocity, usually expessed in units of feet (or meters) per second; f is frequency, generally given in cycles per second; and λ is wavelength, usually expressed in feet (or meters). Light has an extremely high velocity, moving at the rate of 186,282.4 mi/sec.

In geometrical optics, light is considered to travel from a point source through a transmitting medium in straight lines called *light rays.* As illustrated in Fig. 2-2, an

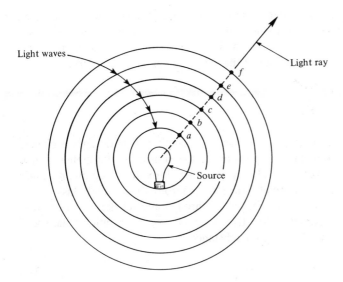

Figure 2-1 Light waves emanating from a point source in accordance with the concept of physical optics.

infinite number of light rays radiate in all directions from any point source. The entire group of radiating lines is called a *bundle of rays*. This concept of radiating light rays develops logically from physical optics if one considers the travel path of any specific point on a light wave as it radiates away from the source. In Fig. 2-1, for example, point *a* radiates to *b, c, d, e, f,* etc., as it travels from the source, thus creating a light ray.

In analyzing and solving photogrammetric problems, rudimentary line diagrams are often necessary. Their preparation generally requires tracing the paths of light rays

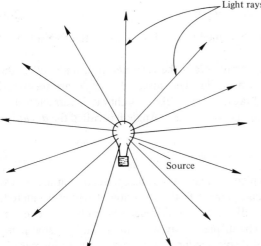

Figure 2-2 Bundle of light rays emanating from a point source in accordance with the concept of geometrical optics.

through air and various optical elements. These same kinds of diagrams are often used as a basis for deriving fundamental photogrammetric equations. For these reasons a basic knowledge of the behavior of light, and especially of *geometrical optics,* is prerequisite to a thorough understanding of the science of photogrammetry.

2-2 REFRACTION OF LIGHT

When light passes from one transmitting material to another, it undergoes a change in velocity in accordance with the composition of the substances through which it travels. Light achieves its maximum velocity traveling through a vacuum, it moves somewhat slower through air, and travels still slower through water and glass.

The measure of the rate at which light travels through any substance is known as the *refractive index* of the material. Refractive index is simply the ratio of the speed of light in a vacuum to its speed through a substance, or

$$n = \frac{c}{V} \tag{2-2}$$

In Eq. (2-2), n is the refractive index of a material, c is the velocity of light in a vacuum, and V is its velocity in the substance. The refractive index for any material is determined through experimental measurement. Typical values for indexes of refraction of common media are vacuum, 1.000; air, 1.0003; water, 1.33; and glass, 1.5 to 1.7.

When light rays pass from one homogeneous transparent medium to a second such medium having a different refractive index, the path of the light ray is bent or *refracted* unless it intersects the second medium normal to the interface. If the intersection occurs obliquely, as shown in Fig. 2-3, then the *angle of incidence* ϕ is related to the *angle of refraction* ϕ' by the law of refraction, frequently called *Snell's law.* This law is stated as follows:

$$n \sin \phi = n' \sin \phi' \tag{2-3}$$

where n is the refractive index of the first medium and n' is the refractive index of the second medium.

In Fig. 2-3, *IA* is the incident light ray, *AR* is the refracted ray, and *NN'* is the normal to the interface between the two media. The angles ϕ and ϕ' are measured from *NN'* to the incident and refracted rays, respectively. A light ray is refracted such that the incident and refracted rays lie in the same plane. Angle θ is called the *deviation angle* and is given by

$$\theta = \phi - \phi' \tag{2-4}$$

If a light ray is directed toward a piece of glass which has perfectly parallel sides, the light ray will emerge from the glass parallel to the direction at which it entered. If the light ray strikes the glass obliquely, it will be displaced laterally, however, as shown by the distance h in Fig. 2-4. The displaced ray will emerge in the same plane as the incident ray, and the amount of displacement will increase with an increase in

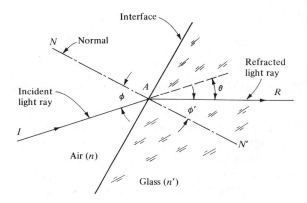

Figure 2-3 Refraction of light rays.

the angle of incidence and an increase in the thickness of the glass. The following equation in terms of the thickness t, the refractive indexes n and n', and the angles ϕ and ϕ' may be used to calculate this lateral displacement:

$$h = t \sin \phi \left(1 - \frac{n \cos \phi}{n' \cos \phi'} \right) \tag{2-5}$$

Example 2-1 Light is directed through air toward a $\frac{1}{4}$-in-thick parallel-sided glass plate at a 60° angle of incidence. What will be the lateral displacement of the light ray as it again emerges into air on the other side of the plate? (Assume n of air $= 1.000$ and n' of crown glass to be 1.520.)

From Eq. (2-3),

$$\sin \phi' = \frac{1.000}{1.520} (0.8660) = 0.5698 \qquad \text{from which } \phi' = 34°44'$$

From Eq. (2-5),

$$h = 0.25(0.8660) \left[1 - \frac{1.000(0.5000)}{1.520(0.8218)} \right] = 0.13 \text{ in}$$

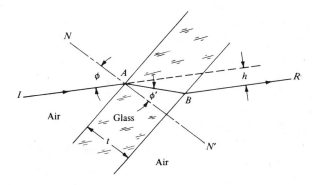

Figure 2-4 Refraction through a parallel-sided glass plate.

2-3 REFLECTION OF LIGHT

When a light ray strikes a smooth surface such as highly polished metal, it is reflected so that the *angle of reflection* ϕ'' is equal to its incidence angle ϕ, as shown in Fig. 2-5. Both angles lie in a common plane and are measured from NN', the normal to the reflecting surface.

Most surfaces partially refract light rays and partially reflect them, as shown in Fig. 2-5. The greater the angle of incidence, the greater the reflected portion. For a light ray traveling from a medium of higher refractive index to a medium of lower refractive index (e.g., from glass to air), there is a particular angle of incidence where the angle of refraction is exactly 90°. The angle of incidence at which this condition first occurs is called the *critical angle* ϕ_c. If the refractive indexes of the two media are known, the critical angle can be calculated using the following modification of Snell's law in which the sine of the 90° refraction angle is 1.000:

$$\sin \phi_c = \frac{n'}{n} \tag{2-6}$$

In Eq. (2-6), ϕ_c is the critical angle of incidence and n and n' are as previously defined.

Example 2-2 Find the critical angle for a light ray traveling from crown glass ($n = 1.520$) to air ($n' = 1.000$).

By Eq. (2-6),

$$\sin \phi_c = \frac{1.000}{1.520} = 0.6579 \qquad \text{from which } \phi_c = 41°08'$$

For angles of incidence greater than the critical angle, there will be total reflection and therefore no light loss due to refraction. As will be explained later, this important factor is the underlying basis for the use of prisms in optical systems.

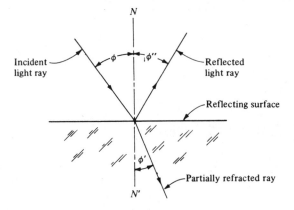

Figure 2-5 Reflection of light rays.

2-4 PLANE MIRRORS

Plane mirrors used for nonscientific purposes generally consist of a plane sheet of glass with a thin reflective coating of silver on the back. The silver coating is usually covered with paint for protection. This type of "back surfaced" mirror is optically undesirable, however, because it reflects rays which interfere with the primary reflected light ray, as shown in Fig. 2-6. These undesirable reflections may be avoided by using *first-surface* mirrors, those which have their silver coating on the front of the glass, as shown in Fig. 2-7. Although mirrors may be made optically acceptable in this manner, they are easily tarnished by fingerprints, smoke, etc., and are not normally used if they can be replaced by prisms.

In certain photogrammetric instruments it is possible to simultaneously view a reflected image and the image of an object which lies behind the reflecting mirror. This may be accomplished through the use of *half-silvered* mirrors. These mirrors consist of a piece of glass surfaced with a light coating of silver such that half the intensity of a light ray carrying an image passes through the mirror and half is reflected. In Fig. 2-8, the light ray from A' is reflected by the mirror and the light ray from A passes through the mirror. Thus a half-silvered mirror enables two different objects to be viewed in superposition at the same time. As described in Chap. 9, this is useful for transferring detail from a photo to a map by direct tracing, e.g., tracing planimetric detail on a map at A' from an aerial photo situated at A. This principle is called *camera lucida*.

2-5 PRISMS

For angles of incidence greater than the critical angle, no refraction occurs, but rather, every light ray undergoes total reflection. This principle is illustrated with the *right-*

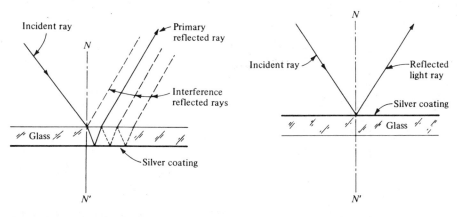

Figure 2-6 Back-surfaced mirror.   **Figure 2-7** First-surface mirror.

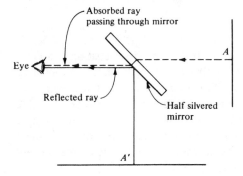

Absorbed ray
passing through mirror

A

Eye

Reflected ray

Half silvered
mirror

A'

Figure 2-8 Use of half-silvered mirror.

angle prism of Fig. 2-9. Light ray *IA* enters the prism normal to one of the faces, undergoes total reflection at *A* (ϕ is 45°, which is greater than $ϕ_c$), and exits along path *AR*. An advantage of using prisms rather than first-surface mirrors is that total reflection may be achieved without the use of silver coatings and therefore no surfaces are subject to tarnishing. Consequently prisms are frequently used as reflecting devices in photogrammetric instruments.

Some common types of prisms are illustrated in Fig. 2-10*a* through *d*. The *right-angle* type of Fig. 2-10*a*, described above, is the most common prism. Its function is to deflect light rays 90°. Figure 2-10*b* illustrates the *porro* prism, a right-angle prism oriented so that reflected light rays are deflected 180° from their incident directions. Note that both prisms *a* and *b* invert their images. The *dove* prism of Fig. 2-10*c* rotates the image 180° about the longitudinal axis of the prism but does not change the direction of the ray paths. The *pentaprism* of Fig. 2-10*d* deflects light rays 90° regardless of the orientation of the prism. All these prisms are commonly used in photogrammetric instruments to change either the direction of a ray path or the orientation of an image.

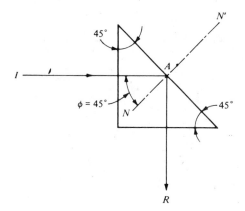

N'

45°

A

I — I

ϕ = 45°

N

45°

R

Figure 2-9 Total reflection through a right-angle prism (ϕ greather than $ϕ_c$).

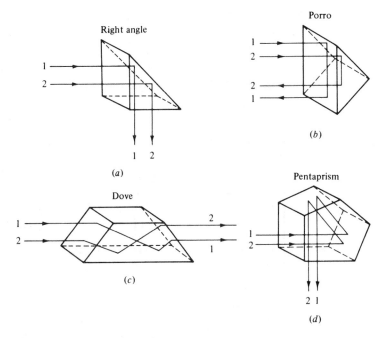

Figure 2-10 Common types of prisms.

2-6 SIMPLE THIN LENSES

A simple thin lens consists of a piece of optical glass which has been ground so that it has either two spherical surfaces or one spherical surface and one flat surface. By definition, the thickness of a thin lens is considered negligible. Various types of thin lenses are shown in Fig. 2-11*a* through *f*. Incident light rays passing through the *converging* (positive) lenses *a* through *c* are refracted toward each other according to Snell's law. The *diverging* (negative) lenses *d* through *f* refract incident light rays in a diverging manner as illustrated.

The primary function of a lens is to gather light rays from object points and bring them to focus at some distance on the opposite side of the lens. A lens accomplishes this function through the principles of refraction. The simplest and most primitive device which performs the functions of a lens is a tiny pinhole which theoretically allows a single light ray from each object point to pass. The tiny hole of diameter d_1 of the pinhole camera illustrated in Fig. 2-12 produces an inverted image of the object. The image is theoretically in focus regardless of the distance from the pinhole to the camera's image plane. Pinholes allow so little light to pass, however, that they are extremely slow and unsuitable for photogrammetric work. For practical reasons they are replaced by glass lenses.

The advantage of a lens over a pinhole is the increased amount of light that is allowed to pass. A lens gathers an entire *pencil of rays* from each object point instead

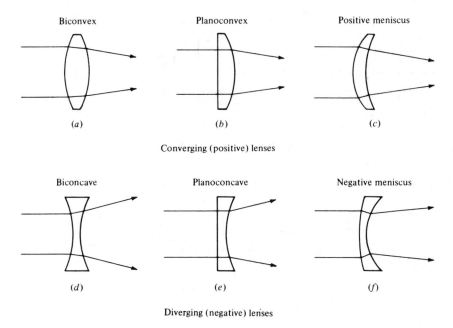

Biconvex   Planoconvex   Positive meniscus

(a)   (b)   (c)

Converging (positive) lenses

Biconcave   Planoconcave   Negative meniscus

(d)   (e)   (f)

Diverging (negative) lenses

Figure 2-11 Various types of thin lenses.

of only a single ray. As discussed earlier and illustrated in Fig. 2-2, when an object is illuminated, each point in the object reflects a bundle of light rays. This condition is also illustrated in Fig. 2-13. A lens placed in front of the object gathers a pencil of light rays from each point's bundle of rays and brings these rays to focus at a point in a plane on the other side of the lens called the *image plane*. An infinite number of image points, focused in the image plane, forms the image of the entire object. Note from Fig. 2-13 that the image is inverted by the simple thin lens.

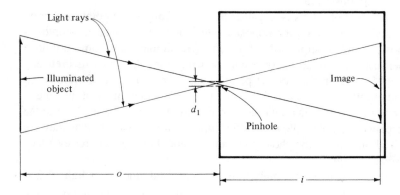

Light rays

Illuminated object

d_1

Pinhole

Image

o

i

Figure 2-12 Pinhole lens of the pinhole camera.

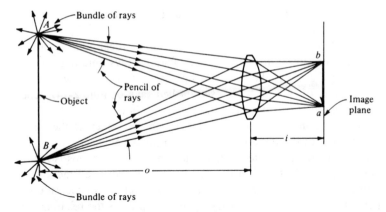

Figure 2-13 Pencils of light rays and image formation.

The *optical axis* of a thin lens is defined as the line joining the centers of curvature of the spherical surfaces of the lens. In Fig. 2-14, O_1 and O_2 are the radius points of the two spherical surfaces of the lens, and R_1 and R_2 are the radii of the surfaces. The optical axis is the line O_1O_2. For a lens having one plane surface, the optical axis is defined as the perpendicular to the plane surface passing through the one center of curvature.

Light rays that are parallel to the optical axis as they enter a lens come to focus at F, the *focal point* of the lens. This is illustrated in Fig. 2-14. As a corollary to that description, the focal point of a lens is the point on the optical axis having the property that any light ray passing through it as it enters a lens travels parallel to the optical axis of the lens after being refracted by the lens. Actually any thin lens has two focal points, one on either side of the lens at equal distances from the lens. If light rays parallel to the optical axis were incident to the lens of Fig. 2-14 from the right, they

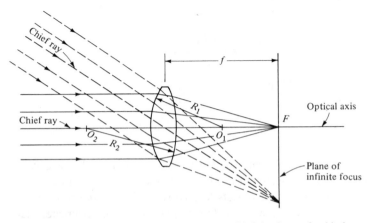

Figure 2-14 Optical axis, focal length, and plane of infinite focus of a thin lens.

would come to focus at a point on the optical axis the same distance to the left of the lens that F is to the right of the lens.

A plane perpendicular to the optical axis passing through the focal point is called the *plane of infinite focus,* or simply the *focal plane.* Parallel rays entering a converging lens, regardless of the angle they make with the optical axis, are ideally brought to focus in the plane of infinite focus (see dashed rays of Fig. 2-14). The point of focus occurs where the *chief ray* strikes the plane of infinite focus. The chief ray is that light ray passing undeviated in direction through the center of the lens. Note that lateral displacement of the ray due to lens thickness is neglected.

The distance from the plane of infinite focus to the center of a thin lens is f, the *focal length* of the lens. This is illustrated in Fig. 2-14. Focal length is a function of the index of refraction of the glass from which the lens is ground and of the radii of curvature of the spherical surfaces. The following equation expresses the relationship of these parameters:

$$\frac{1}{f} = (n - 1) \left(\frac{1}{R_1} + \frac{1}{R_2} \right) \tag{2-7}$$

In Eq. (2-7), f is the lens focal length, n is the index of refraction of the glass from which it is made, and R_1 and R_2 are the radii of the ground surfaces. In using this equation it is assumed that light rays travel from left to right: R_1 applies to the first surface encountered by the ray and is considered positive if its center of curvature lies to the right of the lens and negative if it is left; R_2 applies to the second surface encountered and is considered positive if it lies to the left of the lens and negative if it is right. If the algebraic sign of the focal length is positive, the lens is converging; if it is negative, the lens is diverging. The flat surfaces of planoconvex and planoconcave lenses are considered to have radii of infinity.

Example 2-3 Compute the focal length of a biconvex lens ground from glass having a refraction index of 1.52 and whose left surface has a radius of 50.0 mm and right surface a radius of 75.0 mm.

By Eq. (2-7),

$$\frac{1}{f} = (1.52 - 1.00) \left(\frac{1}{50.0} + \frac{1}{75.0} \right)$$

$$f = 57.7 \text{ mm}$$

2-7 LENS FORMULA

A pencil of incident light rays coming from an object located an infinite distance away from the lens will be parallel, as illustrated in Fig. 2-14, and the image will come to focus in the plane of infinite focus. For objects located some finite distance from the lens, the *image distance* (distance from lens center to plane of focus) is greater than the focal length. It is obvious from Fig. 2-15 that the closer the object is to the lens,

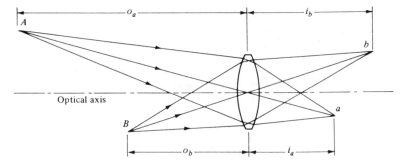

Figure 2-15 Image distance increases with decreased object distance.

the farther behind the lens will be the point of focus. The following equation, called the *lens formula,* expresses the relationship of object distance o and image distance i to the focal length f of a converging lens:

$$\frac{1}{o} + \frac{1}{i} = \frac{1}{f}$$

(2-8)

If the focal length of a lens and the distance to an object are known, the resulting distance to the image plane can be calculated using the lens formula.

Example 2-4 Find the image distance for an object distance of 50.0 m and a focal length of 50.0 cm.
By Eq. (2-8),

$$\frac{1}{50.0} + \frac{1}{i} = \frac{1}{0.50}$$

$$\frac{1}{i} = \frac{1}{0.5} - \frac{1}{50.0}$$

$$i = \frac{1}{1.98} = 0.505 \text{ m} = 50.5 \text{ cm}$$

2-8 SCHEIMPFLUG CONDITION

If the image plane is tilted with respect to the object plane when projecting images through a lens, the *Scheimpflug condition* must be adhered to in order to achieve sharp focus of all images. The Scheimpflug condition, as illustrated in Fig. 2-16, states that for perfect focus, the image plane, object plane, and lens plane (plane through the optical center of the lens and perpendicular to the optical axis) all must intersect along a common line. (When the image and object planes are parallel, the lens plane must also be parallel to these planes, in which case the line of intersection of the three planes theoretically occurs at infinity.)

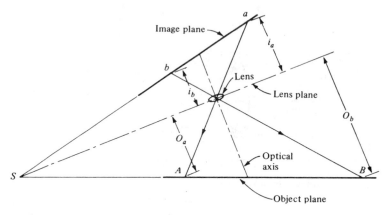

Figure 2-16 Geometry of the Scheimpflug condition.

In Fig. 2-16 the image plane is tilted with respect to the object plane and the two planes intersect along line S. Note that, as a result of tilt, object distances of projected images vary, depending upon their location in the image plane; e.g., object distance O_b corresponding to image b is greater than O_a corresponding to image a. To maintain sharp focus, the lens formula—Eq. (2-8), which relates image distances and object distances for a given lens—must be satisfied for all images. As object distances increase, image distances must be decreased to satisfy the lens formula. This is accomplished by tilting (*canting*) the lens as shown in Fig. 2-16 so that the lens plane passes through S. (Note that image distances and object distances are measured perpendicular to the lens plane.)

Situations in photogrammetry where the Scheimpflug condition must be enforced are in rectification of tilted photographs (see Sec. 11-19) and in orienting oblique photographs in a stereoscopic plotter (see Sec. 17-2). In both of these cases the image plane is tilted with respect to the object plane and object distances are relatively short, so that focus is critical. In taking oblique aerial photos, it is unnecessary to enforce the Scheimpflug condition, because object distances (flying heights) are great with respect to image distances and sharp focus is maintained with image distance set for infinite object distances.

2-9 REAL AND VIRTUAL IMAGES

The image formed by a lens is said to be *real* if it may be made visible by placing a screen in the image plane. Such is the case of the image of Fig. 2-13. The converging lenses of Fig. 2-11a through c all form real images if the objects are located at object distances greater than the focal length of the lens. *Virtual* images cannot be formed on a screen because the rays do not actually come together at a focus point. Instead, the rays must be projected backward to their intersection, as indicated by the dashed

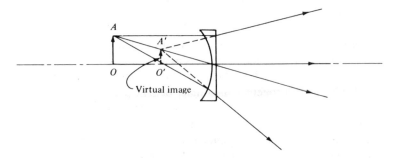

Figure 2-17 Virtual image formed by diverging lens.

lines of Figs. 2-17 and 2-18. All diverging lenses form virtual images; so do converging lenses when the object distance is less than the focal length.

In Fig. 2-18 the object distance is less than the focal length of the converging lens. In this case the virtual image is larger than the real image, which illustrates the use of a lens as a magnifying glass. Light rays emanating from A are refracted in a convergent manner by the lens but not enough to enable them to come to focus. To an observer located on the right, the light rays appear to emanate from O' and A' instead of O and A. Therefore $O'A'$ is the magnified virtual image of the object OA.

2-10 LATERAL MAGNIFICATION

Lateral magnification of a lens is the ratio of the image size to the object size. For the lens of Fig. 2-13, the lateral magnification is the ratio ab/AB. By comparison of similar triangles, the following magnification equation is obtained:

$$M = \frac{i}{o} \tag{2-9}$$

where M is the lateral magnification and o and i are object and image distances, respectively, as previously defined. In Figs. 2-17 and 2-18 lateral magnification is the ratio $A'O'/AO$.

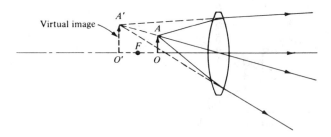

Figure 2-18 Virtual image of a converging lens (magnifying glass).

2-11 THICK LENSES

The analysis of thin lenses in the previous sections was simplified by assuming that their thicknesses were negligible. With thick lenses this assumption is no longer valid. Thick lenses may consist of a single thick element or a combination of two or more elements which are either cemented together in contact or otherwise rigidly held in place with air spaces between the elements. A thick "combination" lens used in an aerial camera is illustrated in Fig. 2-19. Note that it consists of 12 individual elements.

Two points called *nodal points* must be defined for thick lenses. These points, termed the *incident* nodal point and the *emergent* nodal point, lie on the optical axis. They have the property that any light ray directed toward the incident nodal point passes through the emergent nodal point and emerges on the other side of the lens in a direction parallel to the direction of the original incident ray. In Fig. 2-20, for example, rays AN and $N'a$ are parallel, as are rays BN and $N'b$. Points N and N' are the incident and emergent nodal points, respectively, of the thick lens.

If parallel incident light rays (rays from an object at an infinite distance) pass through a thick lens, they will come to focus at the plane of infinite focus. The focal length of a thick lens is the distance from the emergent nodal point N' to this plane of infinite focus.

2-12 LENS QUALITY

It is impossible for a single lens to produce a perfect image; it will, instead, always be somewhat blurred. These imperfections which degrade the sharpness of the image

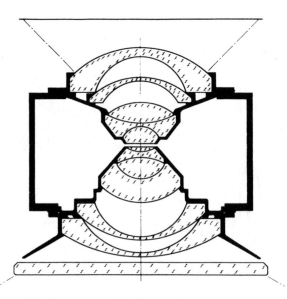

Figure 2-19 Cross section of the Super Aviogon lens. (*Courtesy of Wild Heerbrugg Instruments, Inc.*)

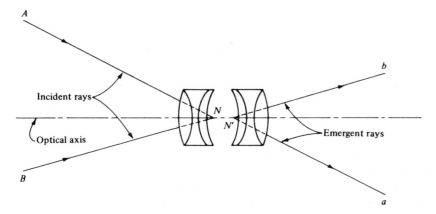

A

Incident rays

Optical axis

B

b

N

N'

Emergent rays

a

Figure 2-20 Nodal points of a thick lens.

are termed *aberrations*. Their presence gives rise to the use of combination lenses discussed in the previous section. Through the use of additional lens elements, lens designers are able to correct for aberrations and bring them within tolerable limits.

The primary lens aberrations are (1) spherical aberration, (2) coma, (3) astigmatism and curvature of field, and (4) chromatic aberration. *Spherical aberration,* as illustrated in Fig. 2-21, is the degradation of the axial image. It is caused by faulty grinding of the spherical surfaces of the lens elements and results in rays entering near the outer edge of the lens being brought to focus nearer the lens than the rays which enter near the center of the lens. The image of a point thus formed is a circle called the *circle of confusion*. If spherical aberration exists for a lens, the size of the circle of confusion may be reduced by adjusting the position of the image plane until the smallest possible circle (*circle of least confusion*) is obtained. The image-plane position that produces the circle of least confusion is illustrated in Fig. 2-21. *Coma* is similar to spherical aberration, except that it applies to the failure of oblique rays, instead of axial rays, to come to focus at a point. Instead of the image being a circle, it is in the shape of a comet.

Astigmatism is a condition in which lines in the object that are perpendicular to each other do not come to focus at the same image distance. As shown in Fig. 2-22,

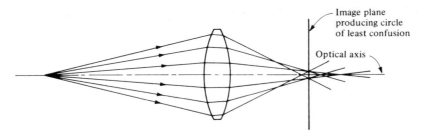

Image plane producing circle of least confusion

Optical axis

Figure 2-21 Spherical aberration.

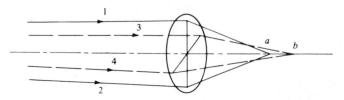

Figure 2-22 Astigmatism.

rays 1 and 2 from a vertical object line are focused at *a,* while rays 3 and 4, which are from a perpendicular line in the object, are focused at *b.* Astigmatism is caused by imperfect grinding of the lens surfaces. It is common in the human eye. With an astigmatic lens, images from points of equal object distances, but whose rays make varying angles with the optical axis, will not come to focus in the same image plane. Rather, they will form a curved surface. This condition is called *curvature of field.* Astigmatism and curvature of field can be minimized by using combination lenses composed of converging and diverging elements.

Chromatic aberration is caused by the different refractive characteristics of the various colors which make up white light. As shown in Fig. 2-23, blue light is refracted more than red light, and therefore these two colors fail to come to focus at a common point. The effects of chromatic aberration can also be compensated for through the use of converging and diverging lenses in combination.

The *aberrations* discussed above degrade the quality or sharpness of the image. Lens *distortions* on the other hand, do not degrade image quality but deteriorate the geometric quality (or positional accuracy) of the image. Lens distortions are classified as either *radial* or *tangential.* Both occur if light rays are bent, or change directions, so that after they pass through the lens they do not emerge parallel to their incoming directions. Radial distortion, as implied by its name, causes imaged points to be distorted along radial lines from the optical axis. It is caused by faulty grinding of the lens elements. Outward radial distortion is considered positive and inward radial distortion is considered negative. Tangential distortion occurs at right angles to radial lines from the optical axis. It is caused by faulty centering of the lens elements of a combination lens. Tangential distortion is generally of much less consequence than radial distortion and can often be disregarded. In Fig. 2-24, instead of the incident light ray being imaged at its correct position *a,* it is imaged at *a'.* The magnitude of radial distortion is Δr (positive in this case because it is outward), and the magnitude of tangential distortion is Δt.

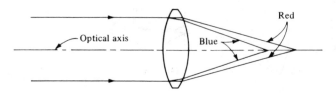

Figure 2-23 Chromatic aberration.

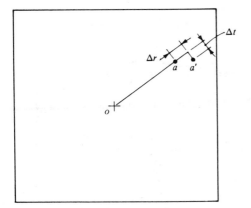

Figure 2-24 Radial and tangential lens distortions.

If distortions do occur because of a camera lens, photogrammetrists will obtain erroneous measurements for image positions in the resulting photographs. Distortions of a particular lens may be determined by calibration tests, as described in Chap. 4, and corrections can be applied to measurements to eliminate their effect. The correction procedure is described in Chap. 5.

Distortions and aberrations cannot be eliminated completely in designing lenses, but they can be minimized through the use of combination lenses. Often the reduction of one type of distortion or aberration affects another type adversely, and therefore lens design is a science of compromise in attempting to obtain the least possible detrimental combination of all distortions and aberrations. The quantity of calculations involved in lens design is enormous if the many possible compromises are to be thoroughly investigated. With the advent of computers, lens designers have performed these computations more readily, and better and better lenses have resulted. Today distortions of many combination lenses have been reduced to such negligible amounts that they are often referred to as *distortion-free lenses*.

Resolution or *resolving power* of a lens is the ability of the lens to show detail. One common method of measuring lens resolution is to count the number of *line pairs* (black lines separated by white line spaces of equal thickness) which can be clearly distinguished within a width of 1 mm in an image produced by the lens. *Modulation transfer function* (MTF) is another way of specifying the resolution characteristics of a lens. Both methods of specifying resolving power are discussed in Sec. 4-16, and a line-pair test pattern is shown in Fig. 4-16. Good resolution is important in photogrammetry because photo images must be sharp and clearly defined for precise measurements and accurate interpretative work. Photographic resolution is not just a function of the camera lens, however, but also depends on film quality and processing (see Sec. 3-5).

2-13 DEPTH OF FIELD

The *depth of field* of a lens is the range in object distance that can be accommodated by a lens without introducing significant image deterioration. In Fig. 2-25a, light rays

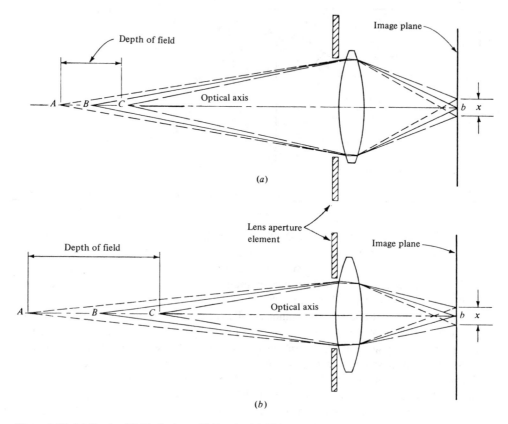

Figure 2-25 (*a*) Depth of field of a lens. (*b*) Depth of field increases with decreased lens aperture.

from object point B are perfectly focused as a point b in the image plane. Light rays from A and C, on the other hand, are imaged as circles of confusion of diameter x. The larger the circles of confusion become, the more deteriorated is the image. By placing a certain limit on the tolerable diameter of the circle of confusion, depth of field is limited. In Fig. 2-25*a*, depth of field is limited to the distance AC by the tolerable diameter x of the circle of confusion.

For a given lens, depth of field can be increased by reducing the size of the lens opening (aperture) as shown in Fig. 2-25*b*. This limits the usable area of the lens to the central portion. For aerial photography, depth of field is seldom of consequence, because variations in the object distance are generally a very small percentage of the total object distance. For close-range photography, however, depth of field is often extremely critical. The shorter the focal length of a lens, the greater its depth of field, and vice versa. Thus, if depth of field is critical, it can be somewhat accommodated through the selection of an appropriate lens.

Depth of focus, which is akin to depth of field, is the range in image distance which can be tolerated without introducing significant image deterioration.

REFERENCES

American Society of Photogrammetry: "Manual of Photogrammetry," 4th ed., Falls Church, Va., 1980, chap. 3.

————: "Manual of Photogrammetry," 3d ed., Falls Church, Va., 1966, chap. 3.

Brown, E. B.: "Modern Optics," Reinhold Publishing Corporation, New York, 1965.

Carlson, F. P.: "Introduction to Applied Optics for Engineers," Academic Press, Inc., New York, 1976.

Fritz, L. W., and H. H. Schmid: Stellar Calibration of the Orbigon Lens, *Photogrammetric Engineering,* vol. 40, no. 2, p. 101, 1974.

Ghatak, A.: "An Introduction to Modern Optics," McGraw-Hill Book Company, New York, 1972.

Greenleaf, A. R.: "Photographic Optics," The Macmillan Company, New York, 1950.

Habell, K. J., and A. Cox: "Engineering Optics," rev. ed., Sir Isaac Pitman & Sons, Ltd., London, 1953.

Jensen, N.: "Optical and Photographic Reconnaissance Systems," John Wiley & Sons, Inc., New York, 1968.

Kissam, P.: "Optical Tooling," McGraw-Hill Book Company, New York, 1962.

Levine, H., and S. Rosin: The Geocon IV Lens, *Photogrammetric Engineering,* vol. 36, no. 4, p. 335, 1970.

Smith, Warren J.: "Modern Optical Engineering," McGraw-Hill Book Company, New York, 1966.

Washer, F. E.: Resolving Power Related to Aberration, *Photogrammetric Engineering,* vol. 32, no. 2, p. 213, 1966.

Wright, R. H.: An Advanced Optical Objective Lens, *Photogrammetric Engineering and Remote Sensing,* vol. 42, no. 8, p. 1049, 1976.

PROBLEMS

2-1 Explain the difference between physical and geometrical optics.

2-2 What is the speed of light in kilometers per second?

2-3 A certain electromagnetic energy is propagated in a vacuum at a frequency of 24,500,000 cycles/sec. What is the wavelength (to the nearest foot) of this energy?

2-4 If a certain type of glass has an index of refraction of 1.550, what is the speed of light through this glass?

2-5 A ray of light enters glass (index 1.570) from air at an incident angle of 25°. Find the angles of refraction and of deviation.

2-6 Repeat Prob. 2-5, except that the incident angle is 38° and the index of refraction of the glass is 1.550.

2-7 A light ray emanating from under water (index 1.333) makes an angle of 40° with the normal to the surface. What is the angle that the refracted ray makes with the normal as it emerges into air (index 1.000)?

2-8 Repeat Prob. 2-7, except that the ray makes an angle of 35° with the normal.

2-9 A light ray is directed through air (index 1.000) at a 25° angle of incidence toward a parallel-sided glass plate (index 1.570) which is 6.35 mm thick. What will be the lateral displacement of the light ray as it again emerges into air?

2-10 Repeat Prob. 2-9, except that the angle of incidence is 35° and the thickness of the glass plate is 0.25 in.

2-11 Find the critical angle for a light ray traveling from glass (index 1.600) to air (index 1.000).

2-12 Repeat Prob. 2-11, except that the glass has a refractive index of 1.520.

2-13 Find the critical angle for a light ray traveling from glass (index 1.750) to water (index 1.333).

2-14 What is the angle of deviation for a light ray traveling from crown glass (index 1.520) to flint glass (index 1.750) if the angle of incidence is 35°?

2-15 Derive Eq. (2-5) for lateral displacement of a light ray passing through a parallel-sided plate.

2-16 Explain the difference between first-surface and back-surface mirrors. What advantage and disadvantage does the first-surface mirror have?

2-17 The two spherical surfaces of a thin biconvex lens such as that of Fig. 2-11a have radii R_1 and R_2 of 10.0 cm and 15.0 cm, respectively. If the refractive index of the glass is 1.650, what is the focal length of the lens?

2-18 Repeat Prob. 2-17, except that the glass has an index of refraction of 1.520.

2-19 Repeat Prob. 2-17, except that the lens is a planoconvex lens such as that of Fig. 2-11b whose one ground spherical surface has a radius of 12.5 cm.

2-20 Repeat Prob. 2-17, except that the lens is a biconcave lens such as that of Fig. 2-11d whose ground spherical surfaces have radii R_1 and R_2 of 100.0 mm and 75.0 mm, respectively. Also, the glass has a refractive index of 1.750.

2-21 To what radius of curvature must the two spherical surfaces of an equiconvex lens (biconvex lens with surfaces of equal radii) be ground so that the focal length will be 85.0 mm if the glass has a refractive index of 1.570?

2-22 An object located 45.0 in in front of a thin lens has its image in focus 10.0 in from the lens on the other side. What is the focal length of the lens?

2-23 A thin lens has a focal length of 100.0 mm. What is the object distance for an image that is perfectly focused at an image distance of 165 mm?

2-24 Prepare a table of image distances (in millimeters) versus object distances of exactly 1, 2, 5, 10, 100, 1,000, and 10,000 feet for a lens having a 210.00-mm focal length.

2-25 An object 10.0 cm high is located 13.8 in in front of a lens whose focal length is 64.00 mm. What is the image distance, lateral magnification, and height of the magnified image?

THREE

PRINCIPLES OF PHOTOGRAPHY†

3-1 INTRODUCTION

Photography, which means "drawing with light," originated long before cameras and light-sensitive photographic films came into use. Ancient Arabs discovered that when inside a dark tent, they could observe inverted images of illuminated outside objects. The images were formed by light rays which passed through tiny holes in the tent. The principle involved was actually that of the pinhole camera of the type shown in Fig. 2-12. In the 1700s French artists used the pinhole principle as an aid in drawing perspective views of illuminated objects. While inside a dark box, they traced the outlines of objects projected onto the wall opposite a pinhole.

In 1839 Louis Daguerre of France developed a photographic film that could capture a permanent record of images which illuminated it. By placing this film inside a dark "pinhole box," a picture could be obtained without the help of an artist.

This box used in conjunction with photographic film became known as a *camera*.

3-2 COMPARISON OF PINHOLE AND LENS CAMERAS

Photographic film has improved a great deal through the years, but it still is not sensitive enough to capture an image projected through a pinhole in a short period of time. Therefore it was necessary to increase the size of the opening to allow more light to pass. If the pinhole is enlarged, however, the resulting image is blurred. By replacing the pinhole with a lens it is possible to enlarge the hole and still retain sharp focus.

The geometry of the lens camera of Fig. 3-1 is identical to the geometry of the pinhole camera of Fig. 2-12 if the thickness t of the lens is disregarded. Note, however, that the lens opening has been increased from pinhole size d_1 in Fig. 2-12 to diameter

---

†By Dr. James P. Scherz, Professor of Civil and Environmental Engineering, University of Wisconsin, Madison, Wis.

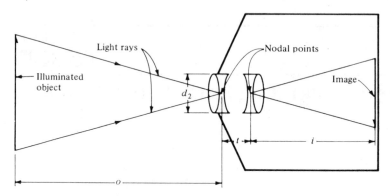

Figure 3-1 Geometry of the lens camera.

d_2 in Fig. 3-1. Instead of object distances and image distances being unrestricted as with the pinhole camera, with the lens camera these distances are governed by the lens formula Eq. (2-8). To satisfy this equation the lens camera must be focused for each different object distance by adjusting the image distance. When object distances approach infinity, as for photographing objects at great distances, the term $1/o$ in Eq. (2-8) approaches zero and image distance i is then equal to f, the focal length of the lens. With aerial photography, object distances are very great with respect to image distances; therefore aerial cameras are manufactured with their focus fixed for infinity. This is accomplished by making the image distance equal to the focal length of the camera lens.

3-3 ILLUMINANCE

Illuminance of any photographic exposure is the brightness or amount of light received per unit area on the image plane surface during exposure. The unit of illuminance is the *meter-candle*. One meter-candle is the illuminance produced by a standard candle at a distance of 1 meter.

Illuminance is proportional to the amount of light passing through the lens opening during exposure, and this, of course, is proportional to the area of the opening. Since the area of the lens opening is $\pi d^2/4$, illuminance is proportional to the variable d^2, the square of the diameter of the lens opening.

Image distance is another factor which affects illuminance. As illustrated in Fig. 3-2, radii of areas in the image plane illuminated by light rays passing through a lens are directly proportional to their image distances. By similar triangles in Fig. 3-2, for example, if i_2 equals $2i_1$, then r_2 equals $2r_1$. Thus, illuminated areas are proportional to the squares of their radii; that is, $A_1 = \pi r_1^2$ and $A_2 = \pi r_2^2$. Illuminance is inversely proportional to illuminated area, and because of the proportionality of i and r, illuminance is proportional to $1/i^2$. Normally in photography, object distances are sufficiently long so that the term $1/o$ in Eq. (2-8) is nearly zero, in which case i is equal to f. Thus, illuminance is proportional to the variable $1/f^2$. Since illuminance is proportional to both d^2 and $1/f^2$, these two quantities may be combined so that

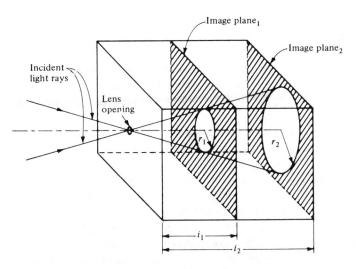

Figure 3-2 Illuminance is a function of image distance.

illuminance is proportional to d^2/f^2. The square root of this term is called the *brightness factor*, or

$$\sqrt{\frac{d^2}{f^2}} = \frac{d}{f} = \text{brightness factor} \tag{3-1}$$

The inverse of Eq. (3-1) is also an inverse expression of illuminance and is the very common term "f-stop," also called "f-number." In equation form,

$$f\text{-stop} = \frac{f}{d} \tag{3-2}$$

According to Eq. (3-2), f-stop is the ratio of focal length to the diameter of the lens opening or *aperture*. As the aperture increases, f-stop numbers decrease and illuminance increases, thus requiring less exposure time, i.e., faster shutter speeds. Because of this correlation between f-stop and shutter speed, f-stop is the term used for expressing *lens speed* or "light-gathering" power of a lens. Illuminance produced by a particular lens is correctly expressed by Eq. (3-2), whether the lens has a very small diameter with short focal length or a very large diameter with a long focal length. If f-stop is the same for two different lenses, the illuminance of each of their images will be the same.

3-4 RELATIONSHIP OF APERTURE AND SHUTTER SPEED

A sunbather gets a suntan or sunburn when exposed to sunshine. The darkness of tan or severity of burn is the product of the sun's illuminance (brightness) and the time of exposure to the sun. *Total exposure* of photographic film likewise is the product of illuminance and time of exposure. Its unit is *meter-candle-seconds*.

In making photographic exposures, the correct amounts of illuminance and time may be correlated using a *light meter*. Illuminance is regulated by varying f-stop settings on the camera, while time of exposure is set by varying shutter speed. Variations in f-stop settings are actually variations in the diameter of the *aperture,* which can be controlled with a *diaphragm*—a circular shield that enlarges or contracts, changing the diameter of the opening of the lens and thus regulating the amount of light that is allowed to pass through it.

With a lens camera, as the diameter of the aperture increases, enabling faster exposures, depth of field becomes less and lens distortions become more severe. There are situations when a small diaphragm opening is desirable, and there are times when the reverse is true. To photograph a scene with great variations in object distances and yet retain sharp focus of all images, a large depth of field is required. In this case, to maximize depth of field, the picture would be taken at a slow shutter speed and large f-stop setting corresponding to a small-diameter lens opening. On the other hand, in photographing rapidly moving objects or in making exposures from a moving vehicle such as an airplane, a fast shutter speed is essential to reduce image motion. In this situation a small f-stop setting corresponding to a large-diameter lens opening would be necessary for sufficient exposure.

From the previous discussion it is apparent that there is an important relationship between f-stop and shutter speed. If exposure time is cut in half, total exposure is also halved. Conversely, if aperture area is doubled, total exposure is also doubled. If shutter time is halved and aperture area is doubled, total exposure remains unchanged.

Except for inexpensive models, cameras are manufactured with the capability of varying both shutter speeds and f-stop settings. The nominal f-stop settings are 1, 1.4, 2.0, 2.8, 4.0, 5.6, 8.0, 11, 16, 22, and 32. Not all cameras will have all these, but the more expensive cameras will have most of them. The camera pictured in Fig. 3-3, for example, has f-stops ranging from f-1.4 to f-16. This camera is also equipped for varying shutter speeds down to $\frac{1}{1000}$ sec. F-stop number 1, given as f-1, occurs, according to Eq. (3-2), when the aperture diameter equals the lens focal length. A setting at f-1.4 halves the aperture area from that of f-1. In fact, each succeeding number of the nominal f-stops listed above halves the aperture area of the preceding one, and it is seen that each succeeding number is obtained by multiplying the preceding one by $\sqrt{2}$. This is illustrated as follows:

Let $d_1 = f$, where d_1 is aperture diameter.
Then $f/d_1 = 1 = f$-stop.
At f-stop $= 1$,

$$\text{Aperture area} = A_1 = \frac{\pi(d_1)^2}{4}$$

If aperture diameter is reduced to d_2, giving a lens opening area half of A_1, then

$$A_2 = \frac{A_1}{2} = \frac{\pi(d_2)^2}{4} = \frac{\pi(d_1)^2}{2(4)}$$

Figure 3-3 Single lens reflex camera having f-stop settings ranging from f-1.4 to f-16, and variable shutter speeds ranging down to 1/1000th second. (*Courtesy Paillard Inc.*)

From the above, $d_2 = d_1/\sqrt{2}$, and the corresponding f-stop number is

$$f\text{-stop} = \frac{f\sqrt{2}}{d_1} = 1\sqrt{2} = 1.4$$

The relationship between f-stop and shutter speed leads to many interesting variations in obtaining correct exposures.

Example 3-1 Suppose that a photographic film is optimally exposed with an f-stop setting of f-4 and a shutter speed of $\frac{1}{500}$ sec. What is the correct f-stop setting if shutter speed is changed to $\frac{1}{1,000}$ sec?

SOLUTION Total exposure is the product of diaphragm area and shutter speed. This product must remain the same for the $\frac{1}{1,000}$ sec shutter speed as it was for the $\frac{1}{500}$ sec shutter speed, or

$$\text{Area}_1 \times \text{time}_1 = \text{area}_2 \times \text{time}_2$$

Rearranging,

$$\text{Area}_2 = \text{area}_1 \times \frac{\text{time}_1}{\text{time}_2} \qquad (a)$$

Let d_1 and d_2 be diaphragm diameters for $\frac{1}{500}$- and $\frac{1}{1,000}$-sec shutter times, respectively. Then the respective diaphragm areas are

$$\text{Area}_1 = \frac{\pi(d_1)^2}{4} \quad \text{and} \quad \text{area}_2 = \frac{\pi(d_2)^2}{4} \qquad (b)$$

By Eq. (3-2),

$$d_1 = \frac{f}{f\text{-stop}_1} \quad \text{and since } f\text{-stop}_1 = 4, \ d_1 = \frac{f}{4} \qquad (c)$$

Substituting (b) and (c) into (a),

$$\frac{\pi(d_2)^2}{4} = \frac{\pi(f)^2}{4(4)^2} \times \frac{\frac{1}{500}}{\frac{1}{1,000}}$$

Reducing,

$$\frac{f}{d_2} = \sqrt{\frac{500 \times 16}{1,000}} = 2.8$$

Hence f-2.8 is the required f-stop. The above is simply computational proof of an earlier statement that each successive nominal f-stop setting halves the aperture area of the previous one, or in this case f-2.8 doubles the aperture area of f-4, which is necessary to retain the same exposure if shutter time is halved.

3-5 CHARACTERISTICS OF PHOTOGRAPHIC EMULSIONS

Photographic films consist of two parts: *emulsion* and *backing* or *support*. The emulsion contains light-sensitive silver halide crystals. These are placed on the backing or support in a thin coat as shown in Fig. 3-4. The support material is usually paper, plastic film, or glass.

Silver halide crystals have the property that when exposed to light the bond between the silver and the halide is weakened. An emulsion that has been exposed to light contains an invisible image of the object called the *latent image*. When the latent image is developed, areas of the emulsion that were exposed to intense light turn to free silver and become black. Areas that received no light become white if the support is white paper. (They become clear if the support is glass or transparent plastic film.) The degree of darkness of developed images is a function of total exposure (product of illuminance and time) which originally sensitized the emulsion to form the latent image. In any photographic exposure there will be variations in illuminance received

Emulsion of silver halide crystals

Support material

Figure 3-4 Cross section of a photographic film.

from different objects in the photographed scene, and therefore between black and white there will exist various tones of gray which result from these variations in illuminance. Actually the crystals turn black, not gray, when exposed to sufficient light. However, if the light received in a particular area is sufficient to sensitize only a portion of the crystals, a gray tone results from a mixture of the resulting black and white. The greater the exposure, the greater the percentage of black in the mixture and hence the darker the shade of gray.

The degree of darkness of a developed emulsion is called its *density*. The greater the density, the darker the emulsion. Density is the measure of the amount of light that can be transmitted through the emulsion; i.e., a black emulsion transmits no light and a clear one transmits almost 100 percent light. Opacity is the reciprocal of transmittance. The unit of density is the common logarithm of opacity. As an example, if 10 percent of the light is transmitted, transmittance is $\frac{1}{10}$, opacity is $\frac{1}{0.10}$ or 10, and density is the common logarithm of 10 or 1.0. The amount of light incident to an emulsion and the amount transmitted can be measured with an instrument called a *densitometer* (see Sec. 5-15).

If exposure is varied for a particular emulsion, corresponding variations in densities will be obtained. A plot of density on the ordinate versus logarithms of exposure on the abscissa for a given emulsion produces a curve called the *characteristic curve,* also known as the *D-Log E curve,* or the *H and D curve.* A typical characteristic curve is shown in Fig. 3-5. Characteristic curves for different emulsions vary somewhat, but they all have the same general shape. The lower part of the curve, which is concave upward, is known as the *toe* region. The upper portion, which is concave downward, is the *shoulder* region. A *straight-line* portion occurs between the toe and shoulder regions.

Characteristic curves are useful in describing the characteristics of photographic emulsions. The slope of the straight-line portion of the curve, for example, is a

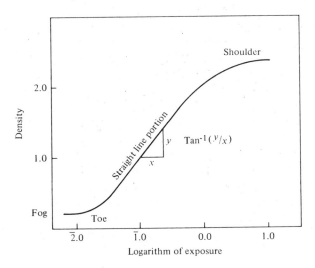

Figure **3-5** Typical characteristic curve of a photographic emulsion.

measure of the contrast of the film. The steeper the slope, the greater the *contrast* (change in density) for a given range of exposure. Contrast of a given film is expressed as *gamma,* the tangent of the angle between the straight-line portion of the curve and the abscissa axis, as shown on Fig. 3-5. From Fig. 3-5 it is evident that for an exposure of zero the film has some density. The density of an unexposed emulsion is called *fog,* and on the curve it is the density corresponding to the low portion of the toe region. It is also apparent from Fig. 3-5 that exposure must exceed a certain minimum before density greater than fog occurs. Also, exposures within the shoulder region affect the density very little, if any. Thus, a properly exposed photograph is one in which the entire range of exposure occurs within the straight-line portion of the curve.

Just as the skin of different people varies in sensitivity to sunlight, so also does the sensitivity of emulsions vary. Light sensitivity of photographic emulsions is a function of the size and number of silver halide crystals or *grains* in the emulsion. When the required amount of light exposes a grain of emulsion, the entire grain becomes exposed regardless of its size. If a certain emulsion is composed of grains smaller than those in another emulsion, such that it requires approximately twice as many grains to cover the film, this emulsion will also require about twice as much light to properly expose it. Conversely, as grain size increases, the total number of grains in the emulsion decreases and the amount of light required to properly expose it decreases. Film is said to be more sensitive and faster when it requires less light for proper exposure. Faster films can be used advantageously in photographing rapidly moving objects.

As sensitivity and grain size increase, the resulting image becomes coarse and *resolution* (sharpness or crispness of the picture) is reduced. Thus, for highest pictorial quality, such as portrait work, slow fine-grained emulsions are preferable. Film resolution can be tested by photographing a standard test pattern. The pattern consists of groups of *line pairs* (parallel lines of varying thickness separated by spaces of width equal to the line thickness). Resolution can then be given either by the count of the maximum number of lines per millimeter in the smallest line pattern that can clearly be discerned on the developed film, or by the *modulation transfer function.* These topics are discussed further in Sec. 4-13.

Photographers have developed exposure guides for films of various sensitivities. In the United States the American Standards Association (ASA) system is used. In this system the ASA number assigned to a film is roughly equal to the inverse of shutter speed (in seconds) required for proper exposure in pure sunlight for a lens opening of f-16. According to this rule of thumb, if a film is properly exposed in pure sunlight at f-16 and $\frac{1}{200}$ sec, it would be classified ASA 200. This rule of thumb is seldom used today because of the availability of light meters which, given the ASA rating of the film being used, automatically yield proper exposures (f-stops and shutter speeds) for particular lighting conditions. In Europe an exposure standard called DIN has been developed. Most light meters today give scales which accommodate both the ASA and DIN systems. Many cameras now have built-in automatic exposure systems. These consist of a light-sensitive cell which measures ambient light conditions and automatically adjusts lens aperture and/or shutter speed to optimum values for the conditions.

3-6 PROCESSING BLACK-AND-WHITE EMULSIONS

The five-step darkroom procedure for processing an exposed black-and-white emulsion is as follows:

1. *Developing*. In this first step the exposed emulsion is placed in a chemical solution called *developer*. The action of the developer causes grains of silver halide which were exposed to light to be reduced to free black silver. The free silver produces the blacks and shades of gray of which the image is composed. Developers vary in strength and other characteristics and must therefore be carefully chosen to produce the desired results. Generally the time of immersion of the film in the developer is between 1 and 15 min, depending upon the particular film and developer used. The contrast of the final image can be changed somewhat by changing development time and temperature of the developer.
2. *Stop Bath*. When proper darkness and contrast of the image has been attained in the developing stage, it is necessary to stop the developing action. This is done with a stop bath—an *acetic* solution which neutralizes the *basic* developer solution. The emulsion is immersed in the stop bath for just a few seconds.
3. *Fixing*. Not all the silver halide grains are turned to free black silver as a result of developing. Instead, there remain many undeveloped grains which would also turn black upon exposure to light if they were not removed. To prevent further developing which would ruin the image, the undeveloped silver halide grains are dissolved out in the fixing solution. The fixing bath also hardens the emulsion. Normal immersion time in the fixing bath is from 10 to 20 min.
4. *Washing*. In this step the emulsion is washed in clear running water to remove any remaining chemicals. If not removed, these chemicals could cause spotting or haziness of the image. Normal washing time is from 10 to 20 min. A detergent may be added to decrease washing time.
5. *Drying*. In this final step the emulsion is dried to remove the water from the emulsion and backing material. This can be done in a variety of ways, from simply air drying to drying in elaborate heated dryers.

Modern equipment is capable of automatically performing the entire five-step darkroom procedure nonstop. The result obtained from developing black-and-white film is a *negative*. It derives its name from the fact that it is reversed in color and geometry from the original scene which was photographed, i.e., black objects appear white and vice versa, and images are inverted.

A *positive* is obtained from the negative by repeating the photographic process. This reverses color and geometry again, thereby producing an image which is true in tone and geometry. In producing a paper print positive from a negative, printing paper which is covered with a layer of emulsion is exposed by passing light through the negative onto the emulsion. Light is transmitted through the various areas of the negative in proportion to the lightness of the negative; e.g., black areas will not transmit light at all and therefore they will not expose their corresponding areas of the printing paper. As a result of exposure through the negative, a latent image is formed on the printing paper. This latent image is processed using the same five-step darkroom procedure outlined above.

Besides using printing paper, positives may also be prepared on plastic film or glass plates. In photogrammetric terminology, positives prepared on glass plates or transparent plastic materials are called *diapositives*.

3-7 SPECTRAL SENSITIVITY OF EMULSIONS

The sun and various artificial sources such as light bulbs emit a wide range of so-called *electromagnetic energy*. The entire range of this electromagnetic energy is called the *electromagnetic spectrum*. X-rays, visible light rays, and radio waves are some familiar examples of energy variations within the electromagnetic spectrum. Electromagnetic energy travels in regular sinusoidal oscillations called *waves*. Variations in electromagnetic energy are classified according to variations in their wavelengths or frequencies of propagation. The velocity of electromagnetic energy in a vacuum is constant and related to frequency and wavelength through the following expression (see also Sec. 2-1):

$$c = f\lambda \tag{3-3}$$

In Eq. (3-3), c is velocity in a vacuum of electromagnetic energy, f is frequency, and λ is wavelength. Figure 3-6 illustrates the wavelength classification of the electromagnetic spectrum. Visible light (that electromagnetic energy to which our eyes are sensitive) is composed of only a very small portion of the spectrum (see Fig. 3-6). It consists of energy with wavelengths in the range of from about 0.4 to 0.7 μm. Energy having wavelengths slightly shorter than 0.4 μm is called *ultraviolet,* and energy with wavelengths slightly longer than 0.7 μm is called *near-infrared.* Ultraviolet and near-infrared cannot be detected by the human eye.

Within the wavelengths of visible light, the human eye is able to distinguish different colors. The primary colors blue, green, and red are composed of slightly different wavelengths: blue is composed of energy having wavelengths of from about 0.4 to 0.5 μm, green from 0.5 to 0.6 μm, and red from 0.6 to 0.7 μm. All other hues are made up of combinations of the primary colors; e.g., yellow is the combi-

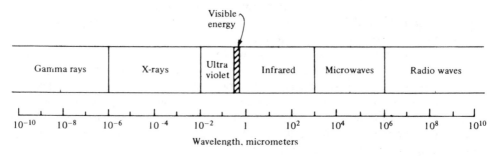

Figure 3-6 Classification of the electromagnetic spectrum by wavelength.

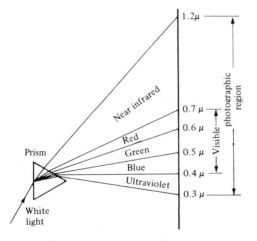

Figure 3-7 White light broken into the individual colors of the visible and near-visible spectrum by means of a prism.

nation of red and green light. There are multitudes of these combination colors. White light is the combination of all of the visible colors. It can be broken down into its component colors by passing it through a prism, as shown in Fig. 3-7. Color separation occurs because of different refractions that result from energy of different wavelengths.

To the human eye, an object appears a certain color because it reflects energy of the wavelengths producing that color. If an object reflects all the visible energy that strikes it, that object will appear white. On the other hand, if an object absorbs all light and reflects none, that object will appear black. If an object absorbs all green and red energy but reflects blue, that object will appear blue.

Just as the retina of the human eye is sensitive to variations in wavelength, photographic emulsions can also be manufactured with variations in wavelength sensitivity. Black-and-white emulsions composed of untreated silver halides are sensitive only to blue and ultraviolet energy. A red object, for example, will not produce an image on such an emulsion. These untreated emulsions are usually used on printing papers for making positives from negatives. When these printing papers are used, red or yellow lights called *safe lights* can conveniently be used to illuminate the darkroom because these colors cannot expose a paper which is sensitive only to blue light.

Black-and-white silver halide emulsions can be treated by use of fluorescent dyes so that they are sensitive to other wavelengths of the spectrum besides blue. Emulsions sensitive to the blue and green range are called *orthochromatic;* those sensitive to blue, green, and red are called *panchromatic.* Emulsions can also be made to respond to energy in the near-infrared range. These emulsions are called *infrared,* or IR. Infrared films make it possible to obtain photographs of energy which is invisible to the human eye. An early application of this type of emulsion was in camouflage detection, where it was found that dead foliation or green netting, which had the same green color as live foliation to the human eye, reflected infrared energy differently. This difference could be detected through infrared photography. Infrared film is now widely used for a variety of applications such as detection of crop stress, tree species mapping, etc. Figure 3-8 illustrates sensitivity differences of various emulsions.

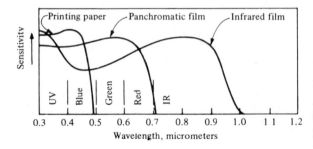

Figure 3-8 Sensitivities of various black-and-white emulsions.

3-8 FILTERS

The red or yellow safe light described in the previous section usually is simply an ordinary white light covered with a red or yellow filter. If the filter is red, it blocks passage of blue and green wavelengths and allows only red to pass. Filters placed in front of camera lenses also allow only certain wavelengths of energy to pass through the lens and expose the film. The use of filters on cameras can be very advantageous for certain types of photography.

Atmospheric haze is largely caused by the scattering of ultraviolet and short blue wavelengths. Pictures which are clear in spite of atmospheric haze can be taken through haze filters. These filters block passage of objectionable scattered short wavelengths (which produce haze) and prevent them from entering the camera and exposing the film. Because of this advantage, haze filters are almost always used on aerial cameras.

Filters for aerial mapping cameras are manufactured from high-quality optical glass. This is the case because light rays which form the image must pass through the filter before entering the camera. In passing through the filter, light rays are subjected to distortions caused by the filter. The camera should therefore be calibrated (see Secs. 4-10 through 4-13), with the filter locked firmly in place; after calibration, the filter should not be removed, for this would upset the calibration.

3-9 PROCESSING COLOR EMULSIONS

Normal color and *color infrared* emulsions are relatively recent advancements which are now finding widespread use in photogrammetry. Color emulsions consist of three layers of silver halides, as shown in Fig. 3-9. The top layer is sensitive to blue light, the second layer is sensitive to green and blue light, and the bottom layer is sensitive to red and blue light. A blue blocking filter is built into the emulsion between the top two layers, thus preventing blue light from exposing the bottom two layers. The result is three layers sensitive to blue, green, and red light, respectively, from top to bottom. The sensitivity of each layer is indicated in Fig. 3-10.

In making a color exposure, light entering the camera sensitizes the layer or layers of the emulsion which correspond to the color or combination of colors of the original

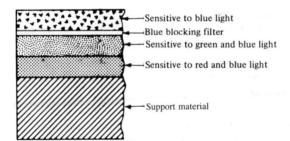

Sensitive to blue light
Blue blocking filter
Sensitive to green and blue light
Sensitive to red and blue light

Support material

Figure 3-9 Cross section of normal color film.

scene. There are a variety of color films available, each requiring a slightly different developing process. The first step of color developing accomplishes essentially the same result as the first step of black-and-white developing. The exposed halides in each layer are turned into black crystals of silver. The remainder of the process depends on whether the film is *color negative* or *color reversal* film. With color negative film, a negative is produced and color prints are made from the negative. Color reversal film produces a true color transparency directly on the film. It is used for making color slides.

In developing color negative film, silver halides in each layer that turned black in the original step are replaced with dyes of the complementary colors to which the layer is sensitive; i.e., the black silver grains of the upper or blue-sensitive layer are replaced with *yellow* dye (yellow is the complementary color of blue and is composed of green and red wavelengths); the black silver crystals in the second or green-sensitive layer are replaced with *magenta* dye (magenta is the complementary color of green and is composed of blue and red wavelengths); and the silver crystals of the third or red-sensitive layer are replaced with *cyan* dye (cyan is the complementary color of red and is composed of blue and green wavelengths).

In producing a color print from the negative, white light (light composed of all

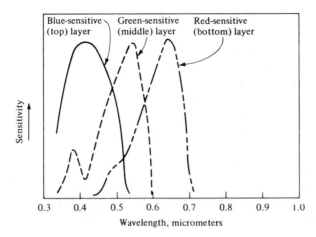

Blue-sensitive (top) layer   Green-sensitive (middle) layer   Red-sensitive (bottom) layer

Sensitivity

0.3   0.4   0.5   0.6   0.7   0.8   0.9   1.0
Wavelength, micrometers

Figure 3-10 Color sensitivity of the three layers of normal color film.

wavelengths in the visible range) is directed through the color negative to expose a three-layered color printing emulsion. The color negative acts as a filter, exposing the three layers to the three colors yellow, magenta, and cyan. In developing the exposed print, the complementary colors of yellow, magenta, and cyan (blue, green, and red, respectively) are produced. This second color reversal yields the original colors of the photographed scene on the processed print.

With color reversal film, the emulsion is first partially developed. It is then exposed again and the developing process repeated to obtain a color reversal. This is the same process as with color negative film except that both processes are done on the original film. The net result is a color transparency composed of the original colors in the photographed scene.

During World War II there was great interest in increasing the effectiveness of films in the infrared region of the spectrum. This interest led to the development of *color infrared* or *false-color* film. The military called it *camouflage detection* film because it allowed photo interpreters to easily differentiate between camouflage and natural foliage. Like normal color film, color IR film also has three emulsion layers, each sensitive to a different part of the spectrum. Fig. 3-11 illustrates the sensitivity curves for each layer of color IR film. The top layer is sensitive to ultraviolet, blue, and green energy. The middle layer has its sensitivity peak in the red portion of the spectrum, but it, too, is sensitive to ultraviolet light. The bottom layer is sensitive to ultraviolet and infrared. Color IR film is commonly used with a yellow filter, which blocks wavelengths shorter than about 0.5 μm. The cross-hatched area of Fig. 3-11 illustrates the blocking effect of a yellow filter.

With color IR film and a yellow filter, any object that reflects infrared energy will appear red on the final processed picture. Objects that radiate red will appear green, and objects radiating green will appear blue. It is this misrepresentation of color which accounts for the name "false color." Although color IR film was developed by the military, it has found a multitude of uses in civilian applications. Some of these applications are described in Chap. 20.

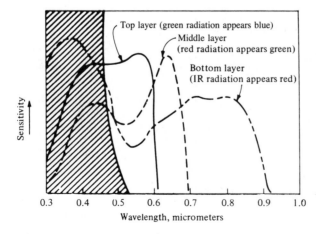

Figure 3-11 Sensitivity of color infrared (false-color) film.

Figure 3-12 Equipment for processing black-and-white aerial film. (*Courtesy Carl Zeiss, Oberkochen.*)

3-10 PROCESSING BLACK-AND-WHITE AERIAL FILM

Black-and-white aerial film is developed directly on the roll. The simplest method of developing aerial film is the *rewind process*. The equipment for this process, shown in Fig. 3-12, consists of three tanks, one each for developer, stop bath, and fixer. In addition there is a rewind apparatus consisting of two reels mounted on a frame. The apparatus just fits into the tanks. The film is wound onto one of the reels and immersed first into the developer tank. By means of either an electric motor or a hand crank, the film is wound back and forth from one reel to the other during developing. This rewind procedure is repeated for the stop bath and fixer. The film is then washed in clear running water and dried. Aerial film is usually developed in complete darkness because it is sensitive to all wavelengths of visible light.

More elaborate continuous processors are available which automatically develop, wash, and dry rolls of black-and-white film. Equipment is also available for the continuous processing of color aerial films.

3-11 CONTACT PRINTING

Contact printing is the direct process of making a photo positive from a negative. The emulsion side of a negative is placed in direct contact with the unexposed emulsion contained on printing paper, plastic material, or glass plate. Together these are placed in a contact printer and exposed with the emulsion of the positive facing the light source. Figure 3-13 illustrates a single-frame contact printer. The instrument is equipped with a timing device which automatically makes the exposure for the desired time. Two reel holders on either side of the stage make it possible to handle spools of aerial film easily. The rubber membrane on the cover of the printer is filled with air, so that when the cover is closed and the exposure made, all parts of the positive and negative

Figure 3-13 KG-30 single-frame contact printer. (*Courtesy Carl Zeiss, Oberkochen.*)

are pressed firmly together onto the glass stage. Contact printers which are capable of automatically and continuously exposing the aerial negatives of an entire roll of film are also available. In contact printing, the positive that is obtained is the same size as the negative from which it was made.

Uniform lighting throughout the negative during exposure will underexpose the emulsion in more dense areas of the negative and overexpose less dense areas. This can be compensated for in a process called *dodging,* which consists of adjusting the amount of light passing through different parts of the negative so that optimum exposure over the entire print is obtained in spite of density variations. The contact printer of Fig. 3-12 has a bank of lights arranged in a rectangular pattern of rows beneath the exposure stage. Dodging is done manually by turning off lamps in various positions until optimum overall lighting is achieved.

Log Etronic contact printers are available which automatically perform dodging. With these instruments a spotlight source from a cathode ray tube makes the exposure by scanning systematically back and forth across the negative. A photo tube monitors the light transmitted through the negative and automatically increases or decreases scanning speed to achieve optimum exposure in spite of varying negative density. Figure 3-14 shows the Log Etronic Mark IV variable dodging contact printer.

3-12 PROJECTION PRINTING

If positives are desired at a scale either enlarged or reduced from original negative size, the *projection printing* process is used. The geometry of projection printing is illustrated in Fig. 3-15. In this process, the negative is placed in the projector of the printer and illuminated from above. Light rays carry images c and d, for example, from the negative, through the projector lens, and finally to their locations C and D on the positive, which is situated on the easel plane beneath the projector. The emulsion of the positive, having been exposed, is then processed in the manner previously described.

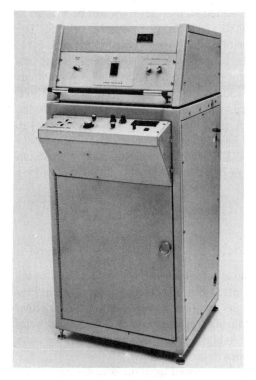

Figure 3-14 Log Etronic variable dodging contact printer. (*Courtesy Log Etronics, Inc.*)

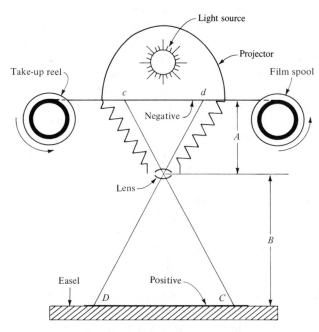

Figure 3-15 Geometry of enlargement with a projection printer.

Distances A and B of Fig. 3-15 can be varied so that positives can be printed at varying scales, and at the same time the lens formula, Eq. (2-8), can be satisfied. The enlargement or reduction ratio from negative to positive size is equal to the ratio B/A.

The easel of a projection printer often has many small holes in it which are connected to a vacuum system. When the exposure is made, this vacuum system holds the positive flat so that distortions from buckling are prevented. The easel and lens of some enlargers are capable of being tilted, which makes it possible to remove the distortions inherent in tilted photographs. A print made from a tilted photograph in which these distortions have been removed has the same geometry as a vertical photograph and is called a *rectified photograph*. Rectification is discussed in more detail in Secs. 11-14 through 11-20. A projection printer capable of enlarging and rectifying is shown in Fig. 11-19.

3-13 HALFTONE PROCESS

When ink reproductions are to be made from an original copy which contains varying gray tones, e.g., aerial photos, mosaics, etc., it is necessary to use the *halftone* process. The need for halftone stems from the fact that in reproduction, ink is either printed or it is not printed; there is no "half-ink" process which can produce tones of gray. If ink is printed, the resulting color is black; and if it is not printed, the color is that of the support material—white if paper is used.

To get varying tones of gray, halftones are prepared from the original copy using a *screen* in conjunction with a projection printer. The screen is a transparent medium upon which fine black lines have been etched to give a quadrille pattern which, except for the closeness of spacing, would look much like a screen door. The screen is placed on the easel over the emulsion of the film which will become the halftone. The screen breaks the light which exposes the film emulsion into small squares, and as a result the image is composed of a mass of small black dots and white spaces. The size of the black dots in any particular area of the halftone depends on the light intensity transmitted through that corresponding part of the negative; i.e., high-intensity light passing through light areas of the negative energizes the emulsion more than low-intensity light and thereby causes larger dots to be formed. In the reproduction process, the dots print ink and the spaces do not. Therefore an area of large dots will reproduce nearly black, while an area of very small dots will appear nearly white, and dots of varying intermediate sizes will produce shades of gray.

The dots of halftone copy are generally so small and close together that the human eye is unaware of them and their overall appearance yields smooth and continuous varying tones of gray. Halftone screens commonly have from about 40 to more than 200 lines per inch. A 100-line screen produces 10,000 small dots per square inch. Coarser screens of about 50 lines per inch are commonly used for rough work such as newspaper pictures, while finer screens such as 100 to 133 are usually used for printing high-quality illustrations, like those in this textbook. Figures 3-12 through 3-14 are but a few of the many examples of the use of halftones in this book. If these figures are examined under magnification, printing dots will actually be seen.

REFERENCES

American Society of Photogrammetry: "Manual of Color Photography," Falls Church, Va., 1968.

――――: "Manual of Photogrammetry," 4th ed., Falls Church, Va., 1980, chap. 6.

――――: "Manual of Photogrammetry," 3d ed., Falls Church, Va., 1966, chap. 6.

Anson, A.: Developments in Aerial Color Photography, *Photogrammetric Engineering,* vol. 34, no. 10, p. 1048, 1968.

Baines, H., and E. S. Bomback: "The Science of Photography," Halsted Press, John Wiley & Sons, Inc., New York, 1974.

Calhoun, J. M., et al.: Physical Properties of Estar Polyester Base Aerial Films for Topographic Mapping, *Photogrammetric Engineering,* vol. 27, no. 3, p. 461, 1961.

Carman, P. D., and S. F. Johnston: Effects of Ambient Conditions on Film Sensitivity, *Canadian Surveyor,* vol. 34, no. 2, p. 131, 1980.

―――― and H. Brown: Resolution of Four Films in a Survey Camera, *Canadian Surveyor,* vol. 24, no. 5, p. 550, 1970.

Craig, D. R.: Logetronics, *Photogrammetric Engineering,* vol. 21, no. 4, p. 556, 1955.

Engels, C. E.: "Photography for the Scientist," Academic Press, Inc., London, 1968.

Fleming, J.: Exploiting the Variability of Aerochrome Infrared Film, *Photogrammetric Engineering and Remote Sensing,* vol. 44, no. 5, p. 601, 1978.

Fritz, N. L.: Available Color Aerial Photographic Materials, *Photogrammetric Engineering and Remote Sensing,* vol. 42, no. 4, p. 525, 1976.

――――: Filters: An Aid in Color Infrared Photography, *Photogrammetric Engineering and Remote Sensing,* vol. 43, no. 1, p. 61, 1977.

Gliatti, E.: Modulation Transfer Analysis of Aerial Imagery, *Photogrammetria,* vol. 33, no. 5, p. 171, 1977.

Harman, W., Jr.: Recent Developments in Aerial Film, *Photogrammetric Engineering,* vol. 27, no. 1, p. 151, 1961.

James, T. H., and G. C. Higgins: "Fundamentals of Photographic Theory," 2d ed., Morgan and Morgan, Inc., New York, 1960.

Larmore, L.: "Introduction to Photographic Principles," 2d ed., Dover Publications, Inc., New York, 1965.

Malan, O. G.: Color Balance of Color-IR Film, *Photogrammetric Engineering,* vol. 40, no. 3, p. 311, 1974.

Mees, C. E. K.: "The Theory of the Photographic Process," rev. ed., The Macmillan Company, New York, 1954.

Michener, B. C.: Drying of Processed Aerial Films, *Photogrammetric Engineering,* vol. 29, no. 2, p. 321, 1963.

Norton, C. L., et al.: Optical and Modulation Transfer Function, *Photogrammetric Engineering and Remote Sensing,* vol. 43, no. 5, p. 613, 1977.

Rosenbruck, K.: Considerations Regarding Image Geometry and Image Quality, *Photogrammetria,* vol. 33, no. 5, p. 155, 1977.

Scarpace, F. L., and G. Friedricks: A Method of Determining Spectral Analytical Dye Densities, *Photogrammetric Engineering and Remote Sensing,* vol. 44, no. 10, p. 1293, 1978.

Schallock, G. W.: Metric Tests of Color Photography, *Photogrammetric Engineering,* vol. 34, no. 10, p. 1063, 1968.

Sorem, A. L.: Principles of Color Photography, *Photogrammetric Engineering,* vol. 33, no. 9, p. 1008, 1967.

Specht, M. R.: IR and Pan Films, *Photogrammetric Engineering,* vol. 36, no. 4, p. 360, 1970.

Stephens, P. R.: Comparison of Color, Color Infrared and Panchromatic Aerial Photography, *Photogrammetric Engineering and Remote Sensing,* vol. 42, no. 10, p. 1273, 1976.

Tarkington, R. G.: Kodak Panchromatic Negative Films for Aerial Photography, *Photogrammetric Engineering,* vol. 25, no. 5, p. 695, 1959.

―――― and A. L. Sorem: Color and False Color Films for Aerial Photography, *Photogrammetric Engineering,* vol. 29, no. 1, p. 88, 1963.

Welch, R.: Photogrammetric Image Evaluation Techniques, *Photogrammetria,* vol. 31, no. 5, p. 161, 1975.

———: Progress in the Specification and Analysis of Image Quality, *Photogrammetric Engineering and Remote Sensing,* vol. 43, no. 6, p. 709, 1977.

Worsfold, R. D.: More on Color Compensating Filters with Infrared Film, *Photogrammetric Engineering and Remote Sensing,* vol. 44, no. 1, p. 97, 1978.

———: Color Compensating Filters with Infrared Film, *Photogrammetric Engineering and Remote Sensing,* vol. 42, no. 11, p. 1385, 1976.

PROBLEMS

3-1 Explain why the lens camera replaced the early pinhole camera.

3-2 Define the photographic terms illuminance, aperture, emulsion, latent image, and fog.

3-3 A camera has a focal length of 55.0 mm. Its f-stop settings range from f-1.4 to f-22. What is the maximum diameter of the aperture? Minimum diameter?

3-4 Prepare a table of lens aperture diameters versus nominal f-stop settings ranging from f-1 to f-32 for an 80.0-mm-focal-length lens.

3-5 An exposure is optimum at a shutter speed of $\frac{1}{250}$ sec and f-4. If it is necessary to change the shutter speed to $\frac{1}{1,000}$ sec, what should be the corresponding f-stop to retain optimum exposure?

3-6 The problem is the same as Prob. 3-5, except that it is desired to expose at $\frac{1}{100}$ sec.

3-7 An exposure is optimum at a shutter speed of $\frac{1}{500}$ sec at f-5.6. To increase depth of field, it is necessary to expose at f-22. What is the required shutter speed to retain optimum exposure?

3-8 A camera has a focal length of 64.0 mm. What image distance is required for perfect focus if the object distance is 5 ft? 10 ft? 15 ft? 20 ft?

3-9 Repeat Prob. 3-8, except that the camera focal length is 28.0 mm.

3-10 What is the relationship between film speed and emulsion grain size?

3-11 What is the relationship between resolution and emulsion grain size?

3-12 What is a characteristic, H and D, or D-Log E curve?

3-13 Discuss the darkroom procedure for processing black-and-white emulsion.

3-14 Describe the electromagnetic spectrum in terms of the wavelengths of the various types of energy.

3-15 What wavelengths of energy form the primary colors?

3-16 What are the complementary colors of yellow, magenta, and cyan?

3-17 Explain when and why a "safe light" can be used in a darkroom.

3-18 Describe the characteristics of color infrared film.

3-19 Explain why a haze filter is used on aerial cameras.

3-20 In Fig. 3-15, assume that negative distance cd and positive distance CD measure 4.28 and 12.05 in, respectively. What is the enlargement factor for this print?

3-21 In Fig. 3-15, assume that distance A is 213 mm and, when the positive is in perfect focus, B is 382 mm. What is the focal length of the enlarger lens? What is the enlargement factor for this positive print?

3-22 Describe the halftone process and why it is needed.

3-23 Examine Fig. 3-3 under magnification and determine the number of lines per inch of the halftone screen used.

FOUR

AERIAL CAMERAS

4-1 INTRODUCTION

There are so many important instruments in photogrammetry that it would be difficult to specify the most significant. Surely, however, the camera is one of the most important since it is used to obtain the photographs upon which much photogrammetry depends. To understand the science of photogrammetry, especially the geometry of photographs, it is essential to have a basic understanding of cameras and how they operate.

The remarkable success of photogrammetry in recent years is due in large part to the progress that has been made in developing precision cameras. Perhaps the most noteworthy among recent camera developments has been the perfection of lenses of extremely high resolving power and almost negligible distortion. This has greatly increased the accuracy of photogrammetry. There have also been many significant improvements in general camera construction and operation.

In Chap. 1, the two different basic classifications of photographs, *terrestrial* and *aerial,* were described and examples of each were shown. Consistent with these photographic categories, cameras also fall into the general classifications of either terrestrial or aerial. Although in recent years a significant influx in the use of terrestrial photos has occurred, and their use continues to expand, applications using aerial photos still dominate the photogrammetric industry. For this reason the subject of aerial cameras is treated separately in this chapter, and various aspects of aerial photogrammetry are discussed in several chapters which follow immediately. Terrestrial cameras are discussed in Chap. 18.

The requirements of aerial mapping cameras are quite different from those of ordinary amateur cameras, such as that shown in Fig. 3-3. The primary requirement of any photogrammetric aerial camera is a lens of high geometric quality. Aerial cameras must be capable of exposing in rapid succession a great number of photographs to exacting specifications. Since they must perform this function while moving in an aircraft at high speed, they must have short cycling times, fast lenses, and efficient shutters. They must be capable of faithful functioning under the most extreme weather conditions and in spite of aircraft vibrations.

Aerial cameras generally use roll film and have magazine capacities of from 200 to 400 ft or more. Cameras have been developed to automatically expose pictures on glass plates, and although this procedure provides highest photogrammetric accuracy, it is not used extensively at this time because it is less convenient and more expensive.

Because the aerial photographic flight mission is fairly expensive and since weather and other conditions may prevent aerial photography for long periods of time, it is extremely important that every precaution be taken in the manufacture of aerial cameras to guarantee the quality and reliability of the photography on each mission.

4-2 TYPES OF AERIAL CAMERAS

Four main types of aerial cameras are (1) *single-lens frame* cameras, (2) *multilens frame* cameras, (3) *strip* cameras, and (4) *panoramic* cameras. These are described as follows:

4-2.1 Single-Lens Frame Cameras

Single-lens frame cameras are by far the most common cameras in use today. They are used almost exclusively in obtaining photographs for mapping purposes because they provide the highest geometric picture quality. With a single-lens frame camera the lens is held fixed relative to the focal plane. The film is generally fixed in position during exposure, although it may be advanced slightly during exposure to compensate for image motion. The entire format is exposed simultaneously with a single click of the shutter.

Single-lens frame cameras are often classified according to their *angular field of view*. Angular field of view, as illustrated in Fig. 4-1, is the angle α subtended at the rear nodal point of the camera lens by the diagonal d of the picture format. [The most common frame or format size of aerial mapping cameras is 9 in (23 cm) square.] Classifications according to angular field of view are

(*a*) Normal angle (up to 75°)
(*b*) Wide angle (75 to 100°)
(*c*) Super-wide angle (greater than 100°)

Angular field of view may be calculated as follows (see Fig. 4-1):

$$\alpha = 2 \tan^{-1} \left(\frac{d}{2f} \right) \tag{4-1}$$

For a nominal 6-in- (152-mm-) focal-length camera with a 9-in-(23-cm-) square format the angular field of view is:

$$\alpha = 2 \tan^{-1} \left(\frac{\sqrt{9^2 + 9^2}}{2(6)} \right) = 93° \text{ (wide angle)}$$

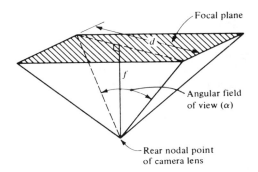

Focal plane

d

f

Angular field
of view (α)

Rear nodal point
of camera lens

Figure 4-1 Angular field view of a camera.

Single-lens frame cameras are available in a variety of lens *focal lengths,* and the choice will depend on the purpose of the photography. The most common one in use today for mapping photography has a 6-in (152-mm) focal length and 9-in-(23-cm-)square format, although $3\frac{1}{2}$-in (89-mm), $8\frac{1}{4}$-in (210-mm) and 12-in (305-mm) focal lengths with 9-in formats are also used. The 6-in focal length with 9-in format provides the best combination of geometric strength and photo scale for mapping. Longer focal lengths such as 12 in are used primarily for obtaining photographs for aerial mosaics and for reconnaissance and interpretation purposes. They enable reasonably large photo scales to be obtained in spite of high flying heights, and reduce image displacements due to relief variations (see Sec. 6-8).

From Eq. (4-1), it is seen that for a particular format size, angular field of view increases as focal length decreases. Short focal lengths, therefore, yield wider ground coverage at a given flying height than longer focal lengths. Short-focal-length cameras with smaller formats have been in common use in the space program, where the physical dimensions of the camera are somewhat limited by available room in the spacecraft.

In Fig. 1-4 the Zeiss RMK 15/23 aerial mapping camera was shown. Figure 4-2 illustrates the Fairchild KC-6A aerial mapping camera, and Fig. 4-3 shows the Wild RC-10. These three cameras, together with a few others, are being used today to take the bulk of aerial photos for mapping purposes. All three are precision single-lens frame cameras having 9-in-(23-cm-)square formats and film capacities of approxi-

Figure 4-2 Fairchild KC-6A aerial mapping camera. (*Courtesy Fairchild Space and Defense Systems.*)

Figure 4-3 Wild RC-10 aerial mapping camera. (*Courtesy Wild Heerbrugg Instruments, Inc.*)

mately 400 ft (120 m). The KC-6A and the RMK 15/23 both have nominal 6-in-(152-mm-)focal-length lenses. The RC-10 is capable of accepting interchangeable cones with lenses having nominal focal lengths of $3\frac{1}{2}$ in (89 mm), 6 in (152 mm), $8\frac{1}{4}$ in (210 mm), or 12 in (305 mm).

Figure 4-4 shows the new ITEK LFC Large Format single-lens frame camera. This camera's lens has a 12-in (305-mm) focal length. Its format is 9 by 18 in (23

Figure 4-4 LFC Large format single-lens frame camera. (*Courtesy ITEK Optical Systems.*)

Figure 4-5 Hasselblad MK-70 camera. (*Courtesy Paillard, Inc.*)

by 46 cm) and its film capacity is 4,000 ft (1,220 m). Designed principally for the space program, it will be carried in orbit by Space Shuttle and other spacecraft, although it can also be used in standard aircraft. The principal uses of the LFC camera, in addition to topographic mapping, will be in environmental monitoring and geologic exploration.

The Hasselblad camera of Fig. 4-5 is a small format single-lens frame camera that has been used extensively for space photography. It uses 70 mm film and can be obtained with various focal-length lenses. An example of a space photograph taken with a Hasselblad camera was shown in Fig. 1-9.

4-2.2 Multilens Frame Cameras

Multilens frame cameras have the basic characteristics of single-lens frame cameras except that they have two or more lenses and expose two or more pictures simultaneously. Figure 4-6 shows a multilens frame camera having six lenses. These types

Figure 4-6 MPC multilens camera. (*Courtesy ITEK Optical Systems.*)

of cameras, which are becoming increasingly popular, are used for environmental monitoring, mapping of natural and cultural resources, etc. All cameras simultaneously expose the same area, but the different cameras contain films with emulsions that are sensitive to different regions of the electromagnetic energy spectrum; hence they are commonly also referred to as *multispectral cameras*. Differences in the resulting photos provide clues that are useful in identifying and interpreting photographed objects.

In years past, multilens cameras were used to obtain greater width of ground coverage than could be obtained with single-lens frame cameras. The so-called *nine-lens camera* of the National Geodetic Survey, for example, exposed a vertical photo which was surrounded by eight low obliques. Its total angular coverage was 130°. *Trimetrogon* photography was used extensively for small-scale charting from about 1940 to 1960. This photography was obtained with a three-camera system. Two cameras exposed high oblique photos aimed at the flanks while a third camera simultaneously took a vertical photo. The three photos provided horizon-to-horizon ground coverage transverse to the direction of flight.

Convergent cameras can also be classified as multilens cameras since they simultaneously expose two photographs. These cameras are described in Sec. 4-3.

4-2.3 Strip Cameras

Strip cameras expose a continuous photograph of a strip of terrain beneath the path of the aircraft. This is accomplished by passing the film over a narrow slit opening in the focal plane of the camera at a rate synchronized with the speed of passage of ground images across the focal plane. Light rays entering the camera lens from a terrain point are therefore focused at a single point on the film during the exposure period. Strip cameras may use a single lens or they may have two lenses—one pointing, say, 20° forward in the direction of flight and the other pointing, say, 20° aft. This arrangement provides stereoscopic coverage. Strip photography, although not applied extensively, is useful in route studies for highways, railroads, pipelines, etc.

4-2.4 Panoramic Cameras

Panoramic cameras photograph a strip of terrain from horizon to horizon; the strip being transverse to the direction of flight. The subject of panoramic photography is discussed in more detail in Secs. 17-12 through 17-17.

4-3 CONVERGENT CAMERAS

A convergent camera, as shown in Fig. 4-7, is actually a special case of two single-lens frame cameras mounted together and operated simultaneously. One camera points forward along the flight line and the other backward. The result is two low oblique photographs taken from the same exposure station. Convergent photography provides a high order of photogrammetric accuracy by enabling a large *base-height* ratio (see

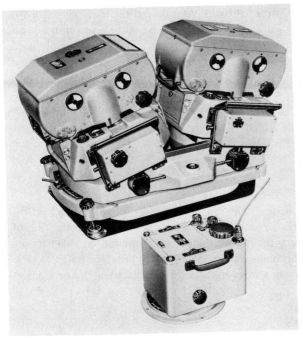

Figure 4-7 Zeiss 2-RMK 21/18 convergent camera. (*Courtesy Carl Zeiss, Oberkochen.*)

Sec. 7-8) to be achieved—a favorable geometric condition for mapping photography. Convergent photography enables up to 100 percent end lap to be attained, a factor which can reduce the amount of ground control required in mapping projects.

4-4 MAIN PARTS OF FRAME AERIAL CAMERAS

Although all frame aerial cameras are somewhat different in construction, they are enough alike so that a general description can be given which adequately encompasses all of them. The three basic components or assemblies of a frame aerial camera, as shown in the generalized cross section of Fig. 4-8, are (1) the *magazine,* (2) the *camera body,* and (3) the *lens cone assembly.*

4-4.1 Camera Magazine

The camera magazine houses the reels which hold exposed and unexposed film, and it also contains the *film advancing* and *film flattening* mechanisms. Film flattening is very important in aerial cameras, for if the film should be buckled during exposure, image positions on the resulting photographs would be incorrect. Film flattening may be accomplished in any of the following four ways: (1) by applying tension to the film during exposure; (2) by pressing the film firmly against a flat focal-plane glass which lies in front of the film; (3) by applying air pressure into the air-tight camera cone, thereby forcing the film against a flat plate lying behind the focal plane; or (4) by drawing the film tightly up against a vacuum plate whose surface lies in the focal plane. The vacuum system has proved most satisfactory and is the most widely used method of film flattening in aerial cameras. A focal-plane glass in front of the film is objectionable because image positions are distorted due to refraction of light rays passing through the glass (see Sec. 2-2). These distortions can be determined through calibration, however, and their effect eliminated in subsequent photogrammetric operations.

4-4.2 Camera Body

The camera body is a one-piece casting which usually houses the drive mechanism. The drive mechanism operates the camera through its cycle; the cycle consisting of (1) advancing the film, (2) flattening the film, (3) cocking the shutter, and (4) tripping the shutter. Power for the drive mechanism is most commonly provided by an electric motor. The camera body also contains carrying handles and electrical connections.

4-4.3 Lens Cone Assembly

The lens cone assembly contains a number of parts and serves several functions. Contained within this assembly are the *lens, filter, shutter,* and *diaphragm* (see Fig. 4-8). With most mapping cameras, the lens cone assembly also contains an *inner cone* or *spider*. The spider rigidly supports the lens assembly and focal plane in a fixed relative position. This fixes the so-called elements of *interior orientation* of the camera. These elements are carefully determined through camera calibration (see Sec. 4-10) so that they are available for photogrammetric calculations. The spider is made of metal having a low coefficient of thermal expansion, so that changes in operating temperatures do not upset the calibration. In some aerial cameras which do not have inner cones, the body and outer lens cone act together to hold the lens with respect to the focal plane. The inner cone of an aerial camera is shown in Fig. 4-9.

The *camera lens* is the most important (and most expensive) part of an aerial camera. It gathers light rays from the object space and brings them to focus in the focal plane behind the lens. Lenses used in aerial cameras are highly corrected compound lenses consisting of several elements. The lens of Fig. 2-19, for example, is the $3\frac{1}{2}$-in-(89-mm-)focal-length *Super Aviogon* used in the super-wide-angle lens cone of the Wild RC-9 and RC-10 aerial mapping cameras.

The *filter* serves three purposes: (1) it reduces the effect of atmospheric haze, (2) it helps provide uniform light distribution over the entire format, and (3) it protects the lens from damage and dust.

The *shutter* and *diaphragm* together regulate the length of time a given amount of light is allowed to pass through the lens to make the exposure. The shutter controls the length of time that light is permitted to pass through the lens. Shutters are discussed

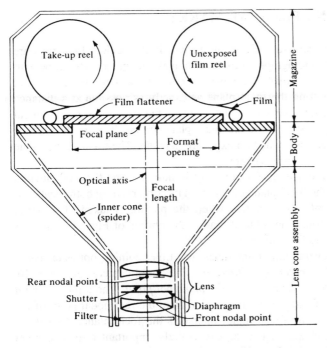

Figure 4-8 Generalized cross section of a frame aerial mapping camera.

in detail in Sec. 4-6. As discussed in Sec. 3-4, the diaphragm regulates the f-stops of the camera by varying the size of the aperture to control the amount of light passing through the lens. F-stops of aerial cameras typically range from about f-4 down to f-22. Thus, for a nominal 6-in-(152-mm-)focal-length lens, the diameter of the aperture would range from about 38 mm at f-4 to about 7 mm at f-22. The diaphragm is normally located in the air space between the lens elements of an aerial camera and consists of a series of leaves which can be rotated to vary the size of the opening.

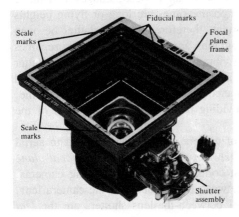

Figure 4-9 Inner cone of an aerial camera. (*Courtesy Fairchild Space and Defense Systems.*)

4-5 FOCAL PLANE AND FIDUCIAL MARKS

The *focal plane* of an aerial camera is the plane in which all incident light rays are brought to focus. In aerial photography object distances are great with respect to image distances. Aerial cameras therefore have their focus fixed for infinite object distances. This is done by setting the focal plane as exactly as possible at a distance equal to the focal length behind the rear nodal point of the camera lens. The focal plane is defined by the upper surface of the focal-plane frame. This is the surface upon which the film emulsion rests when an exposure is made. The focal-plane frame of the lens cone of Fig. 4-9 is clearly visible.

Camera fiducial marks are usually four or eight in number, and they are situated in the middle of the sides of the focal plane opening, in its corners, or in both locations. These marks are exposed onto the negative when the picture is taken. The aerial photograph of Fig. 1-5 has four side fiducial marks; the photos of Figs. 1-7 and 1-8 have four corner and four side fiducial marks.

Fiducial marks serve several important functions. Lines joining opposite marks intersect at a point called the *center of collimation,* and aerial cameras are carefully manufactured so that this occurs very close to the *principal point.* The principal point is defined as the point in the focal plane where a line from the rear nodal point of the camera lens, perpendicular to the focal plane, intersects the focal plane. As will be demonstrated in subsequent chapters, it is an exceedingly important reference point in photogrammetric work. Besides approximately locating the principal point, as described in Sec. 5-2, lines joining opposite fiducial marks also provide a rectangular coordinate axis system for measuring image positions on photographs. In addition to these uses, fiducial marks are also important for making corrections for image deformations due to shrinkage or expansion of photographic materials (see Secs. 5-10 and 5-11).

4-6 SHUTTERS

Because the aircraft carrying a camera moves at a rapid speed, images will move across the focal plane during exposure. If exposure times are long or flying heights low, blurred images may result. It is important, therefore, that the shutter be open for a very short duration when aerial photographs are taken. Short exposure times also reduce the detrimental effects of aircraft vibrations on image quality. The shutter speeds of aerial cameras typically range from about $\frac{1}{100}$ to $\frac{1}{1,000}$ sec. Shutters are designed to operate efficiently so that they open instantaneously, remain open the required time, and then instantaneously close, thereby providing uniform light to all parts of the focal plane.

There are a number of different types of camera shutters. Those used in aerial cameras are generally classified as either *between-the-lens* shutters or *focal-plane* shutters. Between-the-lens shutters are most commonly used in mapping cameras. These shutters are placed in the air space between the elements of the camera lens, as illustrated in Fig. 4-8. Common types of between-the-lens shutters are the *leaf* type, *blade* type, and *rotating disk* type. A schematic diagram of the leaf type is shown in Fig. 4-10. It consists usually of five or more leaves mounted on pivots and spaced around the periphery of the diaphragm. When the shutter is tripped, the leaves rotate about their pivots to the open position of Fig. 4-10*b*, remain open the desired time, and then snap back to the closed position of Fig. 4-10*a*. Some camera shutters use two sets of leaves, one for opening and the other for closing. This increases shutter efficiency, shutter speed, and shutter life.

The blade-type shutter consists of four blades, two for opening and two for closing. Its operation is similar to that of a guillotine. When the shutter is triggered, the two thin "opening" plates or blades move across the diaphragm to open the shutter. When the desired exposure time has elapsed, two "closing" blades close it.

The rotating disk type of shutter consists of a series of continuously rotating disks. Each disk has a cutaway section, and when these cutaways mesh, the exposure is made. The speed of rotation of the disks can be varied so that the desired exposure times are obtained. This type of shutter is very efficient because no starting or stopping of the parts is required, as with other types.

Focal-plane shutters are so named because they are located in front of the focal plane. The most common type of focal-plane shutter, the *curtain* type, consists of a curtain containing a slit. The curtain width equals the width of the focal plane. When the shutter is tripped, the slit moves across the focal plane. Exposure time is varied by varying either the speed at which the curtain moves or the width of the slit. These shutters expose different areas of the focal plane at slightly different times, and this causes relative image position errors in the resulting pictures. They are therefore not suitable for mapping cameras, but are used in reconnaissance cameras.

Another type of focal-plane shutter is the *louver* shutter. It consists of a number of louvers which are operated simultaneously in a manner similar to the operation of venetian blinds. These shutters are not as efficient as other types, and shadows created by the open louvers cause uneven lighting over the focal plane.

Cameras have been perfected to compensate for image motion which occurs during the time that the shutter is open. *Image motion compensation* (IMC) is usually accomplished by moving the film slightly across the focal plane during exposure, in the direction of, and at a rate just equal to, the rate of image movement.

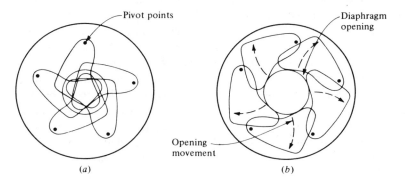

Figure 4-10 Schematic diagram of a leaf-type shutter. (*a*) Shutter closed; (*b*) shutter open.

4-7 CAMERA MOUNTS

The camera mount is the mechanism used to attach the camera to the aircraft. It has dampener devices which prevent aircraft vibrations from being transmitted to the camera. The mount is also usually designed so that the camera can be rotated in azimuth to correct for *crab*. Crab is a disparity in the orientation of the camera in the aircraft with respect to the aircraft's actual travel direction. It is usually the result of side winds which cause the aircraft's direction of heading to deviate from its actual travel direction, as shown in Fig. 4-11*a*. Crab may be of variable amounts, depending on wind velocity and direction. It has the undesirable effect of reducing the stereoscopic ground coverage of aerial photos, as shown in Fig. 4-11*b*. Figure 4-11*c* shows the ground coverage when the camera has been rotated within the mount in the aircraft to make two sides of the format parallel to the actual direction of travel. Aerial cameras are usually equipped with level vials for keeping the camera properly oriented for vertical photos.

4-8 CAMERA CONTROLS

Camera controls are those devices necessary for operating the camera and varying camera settings according to conditions at the time of photography. The *intervalometer* is a device which automatically trips the shutter and actuates the camera cycle at desired times. Older types of intervalometers could be set to automatically make exposures at fixed intervals of time. The time interval could be calculated and depended upon the camera focal length and format size, desired end lap, flying height, and aircraft velocity. The disadvantage of this type of intervalometer is that with fixed time intervals, variations in end lap occur with variations in terrain elevation, flying height, or aircraft velocities.

   Newer intervalometers, such as that shown in Fig. 4-12, make exposures at a desired percent end lap in spite of variations in terrain, flying height, and aircraft velocity. This is done by means of a rotating chain shown in the *viewfinder*. (The viewfinder enables the operator to continually view the terrain beneath the aircraft and to see the ground coverage of each photo.) The chain moves in the viewfinder in the same direction as the passing images. By means of a rheostat, an operator can vary the rate of movement of the chain and make it travel at the same rate as passing images. The desired end lap is set on a dial on the intervalometer, and when the chain has moved the amount which corresponds to the desired end lap, the intervalometer automatically actuates the camera cycle. Of course, an operator can manually actuate the camera cycle at any desired time. The operator can also correct for crab by rotating the viewfinder so that the passing images move parallel to the edges of the picture format. The camera is then given the same amount of rotation so that crab is eliminated.

   Another aerial camera control device is the *exposure control mechanism*. This

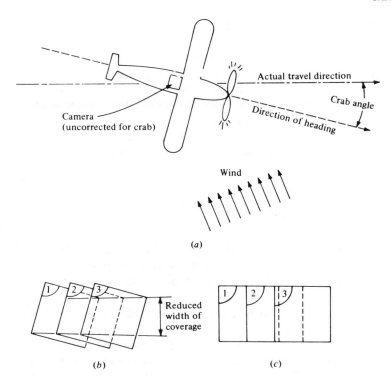

(a)

(b)   *(c)*

Figure 4-11 (*a*) Camera exposing aerial photography with crab present. (*b*) Crabbed overlapping aerial photographs. (*c*) Overlapping aerial photographs with no crab.

consists of an exposure meter which measures terrain brightness and correlates it with the optimum combination of diaphragm opening and shutter speed. Exposure control units are available which operate automatically and constantly vary camera settings to provide optimum exposures.

Figure 4-12 IRU Intervalometer. (*Courtesy Carl Zeiss, Oberkochen.*)

4-9 AUTOMATIC DATA RECORDING

Most modern aerial mapping cameras are equipped with a data recording system which automatically produces pertinent data on the pictures. The usual data consists of date, flying altitude, calibrated focal length of the camera, photo number, job identification, etc. This information is entered on a data block which is exposed onto the film when the photograph is taken. The images of an automatic data recording system are shown along the top border of the aerial photograph in Fig. 1-5. Automatic data recording is a convenience which saves time and prevents mistakes in later use of the photographs.

4-10 CAMERA CALIBRATION

After manufacture and prior to use, aerial cameras are carefully calibrated to determine precise values for a number of constants. These constants, generally referred to as the *elements of interior orientation,* are needed so that accurate data can be determined from photographs.

In general, camera calibration methods may be classified into one of three basic categories: (1) *laboratory* methods, (2) *field* methods, and (3) *stellar* methods. Of these, laboratory methods are most frequently utilized and are normally performed by either camera manufacturers or agencies of the Federal Government. In one of the two methods of laboratory calibration, which uses a *multicollimator,* and in the field and stellar procedures, the general approach consists of photographing an array of targets whose relative positions are accurately known. Elements of interior orientation are then determined by making precise measurements of the target images and comparing their actual imaged locations with the positions they should have occupied had the camera produced a perfect perspective view. In the other laboratory method, which employs a *goniometer,* direct measurements are made of projections through the camera lens of precisely positioned grid points located in the camera focal plane. Comparisons are then made with what the true projections should have been.

The elements of interior orientation that can be determined through camera calibration are

1. *Equivalent focal length*. The focal length which is effective near the center of the camera lens.
2. *Calibrated focal length* (often called the *"camera constant"*). The focal length which produces an overall mean distribution of radial lens distortion.
3. *Average radial lens distortion*. Distortion in image position along radial lines from the principal point.
4. *Tangential lens distortion*. Distortion in image position perpendicular to radial lines from the principal point. (It is normally very small and except for the most precise work, can usually be neglected.)
5. *Principal point location*. Coordinates of the principal point given with respect to the *x* and *y* fiducial axes. (Although it is the intent in camera manufacture to place

the fiducial marks so that lines between opposite pairs intersect at the principal point, there is nearly always some small deviation from this ideal condition.)

6. *Distances between opposite fiducial marks.* (Often given by coordinates of the fiducial marks.)
7. *Angle of intersection of fiducial lines.* (Should be 90° ± 1').
8. *Flatness of focal plane.* [Should not deviate by more than ± 0.0005 in (0.01 mm) from a plane].

In addition to the determination of the above elements of interior orientation, *resolution* (the sharpness or crispness with which a camera can produce an image) is also commonly obtained as a part of camera calibration.

The procedures of determining items 1 through 5 vary somewhat, depending upon whether the calibration method is laboratory, field, or stellar. In any case, the mathematics of this phase of camera calibration can become rather complicated. Items 6 and 7 are obtained from direct precision measurements on an exposed glass plate (*flash plate*) upon which the fiducial marks are imaged. As discussed in Chap. 5, calibrated *x* and *y* fiducial distances are important in correcting photographic measurements for film shrinkage or expansion. Item 8 is measured directly by means of a special gage. Resolution is determined by direct observation and measurement of the images produced by a given lens.

4-11 LABORATORY METHODS OF CAMERA CALIBRATION

As noted in the previous section, the *multicollimator* method and the *goniometer* method are two types of laboratory procedures of camera calibration. The multicollimator method consists of photographing onto a glass plate images projected through a number of individual collimators mounted in a precisely known angular array. A single collimator consists of a lens with a cross mounted in its plane of infinite focus. Therefore, light rays carrying the image of the cross are projected through the collimator lens and emerge parallel. If these light rays are directed toward the lens of an aerial camera, the cross will be perfectly imaged on the camera's focal plane because aerial cameras are focused for parallel light rays (those having infinite object distances).

A multicollimator for camera calibration consists of several individual collimators mounted in two perpendicular vertical planes. One plane of collimators is illustrated in Fig. 4-13. The individual collimators are rigidly mounted so that the optical axes of adjacent collimators intersect precisely at known angles, such as θ of Fig. 4-13. The camera to be calibrated is placed so that its focal plane is perpendicular to the central collimator axis and so that the front nodal point of its lens is at the intersection of all collimator axes. In this orientation, image *g* of the central collimator, which is called the *principal point of autocollimation*, occurs very near the principal point, and also very near the intersection of fiducial lines (center of collimation). The camera is further oriented so that when the calibration exposure is made, the collimator crosses will be imaged along the diagonals of the camera format, as shown in Fig. 4-14.

Collimator crosses A through M of Fig. 4-13, for example, are imaged at a through m of Fig. 4-14. Crosses from the perpendicular plane of collimators are imaged at n through y of Fig. 4-14.

Distances on the exposed plate between imaged crosses are precisely measured. *Equivalent focal length* (EFL) is that focal length which is effective in the essentially distortion-free central area of the camera lens. In the image plane, this area is considered to be within the circle whose center is at the central collimator image g and whose radius is the mean distance to the four collimator images, f, t, h, and s. Equivalent focal length is computed by dividing the average of the four measured distances gf, gh, gs, and gt by the tangent of θ, or

$$\text{EFL} = \frac{gf + gh + gs + gt}{4 \tan \theta} \tag{4-2}$$

Based on EFL, theoretical distances from central collimator image g to all other collimator crosses can be computed. As an example, distances ge, gi, gu, and gr should theoretically be equal to EFL $\times \tan 2\theta$. These four distances are measured, and radial-lens distortion at 2θ is obtained by subtracting the *calculated* theoretical distance from the average of the four *measured* distances. Using this approach, radial-lens distortion can be calculated for each increment of θ.

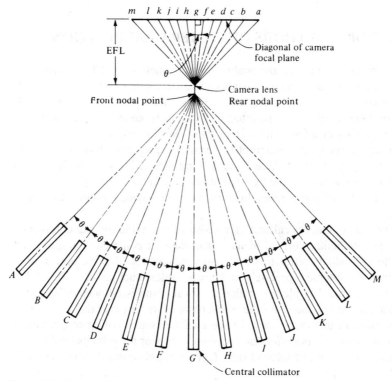

Figure 4-13 Bank of thirteen collimators for camera calibration.

Example 4-1 The following table lists the averages of the four measured distances to collimator crosses versus their angular values from the central cross. Calculate the equivalent focal length (EFL) based on the $7\frac{1}{2}°$ crosses, and determine average radial-lens distortions based upon EFL.

Angle	Average measured distance, mm
$7\frac{1}{2}°$	20.183
$15°$	41.048
$22\frac{1}{2}°$	63.481
$30°$	88.574
$37\frac{1}{2}°$	117.662
$45°$	153.135

SOLUTION From Eq. (4-2),

$$\text{EFL} = \frac{20.183}{\tan 7\frac{1}{2}°} = 153.305 \text{ mm}$$

Based on EFL, the average radial-lens distortion Δr, at the $15°$ crosses, is

$$\Delta r = 41.048 - 153.305 \tan 15° = -0.030$$

This calculation and those for the other crosses are tabulated below.

Angle	Average measured distance, mm	EFL tan angle, mm	Radial lens distortion Δr, mm
$15°$	41.048	41.078	-0.030
$22\frac{1}{2}°$	63.481	63.501	-0.020
$30°$	88.574	88.511	0.063
$37\frac{1}{2}°$	117.662	117.635	0.027
$45°$	153.135	153.305	-0.170

From the preceding discussion, it should be apparent that a focal length other than EFL could be used to calculate radial-lens distortion and that a different focal length would produce different radial-lens distortions. *Calibrated focal length* (CFL) of a lens is a focal length which produces an overall mean distribution of radial distortion, and is selected so that maximum positive radial distortion is equal to maximum negative radial distortion. Calibrated focal length is the value normally used in photogrammetric calculations, since if corrections for radial-lens distortions are neglected, minimum detrimental effects are caused. Henceforth in this text, the term "focal length," symbolized by f, will be understood to mean calibrated focal length.

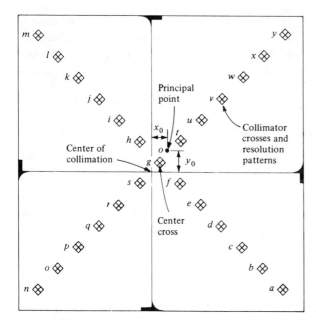

Figure 4-14 Images of photographed collimator targets.

Example 4-2 For Example 4-1, compute the calibrated focal length (CFL) which causes maximum positive and negative radial-lens distortions to be equal. Then calculate radial-lens distortions based upon this CFL.

SOLUTION In Example 4-1, maximum positive radial-lens distortion occurs at 30°, and maximum negative distortion occurs at 45°. With a slight change in focal length, the maximum positive and negative values will still occur at these angles. An equation in terms of the CFL which will cause these two values to be equal is

$$\text{Maximum positive } \Delta r + \text{maximum negative } \Delta r = 0$$
$$88.574 - \text{CFL tan } 30° + 153.135 - \text{CFL tan } 45° = 0$$
$$241.709 = 1.577350 \text{ CFL}$$
$$\text{CFL} = 153.237 \text{ mm}$$

Using this CFL, radial-lens distortions are computed, as before, in tabular fashion.

Angle	Average measured distance, mm	CFL tan angle, mm	Radial-lens distortion Δr, mm
$7\frac{1}{2}°$	20.182	20.174	+0.008
15°	41.048	41.060	−0.012
$22\frac{1}{2}°$	63.481	63.473	+0.008
30°	88.574	88.472	+0.102
$37\frac{1}{2}°$	117.662	117.583	+0.079
45°	153.135	153.237	−0.102

If the measured distances to equal-angle collimator crosses along the four diagonals are of equal magnitude, lens distortions determined in the above manner are symmetrical about the central collimator cross. If the distortions are not symmetrical about the central collimator cross, the point about which they are symmetrical, called the *principal point of symmetry,* can be determined. However, except for the most precise analytical photogrammetry work, the principal point is so close to the central collimator cross, and the distortions are so nearly symmetrical, that the principal point can be considered as the point of symmetry for making distortion corrections. The manner of making radial-lens-distortion corrections is described in Sec. 5-12.

A *radial distortion curve* may be drawn by plotting radial distortion on the ordinate versus radial distance from the central collimator image as the abscissa. An example of such a curve is shown in Fig. 4-15. This curve indicates at a glance the nature of the radial distortion of a particular lens. Note that the curve of Fig. 4-15 is based upon CFL, since the maximum positive and maximum negative distortions are essentially equal.

A paradox associated with the multicollimator method of camera calibration is that targets are dense near the center of the format where lens distortion is practically nonexistent, and sparse in the outer areas of the format where lens distortions are maximum. This problem can be circumvented in either field or stellar calibration.

The goniometer laboratory procedure of camera calibration is very similar to the multicollimator method, but consists of centering a precise grid plate in the camera focal plane. The grid is illuminated from the rear and projected through the camera lens in the reverse direction. The angles at which the projected grid rays emerge are measured with a goniometer. CFL and radial distortions are then determined by comparing actual measured angles with their theoretically true angles.

4-12 FIELD AND STELLAR METHODS OF CAMERA CALIBRATION

Both the multicollimator and goniometer methods of laboratory camera calibration require expensive and precise special equipment. An advantage of field and stellar methods is that this special equipment is not necessary. Several different field and stellar methods of camera calibration have been developed. Field procedures require that an array of targets be established and that their locations with respect to the camera station be measured using precise surveying techniques. The targets are placed far enough from the camera station so that there is no noticeable image degradation. (Recall that an aerial camera is fixed for infinite focus.)

In the stellar method, a target array consisting of identifiable stars is photographed, and the instant of exposure is recorded. Right ascensions and declinations of the stars can be obtained from an ephemeris for the precise instant of exposure so that the angles subtended by the stars at the camera station become known. These then are compared to the angles obtained from precise measurements of the imaged stars.

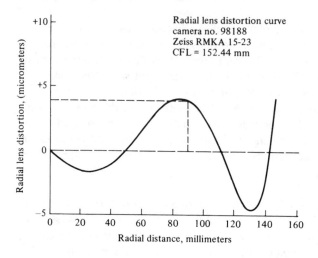

Figure 4-15 Radial-lens distortion curve for the Zeiss Pleogon aerial camera lens. (*Courtesy Wisconsin Department of Transportation.*)

In recent years, researchers have been investigating "in-flight" camera calibration where an array of precisely surveyed ground control targets are photographed. Potential advantages of this method are new calibration constants each time the camera is used and the likelihood of greater accuracy due to calibration under actual operating conditions.

4-13 CALIBRATING THE RESOLUTION OF A CAMERA

In addition to determining interior orientation elements, laboratory methods of camera calibration also provide an evaluation of the camera's resolving power. As noted in Sec. 2-12, there are two common methods of specifying lens resolving power; one is a direct count of the maximum number of lines per millimeter that can be clearly reproduced by a lens; the other is the *modulation transfer function* (MTF) of the lens. The method of calibration employed to determine line-count consists of photographing resolution test patterns using a very-high-resolution emulsion. The test patterns, an example of which is shown in Fig. 4-16, are made up of numerous sets of *line pairs* (parallel black lines of varying thickness separated by white spaces of the same thickness). The measure of line thickness for each set is its number of lines per millimeter. Line-thickness variations in a typical test pattern may range from 10 to 80 or more lines per millimeter. If the multicollimator method is used to calibrate a camera, the test patterns may be projected by the collimators simultaneously with the collimator crosses and imaged on the diagonals of the camera format. After the photograph is made, the resulting images are examined under magnification to determine the finest set of parallel lines which can be clearly resolved. The average of the four resolutions at each angular increment from the central collimator is reported in the calibration certificate.

Whereas the above-described maximum-line-count method appears to be a relatively simple way of quantifying resolving power, it is not without its shortcomings.

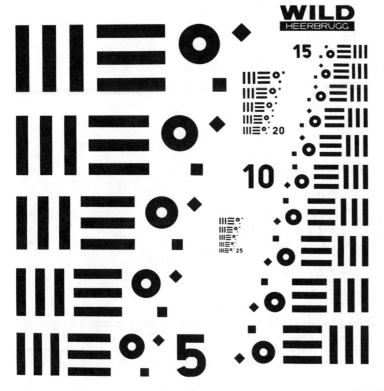

Figure 4-16 Resolution test pattern used in camera calibration. (*Courtesy Wild Heerbrugg Instruments, Inc.*)

In the line-count procedure, with each succeedingly smaller test pattern, the sharpness of distinction between lines and spaces steadily diminishes, and the smallest pattern which can clearly be discerned becomes somewhat subjective. The preferred measure of resolution is by modulation transfer.

To determine modulation transfer, density scans using a microdensitometer (see Sec. 5-15) are taken across test patterns similar to those used in the line-count procedure. For heavy lines with wide spacing, the actual distribution of brightness (density variations) across the object pattern would appear as shown in Fig. 4-17a. However, brightness distributions measured with a densitometer across the image of this pattern would appear as illustrated in Fig. 4-17b. Note that the edges are rounded somewhat in Fig. 4-17b, but the amplitude (or *modulation*) of brightness differences is the same for the image patterns as for the original object. Thus at this spatial frequency of the pattern, modulation transfer is said to be 100 percent. Density scans across the images of successively closer-spaced patterns will produce reduced modulations, as illustrated in Fig. 4-17c. In this case, amplitude is half that of the original object, and the modulation transfer is 50 percent. By measuring densities across many patterns of varying spatial frequencies, and plotting the resulting modulation transfer on the ordinate versus spatial frequency on the abscissa, a curve such as that illustrated in Fig.

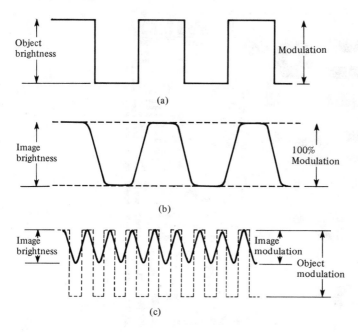

(a)

(b)

(c)

Figure 4-17 (*a*) Modulation of test object. (*b*) Modulation transfer of image of same test object. (*c*) Modulation transfer of image having closer spatial frequency. (Note in (*b*) that 100 percent modulation occurs, but the image shows a reduction in edge sharpness as compared to the test object. In (*c*) edge sharpness is further reduced, and in addition, modulation transfer is reduced to 50 percent of that of the test object.)

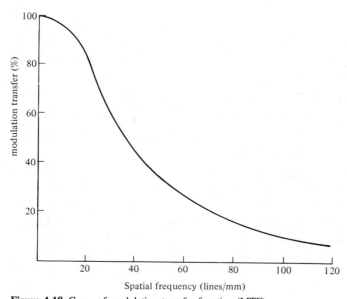

Figure 4-18 Curve of modulation transfer function (MTF).

4-18 is obtained. This curve is the modulation transfer function (MTF). The MTF has a number of advantages. It is a very sensitive indicator of *edge effects* and it also affords the capability of predicting the resolution which may be expected at any given degree of detail. Furthermore, MTF curves can be combined for different lenses, films, and film processes; thus, it is possible to estimate the combined effects of any given imaging system. For these reasons, it is rapidly becoming the preferred method of expressing resolution.

REFERENCES

Abdel-Aziz, Y.: Asymmetrical Lens Distortion, *Photogrammetric Engineering and Remote Sensing,* vol. 41, no. 3, p. 337, 1975.

American Society of Photogrammetry: "Manual of Photogrammetry," 3d ed., Falls Church, Va., 1966, chap. 4

————: "Manual of Photogrammetry," 4th ed., Falls Church, Va., 1980, chap. 4.

Anderson, J. M., and C. Lee: Analytical In-Flight Calibration, *Photogrammetric Engineering and Remote Sensing,* vol. 41, no. 11, p. 1337, 1975.

Bormann, G. E.: The New Wild RC-10 Film Camera, *Photogrammetric Engineering,* vol. 35, no. 10, p. 1033, 1969.

Brock, G. C.: The Possibilities for Higher Resolution in Air Survey Photography, *Photogrammetric Record,* vol. VIII, no. 47, p. 589, 1976.

Carman, P. D.: Camera Calibration Laboratory at N. R. C., *Photogrammetric Engineering,* vol. 35, no. 4, p. 372, 1969.

————: Camera Vibration Measurements, *Canadian Surveyor,* vol. 27, no. 3, p. 208, 1973.

Doyle, F. J.: A Large Format Camera for Shuttle, *Photogrammetric Engineering and Remote Sensing,* vol. 45, no. 1, p. 73, 1979.

Hakkarainen, J.: Image Evaluation of Reseau Cameras, *Photogrammetria,* vol. 33, no. 4, p. 115, 1977.

Hallert, B.: The Method of Least Squares Applied to Multicollimator Camera Calibration, *Photogrammetric Engineering,* vol. 29, no. 5, p. 836, 1963.

Helava, U. V.: New Significance of Errors of Inner Orientation, *Photogrammetric Engineering,* vol. 29, no. 1, p. 126, 1963.

Karren, R. J.: Camera Calibration by the Multicollimator Method, *Photogrammetric Engineering,* vol. 34, no. 7, p. 706, 1968.

Livingston, R. G.: A History of Military Mapping Camera Development, *Photogrammetric Engineering,* vol. 30, no. 1, p. 97, 1964.

Lockwood, H. E., and L. Perry: Shutter/Aperture Settings for Aerial Photography, *Photogrammetric Engineering and Remote Sensing,* vol. 42, no. 2, p. 239, 1976.

McNeil, G. T.: Normal Angle Camera Calibrator, *Photogrammetric Engineering,* vol. 28, no. 4, p. 633, 1962.

Merchant, D. C.: Calibration of the Air Photo System, *Photogrammetric Engineering,* vol. 40, no. 5, p. 605, 1974.

Merritt, E. L.: Methods of Field Camera Calibration, *Photogrammetric Engineering,* vol. 18, no. 4, 1952.

Nielsen, V.: More on Distortions by Focal Plane Shutters, *Photogrammetric Engineering and Remote Sensing,* vol. 41, no. 2, p. 199, 1975.

Rampal, K. K.: System Calibration of Metric Cameras, *ASCE Journal of the Surveying and Mapping Division,* vol. 104, no. SU1, p. 51, 1978.

Rhody, B.: A New Versatile Stereo-Camera System for Large-Scale Helicopter Photography of Forest Resources, *Photogrammetria,* vol. 32, no. 5, p. 183, 1977.

Scholer, H.: On Photogrammetric Distortion, *Photogrammetric Engineering and Remote Sensing,* vol. 41, no. 6, p. 761, 1975.

Tayman, W.: Calibration of Lenses and Cameras at the USGS, *Photogrammetric Engineering,* vol. 40, no. 11, p. 1331, 1974.

Washer, F. E.: The Precise Evaluation of Lens Distortion, *Photogrammetric Engineering,* vol. 29, no. 2, p. 327, 1963.

Welch, R., and J. Halliday: Imaging Characteristics of Photogrammetric Camera Systems, *Photogrammetria,* vol. 29, no. 1, p. 1, 1973.

PROBLEMS

4-1 List the requirements of a precision aerial mapping camera.

4-2 Name and briefly describe the various types of aerial cameras.

4-3 Photography is obtained with a strip camera having a 6-in focal length from an altitude of 5,000 ft above ground. If the aircraft speed is 200 mi/hr, what must be the rate of passage of the film across the focal plane of the camera (in inches per minute) to prevent image blur?

4-4 Repeat Prob. 4-3, except that the altitude is 10,000 ft and the aircraft speed is 600 mi/hr.

4-5 An aerial camera makes an exposure at a shutter speed of $\frac{1}{500}$ sec. If the aircraft speed is 300 mi/hr, how far will the aircraft travel during the exposure?

4-6 Repeat Prob. 4-5, except that the shutter speed is $\frac{1}{1,000}$ sec and the aircraft speed is 550 mi/hr.

4-7 An aerial camera having a $3\frac{1}{2}$-in focal length is carried by an aircraft at 150 mi/hr. If flying height is 3,000 ft above ground and if exposure time is $\frac{1}{200}$ sec, how far will an image move across the focal plane during the exposure?

4-8 Repeat Prob. 4-7, except that the focal length is $8\frac{1}{4}$-in and the exposure time is $\frac{1}{500}$ sec.

4-9 What is the angular field of view of a camera having a 53-mm-square format and a 38-mm focal length? A 60-mm focal length?

4-10 Repeat Prob. 4-9, except that the camera has a 70-mm-square format and 100-mm focal length.

4-11 For a camera having a 9-in-square format, what range of focal lengths could it have to be classified as wide angle?

4-12 Name and briefly describe the main parts of a frame aerial camera.

4-13 Discuss briefly the different types of camera shutters.

4-14 What is the function of the camera mount? The intervalometer?

4-15 What is crab and how may it be caused?

4-16 Why is camera calibration important?

4-17 What are the elements of interior orientation that can be determined in camera calibration?

4-18 Define the term ''principal point.''

4-19 To make a rough determination of a camera's focal length, a 12-ft level rod was held exactly 20 ft from the camera lens, and an exposure made. The rod was held perpendicular to the camera axis, and the exposure made so that the 6-ft marker on the rod was imaged in the center of the photograph. The measured length on the negative of the level rod, from end to end, was 29.95 mm. What was the focal length of the camera lens?

4-20 The following table lists the averages of the four measured distances to collimator crosses versus their angular values from the central cross. Calculate the equivalent focal length based on $7\frac{1}{2}°$ crosses and plot the radial distortion curve based on EFL.

Angle	Averaged measured distance, mm
$7\frac{1}{2}°$	20.081
15°	40.876
$22\frac{1}{2}°$	63.201
30°	88.097
$37\frac{1}{2}°$	117.067
45°	153.482

4-21 Compute the calibrated focal length for the data of Prob. 4-20. (Make maximum positive radial distortion equal to maximum negative radial distortion.) Plot the radial distortion curve based on CFL.

4-22 Repeat Prob. 4-20, except that the following data was obtained:

Angle	Averaged measured distance, mm
5°	18.345
10°	36.970
15°	56.191
20°	76.331
25°	97.787
30°	121.054
35°	146.839

4-23 Repeat Prob. 4-21, except that it applies to the data of Prob. 4-22.

4-24 Name and briefly describe two methods of specifying resolution.

FIVE

PHOTOGRAPHIC MEASUREMENTS AND REFINEMENT

5-1 INTRODUCTION

The solution of photogrammetric problems generally requires some type of photographic measurement. The measurements may be lengths of lines between imaged points, angles between points, or positions of points on photos expressed in terms of rectangular coordinates. Rectangular coordinates are the most common type of photographic measurement, and they are used directly in many photogrammetric equations. Photographic measurements are usually made on positives printed on paper, film, or glass. They could also be made directly on the negatives; however, this is seldom done because this can deface the imagery and it is important to preserve the negatives for making additional prints. Besides length, angle, and coordinate measurements, image densities are also becoming a very common type of photographic measurement and this is discussed in Sec. 5-15.

Equipment used for making photographic measurements varies from inexpensive, simple scales to very elaborate and complex machines which provide automatic digital output. These various types of instruments and the manner in which they are used are described in this chapter. Due to several effects, there will be systematic errors associated with practically all photographic measurements. The sources of these errors and the manners by which they are eliminated are also discussed in this chapter.

5-2 PHOTOGRAPHIC COORDINATE SYSTEM

For cameras with side fiducial marks, the commonly adopted reference system for photographic coordinates is the rectangular axis system formed by joining opposite fiducial marks with straight lines, as shown in Fig. 5-1a. The x axis is usually arbitrarily designated as the fiducial line most nearly parallel with the direction of flight, positive in the direction of flight. The positive y axis is 90° counterclockwise from positive x. The origin of the coordinate system is the intersection of fiducial lines. This point is often called the *center of collimation;* for a precise mapping camera, it is very near the principal point.

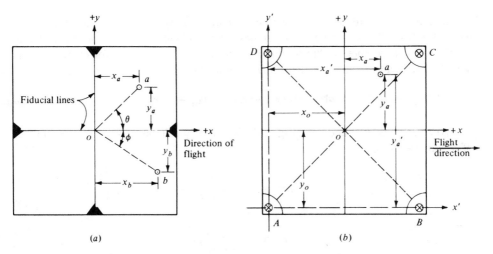

Figure 5-1 Photographic coordinate system. (*a*) side fiducials; (*b*) corner fiducials.

The position of any image on a photograph, such as a of Fig. 5-1a, is given by its rectangular coordinates x_a and y_a, where x_a is the perpendicular distance from the y axis to a and y_a is the perpendicular distance from the x axis to a. Similarly the photographic position of image point b is given by its rectangular coordinates x_b and y_b.

If the camera is equipped with corner fiducials, as is the case with Wild cameras, the reference axes for photocoordinate measurement may be arbitrarily taken as the $x'y'$ system shown in Fig. 5-1b. In this system positive x' is again arbitrarily taken in the direction of flight. The intersection of diagonal lines connecting opposite corner fiducials very nearly defines the principal point. Coordinates measured in the $x'y'$ system are reduced to the conventional xy system with origin at the principal point using the following equations:

$$x_a = x'_a - x_o$$
$$y_a = y'_a - y_o$$

(5-1)

where

$$x_o = \frac{x'_B + x'_C}{4}$$

and

$$y_o = \frac{y'_D + y'_C}{4}$$

It is now quite common for aerial cameras to have eight fiducials installed, in both side and corner locations. Figures 1-7 and 1-8 show this fiducial mark configuration. The photographic coordinate system in this case is defined in Fig. 5-1a. Eight fiducials enable somewhat more accurate corrections to be made for systematic errors in measured image coordinates.

Rectangular coordinates are a very basic and useful type of photographic measurement, for from them distances and angles between points may be calculated using simple analytic geometry. Photographic distance ab of Fig. 5-1a, for example, may be calculated from rectangular coordinates as follows:

$$ab = \sqrt{(x_a - x_b)^2 + (y_a - y_b)^2} \qquad (5\text{-}2)$$

Also, the angles θ and ϕ of Fig. 5-1a are readily calculated from the coordinates of a and b as follows:

$$\theta = \tan^{-1}\left(\frac{y_a}{x_a}\right)$$

$$\phi = \tan^{-1}\left(\frac{y_b}{x_b}\right) \qquad (5\text{-}3)$$

Angle aob is simply the sum of angles θ and ϕ.

5-3 SIMPLE SCALES FOR PHOTOGRAPHIC MEASUREMENTS

There are a variety of simple scales available for photographic measurements. The particular choice of measuring instrument will depend upon the accuracy required for the photogrammetric problem at hand. If a low order of accuracy is acceptable, an ordinary *engineer's scale* may prove satisfactory. With an engineer's scale, accuracy may usually be increased by using the 50 or 60 scale. Where more accuracy is desired, a device such as the metal *microrule* of Fig. 5-2 or the glass scales of Fig. 5-3 may be used. With any of these scales, measured values may be made more exact by taking the mean of several readings.

The main scale of the microrule is graduated at each inch. It has a movable 1-in section at the zero end which is graduated each $\frac{1}{10}$ in. A micrometer having 100 divisions is attached to the movable section, and when it is turned through one complete revolution, the 1-in movable section travels $\frac{1}{10}$ in. Each graduation on the micrometer therefore corresponds to 0.001 in. Accuracy of measurements with the microrule can be increased by using a magnifying glass. The microrule may be used to lay off distances as well as to measure them.

Figure 5-2 Microrule for photographic measurements. (*Courtesy Theo. Alteneder and Sons.*)

Figure 5-3 Glass scales for photographic measurements. (*Courtesy Teledyne-Gurley Co.*)

Glass scales of the type shown in Fig. 5-3 can be obtained in either 6- or 12-in length with either millimeter graduations (least graduations of 0.1 mm) or inch graduations (least graduations of 0.005 in). Readings are taken with the help of magnifying eyepieces which slide along the scale. With a glass scale, readings may be estimated quite readily to one-tenth of the smallest graduated division. These glass scales cannot be used to lay off distances.

5-4 MEASURING PHOTOCOORDINATES WITH SIMPLE SCALES

The conventional procedure for measuring photocoordinates when using an engineer's scale, microrule, or glass scale generally consists of first marking the photocoordinate axis system. This may be done by carefully aligning a straight edge across the fiducial marks and lightly making a line with a razor blade, pin, or very sharp 4h or 5h pencil. Rectangular coordinates are then obtained by direct measurement of the perpendicular distances from these axes.

If the points whose coordinates are to be measured are sharp, distinct points, they may need no further identification. If not, they may be identified with a small pinprick. This should be carefully done under magnification, however, because systematic error will be introduced into measured photocoordinates if points are erroneously marked.

It is important to affix the proper algebraic sign to measured rectangular coordinates; failure to do so will result in frustrating mistakes in solving photogrammetry problems. Points situated to the right of the y axis have positive x coordinates and points to the left have negative x coordinates. Points above the x axis have positive y coordinates, and those below the x axis have negative y coordinates.

5-5 TRILATERATIVE METHOD
OF PHOTOCOORDINATE MEASUREMENT

It is possible to obtain photocoordinates using the simple scales described in Sec. 5-3 but without cutting or scratching fiducial lines. In this procedure, called the *trilaterative method*, distances such as S_a, S_b, S_c, and S_d may be measured from fiducial marks to an image point as illustrated in Fig. 5-4. From photocoordinates of the fiducial marks obtained in camera calibration, coordinates of image points, may then be calculated using trigonometry. The procedure is applicable with corner fiducials as well as side fiducials.

Example 5-1 Suppose that calibrated coordinates of fiducials A and B of Fig. 5-4a are $x_A = -113.00$ mm, $y_A = 0.00$ mm, $x_B = 0.00$ mm, and $y_B = -113.00$ mm. Calculate x_e and y_e if S_a and S_b are measured as 189.89 mm and 100.47 mm, respectively.

SOLUTION By Eq. (5-2),

$$AB = [(113.00)^2 + (113.00)^2]^{1/2} = 159.81 \text{ mm}$$

From the law of cosines,

$$\cos \theta = \frac{S_a^2 + (AB)^2 - S_b^2}{2(S_a)(AB)}$$

$$= \frac{(189.89)^2 + (159.81)^2 - (100.47)^2}{2(189.89)(159.81)} = 0.848591$$

$$\theta = 31° 56' 28''$$

Also

$$\delta = \tan^{-1}\left(\frac{113.00}{113.00}\right) = 45° 00' 00''$$

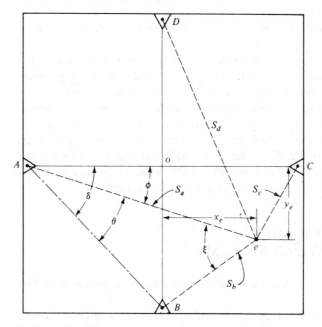

Figure 5-4 Trilaterative method of photocoordinate measurement.

Then

$$\phi = \delta - \theta = (45° \ 00' \ 00'') - (31° \ 56' \ 28'') = 13° \ 03' \ 32''$$

$$x_e = S_a(\cos \phi) + x_A = 189.89(0.974138) - 113.00 = 71.98 \text{ mm}$$

$$y_e = - S_a(\sin \phi) = - 189.89(0.225952) = -42.90 \text{ mm}$$

The trilaterative solution becomes weak as the ξ angle of Fig. 5-4 approaches either 180 or 0°. Strongest solutions are obtained for ξ angles near 90°. Since any two of four possible measured S distances yield a unique solution for the photo coordinates of an image point, the choice should be those two distances yielding an ξ angle nearest 90°. Also, accuracy may be improved by computing photocoordinates in more than one independent solution and then taking the average; e.g., one solution for image coordinates of point e of Fig. 5-4 can be made using S_a and S_b, another can be made using S_a and S_d, etc. A sketch should be used in the calculations to reduce the likelihood of mistakes.

If a computer is available, the trilaterative method can be programmed for solution. As described in a reference cited at the end of this chapter, trilaterative procedures can also be extended to include all measured S distances in a simultaneous least squares solution. Advantages of the trilaterative method over direct measurement from marked fiducial lines are that (1) accuracy is increased, (2) systematic errors of marking fiducial lines are eliminated, and (3) fiducial lines which deface the imagery are not necessary.

5-6 INSTRUMENTS FOR MEASUREMENT OF SHORT DISTANCES

Instruments such as the Zoom Mackroscope shown in Fig. 5-5 make it possible to measure short distances on photographs accurately. This particular instrument has a zoom lens capable of magnifying images from 10 to 30×. A short reticle graduated in units of 0.0001 ft appears superimposed in the field of view. Such instruments are valuable in photo interpretation where short measurements are frequently necessary (e.g., for determining sizes of objects). They are also useful in metric photogrammetry for making corrections for systematic errors that occur in point pricking or in constructing fiducial lines which fail to pass through the fiducial marks.

Faulty fiducial-line marking is illustrated in Fig. 5-6. Photocoordinates to point e have been measured from the faulty axis system. The amounts by which the x axis misses the fiducials are measured as y_a and y_c, and the amounts by which the y axis misses its fiducials are measured as x_b and x_d. The measured x and y fiducial distances are x_m and y_m. General expressions for calculating corrected coordinates x'_e and y'_e of point e are as follows:

$$x'_e = x_e + x_d + (x_b - x_d)\left(\frac{y_m/2 - y_e}{y_m}\right) \tag{5-4}$$

$$y'_e = y_e + y_a + (y_c - y_a)\left(\frac{x_m/2 + x_e}{x_m}\right) \tag{5-5}$$

Figure 5-5 Zoom Mackroscope for precise measurement of short photo distances. (*Courtesy Bausch and Lomb.*)

In Eqs. (5-4) and (5-5), it is necessary to apply correct algebraic signs to y_a, y_c, x_b, and x_d. In the example of Fig. 5-6, these values are all shown positive.

Example 5-2 Suppose that in Fig. 5-6 measured photocoordinates x_e and y_e are 38.27 mm and 49.40 mm, respectively. Measured fiducial distances x_m and y_m are 232.68 mm and 232.43 mm, respectively, and $y_a = 0.54$ mm, $x_b = 0.82$

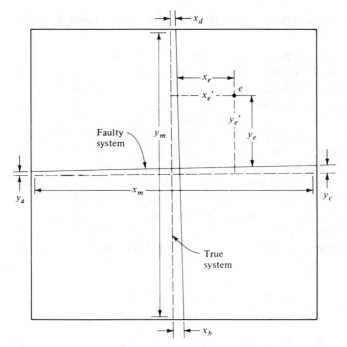

Figure 5-6 Systematic error correction for faulty fiducial axis system.

mm, $y_c = 0.69$ mm, and $x_d = 0.48$ mm. What are photocoordinates x'_e and y'_e corrected for the faulty fiducial axis system?

By Eq. (5-4),

$$x'_e = 38.27 + 0.48 + (0.82 - 0.48)\left(\frac{232.43/2 - 49.40}{232.43}\right)$$

$$= 38.85 \text{ mm}$$

By Eq. (5-5),

$$y'_e = 49.40 + 0.54 + (0.69 - 0.54)\left(\frac{232.68/2 + 38.27}{232.68}\right)$$

$$= 50.04 \text{ mm}$$

5-7 MONOCOMPARATOR MEASUREMENT OF PHOTOCOORDINATES

If the ultimate in photocoordinate measurement accuracy is desired, precise instruments called *comparators* should be used. These instruments are so named because they compare the photographic positions of imaged points with respect to the measurement scales of the devices. There are two basic types of comparators, *monocomparators* and *stereocomparators*. Monocomparators, which are discussed in this section, make measurements on one photograph at a time. With stereocomparators, image positions are measured by simultaneously viewing an overlapping stereo pair of photographs. Stereocomparators are described in Chap. 14. Comparators are used primarily to obtain precise photocoordinates necessary for camera calibration and for analytical photogrammetry.

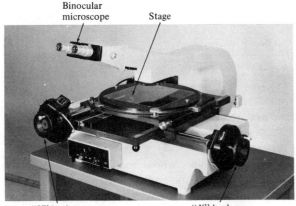

Binocular microscope   Stage

"Y" leadscrew drive wheel   "X" leadscrew drive wheel

Figure 5-7 Mann type 422-F monocomparator. (*Courtesy David W. Mann Co.*)

A common type of monocomparator is shown in Fig. 5-7. This instrument is classified as a *lead-screw* monocomparator. It is capable of making both angle and coordinate measurements on photos having formats as large as 10 in square. The film or diapositive to be measured is first mounted on the stage of the comparator. The stage may be rotated about a vertical axis and is equipped with a slow-motion screw for fine settings to the nearest 20″ of arc. The stage may be moved longitudinally along the X axis of the instrument by means of a lead-screw drive mechanism actuated by the hand wheel on the right side. A similar lead-screw drive and hand wheel on the left side moves the stage transversely along the Y axis of the instrument, the Y axis being perpendicular to the X axis. The instrument can be obtained with small motors for driving the stage along the lead screws. The pitch of the lead screw is 1 mm. A micrometer which records the nearest 0.001 turn of the lead screw makes it possible to read X and Y coordinates to the nearest 0.001 mm (nearest μm). Usually a pair of encoders are attached to the X and Y axes so that the coordinates can be automatically recorded on punched cards, punched paper tape, or magnetic tape. This saves a great deal of time and eliminates costly mistakes in reading and recording. The instrument is equipped with a binocular microscope, having variable magnification of from 10 to 40X, which facilitates setting the reference mark on points at which measurements are to be taken.

There are two basic approaches to measuring photocoordinates with this instrument. In the first of these, the photo is mounted firmly on the stage and the stage is rotated to make the x axis of the photograph parallel with the X axis of the instrument. This is accomplished by trial and error, using the rotary slow-motion screw until the Y coordinates of both side fiducial marks A and C of Fig. 5-8 are equal. Once the stage is set, X and Y readings are taken on the four fiducial marks as well as on all other points whose photographic positions are desired. Owing to setting errors, reading errors, and nonorthogonality of the photo and comparator axes, X readings of fiducials B and D, and Y readings of fiducials A and C of Fig. 5-8 will rarely be equal. Therefore photocoordinates x_e and y_e of a point e are reduced from the comparator axes to the photographic coordinate axes by subtracting the mean of Y coordinates of fiducials A and C from all Y readings, and by subtracting the mean of X coordinates of fiducials B and D from all X readings, or

$$x_e = X_e - \frac{X_B + X_D}{2}$$

$$y_e = Y_e - \frac{Y_A + Y_C}{2}$$

(5-6)

The second basic method of measurement with a lead-screw type of monocomparator is preferred if a computer is available. In this method no attempt is made to orient the plate so that the Y coordinates of fiducials A and C are equal. Instead, the plate is placed on the stage in the approximate orientation of Fig. 5-8. Then measurements are taken to all fiducials and all image points whose coordinates are desired. The coordinates are later reduced numerically from the comparator measurement axis system XY to the conventional photographic xy axis system of Fig. 5-1. One method

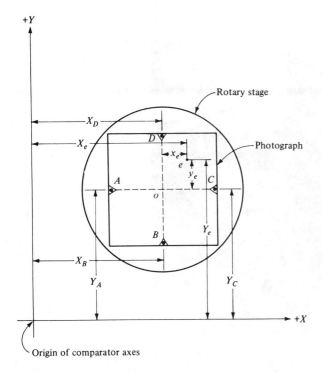

Figure 5-8 Measurement of photo-coordinates with a monocomparator.

of reduction is called *affine coordinate transformation*. The method is discussed in Sec. B-6 of Appendix B and a numerical example is presented.

Another similar type of monocomparator, the Kern MK2, is shown in Fig. 5-9. This instrument has two fixed glass scales mounted orthogonally to each other from which X and Y coordinates are obtained. The photo is mounted on a stage which cannot be rotated but which can be translated in the X and Y directions. Movements of the stage are accomplished freehand to bring the desired image point to the approximate location of the reference measuring reticle. The stage is then clamped, and using slow motion screws, a precise setting can be made while observing under magnification through the eyepiece. By means of a photoelectric cell which monitors the positions of light beams that move along the glass scales when the stage is translated, measurements to the nearest micrometer are obtained and transferred automatically to an electronic digital output device. The usual manner of reducing the XY comparator coordinates to the convential xy photocoordinate system is also by affine coordinate transformation, as discussed in Sec. B-6 of Appendix B.

A considerably different type of monocomparator known as the *multilaterative comparator* is shown in Fig. 5-10. It operates on essentially the same trilaterative principle as that discussed in Sec. 5-5. The measuring apparatus of this comparator consists of a glass-scale measuring arm which is precisely graduated at 1-mm intervals. This measuring arm is free to rotate about a pivot at its base. The instrument is equipped with a microscopic eyepiece having a range of magnification from 10 to $30\times$. Distances are measured from the pivot point to images whose photocoordinates

Figure 5-9 Kern MK2 Monocomparator. (*Courtesy Kern Instruments, Inc.*)

Glass scale   Micrometer
measuring arm   drum

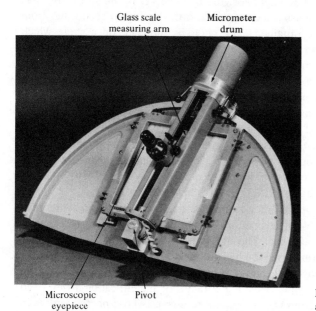

Microscopic   Pivot
eyepiece

Figure 5-10 Multilaterative comparator. (*Courtesy D.B.A. Systems, Inc.*)

are desired. Measurements are made by rotating the measuring arm about the pivot and simultaneously moving the eyepiece so that the point to be measured comes within a circular reticle seen through the eyepiece. Rotation of the micrometer drum at the top of the measuring arm causes the glass scale to translate longitudinally, and once the point is within the circular reticle, the drum is rotated until an even-millimeter graduation bisects the point. The instrument is usually equipped to provide automatic digital output, but it can be read manually. In manual reading, the whole millimeter is obtained directly from the glass scale and the fractional part is taken from the micrometer drum to the nearest half micrometer.

Figure 5-11 illustrates a generalization of the theoretical basis of the multilaterative comparator. Distances R_1, R_2, R_3, and R_4 are measured from the pivot to point e. In the actual measuring process, the pivot remains fixed in position while the measuring stage to which the photograph is attached is rotated four times, each rotation being approximately 90°. Distances are measured at each of these four positions, and thus the geometric equivalent of Fig. 5-11 is created.

From the measured R distances, the coordinates of the four pivot points, fiducial marks, and all image points are computed in the arbitrary XY system shown in Fig. 5-11. The calculations are performed with a computer. After these calculations are made, the coordinates of the fiducial marks are known in both the arbitrary XY comparator system and the calibrated photographic xy system. Using these two sets of coordinates for the fiducial marks, all points may be converted to the fiducial (pho-

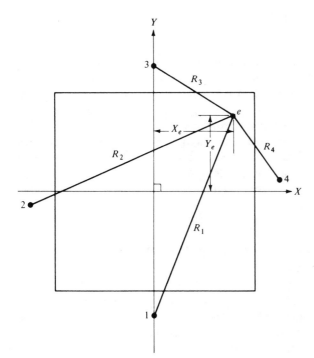

Figure 5-11 Generalized geometric equivalent of the multilaterative comparator measurement system.

tographic) coordinate system using a coordinate transformation such as the two-dimensional affine transformation described in Sec. B-6 of Appendix B.

Although monocomparators are generally very precise, small systematic errors do occur as a result of imperfections in their measurement systems. The magnitudes of these errors can be determined by measuring coordinates of a precise grid plate and then comparing the results with known coordinate values of the grid plate. The overall pattern of differences (errors) can be modeled with polynomials in a manner similar to that described in Sec. A-11 of Appendix A. Measured photocoordinates can then be processed through the polynomial to effectively eliminate the systematic errors of the comparator.

5-8 REFINEMENT OF MEASURED IMAGE COORDINATES

The preceding sections have discussed instruments and techniques for measuring photocoordinates, procedures for eliminating systematic errors in the measurements, and computations for reducing the coordinates to the fiducial axis system. These photocoordinates will still contain systematic errors from various other sources, however. The major sources of these errors are

1. Failure of fiducial axes to intersect at the principal point
2. Shrinkage or expansion of photographic materials
3. Lens distortions
4. Atmospheric refraction distortions
5. Earth curvature distortions

Corrections may be applied to eliminate the effects of these systematic errors. However, all corrections need not be made for all photogrammetric problems; in fact, for crude work they may all be ignored. If, for example, an engineer's scale has been used to make the measurements, uncertainty in the photocoordinates may be so great that the small magnitudes of these systematic errors become insignificant. On the other hand, if precise measurements for an analytical photogrammetry problem have been made with a comparator, all the corrections may be significant. The decision as to which corrections are necessary for a particular photogrammetric problem can be made after considering required accuracy versus magnitude of error caused by neglecting the correction.

5-9 REDUCTION OF COORDINATES TO AN ORIGIN AT THE PRINCIPAL POINT

It has been previously stated that the principal point of a photograph rarely occurs precisely at the intersection of fiducial lines. The actual coordinates of the principal point with respect to the fiducial axes are x_o and y_o, as shown in Fig. 5-12. These coordinates are obtained through camera calibration.

Photogrammetric equations which utilize photocoordinates are based on projective

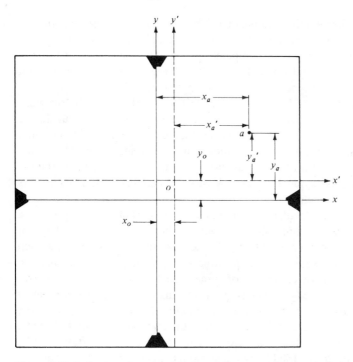

Figure 5-12 Reduction of measured photocoordinates to origin at principal point.

geometry and assume an origin of photocoordinates at the principal point. Therefore it is theoretically correct to reduce photocoordinates from the measurement or fiducial axis system to the axis system whose origin is at the principal point. These principal point axes are shown on Fig. 5-12 as x' and y'. For any image point a, reduction from fiducial axes to principal point axes is simply

$$x'_a = x_a - x_o$$

$$y'_a = y_a - y_o$$

(5-7)

Manufacturers of precision mapping cameras attempt to mount the fiducial marks and camera lens so that the principal point and intersection of fiducial lines coincide. Normally they accomplish this to within a few micrometers, and therefore in ordinary photogrammetric work, especially if paper print positives are being used, this reduction can usually be ignored.

5-10 SHRINKAGE AND EXPANSION OF PHOTOGRAPHIC FILMS AND PAPERS

In photogrammetric work, true positions of images in the picture are required. Photocoordinates measured by any of the previously discussed methods will unavoidably contain small errors due to shrinkage or expansion of the photographic materials which support the emulsion of the negative and positive. Photocoordinates must be corrected for these errors before they are used in photogrammetric calculations; otherwise, errors from this source will be present in the computed results. The magnitude of error in the computed results will depend upon the severity of the shrinkage or expansion error, which depends upon the type of emulsion support materials used.

Most photographic films used to produce negatives for photogrammetric work have excellent dimensional stability, but some small changes in size do occur during processing and storage. Dimensional change during storage may be held to a minimum by maintaining constant temperature and humidity in the storage room. The actual amount of distortion present in a film is a function of several variables, including the type of film and its thickness. Typical values may vary from almost negligible amounts up to approximately 0.20 percent.

A variety of materials are available upon which positive photographic prints are made. Glass is unsurpassed in dimensional stability; when it is used, shrinkage or expansion may be considered nonexistent. If polyester material is used, the same high degree of dimensional stability exists for the positive as for the negative. If the material is paper, however, a much lower degree of dimensional stability exists. For this reason, paper prints are not used for precise photogrammetric work. The amount of paper shrinkage or expansion is a function of temperature, humidity, and paper type and thickness, but to a large degree it is a function of the method of drying the prints. If a hot-drum dryer is used or if the prints are hung to dry, greater distortions can be expected than if the prints are air-dried lying flat at room temperature. Paper shrinkages or expansions of up to 1 percent are not uncommon, and it is sometimes as large as 2 or 3 percent for single-weight papers which are hung to dry. Distortion in the x direction often is markedly different from distortion in y, and this type of "differential" distortion can lead to rather serious consequences if neglected.

5-11 SHRINKAGE CORRECTION

The amount of shrinkage or expansion present in a photograph can be determined by comparing measured photographic distances between opposite fiducial marks with their corresponding values determined in camera calibration. Photocoordinates can be corrected if discrepancies exist. If x_m and y_m are measured fiducial distances on the positive, and x_c and y_c are corresponding calibrated fiducial distances, then the corrected photocoordinates of any point a may be calculated as

$$x_a' = \left(\frac{x_c}{x_m}\right) x_a \qquad (5\text{-}8)$$

$$y_a' = \left(\frac{y_c}{y_m}\right) y_a \qquad (5\text{-}9)$$

In Eqs. (5-8) and (5-9), x_a' and y_a' are corrected photocoordinates and x_a and y_a are measured coordinates. The ratios x_c/x_m and y_c/y_m are simply scale factors in the x and y directions, respectively.

Example 5-3 For a particular photograph, the measured x and y fiducial distances were 233.85 mm and 233.46 mm, respectively. The corresponding x and y calibrated fiducial distances were 232.60 mm and 232.62 mm, respectively. Find the corrected values for the measured photocoordinates which are listed in columns (b) and (c) in the table below.

SOLUTION From Eq. (5-8),

$$x' = \left(\frac{232.60}{233.85}\right)(x) = 0.99465(x)$$

From Eq. (5-9),

$$y' = \left(\frac{232.62}{233.46}\right)(y) = 0.99640(y)$$

Each of the measured values is multiplied by the appropriate constant above and the corrected coordinates are entered in columns (d) and (e) of the table below.

(a)	Measured coordinates		Corrected coordinates	
(a) Point no.	(b) x', mm	(c) y', mm	(d) x', mm	(e) y', mm
1	− 102.57	95.18	− 102.02	94.84
2	− 98.43	− 87.77	− 97.90	− 87.45
3	16.28	− 36.06	16.19	− 35.93
4	65.72	61.84	65.37	61.62
5	104.88	− 73.49	104.32	− 73.23

Shrinkage or expansion corrections may also be applied through the x and y scale factors of a two-dimensional affine coordinate transformation. This method is particularly well suited for analytical photogrammetric calculations and requires a computer. The procedure is described in Sec. B-6 of Appendix B, and a numerical example is presented.

In addition to fiducial marks, or instead of them, some cameras are equipped with a *reseau*. As illustrated in Fig. 5-13, this is usually a glass plate upon which a precise grid of fine crosses has been etched. The reseau plate is mounted in the camera focal plane so that when exposures are made, the grid is imprinted on the negatives, and of course, then appears on all positives made later. The measured positions of the grid marks on the positives can be compared with their precisely known locations in the camera, and if discrepancies exist due to shrinkage or expansion, corrections can be made for them. The usual manner of making these corrections is by affine coordinate transformation as described in Sec. B-6 of Appendix B. The advantage of using a reseau is that the grid pattern is distributed uniformly throughout the entire picture format, and thus corrections can be made for nonuniform shrinkage or expansion that can occur. This is not possible if only side and/or corner fiducials are available, and

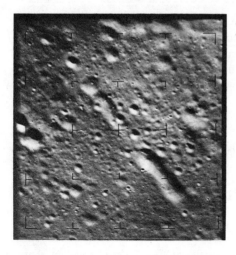

Figure 5-13 Reseau grid superimposed upon the image of a Lunar photograph. (*Courtesy National Space Science Data Center.*)

therefore, for the most precise analytical photogrammetric work, a reseau is preferred. As an alternate to using a glass plate reseau, in some cameras a series of tiny holes in a grid pattern can be drilled through the plate which holds the film firmly against the focal plane during exposure. When the pictures are taken, imprints of these holes appear on the negative.

5-12 CORRECTION FOR RADIAL-LENS DISTORTION

As described in Chaps. 2 and 4, radial-lens distortion causes imaged positions to be distorted along radial lines from the principal point. If the radial distortion characteristics of the camera lens are known through camera calibration, image positions can be corrected.

Three different methods of correcting for radial distortion are (1) reading required corrections from a radial-lens distortion curve, (2) interpolating corrections from a table, and (3) numerical methods in which the radial-lens distoriton curve is approximated by a polynomial. Each of these methods assumes distortions to be symmetric about the principal point. Corrections for radial-lens distortion should theoretically be applied after images have been reduced to the principal point and corrected for shrinkage or expansion.

The following example illustrates the method of reading corrections from a curve:

Example 5-4 Suppose that the photocoordinates of Example 5-3 were measured on a photograph taken with the camera whose radial-lens distortion curve is represented in Fig. 4-15. Calculate coordinates of point no. 4 of Example 5-3 after correcting for radial-lens distortion.

SOLUTION (Calculations are based on shrinkage-corrected coordinates.) The radial distance r from the principal point to point no. 4 is

$$r = \sqrt{(65.37)^2 + (61.62)^2} = 89.83 \text{ mm}$$

Entering the abscissa of Fig. 4-15 with an r value of 89.83 mm, moving vertically to intersect the distortion curve, and then moving leftward horizontally to the ordinate scale (see dashed line of Fig. 4-15), radial-lens distortion Δr of +0.004 mm is read. (Positive values indicate outward distortion.) The corrected radial distance to point no. 4 is obtained by subtracting the positive distortion from the radial distance, or

$$r' = r - \Delta r \tag{5-10}$$

For this example, r' is

$$r'_4 = 89.83 - 0.004 = 89.826 \text{ mm}$$

Now the corrected x' and y' coordinates are calculated in proportion to the ratio of r'/r, or

$$x' = \left(\frac{r'}{r}\right) x \tag{5-11}$$

$$y' = \left(\frac{r'}{r}\right) y \tag{5-12}$$

Applying Eqs. (5-11) and (5-12) to this example,

$$x'_4 = \left(\frac{89.826}{89.83}\right) 65.37 = 65.367 \text{ or } 65.37 \text{ mm (rounded)}$$

$$y'_4 = \left(\frac{89.826}{89.83}\right) 61.62 = 61.617 \text{ or } 61.62 \text{ mm (rounded)}$$

Although radial-lens distortion is essentially negligible for the Zeiss Pleogon lens of the above example, it is quite significant for some camera lenses. The Metrogon lens, for example, has radial-lens distortions in excess of $\pm$ 0.100 mm. Photogrammetrists should therefore ascertain the magnitude of radial-lens distortion for their camera lenses and make corrections if it introduces an intolerable error in any given problem.

The method of approximating radial-lens distortions with a polynomial is the most complex; however, if a computer is available, it is also most convenient. The polynomial method is especially well suited for analytical photogrammetric calculations. The method consists of approximating the radial-lens distortion curve with a polynomial of the form

$$\Delta r = k_1 r + k_2 r^3 + k_3 r^5 + k_4 r^7 \tag{5-13}$$

In Eq. (5-13), Δr is the radial-lens distortion at a radial distance r from the principal point. The k coefficients define the shape of the curve. They are obtained through a least squares curve fitting computation which matches a curve to known radial distortions at varying radial distances as determined through camera calibration. The method of determining the coefficients is demonstrated in an example problem in Sec. A-11 of Appendix A. Once the k's have been determined, radial-lens distortion for any value of r may be readily calculated by back substitution into Eq. (5-13).

The above procedures assume radial-lens distortions to be symmetric about the principal point. *Asymmetric* lens distortions can also be accounted for. The assumption of symmetry, however, accounts satisfactorily for lens distortions for all but the most precise photogrammetric work. *Tangential* lens distortions are generally very small and are seldom corrected for.

5-13 CORRECTION FOR ATMOSPHERIC REFRACTION

It is well known that density (and hence index of refraction) of the atmosphere decreases with increased altitude. Because of this condition, light rays do not travel in straight lines through the atmosphere, but rather are bent according to Snell's law, as shown in Fig. 5-14. The incoming light ray from point A of the figure makes an angle α with the vertical. If refraction were ignored, the light ray would appear to be coming from point B rather than from point A. Photogrammetric equations assume that light rays travel straight paths, and to compensate for the known refracted paths, corrections are applied to the image coordinates.

In Fig. 5-14, if a straight path had been followed by the light ray from object

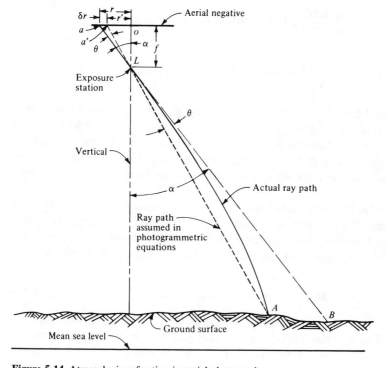

Figure 5-14 Atmospheric refraction in aerial photography.

point A, then its image would have been at a'. The angular distortion due to refraction is θ, and the linear distortion on the photograph is δr. Refraction causes all imaged points to be displaced outward from their correct positions. The magnitude of refraction distortion increases with increasing flying height and also with increasing α angle. Refraction distortion occurs radially from the photographic nadir point (principal point of a vertical photo) and is zero at the nadir point.

From Fig. 5-14 the basic refraction correction equation may be developed as follows:

$$\delta_r = \frac{La\ \theta}{\cos \alpha} \qquad (a)$$

where θ is in radians, and

$$\cos \alpha = \frac{f}{La} \qquad (b)$$

Also,

$$La = \sqrt{r^2 + f^2} \qquad (c)$$

Substituting (b) and (c) into (a) and reducing:

$$\delta_r = \left(\frac{r^2 + f^2}{f}\right) \theta \qquad (5\text{-}14)$$

In Eq. (5-14), r is the radial distance of the image point from the photographic nadir point (principal point for vertical photography), and f is camera focal length. The units of δ_r will be the same as those of r and f.

Assuming a standard atmosphere, angle θ may be calculated for a particular flight altitude, average ground elevation, and α angle. (See references cited at the end of this chapter for the equation.) Figure 5-15 is a plot of θ angles (in seconds) on the ordinate versus α angles (in degrees) on the abscissa. The θ values for plotting this diagram have been calculated for a ground elevation of up to approximately 1,000 ft above MSL using varying values of α and varying flying heights above MSL. For ground elevations higher than about 1,000 ft, slightly different curves, showing smaller refraction angles, would result.

To correct a given imaged point for refraction distortion, the α angle is first calculated based on the measured photocoordinates and camera focal length as follows:

$$\alpha = \tan^{-1} \left(\frac{r}{f}\right) \qquad (5\text{-}15)$$

$$\text{where } r = \sqrt{x^2 + y^2}$$

After calculating θ, or obtaining it from Fig. 5-15, Eq. (5-14) is solved to obtain δr. The correction is applied radially from the photographic nadir point (principal point of a vertical photograph); therefore the corrected radial distance r' to the imaged

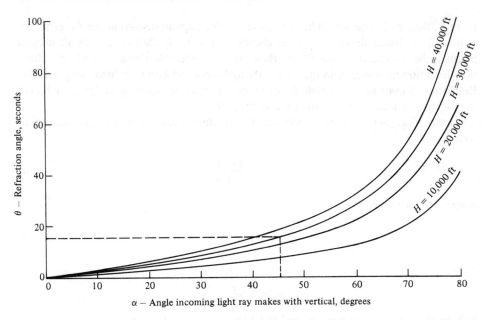

Figure 5-15 Refraction angle, θ versus angle incoming light ray makes with the vertical, α, for ground elevations up to approximately 1000 feet and varying flying heights above mean sea level.

point is obtained using Eq. (5-10), except δ_r is subtracted from r. Corrected x' and y' coordinates may then be obtained by using Eqs. (5-11) and (5-12).

Example 5-5 Assume that the camera used to obtain the photograph of Examples 5-3 and 5-4 had an 88.92-mm focal length, and that the exposure was made from 30,000 ft above MSL. Compute the coordinates of point no. 4 after correcting for atmospheric refraction.

SOLUTION (Calculations are based upon coordinates from Example 5-4 that have been corrected for both shrinkage and expansion, and radial lens distortion.)

$$r = \sqrt{(65.37)^2 + (61.62)^2} = 89.83 \text{ mm}$$

By Eq. (5-15),

$$\alpha = \tan^{-1}\left(\frac{89.83}{88.92}\right) = 45.3°$$

From Fig. 5-15, at $\alpha = 45.3°$ and $H = 30,000$ ft, the value of θ is 16". (This is shown dashed on the figure.)

By Eq. (5-14),

$$\delta_r = \left[\frac{(89.83)^2 + (88.92)^2}{88.92} \right] \frac{16''}{206,265''/\text{rad}} = 0.014 \text{ mm}$$

$$r' = r - \delta_r = 89.83 - 0.014 = 89.816 \text{ mm}$$

$$x' = \left(\frac{89.816}{89.83} \right) 65.37 = 65.360 \text{ mm}$$

$$y' = \left(\frac{89.816}{89.83} \right) 61.62 = 61.610 \text{ mm}$$

5-14 CORRECTION FOR EARTH CURVATURE

If the positions of points in the object space are to be computed in a plane coordinate system, then it may be necessary to account for image distortions caused by curvature of the earth. In Fig. 5-16, A is a ground object point and A' is its position in a map plane which is tangent to the earth at the ground nadir point (point vertically beneath

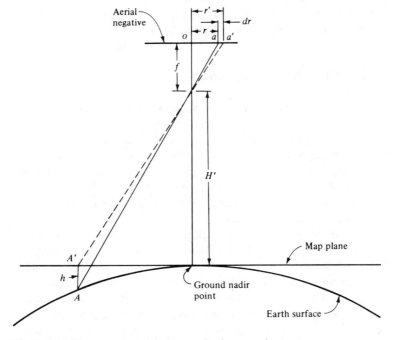

Figure 5-16 Distortions in aerial photography due to earth curvature.

the exposure station). If the map position A' is desired in computation, then it is necessary to use the photo coordinates of the theoretical image position a' instead of the actual image position a. Distance aa' is dr, the distortion for earth curvature. It is radial from the photographic nadir point (principal point of a vertical photo).

From Fig. 5-16 it is seen that the earth curvature correction must be applied outward. The equation for calculating dr is

$$dr = \frac{H'r^3}{2\,Rf^2} \tag{5-16}$$

In Eq. (5-16), H' is flying height above ground, r is radial distance from principal point to image point, R is the earth's radius (20,906,000 ft, or 6,372,200 m), and f is the camera focal length. The units of H' and R must be the same, and the units of dr will be those of r and f. Corrected radial distance r' to an image point is obtained by adding dr to r. As with radial-lens distortion and refraction distortion, corrected photocoordinates may then be obtained using Eqs. (5-11) and (5-12).

Earth curvature distortions become more severe as radial distance from the ground nadir point to the object point increases. This distance increases if radial distance r increases, if flying height is increased, or if focal length is decreased. Figure 5-17 shows radial distortions due to earth curvature for vertical photographs taken with a nominal 6-in- (150 mm-) focal-length camera for various flying heights above ground. As an example of the use of this nomogram, for a flying height of 6,000 ft above ground, an image of a radial distance of 100 mm from the principal point has an earth curvature radial distortion of 6 μm. This example is shown dashed on Fig. 5-17. Note from the figure that the values of dr can become quite large.

5-15 IMAGE DENSITY MEASUREMENTS

As explained in Sec. 3-5, *density* is the degree of darkness or lightness of a film. A photograph usually contains images having a wide range in density variations. In the aerial photo of Fig. 1-8, for example, the river appears as a dark tone, and several buildings are white. Other objects have varying shades of gray between these two extremes of dark and light, e.g., roads, trees, and grassy areas. It is this variation in image densities that enables objects to be identified on photographs. In fact, a photograph is completely defined by the unique spatial arrangement of the infinite number of tonal variations that occur within its borders.

Variations in image densities can be measured using instruments called *microdensitometers*. The measurements are made on diapositives (transparencies printed on glass or film). There are two kinds of microdensitometers: *spot* and *scanning* models. With spot microdensitometers, density measurements at specific points on a photo can be made by manually translating the measurement optics to those locations. Thus, this type of instrument is suitable if only a limited number of points must be measured. If density measurements of entire photos are required, the *scanning* type of microdensitometer is used.

Scanning microdensitometers will systematically scan through a complete pho-

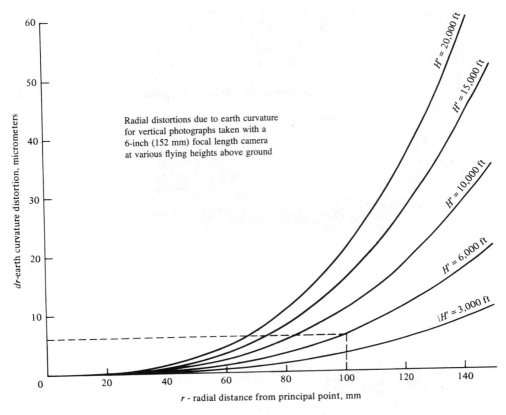

Figure 5-17 Radial distortions due to earth curvature.

tograph, measure density values, and record them on magnetic tape. The contents of a photograph, stored in this way, are available for a variety of automatic data processes that can be performed later using high-speed computers and other equipment.

Scanning microdensitometers are available in two basic types: (1) "drum" designs in which film diapositives are fastened onto a rotating cylinder for measurement, and (2) "flatbed" models which enable a glass or film diapositive to be laid flat on a stage for measurement. The electro-optical components and manner of measurement of both types are basically the same; only the mechanical design differs. Figure 5-18 shows a drum type of scanning microdensitometer, and Fig. 5-19 shows a flatbed type.

The basic operational procedure of a drum type is illustrated in Fig. 5-20. The densitometer consists of a high-speed xy scanning electro-optical system. The film to be scanned is placed over the opening of the cylinder (drum) and forms part of its circumference. In operation the drum rotates while the optics simultaneously translates by means of a lead-screw.

During scanning, the photograph is illuminated from inside the drum and a spot of light is projected through the film onto a square aperture. The size of the aperture, normally referred to as the "spot" or *pixel* size, can be varied. Usually a pixel in the

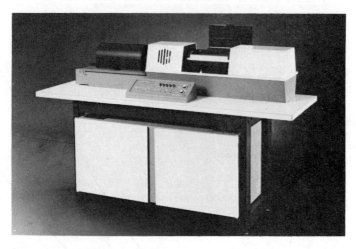

Figure 5-18 Model C-4500 rotating drum scanning microdensitometer. (At a 50 micrometer pixel size, this instrument can scan the entire contents of a 9-inch square photo in less than 20 minutes.) (*Courtesy Optronics International, Inc.*)

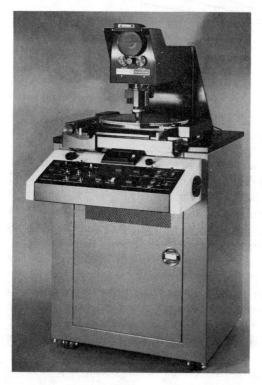

Figure 5-19 Model PDS-1010A flatbet scanning microdensitometer. (*Courtesy Perkin-Elmer, Applied Optics Division.*)

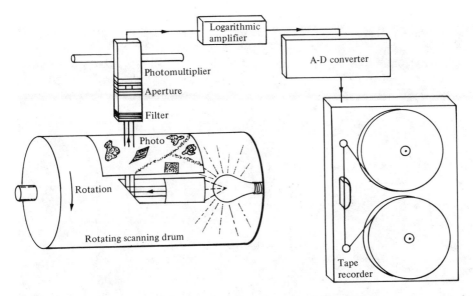

Figure 5-20 Operational procedure of a drum-type scanning microdensitometer.

size range of 25 to 100 μm square is selected, although most scanning microdensitometers have the capability of assigning smaller or larger pixels. As the system scans, the film density for each pixel is obtained by comparing the intensity of the input light signal with the intensity of the light transmitted through the film which has been measured by a *photomultiplier*. An analog signal representing the ratio of the input intensity to transmitted intensity is processed by a logarithmic amplifier, transmitted to an analog-to-digital (A-D) converter, and assigned a density value. (Recall from Sec. 3-5 that density is the common logarithm of the ratio of input intensity to transmitted intensity.) Integer values are assigned to all pixels in proportion to their densities. An operator can select the range of density gradations to be recorded; for example, 256 levels might be selected.

Each scan around the circumference of the film contains as many density sampling points as dictated by the selected pixel size. In the scanning process, a raster of scans is built by *stepping* one interval (equal in width to the pixel size) perpendicular to the previous scan and then sampling the next line of density values around the circumference. As scanning progresses, density value as well as position for each pixel is recorded on magnetic tape. Pixel location along any one circumferential scan provides its scanner system x coordinate, and the count of the number of perpendicular steps that precedes the recording of that pixel gives its scanner system y coordinate. A flatbed scanning microdensitometer works essentially in the same manner except that the input light signal systematically scans back and forth across the diapositive which is laid flat on a stage. Obviously film diapositives must be used on drum types, but either film or glass can be used on a flatbed model.

Figure 5-21*a* shows a black-and-white reproduction of a portion of a color infrared photograph, and Fig. 5-21*b* shows a computer printout which represents density var-

(a)

Figure 5-21 (a) Black and white reproduction of a color infrared photograph. (b) Computer printout of characters representing density variations of the same area shown in Fig. (a). The density data were obtained with a scanning microdensitometer, and represents reflectance in the red portion of the spectrum. The density printout is a negative, e.g., the lake which reflects practically no red shows as a light tone whereas the red seats in the football stadium show as a dark tone.

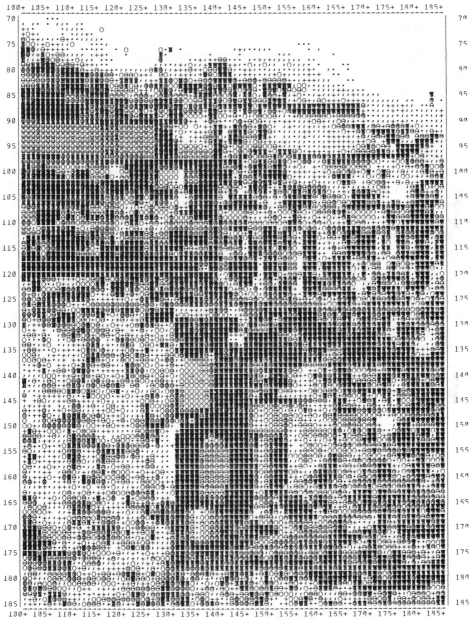

VALUES BELOW 36 ARE BLANK

36 TO 47	48 TO 59	60 TO 71	72 TO 83	84 TO 95	96 TO 107	108 TO 119	120 TO 131	132 AND UP

(b)

iations of the same area obtained using a scanning microdensitometer. The density scan represents red reflectance from the ground, and the computer printout is a negative; e.g., areas that reflected large amounts of red show as dark tones on Fig. 5-21*b*, and those reflecting low amounts of red are light. Note, for example, that the lake in the upper portion of the photo which reflects very little red energy shows light on the printout, while the seats in the football stadium which are red, are dark on the printout. Figure 5-21*b* was obtained by printing various characters according to their row and column locations, and their densities, to create the varying shades of gray. The legend in the lower left of the figure gives a sample of some of the characters which produce the varying gray tones.

The principal utility of the above-described instruments is that they provide an automated system for measuring and digitally recording the contents of a photograph. With this information available, a variety of photogrammetric tasks can be performed automatically by a digital computer, and numerous photogrammetric products can be obtained, basically without the need for human intervention. Some of the tasks that can be performed include automatic image classification and interpretation, which enables compiling information such as the extent of urban areas; the vigor, quantity, and locations of trees of various species; the number of acres of various agricultural crops; and the sizes, shapes and locations of lakes and rivers that appear within the area covered by a given photograph. Products that can be automatically developed include planimetric and topographic maps, digital terrain models, orthophotos, and so forth. Methods of performing these tasks and of developing these products are described in later sections of this book. This general area of automation in photogrammetry, which incorporates high-speed digital computers with scanning microdensitometers and other instrumentation, is a relatively new one. Research is currently extremely active in this area, and new developments loom on the horizon which have a potential of revolutionizing many traditional photogrammetric procedures.

REFERENCES

Abel-Aziz, Y. I.: Asymmetrical Lens Distortion, *Photogrammetric Engineering and Remote Sensing,* vol. 41, no. 3, p. 337, 1975.

American Society of Photogrammetry: "Manual of Photogrammetry," 4th ed., Falls Church, Va., 1980, chap. 9.

Bertram, S.: Atmospheric Refraction, *Photogrammetric Engineering,* vol. 32, no. 1, p. 76, 1966.

Brock, R. H.: Methods for Studying Film Deformation, *Photogrammetric Engineering,* vol. 38, no. 4, p. 399, 1966.

Brown, D. C.: Computational Trade-offs in The Design of a Comparator, *Photogrammetric Engineering,* vol. 35, no. 2, p. 185, 1969.

Bujakiewicz, A.: The Correction of Lens Distortion with Polynomials, *Canadian Surveyor,* vol. 30, no. 2, p. 67, 1976.

Forrest, R. B.: Refraction Compensation, *Photogrammetric Engineering,* vol. 40, no. 5, p. 577, 1974.

Fritz, L. W.: A Complete Comparator Calibration Program, *Photogrammetria,* vol. 29, no. 4, p. 133, 1973.

Gugel, R. A.: Comparator Calibration, *Photogrammetric Engineering,* vol. 31, no. 5, p. 853, 1965.

Jaksic, Z.: Deformations of Estar-base Aerial Films, *Photogrammetric Engineering,* vol. 38, no. 3, p. 285, 1972.

Jeyapalan, K.: Calibration of a Comparator, *Photogrammetric Engineering,* vol. 38, no. 5, p. 472, 1972.

Keating, T. J., and P. R. Wolf: An Improved Method of Digital Image Correlation, *Photogrammetric Engineering and Remote Sensing,* vol. 41, no. 8, p. 993, 1975.

Keller, M., and G. C. Tewinkel: "Aerotriangulation: Image Coordinate Refinement," Technical Bulletin no. 25, U.S. Coast and Geodetic Survey, Washington, D.C., 1965.

Kraus, K.: Film Deformation Correction with Least Squares, *Photogrammetric Engineering,* vol. 38, no. 5, p. 487, 1972.

Kreckel, K. H.: Roll Film Mensuration, *Photogrammetric Engineering,* vol. 31, no. 6, p. 1003, 1965.

Lampton, B. F., and M. J. Umbach: Film Distortion Compensation Effectiveness, *Photogrammetric Engineering,* vol. 32, no. 6, p. 1035, 1966.

Marks, G. W.: Image Error and Photogrammetric Requirements, *ASCE Journal of the Surveying and Mapping Division,* vol. 102, no. SU1, p. 39, 1976.

Scarpace, F. L.: Densitometry on Multi-Emulsion Imagery, *Photogrammetric Engineering and Remote Sensing,* vol. 44, no. 10, p. 1279, 1978.

——— and P. R. Wolf: Atmospheric Refraction, *Photogrammetric Engineering,* vol. 39, no. 5, p. 521, 1973.

Scholer, H.: On Photogrammetric Distortion, *Photogrammetric Engineering and Remote Sensing,* vol. 41, no. 6, p. 761, 1975.

Schut, G. H.: Photogrammetric Refraction, *Photogrammetric Engineering,* vol. 35, no. 1, p. 79, 1969.

Van Roessel, J.: Estimating Lens Distortion with Orthogonal Polynomials, *Photogrammetric Engineering,* vol. 36, no. 6, p. 584, 1970.

Vlcek, J.: Systematic Errors of Image Coordinates, *Photogrammetric Engineering,* vol. 35, no. 6, p. 585, 1969.

Wolf, P. R.: Trilaterated Photo Coordinates, *Photogrammetric Engineering,* vol. 35, no. 6, p. 543, 1969.

———: "Adjustment Computations: Practical Least Squares for Surveyors," PBL Publishers, Madison, Wis., 1980, chap. 18.

——— and R. A. Pearsall: The Kern PG-2 as a Monocomparator, *Photogrammetric Engineering and Remote Sensing,* vol. 42, no. 10, p. 1253, 1976.

Ziemann, H.: Image Deformation and Methods for Its Correction, *Canadian Surveyor,* vol. 25, no. 4, p. 367, 1971.

———: A Coordinate System for Aerial Frame Photography, *Photogrammetric Engineering and Remote Sensing,* vol. 44, no. 5, p. 597, 1978.

PROBLEMS

5-1 Assume that photocoordinates of points a and b of Fig. 5-1 are $x_a = 49.87$ mm, $y_a = 39.24$ mm, $x_b = 79.20$ mm, and $y_b = -62.81$ mm. Calculate photo distance ab, radial distances oa and ob, and the angle (less than 180°) aob.

5-2 Repeat Prob. 5-1, except that the photocoordinates are $x_a = -2.435$ in, $y_a = -3.013$ in, $x_b = -3.985$ in, and $Y_b = 0.946$ in.

5-3 In Fig. 5-4, assume that $x_a = -111.94$ mm, $y_a = 0.000$ mm, $x_b = 0.000$ mm, and $y_b = -111.94$ mm. Calculate x_e and y_e if S_a and S_b are measured as 65.23 mm and 143.91 mm, respectively.

5-4 Repeat Prob. 5-3, except that $x_a = -113.235$ mm, $y_a = 0.000$ mm, $x_d = 0.000$ mm, and $y_d = 113.235$ mm. Also, S_a and S_d are measured as 165.415 mm and 56.725 mm, respectively.

5-5 In Fig. 5-6, assume that a faulty x- and y-axis system was constructed such that $y_a = -0.29$ mm, $x_b = 0.05$ mm, $y_c = 0.15$ mm, and $x_d = 0.40$ mm. Calculate the corrected coordinates of a point e whose coordinates measured with respect to the faulty axis system were $x_e = 95.72$ mm and $y_e = 28.19$ mm. Calibrated fiducial distances x_m and y_m are each 229.00 mm.

5-6 Repeat Prob. 5-5, except that $y_a = 0.25$ mm, $x_b = 0.19$ mm, $y_c = -0.03$ mm, $x_d = -0.43$ mm, $x_e = -75.28$ mm, $y_e = 102.09$ mm, and the calibrated fiducial distances x_m and y_m are 227.420 mm.

5-7 Name and briefly discuss the various systematic errors which may exist in photographic coordinates.

5-8 Calculate the acute angle of intersection of fiducial lines for a camera of the type shown in Fig. 5-1a if comparator measurements of the fiducial marks on a flash plate were as follows:

Mark	X, mm	Y, mm
A	87.294	210.223
B	199.826	96.996
C	313.054	209.555
D	200.512	322.768

5-9 Repeat Prob. 5-8, except that the following flash plate measurements were taken:

Mark	X, mm	Y, mm
A	65.190	215.334
B	178.222	102.107
C	291.950	214.671
D	178.908	327.879

5-10 If the intersection of fiducial lines of the camera of Prob. 5-8 defines the principal point exactly, what are the x and y photocoordinates of the four fiducial marks in the photo system?

5-11 Repeat Prob. 5-10, except that it applies to the data of Prob. 5-9.

5-12 On a paper-print positive, the measured x distance between fiducials (A and C) was 226.38 mm and y between fiducials (B and D) was 225.95 mm. These x and y distances determined in camera calibration were 225.43 mm and 226.70 mm, respectively. Calculate shrinkage-corrected coordinates of points 1, 2, and 3 whose coordinates were measured on the paper print as follows:

Point	X mm	Y mm
1	20.29	− 92.11
2	48.62	85.75
3	− 111.08	− 102.51

5-13 Repeat Prob. 5-12, except that the measured x distance on a paper-print positive between fiducials (A and C) was 8.691 in and y between fiducials (B and D) was 8.978 in; the calibrated distances between these same fiducials were 8.830 in and 8.825 in, respectively; and measured photo coordinates of points 1, 2, and 3 were:

Point	X, in	Y, in
1	0.576	− 2.980
2	− 3.074	1.646
3	− 4.125	− 2.300

5-14 The following $x'y'$ coordinates (see Fig. 5-1b) were determined from measurements for a diapositive having corner fiducials. Calculate coordinates of points 1, 2, and 3 in the conventional xy photocoordinate system. Assume the principal point to be at the intersection of lines joining opposite corner fiducials.

Point	x', mm	y', mm
Fiducial A	0.00	0.00
Fiducial B	211.88	0.00
Fiducial C	211.88	211.96
Fiducial D	0.00	211.96
1	203.28	81.38
2	125.91	107.03
3	57.40	195.82

5-15 Calculate shrinkage-corrected coordinates for points 1, 2, and 3 of Prob. 5-14 if the coordinates of fiducials obtained through camera calibration are as follows:

Fiducial	Calibration x, mm	Calibration y, mm
A	-106.24	-106.19
B	106.24	-106.19
C	106.24	106.19
D	-106.24	106.19

5-16 The photocoordinates of points a, b, and c, after correcting for shrinkage, are as listed below. Find the photocoordinates (to the nearest micrometer) corrected for radial-lens distortion, assuming the photo was taken with a camera whose radial lens distortion curve is shown in Fig. 4-15.

Point	x, mm	y, mm
a	12.723	21.537
b	-64.781	55.231
c	59.260	-115.207

5-17 The photocoordinates listed below have been corrected for shrinkage and radial-lens distortion. Calculate the photocoordinates (to the nearest micrometer) corrected for atmospheric refraction (use Fig. 5-15). Assume the lens of the camera used to take the photo had a focal length of 152.544 mm and that flying height above MSL was 30,000 ft.

Point	x, mm	y, mm
1	28.738	49.211
2	57.820	-93.705
3	-117.232	-102.794

5-18 Repeat Prob. 5-17, except that the camera lens had a focal length of 88.79 mm, flying height above MSL was 40,000 ft, and the photo coordinates were:

Point	x, mm	y, mm
1	59.238	74.281
2	-63.970	-113.444
3	103.296	-98.730

5-19 The photocoordinates listed below have been corrected for shrinkage, radial-lens distortion, and atmospheric refraction distortion. Calculate the photocoordinates (to the nearest micrometer) corrected for earth curvature [use Eq. (5-16) and check with Fig. 5-17]. Assume that the lens of the camera used to take the photo had a 6-in focal length, and that flying height above ground was 15,000 ft.

Point	x, mm	y, mm
a	-64.208	39.820
b	79.413	87.994
c	-106.387	-94.005

5-20 Repeat Prob. 5-19, except that flying height above ground was 10,000 ft and the photocoordinates were:

Point	x, mm	y, mm
a	-88.270	-52.781
b	90.047	82.189
c	-80.725	106.238

5-21 A diapositive was placed on the stage of a monocomparator and oriented so that the Y readings of fiducials A and C of Fig. 5-8 were equal. The following measurements were taken:

Point	X, mm	Y, mm
Fiducial A	43.275	165.319
Fiducial B	156.793	51.807
Fiducial C	270.313	165.319
Fiducial D	156.799	278.831
1	260.648	82.703
2	191.241	228.240
3	87.598	77.461

(a) Reduce the photographic coordinates of points 1, 2, and 3 to the fiducial coordinates axes.

(b) If the calibrated coordinates of the principal point of the camera are $x_o = -0.016$ mm and $y_o = 0.022$ mm, what are the coordinates of points 1, 2, and 3 corrected to the origin at the principal point?

(c) Assume that the calibrated x and y fiducial distances for the camera are 226.705 mm and 226.786 mm, respectively. Using the method described in Sec. 5-11, apply shrinkage corrections to coordinates of points 1, 2, and 3 that were determined in (b).

(d) Assume that the radial-lens distortion curve of the camera is given in Fig. 4-15. Apply radial-lens distortion corrections to the coordinates of points 1, 2, and 3 that were determined in (c).

5-22 Briefly describe the operational procedure of a Drum-type scanning Microdensitometer.

5-23 Compute the number of pixels needed to record the density information of the entire photo using a scanning microdensitometer for the following situations:

(a) Pixel size $= 50$ µm square; photo size $= 9$ in square.

(b) Pixel size $= 100$ µm square; photo size $= 55$ mm square.

(c) Pixel size $= 12.5$ µm square; photo size $= 24$ mm by 36 mm.

SIX

VERTICAL PHOTOGRAPHS

6-1 GEOMETRY OF VERTICAL PHOTOGRAPHS

As described in Chap. 1, photographs taken from an aircraft with the optical axis of the camera vertical or as nearly vertical as possible are called *vertical photographs.* If the optical axis is exactly vertical, the resulting photograph is termed *truly vertical.* In this chapter, equations are developed assuming truly vertical photographs. In spite of precautions taken to keep the camera axis vertical, small tilts are invariably present. For photos intended to be vertical, however, tilts are usually less than 1° and rarely exceed 3°. Photographs containing these small unintentional tilts are called *near-vertical* or *tilted photographs,* and for many practical purposes these photos may be analyzed using the relatively simple "truly vertical" equations of this chapter without serious error.

In this chapter, besides assuming truly vertical photographs, other assumptions are that the photocoordinate axis system has its origin at the photographic principal point and that all photocoordinates have been corrected for shrinkage, lens distortion, atmospheric refraction distortion, and earth curvature distortion.

Figure 6-1 illustrates the geometry of a vertical photograph taken from an exposure station L. The *negative,* which is a reversal in both tone and geometry of the object space, is situated a distance equal to the focal length (distance $o'L$ on Fig. 6-1) behind the rear nodal point of the camera lens. The *positive* may be obtained by direct emulsion-to-emulsion "contact printing" with the negative. This process produces a reversal of tone and geometry from the negative and therefore the tone and geometry of the positive are exactly the same as those of the object space. Geometrically the plane of a contact-printed positive is situated a distance equal to the focal length (distance oL on Fig. 6-1) below the front nodal point of the camera lens. The reversal in geometry from object space to negative is readily seen on Fig. 6-1 by comparing the positions of object points A, B, C, and D with their corresponding negative positions a', b', c', and d'. The correspondence of the geometry of the object space and the positive is also readily apparent. The photographic coordinate axes x and y, as described in Chap. 5, are shown on the positive of Fig. 6-1.

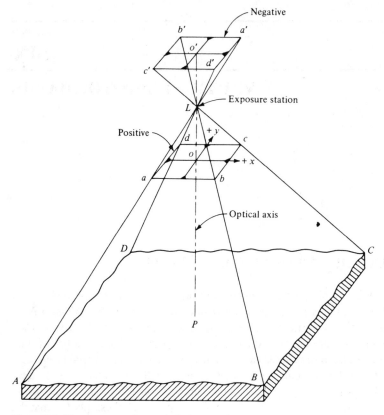

Figure 6-1 The geometry of a vertical photograph.

6-2 SCALE

"Map scale" is ordinarily interpreted as the ratio of a map distance to the corresponding distance on the ground. In similar manner, the scale of a photograph is the ratio of a distance on the photo to that same distance on the ground. On a map, scale is everywhere uniform because a map is an orthographic projection. An aerial photograph, on the other hand, is a perspective projection and, as will be demonstrated herein, its scale varies with variations in terrain elevation.

Scales may be expressed as *unit equivalents, dimensionless representative fractions,* or *dimensionless ratios.* If, for example, 1 in on a map or photo represents 1,000 ft (12,000 in) on the ground, the scale expressed in the aforementioned three ways is

1. Unit equivalents: 1 in = 1,000 ft
2. Dimensionless representative fraction: 1/12,000
3. Dimensionless ratio: 1:12,000

It is helpful to remember that a large number in a scale expression denotes a small scale, and vice versa; e.g., 1 in = 100 ft is a larger scale than 1 in = 1,000 ft.

6-3 SCALE OF A VERTICAL PHOTOGRAPH OVER FLAT TERRAIN

Figure 6-2 shows the side view of a vertical photograph taken over flat terrain. Since measurements are normally taken from photo positives rather than negatives, the negative has been excluded from this and other figures that follow in this text. The scale of a vertical photograph over flat terrain is simply the ratio of the photo distance *ab* to the corresponding ground distance *AB*. That scale may be expressed in terms of camera focal length *f* and flying height above ground *H'* by equating similar triangles *Lab* and *LAB* as follows:

$$S = \frac{ab}{AB} = \frac{f}{H'} \tag{6-1}$$

From Eq. (6-1) it is seen that the scale of a vertical photo is directly proportional to camera focal length (image distance) and inversely proportional to flying height above ground (object distance).

Example 6-1 A vertical aerial photograph is taken over flat terrain with a 6-in-(152.4-mm-) focal-length camera from an altitude of 6,000 ft above ground. What is the photo scale?

By Eq. (6-1),

$$S = \frac{f}{H'} = \frac{6 \text{ in}}{6,000 \text{ ft}} = \frac{1 \text{ in}}{1,000 \text{ ft}} = 1/12,000 = 1:12,000$$

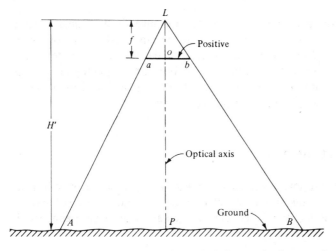

Figure 6-2 Two dimensional view of a vertical photograph taken over flat terrain.

6-4 SCALE OF A VERTICAL PHOTOGRAPH OVER VARIABLE TERRAIN

If the photographed terrain varies in elevation, object distance—or the denominator of Eq. (6-1)—will also be variable and photo scale will likewise vary. Photo scale increases with increasing terrain elevation and decreases with decreasing terrain elevation.

Suppose a vertical aerial photograph is taken over variable terrain from exposure station L of Fig. 6-3. Ground points A and B are imaged on the positive at a and b, respectively. Photographic scale at h, the elevation of points A and B, is equal to the ratio of photo distance ab to ground distance AB. By similar triangles Lab and LAB an expression for photo scale S_{AB} is

$$S_{AB} = \frac{ab}{AB} = \frac{La}{LA} \qquad (a)$$

Also, by similar triangles $LO_A A$ and Loa,

$$\frac{La}{LA} = \frac{f}{H - h} \qquad (b)$$

Substituting Eq. (b) into Eq. (a),

$$S_{AB} = \frac{ab}{AB} = \frac{f}{H - h} \qquad (c)$$

Considering line AB to be infinitesimal, Eq. (c) reduces to an expression for photo scale at a point. In general, by droping subscripts, the scale at any point whose elevation above datum is h may be expressed as

$$S = \frac{f}{H - h} \qquad (6\text{-}2)$$

In Eq. (6-2), the denominator $(H - h)$ is object distance. In this equation as in Eq. (6-1), scale of a vertical photograph is seen to be simply the ratio of image distance to object distance. The shorter the object distance (the closer the terrain to the camera), the greater photo scale, and vice versa. For vertical photographs taken over variable terrain, there are an infinite number of different scales. This is one of the principal differences between a photograph and a map.

6-5 AVERAGE PHOTO SCALE

It is often convenient and desirable to use an *average scale* to define the overall mean scale of a vertical photograph taken over variable terrain. Average scale is the scale at the average elevation of the terrain covered by a particular photograph and is expressed as

$$S_{avg} = \frac{f}{H - h_{avg}} \qquad (6\text{-}3)$$

When an average scale is used, it must be understood that it is exact only at those points which lie at average elevation, and it is an approximate scale for all other areas of the photograph.

Example 6-2 Suppose that highest terrain h_1, average terrain h_{avg}, and lowest terrain h_2 of Fig. 6-3 are 2,000, 1,500, and 1,000 ft above mean sea level, respectively. Calculate maximum scale, minimum scale, and average scale if flying height above mean sea level is 10,000 ft and camera focal length is 6 in (152.4 mm).

By Eq. (6-2) (maximum scale occurs at maximum elevation),

$$S_{max} = \frac{f}{H - h_1} = \frac{6 \text{ in}}{(10,000 - 2,000) \text{ ft}} = \frac{6 \text{ in}}{8,000 \text{ ft}}$$

$$= \frac{1 \text{ in}}{1,333 \text{ ft}} = 1/16,000 = 1:16,000$$

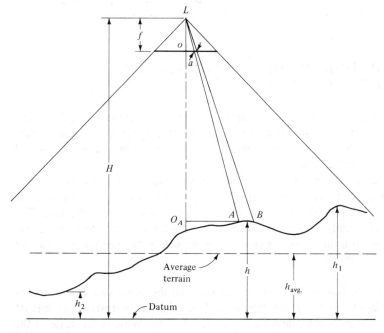

Figure 6-3 Scale of a vertical photograph over variable terrain.

and (minimum scale occurs at minimum elevation)

$$S_{\min} = \frac{f}{H - h_2} = \frac{6 \text{ in}}{(10,000 - 1,000) \text{ ft}} = \frac{6 \text{ in}}{9,000 \text{ ft}}$$

$$= \frac{1 \text{ in}}{1,500 \text{ ft}} = 1/18,000 = 1{:}18,000$$

By Eq. (6-3),

$$S_{\text{avg}} = \frac{f}{H - h_{\text{avg}}} = \frac{6 \text{ in}}{(10,000 - 1,500) \text{ ft}} = \frac{6 \text{ in}}{8,500 \text{ ft}}$$

$$= \frac{1 \text{ in}}{1,417 \text{ ft}} = 1/17,000 = 1{:}17,000$$

In each of the Eqs. (6-1), (6-2), and (6-3), it is noted that flying height appears in the denominator. Thus, for a camera of a given focal length, if flying height increases, object distance $(H - h)$ increases and scale decreases. Figure 6-4a through d illustrate this principle vividly. Each of these vertical photos was exposed using a 9-in (23-cm) format and 6-in-/(152.4-mm-) focal-length camera. The photo of Fig. 6-4a had a flying height of 1,500 ft above average ground, resulting in an average photo scale of 250 ft/in. The photos of Fig. 6-4b, c, and d had flying heights above average ground of 3,000 ft, 6,000 ft, and 12,000 ft, respectively, producing average photo scales of 500 ft/in, 1,000 ft/in, and 2,000 ft/in, respectively.

6-6 OTHER METHODS OF DETERMINING SCALE OF VERTICAL PHOTOGRAPHS

In the previous sections, equations have been developed for calculating scale of vertical aerial photographs in terms of camera focal length, flying height, and terrain elevation. There are, however, other methods of scale determination which do not require knowledge of these values.

A ground distance may be measured in the field between two points whose images appear on the photograph. After the corresponding photo distance is measured, the scale relationship is simply the ratio of the photo distance to the ground distance. The resulting scale is exact only at the elevation of the ground line, and if the line is along sloping ground, the resulting scale applies at approximately the average elevation of the two end points of the line.

Example 6-3 The horizontal distance AB between the centers of two street intersections was measured on the ground as 1,320 ft. The corresponding line ab appears on a vertical photograph and measures 3.77 in. What is the scale of the photo at the average ground elevation of this line?

SOLUTION

$$S = \frac{ab}{AB} = \frac{3.77 \text{ in}}{1,320 \text{ ft}} = \frac{1 \text{ in}}{350 \text{ ft}} = \frac{1}{4,200} = 1\text{:}4,200$$

The scale of a vertical aerial photograph may also be determined if a map covering the same area as the photo is available. In this method it is necessary to measure, on the photograph and on the map, the distances between two well-defined points which can be identified on both photo and map. Photographic scale can then be calculated from the following equation:

$$S = \frac{\text{photo distance}}{\text{map distance}} \times \text{map scale} \tag{6-4}$$

Example 6-4 On a vertical photograph the length of an airport runway measures 6.30 in. On a map which is plotted to a scale of 1:24,000, the runway scales 4.06 in. What is the scale of the photograph at runway elevation?
From Eq. (6-4):

$$S = \frac{6.30 \text{ in}}{4.06 \text{ in}} \times \frac{1}{24,000} = 1/15,470 \text{ or } 1 \text{ in} = 1,290 \text{ ft}$$

The scale of a vertical aerial photograph can also be determined without the aid of a measured ground distance or a map if lines whose lengths are known by common knowledge appear on the photo. "Section lines" of a known 1-mi length, or a football field or baseball diamond, could be measured on the photograph, for example, and photographic scale could be calculated as the ratio of the photo distance to the known ground distance.

Example 6-5 What is the scale of a vertical aerial photograph on which a section line measures 5.93 in?

SOLUTION The length of a section line is assumed to be 5,280 ft. (Actually it can vary considerably from that value.) Photo scale is simply the ratio of the measured photo distance to the ground distance, or

$$S = \frac{5.93 \text{ in}}{5,280 \text{ ft}} = \frac{1 \text{ in}}{890 \text{ ft}} \text{ or } 1\text{:}10,700$$

In each of the methods of scale determination discussed in this section, it must be remembered that the calculated scale applies only at the elevation of the ground line used to determine that scale.

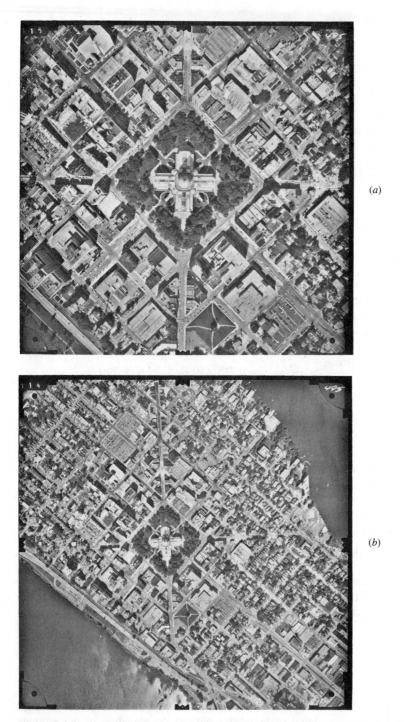

(a)

(b)

Figure 6-4 Four vertical aerial photos taken over Madison, Wisconsin illustrating scale variations due to changing flying heights. Each photo was taken with a camera having a 9-inch (23 cm) square format and 6-inch focal length lens. In Fig. (a), flying height was 1500 ft above average ground and average scale

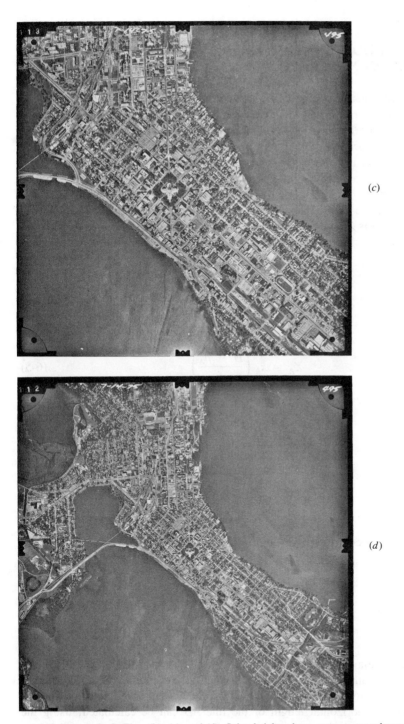

(c)

(d)

was 250 ft per inch. In Figs. (b), (c), and (d), flying heights above average ground were 3000, 6000, and 12,000 ft, respectively, and average scales were 500 ft per inch, 1000 ft per inch, and 2000 ft per inch, respectively. (*Courtesy State of Wisconsin, Department of Transportation.*)

6-7 GROUND COORDINATES FROM A VERTICAL PHOTOGRAPH

The ground coordinates of points whose images appear in a vertical photograph can be determined with respect to an arbitrary ground coordinate system. The arbitrary X and Y ground axes are in the same vertical planes as the photographic x and y axes, respectively, and the origin of the system is at the datum principal point (point in the datum plane vertically beneath the exposure station).

Figure 6-5 shows a vertical photograph taken at a flying height H above datum. Images a and b of the ground points A and B appear on the photograph, and their measured photographic coordinates are x_a, y_a, x_b, and y_b. The arbitrary ground coordinate axis system is X and Y, and the coordinates of points A and B in that system are X_A, Y_A, X_B, and Y_B. From similar triangles $La'o$ and $LA'A_o$, the following equation may be written:

$$\frac{oa'}{A_oA'} = \frac{f}{H - h_A} = \frac{x_a}{X_A}$$

from which

$$X_A = x_a \left(\frac{H - h_A}{f}\right) \tag{6-5}$$

Also, from similar triangles $La''o$ and $LA''A_o$,

$$\frac{oa''}{A_oA''} = \frac{f}{H - h_A} = \frac{y_a}{Y_A}$$

from which

$$Y_A = y_a \left(\frac{H - h_A}{f}\right) \tag{6-6}$$

Similarly, the ground coordinates of point B are

$$X_B = x_b \left(\frac{H - h_B}{f}\right) \tag{6-7}$$

$$Y_B = y_b \left(\frac{H - h_B}{f}\right) \tag{6-8}$$

Upon examination of Eqs. (6-5) through (6-8), it is seen that X and Y ground coordinates of any point are obtained by simply multiplying x and y photocoordinates by the inverse of photo scale at that point. From the ground coordinates of the two points A and B, the horizontal length of the line AB can be calculated, using the pythagorean theorem, as

$$AB = \sqrt{(X_B - X_A)^2 + (Y_B - Y_A)^2} \tag{6-9}$$

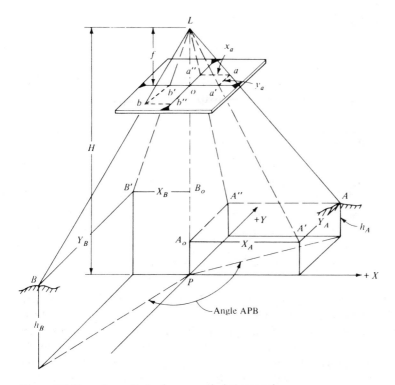

Figure 6-5 Ground coordinates from a vertical photograph.

Also, the horizontal angle APB may be calculated as

$$APB = 90° + \tan^{-1}\left(\frac{X_B}{Y_B}\right) + \tan^{-1}\left(\frac{Y_A}{X_A}\right) \qquad (6\text{-}10)$$

To solve Eqs. (6-5) through (6-8) it is necessary to know the camera focal length, flying height above datum, elevations of the points above datum, and photocoordinates of the points. The photocoordinates are readily measured, camera focal length is commonly known, and flying height above datum is calculated by methods described in Sec. 6-9. Elevations of points may be obtained directly by field measurement, or they may be taken from available topographic maps.

Example 6-6 A vertical aerial photograph was taken with a 6-in-/(152.4-mm-) focal-length camera from a flying height of 4,530 ft above datum. Images a and b of two ground points A and B appear on the photograph, and their measured photocoordinates (corrected for shrinkage and distortions) are $x_a = -52.35$ mm, $y_a = -48.27$ mm, $x_b = 40.64$ mm, and $y_b = 43.88$ mm. Determine the horizontal length of line AB if the elevations of points A and B are 670 and 485 ft above datum, respectively.

SOLUTION From Eqs. (6-5) through (6-8),

$$X_A = \frac{-52.35}{152.4}(4,530 - 670) = -1,326 \text{ ft}$$

$$Y_A = \frac{-48.27}{152.4}(4,530 - 670) = -1,223 \text{ ft}$$

$$X_B = \frac{40.64}{152.4}(4,530 - 485) = 1,079 \text{ ft}$$

$$Y_B = \frac{43.88}{152.4}(4,530 - 485) = 1,165 \text{ ft}$$

From Eq. (6-9)

$$AB = \sqrt{(1,079 + 1,326)^2 + (1,165 + 1,223)^2}$$
$$= \sqrt{(2,405)^2 + (2,388)^2} = 3,389 \text{ ft}$$

Ground coordinates calculated by Eqs. (6-5) through (6-8) are in an arbitrary rectangular coordinate system, as previously described. If arbitrary coordinates are calculated for two or more ''control'' points (points whose coordinates are also known in an absolute ground coordinate system such as the state plane coordinate system), then the arbitrary coordinates of all other points for that photograph can be transformed into the ground system. The method of transformation is discussed in Secs. B-2 through B-5 of Appendix B, and an example problem is given. Using Eqs. (6-5) through (6-8), an entire planimetric survey of the area covered by a photograph can be made.

6-8 RELIEF DISPLACEMENT ON A VERTICAL PHOTOGRAPH

Relief displacement is the shift or displacement in the photographic position of an image caused by the relief of the object, i.e., its elevation above or below a selected datum. With respect to a datum, relief displacement is outward for points whose elevations are above datum and inward for points whose elevations are below datum.

The concept of relief displacement is illustrated in Fig. 6-6, which represents a vertical photograph taken from flying height H above datum. Camera focal length is f, and o is the principal point. The image of terrain point A, which has an elevation h_A above datum, is located at a on the photograph. An imaginary point A' is located vertically beneath A in the datum plane and its corresponding imaginary image position is at a'. On the figure, both $A'A$ and PL are vertical lines, and therefore $A'AaLoP$ is a vertical plane. Plane $A'a'LoP$ is also a vertical plane which is coincident with plane $A'AaLoP$. Since these planes intersect the photo plane along lines oa and oa', respectively, line aa' (relief displacement of point A due to its elevation h_A) is radial from the principal point.

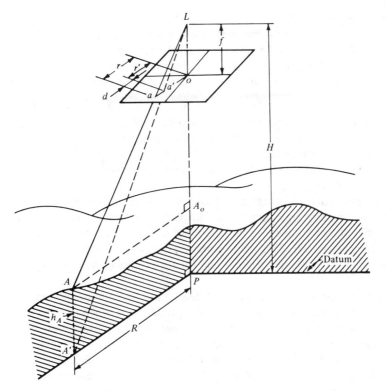

Figure 6-6 Relief displacement on a vertical photograph.

An equation for evaluating relief displacement may be obtained by relating similar triangles Lao and LAA_o in Fig. 6-6 as follows:

$$\frac{r}{R} = \frac{f}{H - h_A} \qquad \text{rearranging: } r(H - h_A) = fR \qquad (d)$$

Also, from similar triangles $La'o$ and $LA'P$,

$$\frac{r'}{R} = \frac{f}{H} \qquad \text{rearranging: } r'H = fR \qquad (e)$$

Equating expressions (d) and (e),

$$r(H - h_A) = r'H$$

Rearranging the above equation, dropping subscripts, and substituting the symbol d for $(r - r')$ gives

$$d = \frac{rh}{H} \qquad (6\text{-}11)$$

where d = relief displacement

$\quad h$ = height above datum of the object point whose image is displaced

$\quad r$ = radial distance on the photograph from the principal point to the displaced image (The units of d and r must be the same.)

$\quad H$ = flying height above the datum selected for measurement of h

Equation (6-11) is the basic relief-displacement equation for vertical photos. Examination of this equation shows that relief displacement increases with increasing radial distance to the image, and it also increases with increased elevation of the object point above the datum. On the other hand, relief displacement decreases with increased flying height above datum. It has also been shown that relief displacement occurs radially from the principal point.

Figure 6-7 is a vertical aerial photograph which vividly illustrates relief displacement. Note in particular the striking effect of relief displacement on the twin chimneys in the upper left of the photo. Notice also that the relief displacement occurs radially from the center of the photograph. This radial pattern is also readily apparent for the relief displacement of all the vertical buildings and for the storage tanks in the lower left of the photo.

Relief displacement often causes straight roads, fence lines, etc., on rolling ground to appear crooked on a vertical photograph. This is especially true when such roads, fences, etc., occur near the edges of the photo. The severity of the crookedness will depend on the amount of terrain variation. Relief displacement causes some imagery to be obscured from view. Several examples of this are seen in Fig. 6-7; e.g., the railroad tracks in the lower left portion of the photo are obscured by the relief displacement of the large storage tanks.

Vertical heights of objects such as buildings, poles, etc., appearing on aerial photographs can be calculated from relief displacements. For this purpose, Eq. (6-11) is rearranged as follows:

$$h = \frac{dH}{r} \qquad (6\text{-}12)$$

To use Eq. (6-12) for height determination, it is necessary that the images of both the top and bottom of the vertical object be visible on the photograph so that d can be measured. Datum is arbitrarily selected at the base of the vertical object. Equation (6-12) is of particular value to the photo interpreter, who is often interested in relative heights of objects rather than absolute elevations.

Example 6-7 The vertical photograph of Fig. 6-7 was taken from an elevation of 1,750 ft above mean sea level. The elevation at the base of the rightmost chimney in the upper left is 850 ft above MSL. The relief displacement d of the chimney was measured as 2.13 in, and the radial distance to the top of the chimney from the photo center was 4.79 in. What is the height of the chimney?

SOLUTION Select datum at the base of the chimney. Then flying height above datum is

$$H = 1,750 - 850 = 900 \text{ ft}$$

Figure 6-7 Vertical photograph illustrating relief displacements. (*Courtesy State of Wisconsin, Department of Transportation.*)

By Eq. (6-12),

$$h = \frac{2.13(900)}{4.79} = 400 \text{ ft}$$

Equation (6-11) may be used to calculate image displacements with respect to datum, and then corrected datum image positions may be located by laying off image displacements along radial lines toward the principal point as shown in Fig. 6-8. These "datum-corrected" images have true relative planimetric positions, just as they would be on a map plotted at photographic datum scale. Datum scale may be calculated by Eq. (6-1) using H above datum in the denominator. From these datum-corrected image positions, angles, lengths, and areas may be measured just as from a map.

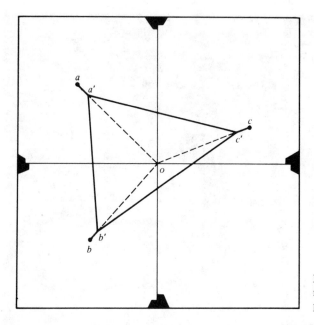

Figure 6-8 Relief displacement corrections laid off radially toward the principal point.

Example 6-8 Figure 6-8 represents a vertical photo taken at a flying height of 6,500 ft above datum with a camera having a 6-in (152.4-mm) focal length. On the photo, points a, b, and c are the imaged positions of the lot corners of a triangular parcel of land. Their radial distances from the principal point are 91.42 mm, 83.50 mm, and 70.06 mm, respectively. Corresponding ground elevations of points A, B, and C are 835 ft, 600 ft, and 450 ft above MSL, respectively. Calculate relief displacements aa', bb', and cc' necessary to locate the datum positions a', b', and c' of the points, and calculate datum scale.

SOLUTION By Eq. (6-11),

$$aa' = \frac{r_a h_A}{H} = \frac{91.42(835)}{6,500} = 11.73 \text{ mm}$$

$$bb' = \frac{r_b h_B}{H} = \frac{83.50(600)}{6,500} = 7.71 \text{ mm}$$

$$cc' = \frac{r_c h_C}{H} = \frac{70.06(450)}{6,500} = 4.85 \text{ mm}$$

By Eq. (6-1),

$$\text{Datum scale} = \frac{f}{H} = \frac{6 \text{ in}}{6,500 \text{ ft}} = \frac{1 \text{ in}}{1,083 \text{ ft}} = 1{:}13{,}000$$

Note: Datum positions a', b', and c' in Fig. 6-8 are obtained by laying off distances aa', bb', and cc' along lines radial to the principal point.

6-9 FLYING HEIGHT OF A VERTICAL PHOTOGRAPH

From previous discussion it is apparent that flying height above datum is an important quantity which is often needed for solving basic photogrammetric equations. Note, for example, that this parameter appears in scale, ground-coordinate, and relief-displacement equations. For rough computations, flying heights may be taken from altimeter readings. Flying heights may also be obtained by using either Eq. (6-1) or (6-2) if a ground line of known length appears on the photograph. This procedure yields exact flying heights for truly vertical photographs if the end points of the ground line lie at equal elevations. In general, the greater the difference in elevation of the end points, the greater the error in computed flying height; therefore the ground line should lie on fairly level terrain. Accurate results can be obtained by this method, however, even though the end points of the ground line are at different elevations, if the images of the end points are approximately equidistant from the principal point of the photograph and on a line through the principal point.

Example 6-9 A certain section line lies on fairly level terrain. Find the approximate flying height above the terrain if the camera focal length is $3\frac{1}{2}$ in (88.9 mm) and the section line scales 3.70 in on the photograph.

SOLUTION (Assuming the section line to be 5,280 ft long.)
   By Eq. (6-1),

$$\frac{3.70 \text{ in}}{5,280 \text{ ft}} = \frac{3\frac{1}{2} \text{ in}}{H'}$$

from which

$$H' = \frac{5,280(3\frac{1}{2})}{3.70} = 5,000 \text{ ft above the terrain}$$

Accurate flying heights can be determined even though the end points of the ground line lie at differing elevations. This procedure requires knowledge of the elevations of the end points of the line as well as of the length of the line. Suppose ground line AB has its end points imaged at a and b on a vertical photograph. Length AB of the ground line may be expressed in terms of ground coordinates, by the pythagorean theorem, as follows:

$$(AB)^2 = (X_B - X_A)^2 + (Y_B - Y_A)^2$$

Substituting Eqs. (6-5) through (6-8) into the above equation gives

$$(AB)^2 = \left[\frac{x_b}{f} (H - h_B) - \frac{x_a}{f} (H - h_A) \right]^2$$

$$+ \left[\frac{y_b}{f} (H - h_B) - \frac{y_a}{f} (H - h_A) \right]^2$$

(6-13)

The only unknown in Eq. (6-13) is flying height, H. When all known values are inserted into the equation, it reduces to the quadratic form of $aH^2 + bH + c = 0$. The direct solution for H in the quadratic is

$$H = \frac{-b \pm \sqrt{b^2 - 4ac}}{2a} \qquad (6\text{-}14)$$

Example 6-10 A vertical photograph was taken with a camera having a focal length of 5.998 in (152.3 mm). Ground points A and B have elevations 1,435 ft and 1,461 ft above sea level, respectively, and the horizontal length of the line AB is 1,919 ft. The images of A and B appear at a and b and their measured photocoordinates are $x_a = 0.717$ in, $y_a = -2.414$ in, $x_b = 4.317$ in, and $y_b = -0.835$ in. Calculate the flying height of the photograph above sea level.

SOLUTION By Eq. (6-13),

$$(1{,}919)^2 = \left[\frac{4.317}{5.998}(H - 1{,}461) - \frac{0.717}{5.998}(H - 1{,}435) \right]^2$$

$$+ \left[\frac{-0.835}{5.998}(H - 1{,}461) + \frac{2.414}{5.998}(H - 1{,}435) \right]^2$$

Reducing,

$$(1{,}919)^2 = (0.6002H - 880)^2 + (0.2633H - 375)^2$$

Squaring terms and arranging in quadratic form,

$$0.4295H^2 - 1{,}253H - 2{,}767{,}536 = 0$$

Solving for H by Eq. (6-14),

$$H = \frac{1{,}253 \pm \sqrt{(1{,}253)^2 + 4(0.4295)(2{,}767{,}536)}}{2(0.4295)}$$

$$= \frac{1{,}253 \pm 2{,}515}{2(0.4295)} = 4{,}387 \text{ ft}$$

Note: Select the positive root, since the negative root yields a ridiculous answer.

6-10 ERROR EVALUATION

Answers obtained in solving the various equations presented in this chapter will inevitably contain errors. It is important to have an awareness of the presence of these errors and to be able to assess their approximate magnitudes. Errors in computed answers are caused partly by random errors in measured quantities which are used in computation and partly by the failure of certain assumptions to be met. Some of the more significant sources of errors in calculated values using the equations of this chapter are

1. Errors in photographic measurements, e.g., line lengths or photocoordinates
2. Errors in ground control
3. Shrinkage and expansion of film and paper
4. Tilted photographs where vertical photographs were assumed

Sources 1 and 2 can be minimized if precise equipment and caution are used in making the measurements. Source 3 can be practically eliminated by making corrections as described in Sec. 5-11. Magnitudes of error introduced by source 4 depend upon the severity of tilt. Generally if the photos were intended to be vertical and if paper prints are being used, these errors are compatible with the other sources. If the photo is severely tilted, or if highest accuracy is desired, the tilted-photo equations of Chap. 11 should be used. For the methods described in this chapter, errors caused by lens distortions, atmospheric refraction, and earth curvature are relatively small and can generally be ignored.

A simple and straightforward approach to calculating the combined effect of several random errors in computed answers is to consider separately the effect on the answer of each error source. This approach involves calculating rates of changes with respect to each variable containing error and requires only simple differential calculus. As an example of this approach, assume that a vertical photograph was taken with a camera having a focal length of 6.000 in. Assume also that a ground distance AB on flat terrain has a length of 5,000 ft and that its corresponding photo distance ab measures 5.00 in. Flying height above ground may be calculated, using Eq. (6-1), as follows:

$$H' = f\frac{AB}{ab} = 6.000\,\frac{5000}{5.00} = 6,000 \text{ ft}$$

Now it is required to calculate the expected error dH' caused by errors in measured quantities AB and ab. This is done by taking derivatives with respect to each of these quantities containing error. Suppose that the error dAB in the ground distance is $\pm$ 1.0 ft and that the error dab in the measured photo distance is $\pm$ 0.01 in. The error in dH' caused by the error dAB in the ground length can be evaluated by taking the derivative dH'/dAB as

$$\frac{dH'}{dAB} = \frac{f}{ab}$$

The value of dH' can now be obtained by substituting numerical values into the above differential equation, as follows:

$$dH' = \frac{f}{ab}\,dAB = \frac{6.000}{5.00}\,1.0 = \pm 1.2 \text{ ft}$$

In similar manner the error dH' caused by error dab is

$$dH' = \frac{-f(AB)}{(ab)^2}\, d_{ab} = \frac{-6.000(5,000)}{(5.00)^2}\, 0.01 = \pm 12.0 \text{ ft}$$

In this example the total error in H' due to the combined effects of the two error sources could be as great as 13.2 ft. Total error is not generally the sum of the individual contributing errors, however, because of the compensating nature of random errors. Rather, the total combined effect would be estimated as the square root of the sum of the squares of the individual errors, or

$$dH'_{\text{total}} = \sqrt{(1.2)^2 + (12.0)^2} = \pm 12.1 \text{ ft}$$

It should be noted that the error in H' caused by the error in the measurement of photo distance ab is the most severe of the two contributing sources. Therefore, to increase the accuracy of the computed value of H', it would be necessary to refine the measured photo distance to a more accurate value. Errors in computed answers using any of the equations presented in this chapter can be analyzed in the manner described above, and the method is valid as long as the contributing errors are small.

REFERENCES

American Society of Photogrammetry: "Manual of Photogrammetry," 3d ed., Falls Church, Va., 1966, chap 2.
————: "Manual of Photogrammetry," 4th ed., Falls Church, Va., 1980, chap. 2.
Hallert, B.: "Photogrammetry," McGraw-Hill Book Company, New York, 1960.
Landis, G. H., and H. A. Meyer: The Accuracy of Scale Determinations on Aerial Photographs, *Journal of Forestry,* vol. 52, p. 863, 1954.
Scherz, J. P.: Errors in Photogrammetry, *Photogrammetric Engineering,* vol. 40, no. 4, p. 493, 1974.

PROBLEMS

6-1 The photo distance between two image points a and b on a vertical photograph is ab and the corresponding ground distance is AB. What is the photographic scale at the elevation of the ground line? (Give the answers both in unit equivalents and dimensionless ratios.)

    (a) $ab = 3.78$ in; $AB = 4,720$ ft
    (b) $ab = 1.24$ in; $AB = 825$ ft
    (c) $ab = 5.83$ in; $AB = 875$ ft
    (d) $ab = 4.65$ in; $AB = 9,305$ ft

6-2 Repeat Prob. 6-1, using values of ab and AB indicated. (Give the answers in dimensionless ratios.)

    (a) $ab = 189.5$ mm; $AB = 429.5$ m
    (b) $ab = 148.3$ mm; $AB = 3,784$ ft
    (c) $ab = 195.7$ mm; $AB = 0.798$ mi
    (d) $ab = 120.0$ mm; $AB = 80.00$ rods

6-3 On a vertical photograph two section corners appear 2.95 in apart. What is the photographic scale at the elevation of the section line?

6-4 On a vertical photograph a college football field measures 0.73 in from goal line to goal line (100 yd). What is the scale of the photograph at the elevation of the football field?

6-5 A semi-tractor and trailer combination which is known to be 55 ft long measures 0.39 in long on a vertical aerial photo. What is the scale of the photo at the elevation of the roadway?

6-6 Repeat Prob. 6-5, except that a railroad boxcar of 80.0-ft known length measures 11.2 mm on the photo.

6-7 An interstate highway pavement of known 24.0-ft width measures 3.40 mm wide on a vertical aerial photo. What is the flying height above the pavement for this photo if the camera focal length was 152.4 mm?

6-8 In the photo of Prob. 6-7, a rectangular-shaped building near the highway has photo dimensions of 0.21 in and 0.34 in. What is the actual size of the structure?

6-9 Repeat Prob. 6-8, except that in the photo of Prob. 6-7 a bridge appears. If its photo length is 26.5 mm, what is the actual length of the bridge?

6-10 A vertical photograph was taken, with a camera having a 6-in focal length, from a flying height 7,500 ft above sea level. What is the scale of the photograph at an elevation of 1,420 ft above mean sea level. What is datum scale?

6-11 Aerial photographs are to be taken for highway planning and design. If a 6-in-focal-length camera is to be used and if an average scale of 1:3,000 is required, what should be the flying height above average terrain?

6-12 A vertical aerial photograph was taken from a flying height of 10,500 ft above datum with a camera having a focal length of 209.45 mm. Highest, lowest, and average terrain appearing in the photograph are at 6,650 ft, 3,085 ft, and 4,800 ft, respectively. Calculate minimum, maximum, and average photographic scale.

6-13 A vertical photograph was taken over the lunar surface from an altitude of 60.0 mi with a 80.20-mm-focal-length camera. What is the actual diameter of a crater whose diameter on the photograph scales 10.63 mm?

6-14 Vertical photography for military reconnaissance is required. If the lowest safe flying altitude over enemy defenses is 15,000 ft, what camera focal length is necessary to achieve a photo scale of 1:24,000?

6-15 A distance ab on a vertical photograph is 1.98 in and corresponding ground distance AB is 4,185 ft. If the camera focal length is 88.95 mm, what is the flying height above the terrain upon which the line AB is located?

6-16 Vertical photography at an average scale of 1:5,000 is to be acquired for the purpose of constructing a mosaic. What is the required flying height above average terrain if the camera focal length is $8\frac{1}{4}$ in?

6-17 The distance on a map between two road intersections in flat terrain measures 1.95 in. The distance between the same two points is 3.48 in on a vertical photograph. If the scale of the map is 1:50,000, what is the scale of the photograph?

6-18 For Prob. 6-17, the intersections occur at an average elevation of 1,250 ft above sea level. If the camera had a focal length of 8.25 in, what was the flying height above sea level for this photo?

6-19 A section line (actual length = 5,280 ft) scales 96.4 mm on a vertical aerial photograph. What is the scale of the photograph?

6-20 For Prob. 6-19, the average elevation of the section line is at 1,590 ft above sea level, and the camera focal length is 152.4 mm. What would be the actual length of a ground line that lies at elevation 950 ft above sea level and measures 57.9 mm on this photo?

6-21 A vertical aerial photo is exposed at 6,400 ft above mean sea level using a camera having a $3\frac{1}{2}$-in focal length. On this photo appears a triangular parcel of land that exists at elevation 850 ft above sea level and has sides that measure 1.43 in, 1.28 in, and 0.95 in, respectively. What is the approximate area of this parcel in acres?

6-22 On a vertical aerial photograph, a line which was measured on the ground to be 1,758 ft long scales 29.4 mm. What is the scale of the photo at the average elevation of this line?

6-23 Points A and B are at elevations 1,288 ft and 1,560 ft above datum, respectively. The photographic coordinates of their images on a vertical photograph are $x_a = 68.27$ mm, $y_a = -32.37$ mm, $x_b = -87.44$ mm, and $y_b = 26.81$ mm. What is the horizontal length of the line AB if the photo was taken from 14,000 ft above datum with a 152.35-mm-focal-length camera?

6-24 Images a, b, and c of ground points A, B, and C appear on a vertical photograph taken from a flying height of 8,350 ft above datum. A 6-in-focal-length camera was used. Points A, B, and C have elevations of 1,725 ft, 1,640 ft, and 2,095 ft above datum, respectively. Measured photocoordinates of the images are $x_a = -2.371$ in, $y_a = 1.864$ in, $x_b = 2.062$ in, $y_b = 3.183$ in, $x_c = 3.704$ in, and $y_c = -3.138$ in. Calculate the lengths of the lines AB, BC, and AC and the area within the triangle ABC.

6-25 The image of a point whose elevation is 1,475 ft above datum appears 53.87 mm from the principal point of a vertical photograph taken from a flying height of 6,000 ft above datum. What would this distance from the principal point be if the point were at datum?

6-26 The images of the top and bottom of a utility pole are 5.11 in and 4.93 in, respectively, from the principal point of a vertical photograph. What is the height of the pole if flying height above the base of the pole is 2,850 ft?

6-27 An area has an average terrain elevation of 1,200 ft above datum. The highest points in the area are 1,850 ft above datum. If the camera focal plane opening is 9 in square, what flying height above datum is required to limit relief displacement with respect to average terrain elevation to 0.20 in? If the camera focal length is $8\frac{1}{4}$ in, what is the resulting average scale of the photography?

6-28 The datum scale of a vertical photograph taken from 3,000 ft above datum is 1:6,000. The diameter of a cylindrical oil storage tank measures 6.87 mm at the base and 7.01 mm at the top. What is the height of the tank if its base lies at 590 ft above datum?

6-29 Assuming that the smallest discernable and measurable relief displacement that is possible on a vertical photo taken from 3,000 ft above ground is 0.5 mm, would it be possible to determine the height of a telephone utility box imaged in the corner of a 9-in-square photo which stands 4 ft high above the ground?

6-30 If the answer to Prob. 6-29 is yes, what is the maximum flying height at which it would be possible to discern the relief displacement of the utility box? If the answer is no, at what flying height would the relief displacement of the box be discernable?

6-31 On a vertical photograph, images a and b of the ground points A and B have photographic coordinates $x_a = -12.68$ mm, $y_a = 70.24$ mm, $x_b = 89.07$ mm, and $y_b = -92.41$ mm. The horizontal distance between A and B is 3,948 ft and the elevations of A and B are 1,283 ft and 1,371 ft above datum, respectively. Calculate the flying height above datum if the camera had a 6-in focal length.

6-32 Repeat Prob. 6-31, except that the horizontal distance AB is 5,258 ft and the camera focal length is 88.92 mm.

6-33 In Prob. 6-13, assume that the values given for focal length, photo distance, and flying height contain random errors of ± 0.1 mm, ± 0.05 mm, and ± 0.2 mi, respectively. What is the expected error in the computed diameter of the crater?

6-34 In Prob. 6-15, assume that the values given for focal length, photo distance, and ground length contain random errors of ± 0.005 mm, ± 0.02 in, and ± 1.0 ft, respectively. What is the expected error in the computed flying height?

6-35 In Prob. 6-26, assume that the random error in each measured photo distance is ± 0.005 in and that the error in the flying height is ± 5 ft. What is the expected error in the computed height of the utility pole?

SEVEN

STEREOSCOPIC VIEWING

7-1 DEPTH PERCEPTION

In our daily activities we unconsciously measure depth or judge distances to a vast number of objects about us through our normal process of vision. Methods of judging depth may be classified as either *stereoscopic* or *monoscopic*. Persons with normal vision (those capable of viewing with both eyes simultaneously) are said to have *binocular* vision, and perception of depth through binocular vision is called stereoscopic viewing. *Monocular* vision is the term applied to viewing with only one eye, and methods of judging distances with one eye are termed monoscopic. A person having normal binocular vision can, of course, view monocularly by covering one eye.

Distances to objects, or depth, can be perceived monoscopically on the basis of (1) relative sizes of objects, (2) hidden objects, (3) shadows, and (4) differences in focusing of the eye required for viewing objects at varying distances. Examples of the first two of these are shown in Fig. 7-1. Depth to the far end of the football field may be perceived, for example, on the basis of the relative sizes of the goal posts. The goal posts are actually the same size, of course, but one appears smaller because it is farther away. Also, the building is quickly judged to be a considerable distance away because it is partially hidden behind the football stadium.

Monoscopic methods of depth perception enable only rough impressions to be gained of distances to objects. With stereoscopic viewing, on the other hand, a much greater degree of accuracy in depth perception can be attained. Stereoscopic depth perception is of fundamental importance in photogrammetry, for it enables the formation of a three-dimensional stereomodel by viewing a pair of overlapping photographs. The stereomodel can then be studied, measured, and mapped. An explanation of how this phenomenon is achieved is the subject of this chapter, and explanations of its use in measuring and mapping are given in the chapters which follow.

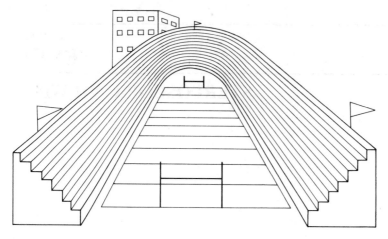

Figure 7-1 Depth perception by relative size and hidden objects.

7-2 THE HUMAN EYE

The phenomenon of stereoscopic depth perception can be more clearly understood with the help of a brief description of the anatomy and physiology of the human eye. The human eye functions in much the same manner as a camera. As shown in Fig. 7-2, the eye is essentially a spherical organ having a circular opening called the *pupil*. The pupil is protected by a transparent coating called the *cornea*. Incident light rays pass through the cornea, enter the eye through the pupil, and strike the *lens,* which is directly behind the pupil. The cornea and lens refract the light rays according to Snell's law.

The lens of the eye is biconvex in shape and is composed of a refractive transparent medium. It is suspended by many muscles which enable the lens to be moved so that the *optical axis* of the eye can be aimed directly at an object to be viewed. As with a camera, the eye must satisfy the lens formula, Eq. (2-8), for each different object distance. The eye's image distance is constant, however; therefore, to satisfy the lens formula for varying object distances, the focal length of the lens changes. When a distant object is viewed, the lens muscles relax, causing the spherical surfaces of the lens to become flatter. This increases the focal length to satisfy the lens formula and accommodate the long object distance. When close objects are viewed, a reverse procedure occurs. The eye's ability to focus for varying object distances is called *accommodation*.

As with a camera, the eye has a diaphragm called the *iris*. The iris (colored part of the eye) automatically contracts or expands to regulate the amount of light entering the eye. When the eye is subjected to intense light, the iris contracts, reducing the pupil aperture. When the intensity of the light lessens, the iris dilates to admit more light.

The cornea partially refracts incident light rays before they encounter the lens.

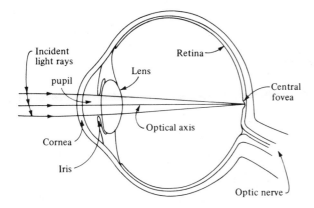

Figure 7-2 Cross section of the human eye.

The lens refracts them further and brings them to focus on the *retina,* thereby forming an image of the viewed object. The retina is composed of very delicate tissue. The most important region of the retina is the *central fovea,* a small pit near the intersection of the optical axis with the retina. The central fovea is the area of sharpest vision. The retina performs a function similar to that performed by the emulsion of photographic film. When it is stimulated by light, the sense of vision is caused, which is transmitted to the brain via the *optic nerve.*

7-3 STEREOSCOPIC DEPTH PERCEPTION

With binocular vision, when the eyes are focused on a certain point, the optical axes of the two eyes converge on that point intersecting at an angle called the *parallactic angle.* The nearer the object, the greater the parallactic angle and vice versa. In Fig. 7-3, the optical axes of the two eyes L and R are separated by a distance b_e, called the *eye base.* For the average adult, this distance is between 63 and 69 mm, or approximately 2.6 in. When the eyes are focused on point A, the optical axes converge, forming parallactic angle ϕ_a. Similarily, when sighting an object at B, the optical axes converge, forming parallactic angle ϕ_b. The brain automatically and unconsciously associates distances D_A and D_B with corresponding parallactic angles ϕ_a and ϕ_b. The depth between objects A and B is $(D_B - D_A)$ and is perceived as the difference in these two parallactic angles.

The ability of human beings to detect changes in parallactic angles, and thus judge differences in depth, is quite remarkable. Although it varies somewhat among individuals, the average person is capable of discerning parallactic angle changes of about 3″ of arc, but some are able to perceive changes as small as 1″. This means that photogrammetric procedures for determining heights of objects and terrain variations based on depth perception by comparisons of parallactic angles can be highly accurate.

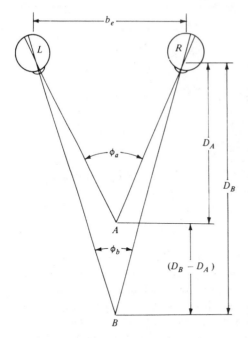

Figure 7-3 Stereoscopic depth perception as a function of parallactic angle.

7-4 VIEWING PHOTOGRAPHS STEREOSCOPICALLY

Suppose that while gazing at object A of Fig. 7-4, a transparent medium containing image marks a_1 and a_2 is placed in front of the eyes as shown. Assume further that the image marks are identical in shape to the object A, and that they are placed on the optical axes so that the eyes are unable to detect whether they are viewing the object or the two marks. Object A could therefore be removed without any noticeable change in the image that is received in the retinas of the eyes. As shown in Fig. 7-4, if the image marks are moved closer together to, say, a_1' and a_2', the parallactic angle increases and the object is perceived to be nearer the eyes at A'. If the marks are moved farther apart to a_1'' and a_2'', the parallactic angle decreases and the brain receives an impression that the object is farther away, at A''.

This phenommenon of creating the three-dimensional or stereoscopic impression of objects by viewing identical images of the objects can be achieved photographically. Suppose that a pair of aerial photographs is taken from exposure stations L_1 and L_2 so that the building appears on both photos, as shown in Fig. 7-5. Flying height above ground is H' and the distance between the two exposures is B, the *air base*. Object points A and B at the top and bottom of the building are imaged at a_1 and b_1 on the left photo and at a_2 and b_2 on the right photo. Now, if the two photos are laid on a table and viewed so that the left eye sees only the left photo and the right eye sees only the right photo, as shown in Fig. 7-6, a three-dimensional impression of the building is obtained. The three-dimensional impression appears to lie below the table top at a distance h from the eyes. The brain judges the height of the building by

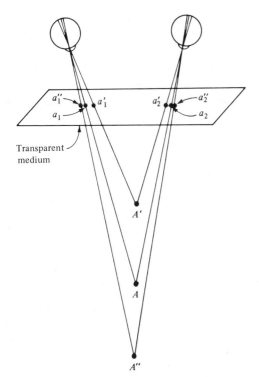

Figure 7-4 The apparent depth to the object, A, can be changed by changing the spacing of the images.

associating depths to points A and B with the parallactic angles ϕ_a and ϕ_b, respectively. When the eyes gaze over the entire overlap area, the brain receives a continuous three-dimensional impression of the terrain. This is acheived by the continuous perception of changing parallactic angles of the infinite number of image points which make up the terrain. The three-dimensional model thus formed is called a *stereoscopic model* or simply a *stereomodel,* and the overlapping pair of photographs is called a *stereopair*.

7-5 STEREOSCOPES

It is quite difficult to view photographs stereoscopically without the aid of optical devices, although some individuals can do it. Besides being an unnatural operation, one of the major problems associated with stereoviewing without optical aids is that the eyes are focused on the photos, while at the same time the brain perceives parallactic angles which tend to form the stereomodel at some depth beyond the photos—a confusing situation to say the least. These difficulties in stereoscopic viewing may be overcome through the use of instruments called *stereoscopes.*

There is a wide selection of stereoscopes serving a variety of special purposes. All operate in essentially the same manner. The *lens* or *pocket* stereoscope shown in

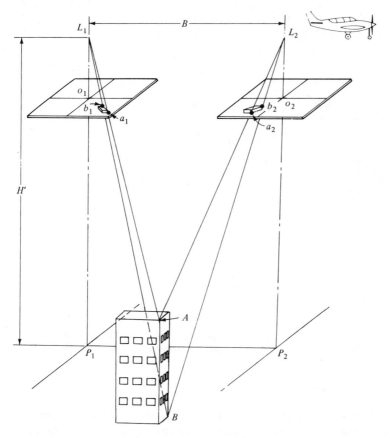

Figure 7-5 Photographs from two exposure stations with building in common overlap area.

Fig. 7-7 is the least expensive and most commonly used stereoscope. It consists of two simple convex lenses mounted on a frame. The spacing between the lenses can be varied to accommodate various eye bases. The legs fold or can be removed so that the instrument is easily stored or carried—a feature which renders the pocket stereoscope ideal for fieldwork. A schematic diagram of the pocket stereoscope is given in Fig. 7-8. The legs of the pocket stereoscope are slightly shorter than the focal length f of the lenses. When the stereoscope is placed over the photos, light rays emanating from points such as a_1 and a_2 on the photos are refracted slightly as they pass through each lens. (Recall from Chap. 2 that a bundle of light rays from a point exactly a distance f from a lens will be refracted and emerge through the lens parallel.) The eyes receive the refracted rays (shown dashed in Fig. 7-8), and on the basis of the eye focusing associated with these incoming rays, the brain receives the impression that they actually originate from a greater distance than that to the table top upon which the photos rest. This overcomes the difficulties noted above. The lenses also serve to magnify the images, thereby enabling detail to be seen more clearly.

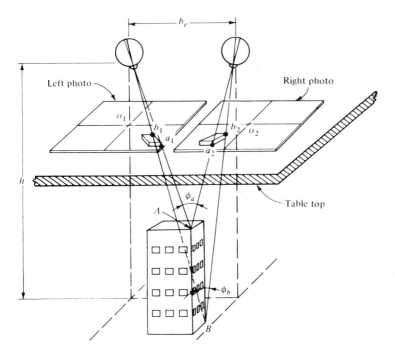

Figure 7-6 Viewing the building stereoscopically.

In using a pocket stereoscope, the photos are placed so that corresponding images are slightly less than the eye base apart; usually about 2 in. For normal 9-in-square format photos taken with 60 percent end lap, the common overlap area of a pair of photos is a rectangular area 5.4 in wide, as shown in Fig. 7-9a. If the photos are separated by 2 in for stereoviewing with a pocket stereoscope, as shown in Fig. 7-9b, there is a rectangular area, shown double crosshatched, in which the top photo obscures the bottom photo, thereby preventing stereoviewing. To overcome this prob-

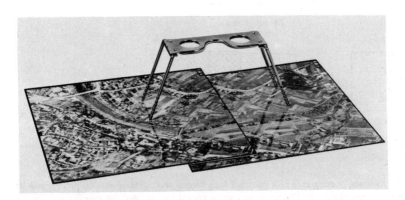

Figure 7-7 Lens or pocket stereoscope. (*Courtesy Carl Zeiss, Oberkochen.*)

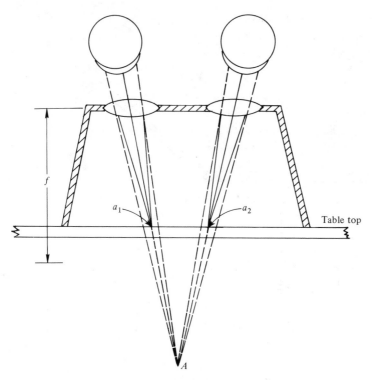

Figure 7-8 Schematic diagram of the pocket stereoscope.

lem, the top photo can be gently rolled up out of the way to enable viewing the corresponding imagery of the obscured area.

The *mirror* stereoscope shown in Fig. 7-10 permits the two photos to be completely separated when viewed stereoscopically. This eliminates the problem of one photo obscuring part of the overlap of the other, and it also enables the entire width of the stereomodel to be viewed simultaneously. The operating principle of the mirror

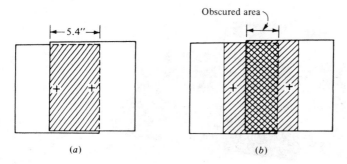

Figure 7-9 (a) The common overlap area of a pair of 9-inch format photos taken with 60 percent end lap (corresponding images coincident). (b) Obscured area when photos are oriented for viewing with pocket stereoscope.

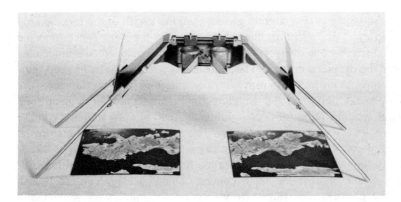

Figure 7-10 ST-4 mirror stereoscope. (*Courtesy Wild Heerbrugg Instruments, Inc.*)

stereoscope is illustrated in Fig. 7-11. The stereoscope has two large wing mirrors and two smaller eyepiece mirrors, all of which are mounted at 45° to the horizontal. Light rays emanating from image points on the photos such as a_1 and a_2 are reflected from the mirror surfaces, according to the principles of reflection discussed in Sec. 2-3, and are received at the eyes forming parallactic angle ϕ_a. The brain automatically associates the depth to point A with that parallactic angle. The stereomodel is thereby created beneath the eyepiece mirrors as illustrated in Fig. 7-11.

Simple lenses are usually placed directly above the eyepiece mirrors as shown in Fig. 7-10. Their separation may be changed to accommodate various eye bases. The

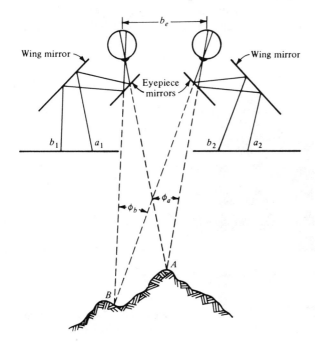

Figure 7-11 Operating principle of the mirror stereoscope.

focal length of these lenses is also slightly greater than the length of the reflected ray path from photo to eyes, and therefore they serve basically the same function as the lenses of the pocket stereoscope. Mirror stereoscopes may be equipped with binoculars which fasten over the eyepiece mirrors. The binoculars, which may be focused individually to accommodate each eye, permit viewing images at high magnification—a factor which is especially important and useful in photo interpretation or in identifying image points. High magnification, of course, limits the field of view so that the entire stereomodel cannot be viewed simultaneously. The stereoscope must therefore be moved about if all parts of the stereomodel are to be seen.

It is extremely important to avoid touching the first-surfaced mirrors of a mirror stereoscope. This is true because the hands contain oils and acids which can tarnish the coatings on the mirrors, rendering them useless. If the mirrors are accidentally fingerprinted, they should be cleaned immediately, using a soft cloth and lens cleaning fluid.

A variation of the mirror stereoscope called the *Old Delft Scanning Stereoscope* has many conveniences. The photos are viewed through oculars via an optical path of prisms and lenses. The oculars can be individually focused and magnification of either 1.5 or 4.5 × can be selected. At either of these magnifications the field of view includes only a limited portion of the stereomodel, but the entire stereomodel may be viewed without moving the stereoscope or photos. This is accomplished by turning knobs which rotate prisms in both optical paths, making it possible to scan the stereomodel in both the X and Y directions. By rotating one prism small residual y *parallaxes* (see Sec. 7-7) can be eliminated.

A different type of stereoscope called the *Zoom stereoscope* is shown in Fig. 7-12. A variety of these instruments are manufactured affording a choice of special features such as continuous zoom magnification up to 120 ×, capability of rotating

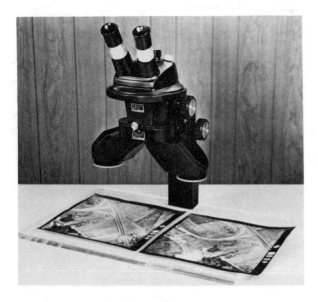

Figure 7-12 Zoom 95 stereoscope. (*Courtesy Bausch and Lomb Co.*)

images optically (which permits convenient correction for crab or alignment of photos), accommodation of various format sizes, and individual focus and enlargement so that two photos of different scales can be viewed stereoscopically. For direct stereoscopic viewing of film negatives, these stereoscopes may be obtained mounted on a light table and equipped with a special scanning mechanism. A reel of film and a takeup reel are mounted on either end of the table. By turning a crank, the frames are brought into position for viewing.

7-6 THE USE OF STEREOSCOPES

Before attempting to use a stereoscope, it is important to study the operator's manual if one is available. This is especially true for stereoscopes having more elaborate optical viewing systems. Also, the lenses and mirrors should be inspected and cleaned if necessary.

In stereoscopic viewing, it is important to orient the photos so that the left and right eyes see the left and right photos, respectively. If the photos are viewed in reverse, a *pseudoscopic view* results in which ups and downs are reversed; e.g., valleys appear as ridges and hills appear as depressions. This can be advantageous for certain work such as tracing drainage patterns, but normally the correct stereoscopic view is desired. The photos should also be oriented so that the shadows appear to fall to the right or toward the observer, and failure to do this can also cause difficulties in stereoscopic viewing.

Accurate and comfortable stereoscopic viewing requires that the eye base, the line joining the centers of the stereoscope lenses, and the flight line all be parallel. Therefore, after the photos have been inspected and laid out so as to prevent a pseudoscopic view, the flight line is marked on both photos. For vertical photographs, the flight line is the line from the center of the left photo to the center of the right photo. In marking the flight line, the photo centers (principal points) are first located by joining opposite fiducial marks with straight lines. Principal points are shown at o_1 and o_2 on Fig. 7-13. *Conjugate principal points* (locations of principal points of adjacent overlapping photos) are marked next. This may be done satisfactorily by carefully observing images immediately surrounding the principal points, finding those corresponding images in the overlap area of the adjacent photos, and then marking the conjugate principal points by estimating their positions with respect to these surrounding images. The conjugate principal points are shown at o_1' and o_2' on Fig. 7-13.

The next step in orienting a pair of photos for stereoscopic viewing is to fasten the left photo down onto the table. Then the right photo is oriented so that the four points defining the flight line (o_1, o_2', o_1', and o_2) all lie along a straight line, as shown in Fig. 7-13. The right photo is retained in this orientation, and while being viewed through the stereoscope, it is moved sideways until the spacing between corresponding images produces a comfortable stereoscopic view. Normally the required spacing between corresponding images is slightly more than 2 in for a pocket stereoscope and about 10 in for a mirror stereoscope.

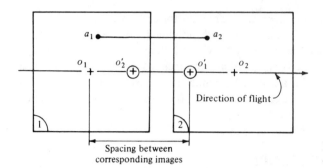

Spacing between
corresponding images

Figure 7-13 Pair of photos properly oriented for stereoscopic viewing.

It is not absolutely necessary to mark the flight lines and orient photos for stereoscopic viewing in the manner outlined above; in fact, for casual stereoviewing, it is normally done by a trial method in which the photos are simply shifted in position until a clear stereoscopic view is obtained. If accuracy and eye comfort is a consideration, however, orientation by the "flight line" procedure is recommended.

As previously stated, comfortable stereoscopic viewing requires that the line joining the stereoscope lens centers be parallel with the flight line. Once the photos are properly oriented, the operator can easily align the stereoscope by simply rotating it slightly until the most comfortable viewing position is obtained. The operator should look directly into the centers of the lenses, thereby holding the eye base parallel with the flight line.

7-7 CAUSES OF Y PARALLAX

An essential condition that must exist for clear and comfortable stereoscopic viewing is that the line joining corresponding images be parallel with the direction of flight. This condition is fulfilled with the corresponding images a_1 and a_2 shown in Fig. 7-13. When corresponding images fail to lie along a line parallel to the flight line, *y parallax* (p_y) is said to exist. Any slight amount of y parallax causes eyestrain, and excessive amounts prevent stereoscopic viewing altogether.

If a pair of truly vertical overlapping photos taken from equal flying heights is oriented perfectly, then no y parallax should exist anywhere in the overlap area. Failure of any of these conditions to be satisfied will cause y parallax. In Fig. 7-14, for example, the photos are improperly oriented and the principal points and conjugate principal points do not lie on a straight line. As a result, y parallax exists at both points *a* and *b*. This condition can be prevented by careful orientation.

In Fig. 7-15 the left photo was exposed from a lower flying height than the right photo, and consequently its scale is larger than the scale of the right photo. Even though the photos are truly vertical and properly oriented, y parallax exists at both points *a* and *b* due to variation in flying heights. To obtain a comfortable stereoscopic view, the y parallax can be eliminated by sliding the right photo upward transverse to the flight line when viewing point *a* and sliding it downward when viewing point *b*.

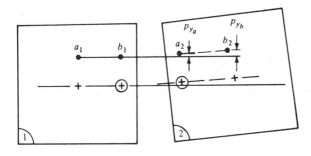

Figure 7-14 *Y* parallax caused by improper orientation of the photos.

The effect of tilted photos is illustrated in Fig. 7-16. The left photo is truly vertical and shows positions of images *a* through *d* of a square parcel of property on flat terrain. The right photo was tilted such that the same parcel appears as a trapezoid. *Y* parallax exists throughout the stereoscopic model as a result of the tilt, as indicated for points *a* and *c*. In practice the direction of tilt is random, and therefore small *y* parallaxes from this source are likely to exist in variable amounts throughout most stereomodels. If it is the intent to obtain vertical photography from a constant flying height, however, these conditions are generally so well controlled that *y* parallaxes from these sources are seldom noticeable. Most serious *y* parallaxes usually occur from improper orientation of the photos, a condition which can easily be corrected.

7-8 VERTICAL EXAGGERATION IN STEREOVIEWING

Under normal conditions the vertical scale of a stereomodel will appear to be greater than the horizontal scale; i.e., an object in the stereomodel will appear to be too tall. This apparent scale disparity is called *vertical exaggeration*. It is usually of greatest concern to photo interpreters, who must take this condition into account when estimating heights of objects, rates of slopes, etc.

Although other factors are involved, vertical exaggeration is caused primarily by the lack of equivalence of the *photographic base-height ratio* (B/H') and the corresponding *stereoviewing base-height ratio* ($b_{e/h}$). B/H' is the ratio of the *air base* (distance between the two exposure stations) to flying height above average ground, and $b_{e/h}$ is the ratio of the *eye base* (distance between the two eyes) to the distance from the eyes at which the stereomodel is perceived. Figure 7-17*a* and Fig. 17-7 *b*

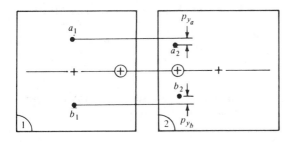

Figure 7-15 *Y* parallax caused by variation in flying height.

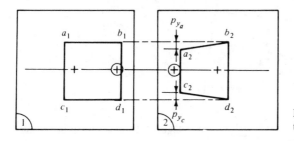

Figure 7-16 Y parallax caused by tilt of the photos.

depict, respectively, the taking of a pair of vertical overlapping photographs and the stereoscopic viewing of those photos. In Fig. 7-17a, f is the camera focal length, B is the air base, H' is the flying height above ground, Y is the height of the ground object AC, and D is the horizontal ground distance KC. In Fig. 7-17a, assume that Y is equal to D. In Fig. 7-17b, i is the image distance from the eyes to the photos, b_e is the eyebase, h is the distance from the eyes to the perceived stereomodel, y is the stereomodel height of object $A'C'$, and d is horizontal stereomodel distance $K'C'$.

An equation for calculating vertical exaggeration can be developed with reference to these figures. From similar triangles of Fig. 7-17a,

$$\frac{x_a}{B} = \frac{f}{H' - Y} \quad \text{from which} \quad x_a = \frac{Bf}{H' - Y} \tag{a}$$

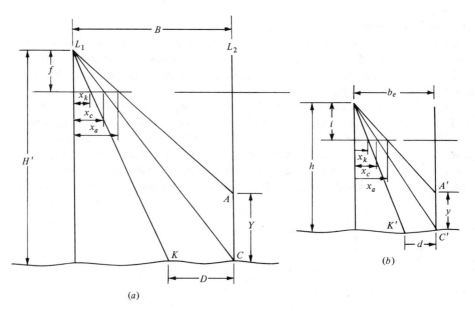

(a)

Figure 7-17 Simplistic diagrams for analyzing vertical exaggeration. (a) Geometry of overlapping vertical aerial photography. (b) Geometry of stereoscopic viewing of the photos of Fig. (a).

Also,

$$\frac{x_c}{B} = \frac{f}{H'} \quad \text{from which } x_c = \frac{Bf}{H'} \tag{b}$$

Subtracting (b) from (a) and reducing,

$$x_a - x_c = Bf \frac{Y}{(H')^2 - H'Y} \tag{c}$$

Also from similar triangles of Fig. 7-17(b),

$$\frac{x_a}{b_e} = \frac{i}{h - y} \quad \text{from which } x_a = \frac{b_e i}{h - y} \tag{d}$$

and $\dfrac{x_c}{b_e} = \dfrac{i}{h}$ from which $x_c = \dfrac{b_e i}{h}$ (e)

Subtracting (e) from (d) and reducing,

$$x_a - x_c = b_e i \frac{y}{h^2 - hy} \tag{f}$$

Equating (c) and (f),

$$Bf \frac{Y}{(H')^2 - H'Y} = b_e i \frac{y}{h^2 - hy}$$

In the above equation the values of Y and y are normally considerably smaller than the values of H' and h, respectively; thus

$$\frac{BfY}{(H')^2} \simeq \frac{b_e iy}{h^2} \quad \text{from which } \frac{y}{Y} = \frac{fh}{H'i} \frac{Bh}{H'b_e} \tag{g}$$

Also from similar triangles of Fig. 7-17a and b,

$$\frac{x_c - x_k}{D} = \frac{f}{H'} \quad \text{from which } D = (x_c - x_k) \frac{H'}{f} \tag{h}$$

and $\dfrac{x_c - x_k}{d} = \dfrac{i}{h}$ from which $d = (x_c - x_k) \dfrac{h}{i}$ (i)

Dividing (i) by (h) and reducing,

$$\frac{d}{D} = \frac{fh}{H'i} \tag{j}$$

Substituting (j) into (g) and reducing,

$$\frac{y}{Y} = \frac{d}{D} \frac{Bh}{H'b_e} \tag{k}$$

In Eq. (k), if the term $Bh/H'b_e$ is equal to 1, there is no vertical exaggeration of the stereomodel. (Recall that Y is equal to D.) Thus an expression for the magnitude of vertical exaggeration, V, is given by

$$V = \frac{B}{H'} \frac{h}{b_e} \quad \text{(approx)} \tag{7-1}$$

From Eq. (7-1) it is seen that the magnitude of vertical exaggeration in stereoscopic viewing can be approximated by multiplying the B/H' ratio by the inverse of the b_e/h ratio. An expression for the B/H' ratio can be developed with reference to Fig. 7-18. In this figure, G represents the total ground coverage of a vertical photo taken from an altitude of H' above ground. Air base B is the distance between exposures. From the figure,

$$B = G - G\frac{PE}{100} = G\left(1 - \frac{PE}{100}\right) \tag{l}$$

In Eq. (l), PE is the percent end lap which gives the amount that the second photo overlaps the first. Also by similar triangles of the figure,

$$\frac{H'}{G} = \frac{f}{d} \quad \text{from which } H' = \frac{fG}{d} \tag{m}$$

In Eq. (m), f is the camera focal length and d is its format dimension. Dividing (l) by (m) and reducing,

$$\frac{B}{H'} = \left(1 - \frac{PE}{100}\right)\frac{d}{f} \tag{7-2}$$

The stereoviewing base-height ratio b_e/h is a somewhat difficult variable to measure, and it differs slightly among individuals. Repeated tests, however, indicate that its value is approximately 0.15.

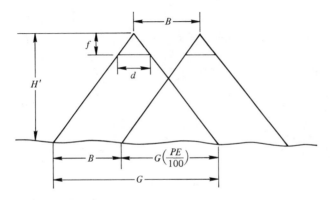

Figure 7-18 Base-height (B/H') ratio.

Example 7-1 Calculate the approximate vertical exaggeration for vertical aerial photos taken with a 6-in-(152.4-mm-)-focal-length camera having a 9-in-(23-cm-) square format if the photos were taken with 60 percent end lap.

SOLUTION By Eq. (7-2),

$$\frac{B}{H'} = \left(1 - \frac{60}{100}\right)\frac{9}{6} = 0.60$$

By Eq. (7-1), assuming b_e/h to be 0.15,

$$V = 0.60\,\frac{1}{0.15} = 4.0 \qquad \text{(approx.)}$$

Note: If a 12-in-focal-length camera had been used, the B/H' ratio would have been 0.30 and vertical exaggeration would have been reduced to 2.

REFERENCES

Ambrose, W. R.: Stereoscopes with High Performance, *Photogrammetric Engineering,* vol. 31, no. 5, p. 822, 1965.

American Society of Photogrammetry: "Manual of Photogrammetry," 4th ed., Falls Church, Va., 1980, chap. 10.

Anson, A.: Significant Findings of a Stereoscopic Acuity Study, *Photogrammetric Engineering,* vol. 25, no. 4, p. 607, 1959.

Collins, S. H.: Stereoscopic Depth Perception, *Photogrammetric Engineering and Remote Sensing,* vol. 47, no. 1, p. 45, 1981.

Dalsgaard, J.: Stereoscopic Vision—A Problem in Terrestrial Photogrammetry, *Photogrammetria,* vol. 34, no. 1, p. 3, 1978.

Goodale, E. R.: An Equation for Approximating the Vertical Exaggeration Ratio of a Stereoscopic View, *Photogrammetric Engineering,* vol. 19, no. 4, p. 607, 1953.

Gumbel, E. J.: The Effect of the Pocket Stereoscope on Refractive Anomalies of the Eyes, *Photogrammetric Engineering,* vol. 30, no. 5, p. 795, 1964.

Howard, A. D.: The Fichter Equation for Correcting Stereoscopic Slopes, *Photogrammetric Engineering,* vol. 34, no. 4, p. 386, 1968.

Jackson, K. B.: Some Factors Affecting the Interpretability of Air Photos, *Canadian Surveyor,* vol. 14, no. 10, p. 454, 1959.

LaPrade, G. L.: Stereoscopy—A More General Theory, *Photogrammetric Engineering,* vol. 38, no. 12, p. 1177, 1972.

———: Stereoscopy—Will Dogma or Data Prevail?, *Photogrammetric Engineering,* vol. 39, no. 12, p. 1271, 1973.

Miller, C. I.: Vertical Exaggeration in the Stereo Space Image and its Use, *Photogrammetric Engineering,* vol. 26, no. 5, p. 815, 1960.

Myers, B. J., and F. P. Van der Duys: A Stereoscopic Field Viewer, *Photogrammetric Engineering and Remote Sensing,* vol. 41, no. 12, p. 1477, 1975.

Nicholas, G., and J. T. McCrickerd: Holography and Stereoscopy: The Holographic Stereogram, *Photographic Science and Engineering,* vol. 13, no. 6, p. 342, 1969.

Palmer, D. A.: Stereoscopy and Photogrammetry, *Photogrammetric Record,* vol. 4, p. 391, 1964.

Raasveldt, H. C.: The Stereomodel, How It Is Formed and Deformed, *Photogrammetric Engineering,* vol. 22, no. 4, p. 708, 1956.

Raju, A. V., and E. Parthasarathi: Stereoscopic Viewing of Landsat Imagery, *Photogrammetric Engineering and Remote Sensing,* vol. 43, no. 10, p. 1243, 1977.

Scheaffer, C. E.: Stereoscope for Strips, *Photogrammetric Engineering,* vol. 34, no. 10, p. 1044, 1968.

Singleton, R.: Vertical Exaggeration and Perceptual Models, *Photogrammetric Engineering,* vol. 22, no. 9, p. 175, 1956.

Thayer, T. P.: The Magnifying Single Prism Stereoscope: A New Field Instrument, *Journal of Forestry,* vol. 61, p. 381, 1963.

Thurrell, R. F., Jr.: Vertical Exaggeration in Stereoscopic Models, *Photogrammetric Engineering,* vol. 19, no. 4, p. 579, 1953.

Treece, W. A.: Estimation of Vertical Exaggeration in Stereoscopic Viewing of Aerial Photographs, *Photogrammetric Engineering,* vol. 21, no. 4, p. 518, 1955.

Yacoumelos, N.: The Geometry of the Stereomodel, *Photogrammetric Engineering,* vol. 38, no. 8, p. 791, 1972.

PROBLEMS

7-1 What are some of the monocular methods of perceiving depth?

7-2 What is a parallactic angle?

7-3 Compare the advantages and disadvantages of the pocket and mirror stereoscopes.

7-4 Give a step-by-step procedure for orienting photos for stereoscopic viewing.

7-5 What is y parallax? What are the causes of y parallax in a stereomodel?

7-6 Prepare a table of B/H' ratios for camera focal lengths of $3\frac{1}{2}$, 6, $8\frac{1}{4}$, and 12 in, camera format of 9-in square, and percent end laps of 55, 60, and 65.

7-7 Calculate the approximate vertical exaggeration in a stereomodel from photos taken with a 6-in-focal-length camera having a 9-in-square format if the photos are taken at 55 percent end lap.

7-8 Repeat Prob. 7-7, except that an $8\frac{1}{4}$-in-focal-length camera was used, and end lap was 65 percent.

EIGHT
STEREOSCOPIC PARALLAX

8-1 INTRODUCTION

Parallax is the apparent displacement in the position of an object, with respect to a frame of reference, caused by a shift in the position of observation. A simple experiment will serve to illustrate parallax. If a finger is held in front of one's eyes and while gazing at the finger the head is quickly shifted from side to side without moving the finger, the finger will appear to move from side to side with respect to objects beyond the finger such as pictures on the wall. Rather than shifting the head, the same effect can be created by alternately blinking one's eyes. The closer the finger is held to the eyes, the greater will be its apparent shift. This apparent motion of the finger is parallax, and it is due to the shift in the position of observation.

If a person looked through the viewfinder of an aerial camera as the aircraft moved forward, images of objects would be seen to move across the field of view. This image motion is another example of parallax caused by shifting the location of the observation point. Again, the closer an object is to the camera, the more its image will appear to move.

An aerial camera exposing overlapping photographs at regular intervals of time obtains a record of positions of images at the instants of exposure. The change in position of an image from one photograph to the next caused by the aircraft's motion is termed *stereoscopic parallax, x parallax,* or simply *parallax.* Parallax exists for all images appearing on successive overlapping photographs. In Fig. 8-1, for example, images of object points A and B appear on a pair of overlapping vertical aerial photographs which were taken from exposure stations L_1 and L_2. Points A and B are imaged at a and b on the left-hand photograph. Forward motion of the aircraft between exposures, however, caused the images to move laterally across the camera focal plane parallel to the flight line, so that on the right-hand photo they appear at a' and b'. Because point A is higher (closer to the camera) than point B, the movement of image a across the focal plane was greater than the movement of image b; in other words, the parallax of point A is greater than the parallax of point B. This calls attention to two important aspects of stereoscopic parallax: (1) the parallax of any

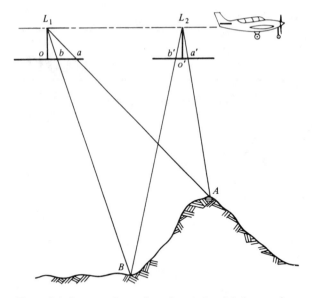

Figure 8-1 Stereoscopic parallax of vertical aerial photographs.

point is directly related to the elevation of the point, and (2) parallax is greater for high points than for low points. Variation of parallax with elevation provides the fundamental basis for determining elevations of points from photographic measurements. In fact, X, Y, and Z ground coordinates can be calculated for points based upon their parallaxes. Equations for doing this are presented in Sec. 8-6.

Figure 8-2 shows the two photographs of Fig. 8-1 in superposition. Parallaxes of object points A and B are p_a and p_b, respectively. Stereoscopic parallax for any point such as A whose images appear on two photos of a stereopair, expressed in terms of "flight line" photographic coordinates, is

$$p_a = x_a - x_a' \tag{8-1}$$

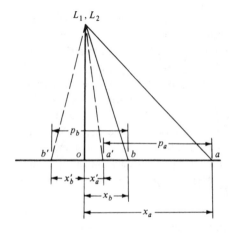

Figure 8-2 The two photographs of Fig. 8-1 shown in superposition.

In Eq. (8-1), p_a is the stereoscopic parallax of object point A, x_a is the measured photocoordinate of image a on the left photograph of the stereopair, and x'_a is the photocoordinate of image a' on the right photo. These photocoordinates *are not* measured with respect to the fiducial axis system. Rather, they are measured with respect to the ''flight line'' axis system described in Sec. 8-2. In Eq. (8-1) it is imperative that proper algebraic signs be given to measured photocoordinates to obtain correct values for stereoscopic parallax.

Figure 8-3 is a portion of a stereopair of vertical photographs taken over Washington, D.C., with a 6-in-focal-length camera at a flying height of 6,000 ft above ground. On these photos, note how all images moved laterally with respect to the y axis from their positions on the left photo to their positions on the right photo. Note also how vividly the Washington Monument illustrates the increase in parallax with higher points; i.e., the top of the monument has moved farther across the focal plane than the bottom of the monument.

In Fig. 8-3 the Washington Monument affords an excellent example for demonstrating the use of Eq. (8-1) for finding parallaxes. The top of the monument has an x coordinate ($x_t = 4.32$ in) and an x' coordinate ($x'_t = 0.72$ in). By Eq. (8-1), the parallax $p_t = 4.32 - 0.72 = 3.60$ in. Also, the bottom of the monument has an x coordinate ($x_b = 3.96$ in) and an x' coordinate ($x'_b = 0.67$ in). Again by Eq. (8-1), $p_b = 3.96 - 0.67 = 3.29$ in. In Sec. 8-8 it will be demonstrated how these parallaxes can be used to compute the height of the Washington Monument.

8-2 PHOTOGRAPHIC ''FLIGHT LINE'' AXES FOR PARALLAX MEASUREMENT

Since parallax occurs parallel to the direction of flight, the photographic x and x' axes for parallax measurement must be parallel with the flight line for each of the photographs of a stereopair. (Primed values denote the right-hand photo of a stereopair.) For a vertical photograph of a stereopair, the flight line is the line connecting principal point and conjugate principal point. Principal points are located in the usual manner by intersecting the x and y fiducial lines. A monoscopic method of establishing conjugate principal points was described in Sec. 7-6. Stereoscopic methods are discussed in Sec. 8-4. The y and y' axes for parallax measurement pass through their respective principal points and are perpendicular to the flight line.

All photographs except those on the ends of a flight strip may have two sets of flight axes for parallax measurements—one to be used when the photo is the left photo of the stereopair and one when it is the right photo. An example is shown in Fig. 8-4, where photographs 1 through 3 were exposed as shown. Parallax measurements in the overlap area of photos 1 and 2 are made with respect to the solid xy axis system of photo 1 and the solid $x'y'$ system of photo 2. However, due to the aircraft's curved path of travel, the flight line of photos 2 and 3 is not in the same direction as the flight line of photos 1 and 2. Therefore, parallax measurements in the overlap area of photos 2 and 3 must be made with respect to the dashed xy axis system on photo 2 and the dashed $x'y'$ system of photo 3. It is possible for the two axis systems to be

Figure 8-3 Overlapping vertical photographs taken over Washington, D.C., illustrating stereoscopic parallax. (*Photos courtesy Owen Ayres and Associates, Inc.*)

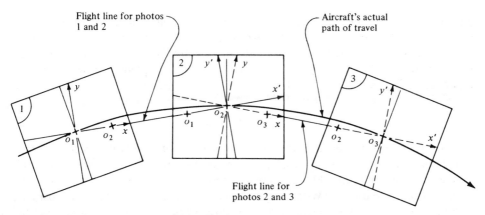

Figure 8-4 Flight line axes for measurement of stereoscopic parallax.

coincident; however, this does not generally occur in practice. Henceforth in this chapter it shall be understood that photographic coordinates for parallax determination are measured with respect to the flight line axis system.

8-3 MONOSCOPIC METHODS OF PARALLAX MEASUREMENT

Parallaxes of points in a stereopair may be measured either monoscopically or stereoscopically. There are certain advantages and disadvantages associated with each method. In either method the photographic flight line axes must first be carefully located by marking principal points and conjugate principal points.

The simplest method of parallax measurement is the monoscopic approach in which Eq. (8-1) is solved after direct measuremnt of x and x' on the left and right photos, respectively. A disadvantage of this method is that two measurements are required for each point.

Another monoscopic approach to parallax measurement is to fasten the photographs down on a table or base material as shown in Fig. 8-5. In this method the

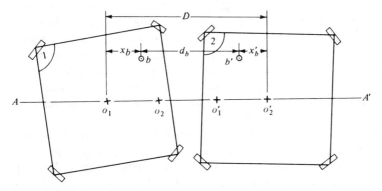

Figure 8-5 Parallax measurement using a simple scale.

photographic flight lines o_1o_2 and $o_1'o_2'$ are marked as usual. A long straight line, AA', is drawn on the base material and the two photos are *carefully* mounted as shown so that the photographic flight lines are coincident with this line. Having fastened the photos down, the distance D between principal points is a constant which can be measured. The parallax of point B is $p_b = x_b - (-x_b')$. However, by examining the figure, it is seen that parallax is also

$$p_b = D - d_b \qquad (8\text{-}2)$$

With D known in Eq. (8-2), to obtain the parallax of a point it is necessary only to measure the distance d between its images on the left and right photos. The advantage is that for each additional point whose parallax is desired, only a single measurement is required. With either of these monoscopic methods of parallax measurement, any of the simple scales described in Sec. 5-3 may be used, and the choice will depend upon the desired accuracy.

8-4 PRINCIPLE OF THE FLOATING MARK

Parallaxes of points can be measured while viewing stereoscopically with the advantages of speed and accuracy. Stereoscopic measurement of parallax makes use of the principle of *floating mark*. When a stereomodel is viewed through a stereoscope, two small identical marks etched on clear glass called *half marks* may be placed over the photographs—one on the left photo and one on the right photo, as illustrated in Fig. 8-6. The left mark is seen with the left eye and the right mark with the right eye. The half marks may be shifted in position until they fuse together into a single mark which appears to exist in the stereomodel and to lie at a particular elevation. If the half marks are moved closer together, the parallax of the half marks is increased and the fused mark will therefore appear to rise. Conversely, if the half marks are moved apart, parallax is decreased and the fused mark appears to fall. This apparent variation in the elevation of the mark as the spacing of half marks is varied is the basis for the term "floating mark."

The spacing of the half marks (parallax of the half marks) may be varied so that the floating mark appears to rest exactly on the terrain. This produces the same effect as though an object of the shape of the half marks had existed on the terrain when the photos were originally taken. The floating mark may be moved about the stereomodel from point to point, and as the terrain varies in elevation, the spacing of the half marks may be varied to make the floating mark rest exactly on the terrain. Figure 8-6 demonstrates the principle of the floating mark and illustrates how the mark may be set exactly on particular points such as A, B, and C by placing the half marks at a and a', b and b', and c and c', respectively.

The principle of the floating mark can be used to transfer principal points to their conjugate locations, thereby marking the flight line axes. In this procedure the principal points are first located as usual at the intersection of fiducial lines. Then by means of a point transfer device such as that shown in Fig. 8-7, these principal points are transferred to their conjugate locations. The point transfer device consists of two

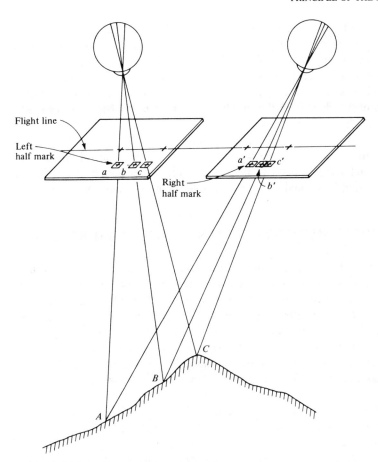

Figure 8-6 The principle of the floating mark.

separate pieces containing identical half marks etched on glass. The left half mark of Fig. 8-7 is placed over one of the principal points, say, the left point o_1, for example. Using a mirror stereoscope for viewing, the right half mark is placed on the right photo and moved about until a clear stereoscopic view of the floating mark is obtained and the fused mark appears to rest exactly on the ground. The right half mark, which is hinged, is then raised to make way for lowering a hinged arm containing a pin.

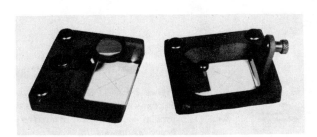

Figure 8-7 Simpson point transfer device. (*Courtesy Alan Gordon Enterprises, Inc.*)

The pin comes down exactly on the position occupied by the right half mark. It is pressed into the photograph, thereby marking the conjugate principal point. This stereoscopic procedure is very accurate if carefully performed and it has the advantage that discrete images near the principal points are not necessary, as they are with the monoscopic method. Imagine, for example, the difficulty of monoscopically transferring a principal point which falls in the middle of a wheat field. This transfer could be done easily by the stereoscopic method, however.

A homemade version of the stereoscopic point-transfer device described above consists of two small pieces of transparent plastic upon which identical crosses are inked. When a half mark is stereoscopically located over a conjugate principal point, its photographic position is marked by pinpricking through the center of the cross.

8-5 STEREOSCOPIC METHODS OF PARALLAX MEASUREMENT

Through the principle of the floating mark, parallaxes of points may be measured stereoscopically. This method employs a stereoscope in conjunction with an instrument called a *parallax bar,* also frequently called a *stereometer.* A parallax bar consists of a metal bar to which are fastened two half marks. The right half mark may be moved with respect to the left mark by turning a micrometer screw. Readings from the micrometer are taken with the floating mark set exactly on points whose parallaxes are desired. From the micrometer readings, parallaxes or differences in parallax are obtained. A parallax bar is shown lying on the photos beneath a mirror stereoscope in Fig. 8-8.

When a parallax bar is used, the two photos of a stereopair are first *carefully* oriented for comfortable stereoscopic viewing, in such a way that the flight line of each photo lies along a common straight line, as shown in Fig. 8-5. The photos are then fastened securely and the parallax bar is placed on the photos. The left half mark, called the *fixed mark,* is unclamped and moved so that, when the floating mark is fused on a terrain point of average elevation, the parallax bar reading is approximately in the middle of the run of the graduations. The fixed mark is then clamped, where it will remain for all subsequent parallax measurements on that particular stereopair. After the fixed mark is positioned in this manner, the right half mark or *movable mark*

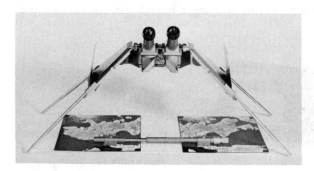

Figure 8-8 Wild ST-4 mirror stereoscope with binocular attachment and parallax bar. (*Courtesy Wild Heerbrugg Instruments, Inc.*)

may be moved left or right with respect to the fixed mark (increasing or decreasing the parallax) as required to accommodate high points or low points without exceeding the run of the parallax bar graduations.

Figure 8-9 is a schematic diagram of the operating principle of the parallax bar. After the photos have been oriented and the fixed half mark positioned as just described, the *parallax bar constant C* for the setup is determined. For the setup, the spacing between principal points is a constant, D. Once the fixed mark is clamped, the distance from the fixed mark to the index mark of the parallax bar is also a constant, K. From Fig. 8-9, the parallax of point A is

$$p_a = x_a - x_a' = D - (K - r_a) = (D - K) + r_a$$

The term $D - K$ is C, the parallax bar constant for the setup. Also, r_a is the micrometer reading. Substituting C into the above equation, the expression becomes

$$p_a = C + r_a \qquad (8\text{-}3)$$

Equation (8-3) assumes the parallax bar micrometer to be "forward-reading"; i.e., readings increase with increasing parallaxes. Should the readings decrease with increasing parallax, the parallax bar is called "backward-reading" and the algebraic sign of r must be reversed.

To calculate the parallax bar constant, a micrometer reading is taken with the floating mark set on a point. The parallax of that particular point is also directly measured monoscopically and calculated using Eq. (8-1). Then with p and r for that point known, the value of C is calculated, using Eq. (8-3), as

$$C = p - r \qquad (8\text{-}4)$$

The parallax bar constant should be determined on the basis of micrometer readings and parallax measurements for two points. Then the mean of the two values of C may be adopted. Any two points may be selected for this purpose; however, the two principal points are convenient and often used. Figure 8-10 is a section through

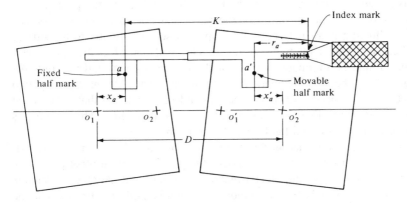

Figure 8-9 Schematic diagram of the parallax bar.

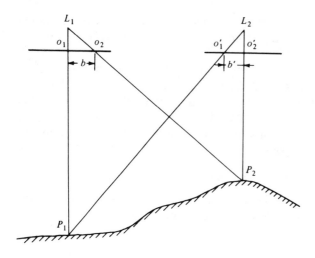

Figure 8-10 Parallax of the principal points.

a pair of overlapping vertical photos. By Eq. (8-1), the parallax of the left-photo ground principal point O_1 is $p_{o_1} = x_{o_1} - (-x'_{o_1}) = 0 - (-b') = b'$. (The x coordinate of o_1 on the left photo is zero.) Also, the parallax of the right-photo ground principal point O_2 is $p_{o_2} = x_{o_2} - x'_{o_2} = b - 0 = b$. From the foregoing, it is seen that the parallax of the left ground principal point is photo base b' measured on the right photo and the parallax of the right ground principal point is photo base b measured on the left photo.

To determine the parallax bar constant using principal points, distances b and b' are first measured. Then the floating mark is stereoscopically fused on the left principal point o_1 and micrometer value r_{o_1} is read, from which we obtain $C_1 = b' - r_{o_1}$. The floating mark is then stereoscopically placed on the right principal point o_2 and micrometer value r_{o_2} is read, from which we obtain $C_2 = b - r_{o_2}$. Errors due to tilt in the photography, unequal flying heights, paper shrinkage, and measurement will generally give two slightly different values of C. The mean value is normally adopted.

In practice, the principal points sometimes become defaced so that stereoscopic measurements r_{o_1} and r_{o_2} are difficult to obtain. In that case, since the parallax of a point depends only upon the elevation of the point and not on its location on the photograph, the floating mark may be stereoscopically set just slightly to the side of them when their micrometer readings are taken. This will not affect the micrometer readings if the ground is reasonably level near the principal point, and it avoids the inaccuracy of setting the floating mark on defaced imagery. Alternately, any two discrete images may be used. However, they should be chosen so that they lie on opposite sides of the flight line and equidistant from the flight line. This minimizes error in parallaxes due to tilt and faulty orientation of the photos.

One of the advantages of measuring parallax stereoscopically is increased speed, for once the parallax bar constant is determined, the parallaxes of all other points are quickly obtained with a single micrometer reading for each point. Another advantage is increased accuracy, not only because binocular vision is used but also because the least reading of most parallax bar micrometer scales is 0.01 mm. A person's ability

Figure 8-11 N2 mirror stereoscope with binoculars and attached parallax bar. (*Courtesy Carl Zeiss, Oberkochen.*)

to set the floating mark, like learning to swim, improves with practice. An experienced person using quality equipment and clear photographs is generally able to obtain parallaxes to within approximately $\pm$ 0.03 mm of their correct values.

Instruments are available which incorporate a stereoscope and parallax bar into a single unit. Figures 8-11 and 8-12 illustrate, respectively, a mirror stereoscope and pocket stereoscope with attached parallax bars. Note the binocular attachment, for greater accuracy, on the unit of Fig. 8-11. The pocket stereoscope unit is especially handy for quick determinations of tree heights, building heights, etc. Although it has the advantage of simplicity and compactness, it does of course retain the inconvenience common to pocket stereoscopes that the top photo obscures part of the overlap area of the stereopair.

8-6 PARALLAX EQUATIONS

As noted earlier, X, Y, and Z ground coordinates can be calculated for points based upon the measurements of their parallaxes. Figure 8-13 illustrates an overlapping pair of vertical photographs which have been exposed at equal flying heights above datum. Images of an object point A appear on the left and right photos at a and a', respectively.

Figure 8-12 TM pocket measuring stereoscope. (*Courtesy Carl Zeiss, Oberkochen.*)

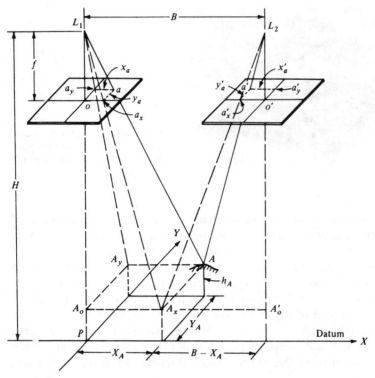

Figure 8-13 Geometry of an overlapping pair of vertical photographs.

The planimetric position of point A on the ground is given in terms of ground coordinates X_A and Y_A. Its elevation above datum is h_A. The XY ground axis system has its origin at the datum principal point P of the left-hand photograph; the X axis is in the same vertical plane as the photographic x and x' flight axes; and the Y axis passes through the datum principal point of the left photo and is perpendicular to the X axis. According to this definition, each stereopair of photographs has its own unique ground coordinate system.

By equating similar triangles of Fig. 8-13, formulas for calculating h_A, X_A, and Y_A may be derived. First of all, from similar triangles L_1oa_y and $L_1A_oA_y$,

$$\frac{Y_A}{H - h_A} = \frac{y_a}{f}$$

$$\text{from which } Y_A = \frac{y_a}{f}(H - h_A) \tag{a}$$

And equating similar triangles L_1oa_x and $L_1A_oA_x$,

$$\frac{X_A}{H - h_A} = \frac{x_a}{f}$$

$$\text{from which } X_A = \frac{x_a}{f}(H - h_A) \tag{b}$$

Also from similar triangles $L_2o'a'_x$ and $L_2A'_oA_x$,

$$\frac{B - X_A}{H - h_A} = \frac{-x'_a}{f}$$

$$\text{from which } X_A = B + \frac{x'_a}{f}(H - h_A) \tag{c}$$

Equating (b) and (c) and reducing,

$$h_A = H - \frac{Bf}{x_a - x'_a}$$

Substituting p_a for $x_a - x'_a$ into the above,

$$h_A = H - \frac{Bf}{p_a} \tag{8-5}$$

Now substituting Eq. (8-5) into each of Eqs. (b) and (a) and reducing,

$$X_A = B\frac{x_a}{p_a} \tag{8-6}$$

$$Y_A = B\frac{y_a}{p_a} \tag{8-7}$$

In Eqs. (8-5), (8-6), and (8-7), h_A is the elevation of point A above datum, H is the flying height above datum, B is the air base, f is the focal length of the camera, p_a is the parallax of point A, X_A and Y_A are ground coordinates of point A in the previously defined unique arbitrary coordinate system, and x_a and y_a are the photocoordinates of point a measured with respect to the flight line axes on the left photo.

Equations (8-5), (8-6), and (8-7) are commonly called the *parallax equations*. They are among the most useful equations to the photogrammetrist. These equations enable a complete survey of the overlap area of a stereopair to be made, provided the focal length is known and sufficient ground control is available so the airbase B and flying height H can be calculated.

Equations (8-6) and (8-7) yield X and Y ground coordinates in the unique arbitrary coordinate system of the stereopair, which is not related to the true ground coordinate system. However, if arbitrary coordinates are determined using these equations for two points whose true ground coordinates are known, then the arbitrary coordinates of all other points can be transformed into the true ground system through a two-dimensional conformal coordinate transformation, as described in Appendix B.

Example 8-1 A pair of overlapping vertical photographs were taken from a flying height of 4,045 ft above sea level with a 152.4-mm-focal-length camera. The air base was 1,280 ft. With the photos properly oriented, parallax bar readings of 12.57 mm and 13.04 mm were obtained with the floating mark set on principal points o_1 and o_2, respectively. On the left photo b was measured as 93.73 mm and on the right photo b' was measured as 93.30 mm. Parallax bar readings of

10.96 mm and 15.27 mm were taken on points A and B. Also, the x and y photocoordinates of points A and B measured with respect to the flight axes on the left photo were $x_a = 53.41$ mm, $y_a = 50.84$ mm, $x_b = 88.92$ mm, and $y_b = -46.69$ mm. Calculate the elevations of points A and B and the horizontal length of line AB.

SOLUTION By Eq. (8-4),

$$C_1 = b' - r_{o_1} = 93.30 - 12.57 = 80.73 \text{ mm}$$

$$C_2 = b - r_{o_2} = 93.73 - 13.04 = 80.69 \text{ mm}$$

$$C = \frac{80.73 + 80.69}{2} = 80.71 \text{ mm}$$

By Eq. (8-3),

$$p_a = C + r_a = 80.71 + 10.96 = 91.67 \text{ mm}$$

$$p_b = C + r_b = 80.71 + 15.27 = 95.98 \text{ mm}$$

By Eq. (8-5),

$$h_A = H - \frac{Bf}{p_a} = 4{,}045 - \frac{1{,}280(152.4)}{91.67} = 1{,}917 \text{ ft above sea level}$$

$$h_B = H - \frac{Bf}{p_b} = 4{,}045 - \frac{1{,}280(152.4)}{95.98} = 2{,}012 \text{ ft above sea level}$$

By Eqs. (8-6) and (8-7),

$$X_a = B \frac{x_a}{p_a} = \frac{53.41(1{,}280)}{91.67} = 746 \text{ ft}$$

$$X_B = B \frac{x_b}{p_b} = \frac{88.92(1{,}280)}{95.98} = 1{,}186 \text{ ft}$$

$$Y_A = B \frac{y_a}{p_a} = \frac{50.84(1{,}280)}{91.67} = 710 \text{ ft}$$

$$Y_B = B \frac{y_b}{p_b} = \frac{-46.69(1{,}280)}{95.98} = -623 \text{ ft}$$

The horizontal length of line AB is,

$$AB = \sqrt{(X_B - X_A)^2 + (Y_B - Y_A)^2}$$

$$= \sqrt{(1{,}186 - 746)^2 + (-623 - 710)^2} = 1{,}404 \text{ ft}$$

8-7 ELEVATIONS BY PARALLAX DIFFERENCES

Parallax differences between one point and another are caused by different elevations of the two points. While parallax Eq. (8-5) serves to define the relationship of stereoscopic parallax to flying height, elevation, air base, and camera focal length, parallax differences are more convenient for determining elevations. In Fig. 8-14, object point C is a control point whose elevation h_C above datum is known. The elevation of object point A is desired. Rearranging Eq. (8-5), parallaxes of both points can be expressed as

$$p_c = \frac{fB}{H - h_C} \tag{d}$$

$$p_a = \frac{fB}{H - h_A} \tag{e}$$

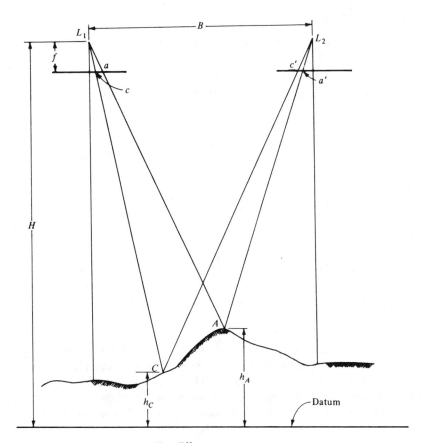

Figure 8-14 Elevations by parallax differences.

The difference in parallax, $p_a - p_c$, obtained by subtracting Eq. (d) from Eq. (e) and rearranging is

$$p_a - p_c = \frac{fB(h_A - h_C)}{(H - h_A)(H - h_C)} \qquad (f)$$

Let $p_a - p_c$ equal Δp, the difference in parallax. Substituting $(H - h_A)$ from Eq. (e) and Δp into Eq. (f) and reducing, the following expression for elevation h_A is obtained:

$$h_A = h_C + \frac{\Delta p(H - h_C)}{p_a} \qquad (8\text{-}8)$$

Example 8-2 In Example 8-1, an additional parallax bar reading of 11.89 was taken on control point C, whose elevation is 1,938 ft above sea level. Calculate the elevations of points A and B of that example using parallax difference Eq. (8-8).

SOLUTION By Eq. (8-3),

$$p_c = C + r_c = 80.71 + 11.89 = 92.60 \text{ mm}$$

For point A,

$$\Delta p = p_a - p_c = 91.67 - 92.60 = -0.93 \text{ mm}$$

By Eq. (8-8),

$$h_A = 1,938 + \frac{(-0.93)(4,045 - 1,938)}{91.67} = 1,917 \text{ ft above sea level}$$

For point B,

$$\Delta p = p_b - p_c = 95.98 - 92.60 = 3.38 \text{ mm}$$

By Eq. (8-8),

$$h_B = 1,938 + \frac{3.38(4,045 - 1,938)}{95.98} = 2,012 \text{ ft above sea level}$$

Note that these answers check the values computed in Example 8-1.

If a number of control points are located throughout the overlap area, use of Eq. (8-8) permits elevations of unknown points to be most accurately determined from the parallax difference of the nearest control point. This minimizes the effects of many errors including photographic tilt, imperfect alignment of the photos for parallax measurement, shrinkage and expansion of the photo papers, and camera-lens distortions.

8-8 APPROXIMATE EQUATION FOR ELEVATIONS FROM PARALLAX DIFFERENCES

The following approximate equation for elevation differences is obtained from Eq. (8-8) by (1) substituting the photo base b of the stereopair for p_a; (2) substituting average flying height above ground, H', for $(H - h_C)$; and (3) letting $\Delta h = h_A - h_C$:

$$\Delta h = \frac{\Delta p H'}{b} \tag{8-9}$$

In Eq. (8-9), Δh is the difference in elevation between two points whose parallax difference is Δp. For photography in a 9-in-square format taken with 60 percent end lap, the photo base b is approximately 90 mm. For moderate relief, parallaxes for all points are approximately equal to b, so that the substitution of b for p_a is valid. Furthermore, if flying height is not extremely low and if relief is moderate, the substitution of average flying height above ground, H' for $H - h_C$ is valid. For very low flying heights or in areas of significant relief, or both, the assumptions of Eq. (8-9) are not met; in these cases Eq. (8-8) should be used. Equation (8-9) is especially convenient in photo interpretation where rough elevations, building and tree heights, etc., are often needed.

Example 8-3 The parallax difference between the top and bottom of a tree is measured as 1.32 mm on a stereopair of photos taken at 3,000 ft above ground. Average photo base is 88 mm. How tall is the tree?

SOLUTION By Eq. (8-9),

$$\Delta h = \frac{1.32 \times 3,000}{88} = 45 \text{ ft}$$

Example 8-4 Using parallax difference Eq. (8-9), determine the height of the Washington Monument from parallax measurements on Fig. 8-3. Flying height was 6,000 ft above ground and the photo base b was measured as 3.25 in.

SOLUTION On Fig. 8-3 the parallax of the top of the monument was measured as 3.60 in and the parallax of the bottom was 3.29 in. Parallax difference is

$$\Delta p = 3.60 - 3.29 = 0.31 \text{ in}$$

By Eq. (8-9) the monument height is

$$\Delta h = \frac{6,000 \times 0.31}{3.25} = 572 \text{ ft}$$

(This is within 3 percent of its true height of 555 ft.)

8-9 MEASUREMENT OF PARALLAX DIFFERENCES

Parallax differences may be determined in any of the following ways:

1. By monoscopic measurement of parallaxes followed by subtraction.
2. By taking differences in parallax bar readings. The validity of this approach is seen if parallaxes determined from parallax bar readings are subtracted, as

$$\Delta p = p - p_c = (C + r) - (C + r_c) = r - r_c$$

3. By *parallax wedge*.

A parallax wedge, as illustrated in Fig. 8-15, consists of a piece of transparent film upon which are drawn two converging lines. The left line is a reference line while the line on the right contains graduations from which readings can be made. The spacing of the two lines depends on whether the parallax wedge will be used with a mirror stereoscope or a pocket stereoscope. For a pocket stereoscope the spacing should vary from about $2\frac{1}{2}$ inches at the bottom to about $1\frac{3}{4}$ inches at the top. This spacing accommodates the usual spacing between corresponding images when a stereopair is oriented for viewing with a pocket stereoscope, and it gives a possible range of about $\frac{3}{4}$ inch in parallax differences that can be measured.

Suppose line spacings of a parallax wedge were exactly 2.50 in at the bottom and 1.70 in at the top as shown in Fig. 8-15. If the total height y of the graduations was exactly 8 in, then for graduations spaced at $\frac{1}{10}$-in intervals along the line, each of the 80 graduations proceeding upward on the scale is 0.01 in closer to the reference line than the next lower graduation. Graduations numbered from 70 to 00 to 50 on the

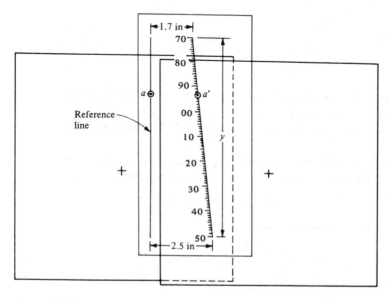

Figure 8-15 Parallax wedge.

parallax wedge therefore represent parallax differences in $\frac{1}{100}$-in units, and with this particular parallax wedge a range of 0.80-in of parallax difference can be accommodated.

When a parallax wedge is used, the photos are first carefully oriented as usual and secured. The parallax wedge is placed in the overlap area, and viewed stereoscopically, the two lines of the parallax wedge will fuse and appear as a single *floating line* in areas where the spacing of the lines is slightly less than the spacing of corresponding photo images. The floating line will appear to split where the parallax of the lines is equal to that of photo images. The position of the parallax wedge can be adjusted so that the floating line splits forming a wedge exactly at a point whose parallax is desired, and at that point a reading is taken from the scale. The parallax wedge reading at point *a* of Fig. 8-15, for example, is 1.93 in. Parallax differences are obtained by simply taking differences in parallax wedge readings for different points.

An expedient means of producing a parallax wedge is to prepare an ink drawing on white paper, photograph it, and then prepare a film positive from the negative. The size of the positive must be carefully produced to obtain proper dimensions. To increase accuracy during preparation of a parallax wedge, the ink drawing can be made at a 2- or 4-times enlargement, and then be photographically reduced to the correct size.

8-10 PARALLAX CORRECTION GRAPH

Errors in measured parallaxes due to photographic tilt, imperfect alignment of the photos for parallax measurement, shrinkage or expansion of the photo papers, and camera lens distortions can be effectively compensated for by constructing a *parallax correction graph*. To construct such a graph it is necessary to have a number of vertical control points evenly distributed throughout the stereo overlap area. Assume, for example, that six well-distributed control points *A* through *F* exist in the overlap area of a stereopair exposed from a flying height of 6,900 ft above datum. The elevations of the six points are given in column (2) of Table 8-1, and parallax bar readings on each of these points are listed in column (3).

Based upon one selected *reference* control point, say, point *C* for this example, parallax differences $\Delta p'$ that result from the measured parallax bar readings are calculated by subtracting the reading on point *C* from each of the other readings. These values are listed in column (4) of Table 8-1. Because the elevations of the six points are all known, the theoretically correct parallax differences Δp that should exist between these points can also be calculated. Discrepancies between $\Delta p'$ and Δp thus represent corrections that must be applied to account for the errors noted above.

To develop the equation for calculating Δp, assume that point *C* has been selected as the reference control point. Then by Eq. (8-5),

$$H - h_c = \frac{Bf}{p_c} \qquad (g)$$

Table 8-1 Parallax corrections

(1) Control point	(2) Elevation, ft	(3) Parallax bar reading, mm	(4) Measured parallax difference, $\Delta'p = r - r_c$, mm	(5) Calculated parallax difference,† $\Delta p = \dfrac{h - h_C}{H - h} p_c$, mm	(6) Parallax correction, $c_p = \Delta'p - \Delta p$, mm
A	1,071	19.86	0.64	0.55	0.09
B	1,135	20.81	1.59	1.46	0.13
C	1,032	19.22	—	—	—
D	1,100	20.27	1.05	0.96	0.09
E	1,184	21.65	2.43	2.17	0.26
F	1,116	20.57	1.35	1.19	0.16

† Parallax p_c of reference control point C is 81.62 mm.

Also by Eq. (8-5), the parallax of any other control point, such as A, is

$$p_a = \frac{Bf}{H - h_A} \tag{h}$$

Substituting (g) and (h) into Eq. (8-8) and reducing and dropping subscripts,

$$\Delta p = \left(\frac{h - h_C}{H - h}\right) p_c \tag{8-10}$$

In Eq. (8-10), h_C and p_c are the elevation and parallax, respectively, of the selected reference control point, and h is the elevation of any other control point. Flying height above datum is H. The reference control point may be arbitrarily selected, but it is convenient to choose the point of lowest elevation, since parallax differences calculated from Eq. (8-10) are then all positive. From Eq. (8-10), parallax differences Δp between the reference control point and the other control points are computed. These are listed in column (5) of Table 8-1. Note that for this calculation the parallax bar constant C must have been determined in accordance with Eq. (8-4) so that parallax p_c could be calculated using Eq. (8-3). For this example, C is 62.40 mm, and the resulting value of p_c is 81.62 mm.

After $\Delta p'$ and Δp have been determined, corrections c_p to be applied to the parallax differences obtained by measurement for each of the other control points can be calculated by simply subtracting Δp from $\Delta p'$. These values, which enable the parallax correction graph to be prepared, are listed in column (6) of Table 8-1.

To construct the parallax correction graph, a transparent overlay is placed over one of the photos of the stereopair and the positions of all vertical control points are marked. The overlay for this example is shown in Fig. 8-16. Parallax corrections of all vertical control points are noted beside each point, and by interpolating between points, isolines connecting points of equal parallax correction are drawn in the same manner that contours of equal elevation are drawn.

Table 8-2 Elevations from corrected parallaxes

(1) Point	(2) Parallax bar reading, mm	(3) Measured parallax difference, $\Delta'p = r - r_c$, mm	(4) Parallax correction (from graph), mm	(5) Corrected parallax difference, $\Delta p = \Delta'p - c_p$, mm	(6) Elevation,† $h = h_C + \dfrac{\Delta p\,(H - h_C)}{p}$, ft
1	19.82	0.60	0.07	0.53	1,070
2	19.32	0.10	0.03	0.07	1,037
3	18.63	−0.59	0.05	−0.64	986
4	19.01	−0.21	0.11	−0.32	1,009
5	21.67	2.45	0.13	2.32	1,194
6	20.60	1.38	0.17	1.21	1,118
7	21.01	1.79	0.21	1.58	1,143
8	21.75	2.53	0.18	2.35	1,196

† Parallax of reference control point C is 81.62 mm.

Locations of points whose elevations are to be determined are also plotted on the transparent overlay, for example, points 1 through 8 of Fig. 8-16. Corrections to be applied to measured parallaxes of each point are read from the parallax correction graph based upon the locations of the points in the overlap area. Elevations of the points are then calculated from the corrected parallaxes using Eq. (8-8). Parallax bar readings, parallax corrections, and computed elevations for points 1 through 8 are given in Table 8-2.

8-11 COMPUTING FLYING HEIGHT AND AIR BASE

To use parallax equations, it is generally necessary to compute flying height and air base. Flying height may be calculated using the methods described in Sec. 6-9. For best results, the average of flying heights for the two photos of a stereopair should be used.

If the air base is known and if one vertical control point is available in the overlap area, flying height for the stereopair may be calculated using Eq. (8-5).

Example 8-5 An overlapping pair of vertical photographs taken with a 152.4-mm-focal-length camera has an air base of 2,125 ft. The elevation of control point A is 927 ft above sea level and the parallax of point A is 89.40 mm. What is the flying height above sea level for this stereopair?

SOLUTION By Eq. (8-5):

$$H = h + \frac{Bf}{p} = 927 + \frac{2,125(152.4)}{89.40} = 4,550 \text{ ft above sea level}$$

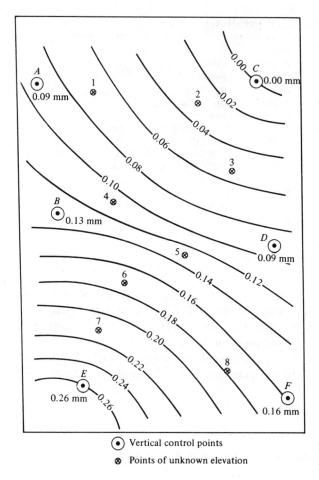

⊙ Vertical control points

⊗ Points of unknown elevation

Figure 8-16 Parallax correction graph.

If the flying height above datum is known and if one vertical control point is available in the overlap area, the air base for the stereopair may be calculated using Eq. (8-5).

Example 8-6 An overlapping pair of vertical photos was exposed with a 152.4-mm-focal-length camera from a flying height of 5,320 ft above datum. Control point C has an elevation of 865 ft above datum and the parallax of its images on the stereopair is 86.27 mm. Calculate the air base.

SOLUTION By rearranged Eq. (8-5),

$$B = (H - h)\frac{p}{f} = \frac{(5{,}320 - 865)(86.27)}{152.4} = 2{,}522 \text{ ft}$$

If a line of known horizontal length appears in the overlap area, then the air base can be readily calculated. The horizontal length of a line may be expressed in terms of rectangular coordinates, according to the pythagorean theorem, as

$$AB = \sqrt{(X_B - X_A)^2 + (Y_B - Y_A)^2}$$

Substituting Eqs. (8-6) and (8-7) into the above for the rectangular coordinates,

$$AB = \left[\left(\frac{Bx_b}{p_b} - \frac{Bx_a}{p_a} \right)^2 + \left(\frac{By_b}{p_b} - \frac{By_a}{p_a} \right)^2 \right]^{1/2}$$

Solving the above equation for B,

$$B = \left[\frac{(AB)^2}{\left(\dfrac{x_b}{p_b} - \dfrac{x_a}{p_a} \right)^2 + \left(\dfrac{y_b}{p_b} - \dfrac{y_a}{p_a} \right)^2} \right]^{1/2} \tag{8-11}$$

Example 8-7 Images of the end points of ground line AB, whose horizontal length is 2,134.1 ft, appear on a pair of overlapping vertical photographs. Photocoordinates measured with respect to the flight axis on the left photo were $x_a = 33.29$ mm, $y_a = 13.46$ mm, $x_b = 41.76$ mm, and $y_b = -95.76$ mm. Photocoordinates measured on the right photo were $x'_a = -52.32$ mm and $x'_b = -44.96$ mm. Calculate the air base for this stereopair.

SOLUTION By Eq. (8-1),

$$p_a = x_a - x'_a = 33.29 - (-52.32) = 85.61 \text{ mm}$$

$$p_b = x_b - x'_b = 41.76 - (-44.96) = 86.72 \text{ mm}$$

By Eq. (8-11),

$$B = \left[\frac{(2,134.1)^2}{\left(\dfrac{33.29}{85.61} - \dfrac{41.76}{86.72} \right)^2 + \left(\dfrac{13.46}{85.61} + \dfrac{95.76}{86.72} \right)^2} \right]^{1/2} = 1,687.2 \text{ ft}$$

If at least two control points are available in the overlap area, the air base may also be determined by radial triangulation. These principles are discussed in Chap. 9.

8-12 MAPPING WITH STEREOSCOPE AND PARALLAX BAR

When accuracy requirements are low, suitable topographic maps may be compiled using a stereoscope and parallax bar. One method of accomplishing this is to use Eqs. (8-6) and (8-7) to compute planimetric positions of all map detail and "contour points" necessary for drawing elevation contours. Elevations of the contour points are preferably determined using parallax difference Eq. (8-8). These points can all be plotted in a planimetrically correct manner according to their coordinates. Elevation contours can then be interpolated between contour points.

The mapping procedure described above requires considerable time and work. A much faster method is simply to trace planimetric positions of points directly from the left photograph of the stereopair on a transparent overlay. Elevations of all necessary points may then be determined by measuring parallax differences. The contours can be interpolated and drawn directly on the overlay while viewing the stereomodel through the stereoscope. This affords the operator advantage of being able to actually see the terrain in three dimensions while drawing the contours. Since planimetric positions of points are directly traced from a photograph, however, this method has the disadvantage that the resulting map is a perspective projection. It varies in scale with terrain elevation and contains all the image positional errors of the left photo.

Another method of map compilation with stereoscope and parallax bar is to use one of the instruments designed specifically for *direct tracing* of topographic features and elevation contours. One such instrument is the *stereocomparagraph* shown in Fig. 8-17. It consists of a mirror stereoscope and parallax bar mounted together as a unit. The photos are oriented for clear stereoviewing, using the ''flight line'' method described in Sec. 7-6, and then they are taped securely. The stereocomparagraph may be attached to a parallel drafting arm and oriented so that the line through the half marks of the parallax bar is parallel to the flight line. The stereocomparagraph may then be moved about the stereomodel—tracing topographic features such as roads, fences, rivers, etc.—by keeping the floating mark constantly in contact with features being traced. A metal arm (not shown in Fig. 8-17) attached to the stereocomparagraph holds a pencil which enables direct tracing of the features onto a map sheet. With the proper parallax bar micrometer readings set, contours may be traced directly by keeping the floating mark in contact with the ground as it is moved about the stereomodel. Parallax bar settings required to trace each contour are readily computed.

Figure 8-17 Fairchild stereocomparagraph. (*Courtesy Alan Gordon Enterprises, Inc.*)

Example 8-8 A pair of overlapping vertical photographs were taken with a 152.00-mm-focal-length camera from a flying height of 6,885 ft above sea level. The air base was 3,240 ft. The stereopair was oriented for parallax measurements and the parallax bar constant was determined as $C = 67.45$ mm. Calculate the parallax bar micrometer settings necessary to trace the 750-, 800-, 850-, 900-, 950- and 1,000-ft contours.

SOLUTION Eq. (8-5) is solved in tabular form as follows:

Contour, ft	$H - h$ ft	$p = \dfrac{Bf}{H - h}$, mm	$r = p - C$, mm
750	6,135	80.27	12.82
800	6,085	80.93	13.48
850	6,035	81.60	14.15
900	5,985	82.29	14.84
950	5,935	82.98	15.53
1,000	5,885	83.68	16.23

With parallax bar instruments such as the stereocomparagraph, the map substitute which results is a perspective projection containing the tilt and relief displacements of the left photo of the stereopair. In areas of little relief it may serve as a map, but in areas of accented terrain, considerable correction would be necessary to develop the map substitute into a map.

There are parallax-bar-type instruments available for mapping which correct for tilt and relief displacements, thereby providing an orthographic projection. Methods employed for making these corrections are ingenious and vary for different instruments. The K.E.K. plotter of Fig. 8-18 is an example of such an instrument. It is a portable instrument consisting basically of a mirror stereoscope, measuring half marks, and a pair of photo carriers. Each photo carrier may be rotated and tilted so that photographic tilt can be accounted for. Accommodations for scale variations are made by varying the spacing of the half marks. This raises the height of the vertical datum plane, the stereomodel being adjusted accordingly to that datum by raising or lowering the photo carriers. When the instrument is properly oriented to ground control, planimetry and contours can be traced by keeping the floating mark in contact with features being traced. A tracing pencil attached to the reference mark records the movement. A pantograph enables some enlargement or reduction from photo scale to map scale.

8-13 ERROR EVALUATION

Answers obtained using the various equations presented in this chapter will inevitably contain errors. It is important to be aware of the presence of these errors and to be able to assess their magnitudes. Some of the sources of error in computed answers using parallax equations are as follows:

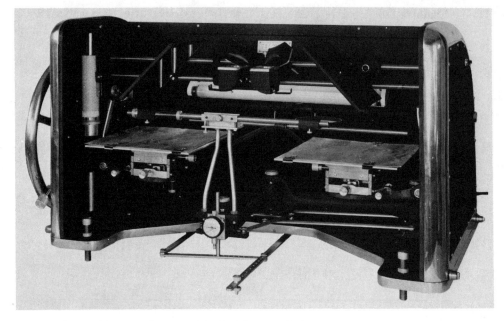

Figure 8-18 K.E.K. plotter. (*Courtesy Philip B Kail Assoc., Inc.*)

1. Locating and marking the flight line on photos
2. Orienting the stereopair for parallax measurement
3. Parallax and photocoordinate measurements
4. Shrinkage or expansion of photographs
5. Unequal flying heights for the two photos of the stereopair
6. Tilted photographs
7. Errors in ground control
8. Other errors of lesser consequence such as camera-lens distortion, atmospheric refraction distortion, etc.

A general approach for determining the combined effect of several random errors in computed answers is presented in Sec. 6-10. The same simple and straightforward method is demonstrated in the following example.

Example 8-9 In the computation of the elevation of point A in Example 8-1, suppose that the random errors were ± 5 ft in H, ± 5 ft in B, and ± 0.03 mm in p_a. Compute the resulting error in h_A due to each of these error sources and compute the total combined effect of all three errors.

SOLUTION The basic equation used was Eq. (8-5), and the derivatives of h_A in that equation taken with respect to each of the three error sources are

1. $\dfrac{dh_A}{dH} = 1$ from which $dh_A = dH$

Therefore dh_A, the error in h_A, caused by an error dH in the flying height, is dH, or ± 5 ft.

2. $\dfrac{dh_A}{dB} = \dfrac{f}{p_a}$ from which $dh_A = \dfrac{f}{p_a} dB$

Substituting numerical values into the above,

$$dh_A = \frac{152.4}{91.67} \pm 5 = \pm 8.3 \text{ ft}$$

3. $\dfrac{dh_A}{dp_a} = \dfrac{Bf}{(p_a)^2}$ from which $dh_A = \dfrac{Bf}{(p_a)^2} dp_a$

Substituting numerical values into the above,

$$dh_A = \frac{1280(152.4)}{(91.67)^2} (\pm 0.03) = \pm 0.7 \text{ ft}$$

The combined effect in the computed elevation of point A of all three random errors is the square root of the sum of the squares of the individual contributions, or

$$dh_{A_{(total)}} = \sqrt{5^2 + 8.3^2 + 0.7^2} = \pm 9.7 \text{ ft}$$

Errors in computed answers using any of the equations of this chapter can be analyzed in the fashion described above. It is, of course, necessary to estimate the magnitude of the random errors. It is more difficult to analyze errors caused by tilt in the photographs. The subject of tilted photographs is discussed in Chap. 11. For the present, however, suffice it to say that for normal photography intended to be vertical, errors in parallax equation answers due to tilt are compatible with errors from the other sources that have been considered.

REFERENCES

Aldred, A. H.: Wind-Sway Error in Parallax Measurements of Tree Height, *Photogrammetric Engineering,* vol. 30, no. 5, p. 732, 1964.

American Society of Photogrammetry: "Manual of Photogrammetry," 3d ed., Falls Church, Va., 1966, chap. 2.

————: "Manual of Photogrammetry," 4th ed., Falls Church, Va., 1980, chap. 2.

Avery, T. E.: Two Cameras for Parallax Height Measurements, *Photogrammetric Engineering,* vol. 32, no. 6, p. 576, 1971.

Bender, L. U.: Derivation of Parallax Equation, *Photogrammetric Engineering,* vol. 33, no. 10, p. 1175, 1967.

Hackman, R. J.: The Isopachometer—A New Type Parallax Bar, *Photogrammetric Engineering,* vol. 26, no. 3, p. 457, 1960.

Hadjitheodorou, C.: Elevation from Parallax Measurements, *Photogrammetric Engineering,* vol. 29, no. 5, p. 840, 1963.

Johnson, E. W.: The Limit of Parallax Perception, *Photogrammetric Engineering,* vol. 23, no. 5, p. 933, 1957.

Moessner, K. E.: Comparative Usefulness of Three Parallax Measuring Instruments in the Measurement and Interpretation of Forest Stands, *Photogrammetric Engineering,* vol. 27, no. 5, p. 705, 1961.

Nash, A. J.: Use a Mirror Stereoscope Correctly, *Photogrammetric Engineering,* vol. 38, no. 12, p. 1192, 1972.

Porter, Goff R.: Errors in Parallax Measurements and Their Assessment in Student Exercises, *Photogrammetric Record,* vol. VIII, no. 46, p. 528, 1975.

Schut, G. H.: The Determination of Tree Heights From Parallax Measurements, *Canadian Surveyor,* vol. 19, p. 415, 1965.

PROBLEMS

8-1 Calculate the stereoscopic parallaxes of points *a* through *d*, given the following measured coordinates:

Point	x(left photo)	x'(right photo)
a	2.36 in	− 1.07 in
b	68.05 mm	−21.61 mm
c	3.92 in	0.39 in
d	100.37 mm	8.52 mm

Which point is the highest in elevation? Which is lowest?

8-2 Calculate the elevations of points *a* through *d* of Prob. 8-1 if the camera focal length is 6 in, flying height above datum is 8,100 ft, and the air base is 4,450 ft.

8-3 A pair of overlapping vertical photographs are mounted for parallax measurement, as illustrated in Fig. 8-5. Distance *D* is measured as 10.37 in. Calculate the stereoscopic parallaxes of the following points whose measured *d* values are as follows:

Point	d
a	6.79 in
b	7.07 in
c	6.35 in
d	6.60 in

Which point is highest in elevation? Which is lowest?

8-4 Repeat Prob. 8-3, except *D* was measured as 266.55 mm, and measured *d* values are as follows:

Point	d
a	170.18 mm
b	164.29 mm
c	176.03 mm
d	166.46 mm

8-5 Assume that point A of Prob. 8-3 has an elevation of 1,190 ft above datum and that the photos were taken with a $3\frac{1}{2}$-in-focal-length camera. If the air base is 3,690 ft, what are the elevations of points B, C, and D?

8-6 Assume that point A of Prob. 8-4 has an elevation of 375 m above datum and that the photos were taken with a camera having a 152.40-mm focal length. If the air base is 1,830 m, what are the elevations of points B, C, and D?

8-7 A pair of overlapping vertical photos are oriented for parallax measurement with stereoscope and forward-reading parallax bar. On the left photo, b measures 82.61 mm and on the right photo b' is 83.06 mm. The parallax bar readings on o_1 and o_2 were 20.82 mm and 20.33 mm, respectively. (*a*) Calculate the parallax bar constant C based on the average for the two principal points. (*b*) Which principal point is higher in elevation? (*c*) Calculate the parallaxes of points a through d, given the following micrometer readings:

Point	Micrometer reading, mm
a	19.91
b	21.08
c	20.84
d	18.67

8-8 Assume that the photos of Prob. 8-7 were taken from a flying height of 8,950 ft above control point A whose elevation is 721 ft above datum. Calculate the elevations of points B, C, and D using Eqs. (8-8) and (8-9) and compare the results of the two equations. (Assume average flying height above ground to be 8,950 ft.)

8-9 From the information given for Probs. 8-1 and 8-2, calculate the horizontal ground length of line AC. Measured y coordinates on the left photo are $y_a = -2.33$ in and $y_c = 4.01$ in.

8-10 Repeat Prob. 8-9, except that the computations are for line BD. Measured y coordinates on the left photo are $y_b = 1.67$ in and $y_d = -3.02$ in.

8-11 From the data of Probs. 8-3 and 8-5, calculate the area on the ground contained within triangle ABC. Measured x and y photocoordinates of a, b, and c on the left photo were $x_a = -0.373$ in, $y_a = 4.370$ in, $x_b = 0.587$ in, $y_b = -4.410$ in, $x_c = 4.823$ in, and $y_c = 1.871$ in.

8-12 The air base of a pair of overlapping vertical photos was determined to be 2,485 ft. The focal length of the camera was 152.35 mm. The image coordinates of point A, whose elevation is 925 ft above datum, was determined on the left photo as $x_a = 3.29$ mm and on the right photo as $x'_a = -84.98$ mm. What is the flying height above datum for the stereopair?

8-13 Repeat Prob. 8-12, except that the air base was 1,055 m, the camera focal length was 209.60 mm, and point A, whose elevation was 283.5 m above datum, had image coordinates of $x_a = 42.93$ mm on the left photo and $x'_a = -47.28$ mm on the right photo.

8-14 The images of two control points A and B appear in the overlap area of a pair of vertical photographs. The following photocoordinates and ground coordinates apply to points A and B. Calculate the air base of the stereopair using Eq. (8-11).

Point	Left photocoordinates x, in	y, in	Right photocoordinates x', in	y', in	Ground coordinates X, ft	Y, ft
A	1.040	-3.827	-2.562	-3.831	256,445.4	91,851.6
B	-0.765	1.346	-3.655	1.344	256,726.4	89,736.1

8-15 Repeat Prob. 8-14, except that the photo coordinates and ground coordinates for points A and B were as follows:

Point	Left photocoordinates		Right photocoordinates		Ground coordinates	
	x, mm	y, mm	x', mm	y', mm	X, m	Y, m
A	65.78	82.71	-33.87	82.70	102,055.75	35,781.09
B	41.82	-76.29	-50.24	-76.31	100,989.84	34,196.60

8-16 Distances b on the left photo and b' on the right photo of a pair of overlapping vertical photos are 90.26 mm and 89.85 mm, respectively. If the air base is 562.5 m and the camera focal length is 88.78 mm, which ground principal point is higher and by how much?

8-17 Repeat Prob. 8-16, except that b and b' are 3.652 in and 3.594 in, respectively, the air base is 2,085 ft, and the camera focal length is 6.008 in.

8-18 A pair of overlapping vertical photos is taken from a flying height of 3,550 ft above ground with a 6-in-focal-length camera. The x coordinates on the left photo of the base and top of a certain tree are 3.21 in and 3.32 in, respectively. On the right photo these x' coordinates are -0.49 in and -0.56 in, respectively. Determine the height of the tree.

8-19 A pair of overlapping vertical photos is taken from a flying height of 6,020 ft above the base of a radio tower. The x coordinates on the left photo of the top and base of the tower were 96.52 mm and 90.49 mm, respectively. On the right photo these x' coordinates were -1.05 mm and -0.98 mm, respectively. What is the approximate height of the tower?

8-20 A pair of overlapping vertical photos were taken with a 6-in-focal-length camera from a flying height of 9,545 ft above sea level. The air base was 4,044 ft. This pair was oriented for parallax measurements with a stereocomparagraph having a forward-reading parallax bar, and the parallax bar constant C was determined as 71.55 mm. Calculate the required micrometer settings for tracing the 2,400-, 2,500-, 2,600-, and 2,700-ft contours.

8-21 A pair of overlapping vertical photos were exposed with a camera having a 209.80-mm focal length. Calculate B and H from the following information on the ground points D and E:

Point	Elevation, m	Left photocoordinates	Right photocoordinates
D	587	$x_d = -17.39$ mm	$x'_d = -111.05$ mm
E	729	$x_e = 99.17$ mm	$x'_e = 1.63$ mm

8-22 Repeat Prob. 8-21, except that the camera focal length is 6.005 in and the following information applies to points D and E:

Point	Elevation, ft	Left photocoordinates	Right photocoordinates
D	1,795	$x_d = 2.79$ in	$x'_d = -0.90$ in
E	1,570	$x_e = 1.03$ in	$x'_e = -2.52$ in

8-23 A pair of overlapping vertical photos was exposed from a flying height of 10,280 ft above datum using a 3.502-in-focal-length camera. The air base was determined as 8,745 ft. The following micrometer readings were taken (forward-reading parallax bar). Calculate the elevations of points 1 through 4 if the elevation of point *A* is 1,525 ft above datum.

Point	Micrometer reading, mm
A	13.29
1	17.86
2	20.21
3	11.40
4	18.65

8-24 Repeat Prob. 8-23, except flying height above datum was 3,750 m, camera focal length was 152.44 mm, air base was 1,815 m, the elevation of point *A* was 765 m above datum, and parallax bar micrometer readings were as follows:

Point	Micrometer reading, mm
A	12.85
1	17.92
2	15.71
3	21.28
4	14.97

8-25 A parallax wedge for use with a pocket stereoscope similar to that shown in Fig. 8-15 has a height of graduations, *y*, equal to 5.00 in. The lateral spacing between reference line and the graduated line is 1.80 in at the top and 2.40 in at the bottom. What is the vertical spacing of reference marks on the graduated line if the difference in parallax between adjacent graduations is 0.01 in?

8-26 A vertical stereopair of aerial photos was exposed from 4,375 ft above datum. Five vertical control points whose elevations are given below appear in the stereo overlap area. Construct a parallax correction graph based on the parallax bar readings given below. Use point *E* as the reference control point. The parallax bar was forward-reading and the constant was 74.55 mm. (Plot the positions of the control points in the overlap area using the given *x* and *y* photocoordinates taken from the left photo of the stereopair.)

Point	Elevation, ft	Parallax bar reading, mm	Photocoordinates left photo	
			x, mm	*y*, mm
A	1,395	28.63	12.3	94.1
B	1,410	29.11	98.2	90.3
C	1,178	21.56	80.1	11.6
D	1,253	23.82	− 6.2	− 90.8
E	1,042	17.55	89.9	− 87.4

8-27 Using the parallax correction graph constructed for Prob. 8-26, calculate elevations for points 1 through 5 whose parallax bar readings are given below. (To obtain parallax corrections, plot the positions of points 1 through 5 on the parallax correction graph from the given photocoordinates measured on the left photo of the stereopair.)

| Point | Parallax bar reading, mm | Photocoordinates left photo | |
		x, mm	y, mm
1	25.49	25.8	70.9
2	18.36	64.0	41.7
3	24.72	10.6	− 1.3
4	29.23	57.2	− 32.4
5	27.40	75.6	− 91.9

8-28 In Prob. 8-12, suppose that random errors were ± 2 ft in h and B, and ± 0.03 mm in each of x_a and x_a'. What is the expected resultant error in the calculated value of H due to these random errors? (Assume the focal length to be error-free.)

8-29 In Prob. 8-13, suppose that random errors were ± 1 m in $h_{A'}$, ± 2 m in B, and ± 0.05 mm in both x_a and x_a'. What is the expected error in the calculated value of H due to these errors? (Assume the focal length to be error-free.)

8-30 In Prob. 8-18, assume that random errors existed in the amounts of ± 5 ft in H and ± 0.01 in for each of the measured photocoordinates. What is the expected error in the calculated height of the tree due to these random errors?

NINE

ELEMENTARY METHODS OF PLANIMETRIC MAPPING WITH VERTICAL PHOTOGRAPHS

9-1 INTRODUCTION

This chapter describes some elementary methods that can be used for compiling planimetric maps from information contained on vertical photos. These include (1) direct tracing of planimetric features from aerial photos or photo enlargements, (2) tracing with the use of reflection or projection instruments, and (3) radial-line triangulation methods.† Each of these techniques is relatively uncomplicated to perform and requires only simple and inexpensive equipment. However, they can have definite utility, depending upon the nature and extent of planimetric mapping to be accomplished.

The above cited methods are quite suitable for planimetric mapping of areas of limited size, and they are especially convenient for revising portions of existing maps. As an example, it may be necessary to include a recently constructed road or shopping center which is not shown on an otherwise satisfactory existing planimetric map. It would be expensive, and unnecessary, to prepare a new map of the area if these features could be satisfactorily superimposed onto the existing map. This type of planimetric map revision can readily be done using procedures described in this chapter.

The accuracies that can be achieved in planimetric mapping using these methods are generally of a lower order than those attainable with stereoplotting instruments (see Chap. 12) or orthophoto equipment (see Chap. 13). However, for some work, especially if ample care is exercised, suitable results can be achieved.

---

†In this context, compiling planimetric maps denotes the drawing of scaled diagrams which show planimetric features by means of lines and symbols. Planimetric maps portray only horizontal position and give no information concerning elevations. The maps are prepared to some designated scale; thus, all features are presumed to be shown in their true relative positions. Various "photomap" products such as enlarged prints of aerial photos, mosaics, and orthophotos are also used to show planimetry. With these, however, rather than using lines and symbols, photo images of objects depict relative locations of features. Chapter 10 discusses photomaps and mosaics, and Chap. 13 describes orthophotos.

9-2 PLANIMETRIC MAPPING BY DIRECT TRACING

The simplest method of planimetric mapping consists in tracing features directly onto a transparent sheet of mylar which has been placed on a light table superimposed over a vertical photo. To maintain a satisfactory level of accuracy, a number of photo control points, uniformly distributed throughout the photo, should be plotted on the mylar overlay at a scale equal to the average scale of the photo.† Before tracing, the mylar should be adjusted in position so that the plotted control point nearest the features to be traced matches its corresponding photo image. Other nearby control should also be fitted as well as possible. As noted in Chap. 6, a vertical photo is not a map and contains scale variations and relief displacements. Also, if the photo is not truly vertical, there will be image distortions from this source. However, by matching control in localized areas as described above, errors from these sources are minimized. If terrain variations are moderate and the photo is nearly vertical, fairly accurate results can be achieved.

A map manuscript produced using the direct tracing procedure will of course have a scale equal to that of the photo. However, the manuscript can be enlarged or reduced to make it correspond to desired final map scale. As an alternative procedure, the photo can first be enlarged or reduced to an average scale which equals final map scale, and then the tracing done directly at that scale.

9-3 PLANIMETRIC MAPPING WITH REFLECTION OR PROJECTION INSTRUMENTS

A preferred approach in the direct-tracing method of planimetric mapping is to utilize reflection or projection instruments, if they are available. With the use of these instruments, photo images can be made to match corresponding map images even though their scales are unequal.

A reflection instrument for planimetric map revision called a *vertical sketchmaster* is shown in Fig. 9-1. Its operating principle is illustrated in Fig. 9-2. A light ray from photo image a is reflected by a large mirror at a_1, reflected again by an eyepiece mirror at a_2, and received by an observer at o. Because the eyepiece mirror is half-silvered, the observer can also simultaneously view the map. By adjusting the position of the map, photo image a and its corresponding map point A can be made to coincide. This principle of operation is known as *camera lucida*. In planimetric map revision, points such as A may be any features shown on the existing map which also appear on the photo. For new planimetric mapping, these may be control points as described in the preceding section.

---

†Photo control, as described in detail in Chap. 15, consists of any discrete objects whose images appear on the photo and whose ground coordinates are known. There are several methods for determining the ground coordinates, including scaling from an existing map, direct field survey, or photogrammetric techniques such as radial-line triangulation (described later in this chapter) and aerotriangulation (described in Chap. 14).

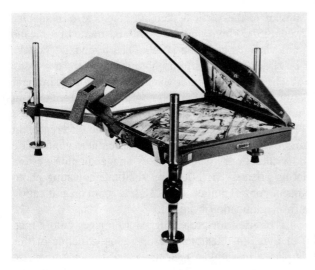

Figure 9-1 Vertical sketchmaster. (*Courtesy Keuffel & Esser Co.*)

The ratio of scales between photo and map is the ratio of ray path lengths aa_1a_2o and Aa_2o. This ratio can be changed using the adjustable legs of the instrument, although the range of scale change is quite limited. Each of the three legs may be adjusted to different lengths, thereby creating a tilted relationship between photo plane and map plane. By adjusting the legs, three photo images forming a triangle can be made coincident with their corresponding map images, and details in that triangular area can be traced on the map. The tilt of the photo plane does not represent the actual tilt of the photograph and does not completely eliminate errors from tilt displacement. It does, however, reduce the amount of tilt distortion in plotted planimetric detail. Of course, local relief displacements are still present.

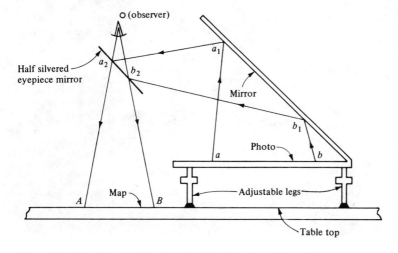

Figure 9-2 Operating principle of vertical sketchmaster.

A horizontal sketchmaster similar to the vertical sketchmaster is also available. With this instrument the easel holding the photo may be tilted and rotated by means of a ball and socket to match photo images and map points. The horizontal sketchmaster enables making a rather large-scale change from photo to map.

The Zoom Transfer Scope (ZTS) of Fig. 9-3 is a versatile reflection instrument for planimetric map revision. With the ZTS an aerial photograph can be viewed in superposition with a map, and information from the photo may be readily transferred onto the map by direct tracing. The instrument can accommodate large differences in scale from photo to map. Its anamorphic optical system enables different magnification ratios to be applied in the x and y directions. It also has an image rotation system which operates by means of rotating prisms. These features facilitate adjusting photo images to coincide with map points or control points. The ZTS has zoom magnification providing continuous variations in magnification from 1 to $7\times$.

A *reflecting projector* may also be used for projecting photo images onto a map for direct tracing. The diagram of Fig. 9-4 illustrates the operating principle of this type of instrument. The photo is placed in a holder near the top of the instrument so that the imagery faces the mirror. Light rays carrying photo images are reflected from the mirror through a lens and projected onto the map sheet on the projection table below. The magnification ratio from photo to map is the ratio of image distance aa_1L to object distance LA. When the magnification ratio is changed, a special mechanical

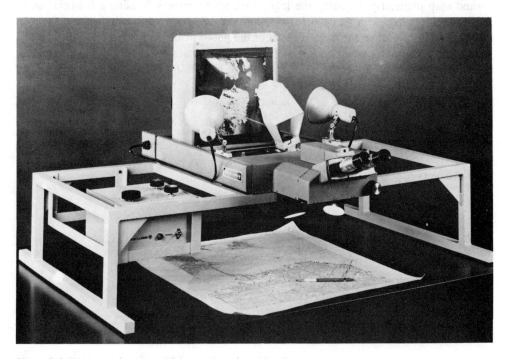

Figure 9-3 Zoom transfer scope. (*Courtesy Bausch and Lomb.*)

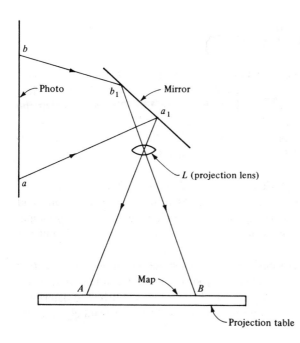

Figure 9-4 Operating principle of a reflecting projector.

arrangement automatically changes image and object distances in the required ratio to satisfy the lens formula, Eq. (2-8), and maintain focus.

9-4 RADIAL-LINE TRIANGULATION

Radial-line triangulation was one of the earliest photogrammetric mapping procedures, done initially by graphical methods called *radial-line plotting*. Slotted templates and mechanical devices were developed later which duplicated the graphical procedures. Still later radial-line triangulation was performed numerically using computers. These procedures are not used extensively today, having given way to other more accurate and convenient methods. Nevertheless, radial-line triangulation affords an opportunity to present in an uncomplicated manner, some otherwise rather complex fundamentals of photogrammetry. Further, this procedure still has practical value in special situations involving only a few photos where a lower order of accuracy will suffice.

In addition to its use for planimetric mapping, radial-line triangulation can also be used to extend or supplement horizontal control; thus it could be used to provide the control needed for the direct tracing methods of planimetric mapping described in the preceding sections. In the sections that follow, various procedures for extending horizontal control by radial-line triangulation are presented first. This is followed by descriptions of the techniques of planimetric mapping using radial-line methods.

9-5 FUNDAMENTAL PRINCIPLE
OF RADIAL-LINE TRIANGULATION

The fundamental principle upon which radial-line triangulation is based is that angles with vertexes at the principal point of a vertical photograph are true horizontal angles. The procedure therefore assumes truly vertical photos. In Fig. 9-5 the plane of the vertical photo is horizontal and parallel to the datum plane. Points A' and B' in the datum plane are vertically beneath object points A and B. Planes $LAA'P'$ and $LBB'P'$ are vertical, and therefore angle aob on the photo is equal to the true horizontal angle $A'P'B'$.

On a vertical photograph, relief displacement, radial-lens distortion, and atmospheric refraction all displace images along radial lines from the principal point and therefore do not affect sizes of photographic angles with vertexes at the principal point. Also variations in flying heights of vertical photos affect photo scale but not angle sizes.

Radial-line triangulation consists basically of two distinct operations: (1) *resection to determine the planimetric* positions of photo exposure stations and (2) *intersection* from two or more photos whose exposure stations have been established to determine the positions of new points. These operations may be performed separately, but are more often done simultaneously.

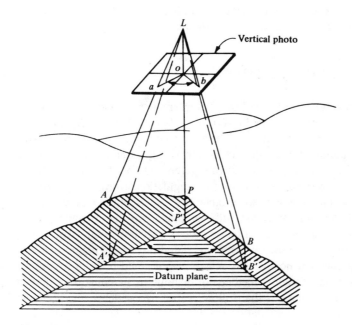

Figure 9-5 Horizontal angles on a vertical photograph.

9-6 GRAPHICAL METHODS OF RADIAL-LINE TRIANGULATION

Radial-line triangulation by graphical techniques is an easily visualized procedure and will be used to introduce the subject. In Fig. 9-6, five photos of a flight strip are laid out in their overlapping positions. Principal points and conjugate principal points have been marked on the photos. Points a and b are images of two horizontal ground control points A and B. A transparent overlay is placed over photo no. 1 and a template is prepared, as shown in Fig. 9-7a, by ruling lines on the overlay from the principal point o_1 through points a, b, c, d, and conjugate principal point o_2. Angles with vertexes at o_1 are true horizontal angles on the template. A second, similar template is prepared for photo no. 2, as shown in Fig. 9-7b. In drawing the rays it is necessary to hold the overlays firmly and use a sharp, hard drawing pencil so that true angles are obtained between rays.

A base map upon which the radial-line triangulation will be performed is prepared next and ground control points A and B are plotted thereon, (Fig. 9-8). The scale of the base map is chosen quite arbitrarily, but it should not normally differ greatly from photo scale. If map scale is chosen larger than photo scale, then the templates must be prepared larger in size than the photos to make possible an increase in scale from photo to map. Template no. 1 is oriented on the base map so that rays o_1a and o_1b simultaneously pass through their respective plotted control points A and B. At the same time template no. 2 is oriented on the map to make rays o_2a and o_2b pass through their respective plotted control points and, in addition, rays o_1o_2 on template no. 1 and o_2o_1 on template no. 2 are made to coincide. With these conditions established, the locations of o_1 and o_2 define the true planimetric map positions of ground principal points (exposure stations) P_1 and P_2. Their positions are marked on the map by pricking through the templates with a pin. The direction of P_1P_2 on the map represents the flight direction for that stereopair, and the distance from P_1 to P_2 represents the air base. This procedure for locating exposure station positions is called *resection—* more specifically *two-point resection,* since it requires two control points.

With the exposure stations of a pair of overlapping photographs fixed on the map, any number of other points whose images appear in the overlap area of the stereopair

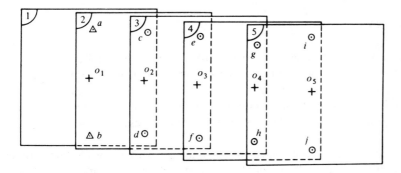

Figure 9-6 Five photos of a flight strip.

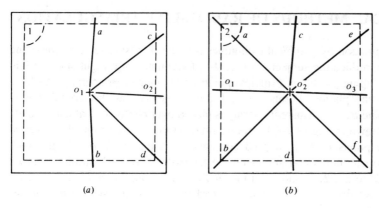

(a)   (b)

Figure 9-7 (a) Template for photo no. 1. (b) Template for photo no. 2.

can be established by *intersection*. When the two templates were originally prepared, rays through new points c and d were also drawn. With the two templates oriented on the map as previously described, the intersection of rays o_1c and o_2c fixes the true planimetric position of point C. Likewise, the intersection of rays o_1d and o_2d locates point D. Map locations of any points in the overlap area can be established by this procedure.

Once points C and D are established on the map, they beome new horizontal control points. These points should have been carefully selected so that their images appeared not only on photos no. 1 and 2 but also on no. 3. (This condition requires greater than 50 percent end lap in the photos.) If C and D are treated as control points, a template can be prepared for photo no. 3 and resection performed, as previously described to locate exposure station P_3. With P_3 located, new points E and F can be fixed on the map by intersection from photos no. 2 and 3. Carefully selecting images e and f so that their images appear on photo no. 4 enables exposure station P_4 to be located. This procedure of successive resection and intersection may be continued through the entire strip of photos. The completed radial-line triangulation for the five-photo strip is shown in Fig. 9-8.

9-7 PASS POINTS

Points C through J of the triangulated strip of Fig. 9-8 are points of extended horizontal control. They may be used for controlling subsequent photogrammetric procedures such as planimetric mapping or mosaic construction. These points were necessary, however, to continue the radial triangulation through the strip, and therefore they are called *pass points* because they enabled the triangulation to pass from one photo to the next.

To satisfactorily serve as pass points, images must be sharp and well defined on all photos in which they appear. They must be located in desirable positions on three successive overlapping photographs. The most ideal positions are opposite the principal points and conjugate principal points, as illustrated in Fig. 9-6. This placement

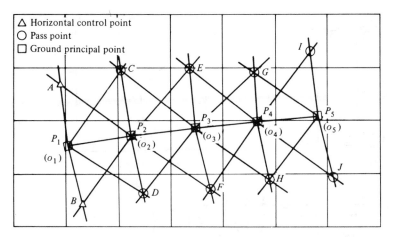

Figure 9-8 Radial-line triangulation assembly of five-photo strip.

creates the strongest geometrical strength and yields highest accuracy. For radial-line triangulation of a block of two or more strips, pass points common to two strips should be chosen in the center of the side lap area.

Careful overall planning should precede a radial-line triangulation project. The photographs should be carefully studied and all pass points selected and labeled prior to constructing templates. When the templates are prepared, all rays should be labeled on the templates to prevent confusion when assembling them on the map.

9-8 THREE-POINT RESECTION

The two-point resection procedure described in Sec. 9-6 requires that a *stereopair* be resected simultaneously and that two horizontal control points appear in the overlap area of the stereopair. If three or more horizontal control points appear anywhere in a single vertical photo, its exposure station can be located by *three-point resection*. In Fig. 9-9a, for example, images *a, b,* and *c* of horizontal control points, *A, B,* and *C* appear in vertical photo no. 1. A template is prepared for that photo by drawing rays from the principal point through the three image points, as shown in Fig. 9-9b. The template is placed on a base map upon which the three control points have been plotted, and it is oriented so that the three rays simultaneously pass through their respective plotted control points, as shown in Fig. 9-9c. This locates exposure station P_1, which is marked on the map by pinpricking. Greater accuracy is achieved if more than three points are used, and in that case all rays must simultaneously pass as nearly as possible through their respective plotted control.

If images of three or more horizontal control points also appear somewhere on adjacent photo no. 2, then its exposure station can also be located by three-point resection. Points *b, c,* and *d* of Fig. 9-9a, for example, could be used for that photo. With the two exposure stations of the stereopair located, new pass points can be positioned by intersection. If pass points are selected whose images appear on photo

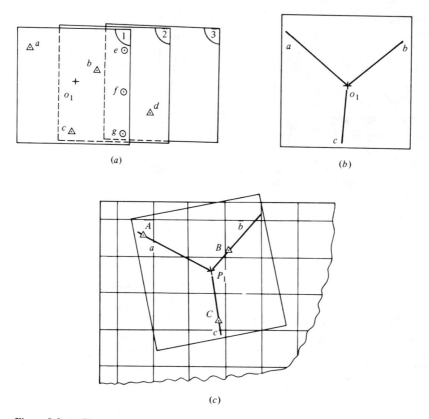

(a)

(b)

(c)

Figure 9-9 (a) Photos with ample control for three-point resection. (b) Template prepared for three-point resection. (c) Locating an exposure station by three-point resection.

no. 3—for example, at positions e, f, and g of Fig. 9-9a—then, with their positions known, they may be used in three-point resection to locate exposure station P_3. In this manner radial-line triangulation by three-point resection may be extended.

Radial-line triangulation by three-point resection has the advantage that conjugate principal points need not be used. A disadvantage of the method is the greater amount of ground control needed. Two-point resection is more convenient for graphical radial-line triangulation or when using slotted or mechanical templates (see secs. 9-11 and 9-12). Three-point resection is useful in locating single photo exposure stations and planimetric mapping of limited areas as described in Sec. 9-15.

9-9 RADIAL-LINE TRIANGULATION OF A BLOCK OF PHOTOS

A block of photos may be assembled in a simultaneous radial-line triangulation by means of pass points located in the side lap of adjacent strips. If another strip below the strip of Fig. 9-8 were to be triangulated simultaneously with the upper strip, for example, then pass points *d, f, h,* and *j* would be chosen in the side lap area common to both strips. These pass points would also be chosen very carefully because their images may appear on as many as six photos.

Pass points which tie strips together are called *tie points*. In assembling templates of two or more strips, not only must the previously described ray intersections be adhered to, but rays from both strips common to tie points must also intersect simultaneously. Because of drafting errors in preparing templates and because certain assumptions such as vertical photos are not actually met, angles on templates may not be true horizontal angles. Therefore, some difficulties may be encountered in assembling templates of a block. In these cases, a best fit is forced which tends to adjust for discrepancies.

9-10 CONTROL POINT LOCATIONS

From foregoing discussions it should be apparent that a minimum of two horizontal control points are necessary for radial-line triangulation. More than two are desirable, however, for providing checks and improving accuracy. In the example of Fig. 9-8, control points *A* and *B* existed in the overlap area of a stereopair. Better overall stability in controlling that triangulation would have been obtained had more than two control points existed or had the two points been as widely spaced in the strip as possible, say, at *A* and *J*. In general, best network stability is obtained with a uniformly distributed control network.

In graphical radial-line triangulation, it is convenient but not necessary for two control points to occur within the overlap of one stereopair of the block. If there are no stereopairs in which this condition exists, graphical methods may still be performed, but either of two slightly different approaches must be taken. In the first approach, the assembly is begun at one control point. Scale and orientation are estimated, and the entire assembly is constructed so that, it is hoped, other control point ray intersections of the finished assembly occur near their respective plotted map positions. Based on discrepancies of plotted control points and their corresponding ray intersections, adjustments are made by trial and error to change scale and orientation until the assembly is fitted as closely as possible to all control.

In the second technique, the first two templates are oriented in an arbitrary location on the map by simply making lines o_1o_2 and o_2o_1 coincide. Scale is arbitrarily set by making map distance P_1P_2 some convenient value. Successive templates are oriented to this arbitrarily located stereopair until the entire network is assembled. Then the arbitrary positions of all points—including exposure stations, pass points, and control points—are pinpricked and their coordinates are scaled from the map. With coordinates of the control points known in both the ground system and arbitrary system, a two-dimensional conformal coordinate transformation may be performed as described in Appendix B to obtain ground coordinates of all exposure stations and pass points.

9-11 SLOTTED TEMPLATES

The slotted template method of radial-line triangulation is similar to the graphical method except that templates are prepared differently. In this method, long narrow slots are cut in a template material to represent rays radiating from the center of the photo. The template material is usually thin stiff cardboard, plastic, or metal. Figure 9-10 shows a cardboard template being prepared in a slotted template cutter.

When a cardboard template is cut with the instrument in Fig. 9-10, the photo is first placed over the cardboard and, while both photo and cardboard are firmly held together, pinpricks into the underlying cardboard are made through the photo to mark principal point, conjugate principal points, all pass points, and any control points that may exist in the photo. A hole centered on the pinprick of the principal point is made in the cardboard using a special punch. This hole serves as the center of rays. The cardboard is placed on the table of the template cutter with the center punch hole over the upright stud, as shown in Fig. 9-10. By means of a sighting device, the cardboard is rotated and aligned so that each slot is cut in its correct position. Imaginary lines from the center of the principal point punch hole through the centers of the elongated slots represent rays which define horizontal angles. All slots should be labeled to avoid confusion in assembling the network.

When templates have been prepared for each photo, the network is assembled on a base map which has been prepared to desired scale. All available horizontal control is first plotted on the map. The map is placed over a softwood board and headless pins called *control pins* are driven firmly into the wood through each plotted control point. The shafts of the pins should be perpendicular to the map surface. The templates are assembled by making all slots (rays) to common points intersect. The network is held together by means of hollow shafted studs which are placed at the intersections of slots to common points. To prevent play in the assembly, the shafts of the studs are the same diameter as the slot widths. An assembly of a small block of seven photos is shown in Fig. 9-11.

The scale of an entire assembly may be made larger or smaller by gently pulling apart or compressing the edges of the assembly. In this manner the scale of the assembly is adjusted until it is equal to map scale—a condition which exists when the hollow shafts of the control point studs fit over their respective control pins. In the

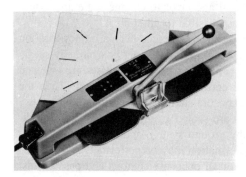

Figure 9-10 Slotted template cutter. (*Courtesy Carl Zeiss, Oberkochen.*)

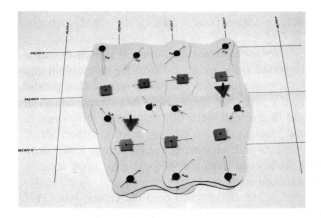

Figure 9-11 An assembly of slotted templates.

assembly of Fig. 9-11, the centers of hollow studs capped with square washers mark the seven exposure stations, centers of small circular washers are pass points, and the two triangles are control points. Once the network has been fitted to the control, map positions of all exposure stations and pass points are marked by pinpricking through the centers of the hollow-shafted studs.

In the example of Fig. 9-11, two control points were used to control the seven-photo assembly. The two points do not exist in the overlap of a stereopair, but this causes no difficulty. Any number of control points can be conveniently accommodated in the slotted template method by simply fitting all control point studs to their respective control pins simultaneously. Because of errors in template construction, non-vertical photos, etc., some difficulties may be encountered in assembling the network. Normally these errors are adjusted for by gently vibrating the network and forcing all intersection and control conditions.

9-12 MECHANICAL TEMPLATES

Radial-line triangulation may also be performed using mechanical templates (known as *spiders*) constructed from kits called *lazy daisys*. Except for the manner of constructing templates, the procedure is the same as slotted template radial-line triangulation.

To construct mechanical templates, each photo is placed individually on a softwood board and a headless pin is carefully driven through the principal point firmly into the underlying wood. The shaft of the pin should be perpendicular to the photo surface. A hollow-shafted small stud is placed over the headless pin, and a threaded bolt which serves as the hub of template rays is mounted over the stud. Headless pins are also carefully driven into the board through the two conjugate principal points, all pass points, and any control points that may exist in that photo. Studs are then placed over them. A line connecting the pin at the principal point and the pin at any pass point, conjugate principal point, or control point represents the ray through that point.

Metal arms having a round hole on one end and an elongated slot on the other are used to construct the rays. The round holes are placed over the central bolt and the slots over the studs at pass points, conjugate principal points, and control points. The arm for each ray is chosen from a variety of lengths with two considerations in mind: first, excessive length should be avoided since this may cause difficulties in assembling the network, and second, enough slot length must exist on either side of all studs to allow for relief displacements and for making a scale change from photo to map. The necessity of meeting these conditions will become clear upon actually constructing a triangulation network from spiders.

Once all arms have been placed, a nut is tightened onto the central bolt. This must be done carefully to prevent buckling of the arms, which would result in erroneous angles between rays. When all spiders have been prepared, they are assembled into a network in exactly the same manner as that described for slotted templates. An assembly of a small block of photos is shown in Fig. 9-12. In this network, the control is again indicated with triangles. Exposure stations and pass points occur at the positions of their respective studs, and they are marked on the map by pinpricking through the studs.

9-13 ERRORS IN RADIAL-LINE TRIANGULATION

Some of the principal sources of error in radial-line triangulation are (1) differential paper shrinkage, (2) graphical or mechanical construction errors, (3) erroneous location of principal points, (4) faulty transfer of conjugate principal points, and (5) tilt in the photos. Differential paper shrinkage causes small errors, but these can be eliminated by using polyester-base materials for both photos and base map. Graphical errors can be minimized by using a sharp, hard drawing pencil and good straightedge.

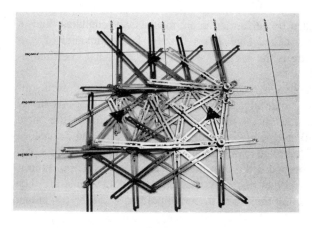

Figure 9-12 An assembly of spiders.

They cannot be entirely eliminated, however, for even a pencil line as thin as $\frac{1}{100}$ in represents 10 ft on a map plotted at a scale of 1,000 ft per in. Construction errors in preparing slotted templates and lazy-daisy spiders can be minimized only by exercising caution in preparing them.

If the terrain is flat, errors caused by faulty location of principal points is insignificant. If the terrain is rugged or rolling, however, principal points should be carefully located by intersecting very fine fiducial lines, and conjugate principal points should be carefully located. On a tilted photo, as discussed in Chap. 11, relief displacement is radial from the nadir point and tilt displacement is radial from the isocenter. Thus if a vertical photo is assumed when in fact it is tilted, angles between rays will not be true horizontal angles. However, for tilts up to about 3° in photos taken over terrain with moderate relief, errors from this source are quite small. The errors become more severe as relief variations and tilts increase. Errors caused by tilted photography can be removed by rectifying the photos, but for intended vertical photography, errors caused by tilt are usually no more serious than graphical or construction errors.

Accuracy in extending control by radial-line triangulation is very dependent upon the density and distribution of existing ground control. On the basis of empirical tests, the following equation was developed which relates expected average error in pass point location to the number of photos and number of ground control points:

$$e = k\left(\frac{t}{c}\right)^{1/2} \tag{9-1}$$

In Eq. (9-1), e is the average error expected in pass point location in millimeters at compilation scale, k is a constant which has been evaluated at 0.16 for careful slotted template radial triangulation, t is the total number of photos in the assembly, and c is the total number of ground control points used in the assembly. The equation is applicable only for a uniform distribution of control points throughout a block of photos.

Although Eq. (9-1) was based on slotted template assemblies, it is plausible to also apply it to carefully constructed graphical and lazy-daisy assemblies. The form of Eq. (9-1) may be changed as follows in order to calculate the number of well-distributed ground control points necessary in a block to achieve a certain accuracy in locating points with radial triangulation:

$$c = t\left(\frac{k}{e}\right)^{2} \tag{9-2}$$

Example 9-1 Suppose it is required that pass points be located within an average accuracy of 10 ft from a block of 40 photos whose average scale is 500 ft /in. How many well-distributed ground control points must be established by field survey to achieve the desired accuracy?

SOLUTION (Assume map compilation is at photo scale.) At compilation scale, the acceptable error e in millimeters is

$$e = \frac{10 \text{ ft } (25.4 \text{ mm/in})}{500 \text{ ft/in}} = 0.508 \text{ mm}$$

By Eq. (9-2),

$$c = 40\left(\frac{0.16}{0.508}\right)^2 = 3.9 \quad \text{(4 uniformly distributed control points required)}$$

9-14 NUMERICAL METHODS OF RADIAL-LINE TRIANGULATION

Several methods have been developed for performing radial-line triangulation numerically. All these methods involve measurement of photocoordinates followed by the formation and solution of a mathematical model which duplicates the manual procedures. Regardless of the numerical method performed, the lengthy nature of the computations make this procedure most feasible when performed on a computer. Descriptions of various numerical methods are given in references cited at the end of this chapter.

The primary advantage in numerical radial-line triangulation is an increase in accuracy which results from the elimination of graphical and mechanical construction errors. The accuracy with which ground control can be extended by numerical procedures depends upon several variables such as tilt in the photos, flying height, and terrain relief. With vertical photography taken over moderate terrain having a uniform distribution of horizontal control points, an accuracy in computed X and Y ground coordinates of approximately $\frac{1}{500}$ of the flying height is generally attainable.

9-15 PLANIMETRIC MAPPING BY RADIAL-LINE PLOTTING

Radial-line methods are useful in planimetric mapping and map revision, especially where only a limited number of features must be located. Assume, for example, that it is desired to plot the map position of the building appearing in the stereopair of Fig. 9-13a. To accomplish this by radial-line methods, templates are prepared for each photo by ruling lines from principal points through conjugate principal points and through the corners of the building, as shown in Fig. 9-13a. All rays through building corners are labeled to avoid confusion in plotting. The template overlays are then placed on the map and oriented so that o_1 and o_2 are coincident with the map locations of exposure stations P_1 and P_2, respectively, and so that line o_1o_2 of overlay no. 1 coincides with line o_2o_1 of overlay no. 2. In this orientation, intersections of lines to common points locate the building corners. Their positions are pinpricked and the building is drawn on the map by connecting adjacent corners with straight lines, as shown in Fig. 9-13b. This procedure yields correct planimetric positions in spite of relief displacements and scale variations of the photos.

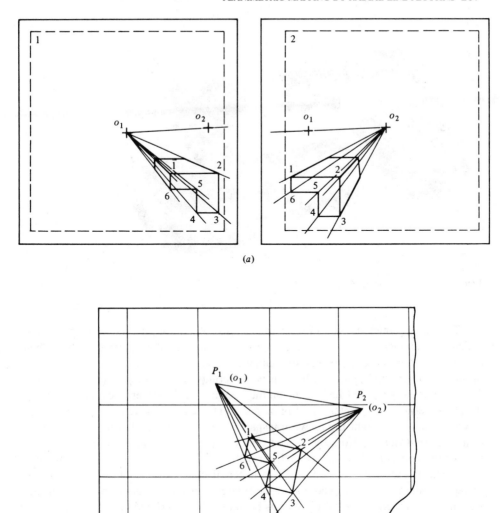

Figure 9-13 (*a*) Templates for locating a building by radial-line intersection. (*b*) Intersection on the base map.

The above procedure presumes known map locations for the exposure stations of the stereopair. These may have been located through radial-line triangulation as previously described. On the other hand, it is often possible to locate exposure stations by either two-point or three-point resection, using as control, discrete points that are common to both the existing map and the photos. Buildings, road intersections, or intersections of streams or power lines with roads are some points that may suffice for control points.

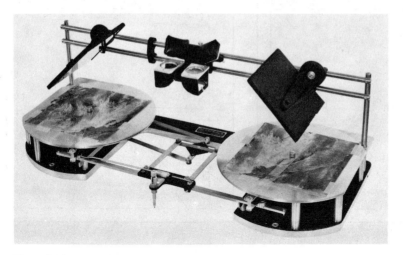

Figure 9-14 Radial plotter. (*Courtesy Philip B. Kail Assoc., Inc.*)

The radial plotter shown in Fig. 9-14 is an instrument for planimetric map revision which continuously solves the radial-line intersection problem mechanically. When the radial plotter is used, a pair of overlapping vertical photos are placed on the stages of the instrument and pinned at their principal points. The stereopair is then oriented for stereoscopic viewing (the instrument incorporates a mirror stereoscope) by rotating the photos about their centers until principal points and conjugate principal points lie along a common straight line. Two transparent arms, upon which a fine black line is etched, extend radially from the pinned principal points. The apparent intersection in the stereomodel of the black lines on the arms forms a cross which serves as a reference mark. The radial arms are connected to a pencil chuck by means of a mechanical linkage; as the pencil is moved about, the reference mark appears to move about the stereomodel. When the radial plotter has been oriented to ground control, planimetric detail can be traced directly on a map beneath the instrument by moving the pencil so as to keep the reference mark in contact with objects being traced.

REFERENCES

American Society of Photogrammetry: "Manual of Photogrammetry," 3d ed., Falls Church, Va., 1966, chap. 9.
———: "Manual of Photogrammetry," 4th ed., Falls Church, Va., 1980, chap. 9.
Kail, P.: The Radial Planimetric Plotter, *Photogrammetric Engineering*, vol. 15, no. 3, p. 402, 1949.
Mikhail, E. M.: A Study in Numerical Radial Triangulation, *Photogrammetric Engineering*, vol. 34, no. 4, p. 358, 1968.
Trorey, L. G.: Slotted Template Error, *Photogrammetric Engineering*, vol. 13, no. 2, p. 227, 1947.
Turpin, R. D.: Numerical Radial Triangulation, *Photogrammetric Engineering*, vol. 32, no. 6, p. 1041, 1966.
Wolf, P. R.: Analytical Radial Triangulation, *Photogrammetric Engineering*, vol. 33, no. 1, p. 109, 1967.

PROBLEMS

9-1 Describe the various techniques available for planimetric mapping using vertical photographs.

9-2 Discuss the sources of error in the direct-tracing methods of planimetric mapping.

9-3 Describe the zoom transfer scope and its advantages in planimetric mapping by direct tracing.

9-4 Describe the different methods available for performing radial-line triangulation.

9-5 List the characteristics of good pass points for radial-line triangulation.

9-6 Prepare tracing-paper templates and perform a graphical two-point resection on the basis of the following ground coordinates and photocoordinates (with respect to the flight line axis) of control points A and B. Determine the ground length of the air base and the azimuth of the flight line. Use a base map scale of 1 in = 1,000 ft. (*Note:* Point A is west of the flight line.)

| | Photocoordinates | | | | Ground coordinates | |
| | Left photo | | Right photo | | | |
Point	x, mm	y, mm	x', mm	y', mm	X, ft	Y, ft
A	42.43	81.90	-47.24	82.01	2,169,818	89,189
B	51.28	-69.71	-37.57	-69.65	2,175,962	91,731

9-7 For the stereopair of Prob. 9-6, perform a graphical intersection to determine the ground coordinates of pass points C and D given the following photocoordinates (with respect to the flight line axis):

| | Photocoordinates | | | |
| | Left photo | | Right photo | |
Point	x, mm	y, mm	x', mm	y', mm
C	71.29	82.05	-8.19	82.16
D	50.83	-39.27	-37.25	-39.20

9-8 Using a base map of scale 1 in = 500 ft, prepare a tracing-paper template and perform a graphical three-point resection to determine the azimuth of the flight line and the ground coordinates of the exposure station of a vertical photograph. Three control points appear in the photo and their ground coordinates and photocoordinates (with respect to flight line axes) are as follows. (*Note:* Exposure station is west of control points.)

| | Photocoordinates | | Ground coordinates | |
Point	x, mm	y, mm	X, ft	Y, ft
H_1	63.25	90.11	2,165,583	128,103
H_2	43.07	15.88	2,165,706	126,442
H_3	82.90	-86.30	2,167,207	124,701

9-9 Repeat Prob. 9-8, except that the three horizontal control points are as follows: (*Note:* Exposure station is east of control points.)

Point	Photocoordinates		Ground coordinates	
	x, mm	y, mm	X, ft	Y, ft
A	−74.96	91.25	1,903,318	90,924
B	−60.74	17.34	1,902,660	89,733
C	−44.99	−79.84	1,901,743	88,183

9-10 Discuss the sources of error in radial-line triangulation.

9-11 Radial-line triangulation is to be performed to extend control through a block of photographs which consists of 10 flight strips with 20 photos per strip. Map compilation scale is 1 in = 1,800 ft. If there are 16 well-distributed control points in the block, what error (in feet) can be expected in the map positions of the pass points?

9-12 Repeat Prob. 9-11, except that the block consists of 6 flight strips of 12 photos per strip, map compilation scale is 1:6,000, and there are 10 well-distributed ground control points in the block.

9-13 It is required to locate pass points on a map to an average accuracy of 10 ft from a block of 60 photographs. If map compilation scale is 1:6,000, how many well-distributed ground control points are needed to achieve the required accuracy?

9-14 Repeat Prob. 9-13, except that required accuracy is 20 ft, the block contains 100 photographs, and map compilation scale is 1:12,000.

TEN
PHOTOMAPS AND MOSAICS

10-1 INTRODUCTION

Photomaps are simply aerial photos which are used directly as planimetric map substitutes. The photos are usually brought to some desired average scale by enlargement or reduction (see Sec. 3-12). Title information, place names, and other data may be superimposed on the photos in the same way that it is done on maps. Photomaps may be prepared from single aerial photos, or they may be made by piecing together two or more individual overlapping photos to form a single continuous composite picture. These composites are commonly referred to as *mosaics*.

A single photo can be used to prepare a photomap if it covers the entire area to be shown. In years past, because of the limited flying heights (and thus restricted photo coverages) that were attainable with older aircraft, mosaics frequently needed to be prepared. Recently, new aircraft such as the Lear jet, have enabled flying heights of 50,000 ft or more to be reached, so that larger areas can now be covered in one photo. This has decreased the need for mosaics somewhat, but if large-scale photomaps of extensive areas are needed, or if desired photomap size and scale requirements exceed the size capabilities of available enlargers, they still provide the solution to the problem.

10-2 ADVANTAGES AND DISADVANTAGES OF PHOTOMAPS AND MOSAICS

Photomaps and/or mosaics are similar to maps in many respects, but they have a number of definite advantages over maps. They show relative planimetric locations of an infinite number of objects, whereas features on maps—which are shown with lines and symbols—must be limited in number. Photomaps or mosaics of large areas can be prepared in much less time and at considerably lower cost than maps. They are easily understood and interpreted by people without photogrammetrey or engineering backgrounds because objects are shown by their images. For this reason they are very useful in describing proposed construction or existing conditions to members of the general public, who would probably be confused by the same representations on a map.

Photomaps and mosaics have the one serious disadvantage that they are not true planimetric representations. Rather, they are constructed from perspective photographs, which are subject to image displacements and scale variations. The most serious image displacements and scale variations are caused by variations in the terrain elevation, tilting of the camera axis, and variations in the flying heights. Of course, some small distortions result from shrinkage or expansion of the photo papers and camera-lens imperfections, but these are generally negligible.

The effects of tilt and flying height variations can be eliminated by rectifying the photographs and ratioing them to a common scale. (This procedure is described in Secs. 11-14 through 11-20.) Rectification does not remove effects of topographic relief, however, and as a result the scale of a photomap or mosaic is never constant throughout unless the area shown is perfectly flat. Relief displacements can be minimized by using a high flying height, while at the same time the decrease in scale is compensated for by using a longer-focal-length camera. In measuring distances or directions from a photomap or mosaic, it must be remembered that, due to image displacements, the scaled values will not be true. They are often used for qualitative studies only, and in that case slight planimetric inaccuracies caused by image displacements are of little consequence.

10-3 USES OF PHOTOMAPS AND MOSAICS

Because of their many advantages, photomaps and mosaics are quite widely used. Their value is perhaps most appreciated in the field of planning, both in land-use planning and in planning for engineering projects. A photomap or mosaic which shows an area completely and comprehensively can be rapidly and economically prepared. All critical features in the area which could affect the project can then be interpreted and taken into account. Alternative plans can be conveniently investigated, including considerations of soil types, drainage patterns, land-use and associated right of way costs, etc. As a result of this type of detailed study, the best overall plan is finally adopted.

Photomaps and mosaics are valuable in numerous other miscellaneous areas besides planning. They are used to study geologic features, to inventory natural resources, to record growth of cities and large institutions, to monitor construction activities at intervals of time, to record property boundaries, etc. They are used as planimetric map substitutes for many engineering projects. Highway departments, for example, that are engaged in preparing plans for extensive construction projects frequently use photomaps to replace planimetric surveys. This not only eliminates most of the ground surveying but also does away with the painstaking procedure of plotting the planimetry in the office. Design drawings and construction specifications are superimposed directly over the photomap. Used in this manner, these products have resulted in a tremendous saving in time and cost with no significant loss of accuracy.

10-4 KINDS OF MOSAICS

If a single photo does not contain an extensive enough coverage, or if it cannot be enlarged to the required scale, a mosaic must be prepared. Aerial mosaics generally fall into three classes: (1) *controlled,* (2) *semicontrolled,* and (3) *uncontrolled.* A controlled mosaic is the most accurate of the three classes. It is prepared from photographs which have been rectified and ratioed; i.e., all prints are made into equivalent vertical photographs which have the same scale. On the base board upon which the assembly will be laid, the horizontal positions of control points are plotted at the same scale as ratioed photo scale. The images of the control points must be recognizable on the photos. The coordinates of the ground control points may be obtained by field survey (see Chap. 15), by radial-line triangulation (see Chap. 9), or by aerotriangulation (see Chap. 14). The mosaic is laid by matching the control point images to their respective plotted positions on the base board. In spite of the precautions taken in preparing controlled mosaics, images of adjacent photographs will not match perfectly, nor will the scale of the mosaic be constant. Relief displacement is the chief reason for this condition. For certain small-scale controlled mosaic preparation, existing maps such as U.S. Geological Survey quadrangle maps may be used to provide the control. In this procedure the photos are ratioed to map scale. (If desired, map scale can first be altered to a convenient size by photographic enlargement or reduction.) The map is pasted to the mounting board and the mosaic is assembled by mounting the photos so that recognizable photo images coincide with their corresponding map positions.

An uncontrolled mosaic is prepared by simply matching the image details of adjacent photos. There is no ground control, and vertical photographs which have not been rectified or ratioed are used. Uncontrolled mosaics are more easily and quickly prepared than controlled mosaics. They are not as accurate as controlled mosaics, but for many qualitative uses they are completely satisfactory.

Semicontrolled mosaics are assembled utilizing some combinations of the specifications for controlled and uncontrolled mosaics. A semicontrolled mosaic may be prepared, for example, by using ground control but using photos that have not been rectified or ratioed. The other combination would be to use rectified and ratioed photos but no ground control. Semicontrolled mosaics are a compromise between economy and accuracy. The mosaic of Fig. 10-1 is a semicontrolled mosaic prepared from unrectified photos but assembled to fit U.S. Geological Survey quadrangle maps.

Mosaics are frequently categorized according to their use. In this type of classification are the *index* mosaic and *strip* mosaic. An index mosaic, or *photo index* as it is sometimes called, is illustrated in Fig. 10-2. It is an uncontrolled mosaic which has been laid to very rough specifications. Its purpose is to serve as an index for correlating photo numbers and photo coverages. A quick check of the index mosaic makes it possible to determine which photos must be retrieved from the files to cover a particular area of interest. An index mosaic is assembled as soon as possible after the flight. In preparing them, no cutting or trimming of the photos is necessary. The assembly is made by matching images, being careful to keep the numbers of all photos clearly visible. A convenient method of assembling an index mosaic is to staple the photos

Figure 10-1 Semicontrolled mosaic showing entire San Francisco Bay area. This mosaic consists of over 2,000 individual aerial photographs. (*Copyright photo, courtesy Pacific Resources, Inc.*)

Figure 10-2 Example of a "photo-index" or index mosaic.

to a fiber board such as Celotex. Upon completion, the assembly is photographed, and reduced prints are made for use. By assembling an index mosaic as soon as possible after the project, the photographic coverage of the project can be checked. Any gaps or missed areas will be visible, and arrangements can be made immediately for reflights if necessary.

A strip mosaic is the assembly of a series of photographs along a single flight strip. Strip mosaics are extremely useful in planning and designing linear engineering projects such as roads, railroads, pipelines, transmission lines, aqueducts, etc. Strip mosaics may be controlled, uncontrolled, or semicontrolled.

10-5 MATERIALS FOR PREPARING MOSAICS

The following materials are suggested for preparing mosaics:

Photographs. For mosaic work, photographs should be taken with a minimum of about 60 percent end lap and 30 percent side lap. This makes it possible to use only the central portions of the photos, thereby reducing distortions due to relief and tilt. If the area is very flat, these percentages may be reduced; and if the area is rugged, they may be increased somewhat. The photographs should be printed on single-weight paper, and special care should be exercised in developing and printing to obtain uniform tones on all photos. This will result in an overall uniform tone on the mosaic.

Mounting boards. These are the surfaces upon which the mosaic is to be laid. Mounting boards should have a smooth, hard, nonporous surface. Masonite makes an excellent mounting surface. Plywood may also be used, although it should be thoroughly moistened prior to laying the mosaic. Mounting boards should be large enough to accommodate the entire area of the assembly. After the mosaic has been completed and photographed for reproduction, the mounting board may be cleaned by soaking in water. In this manner it may be used repeatedly.

Adhesive. Gum arabic is a commonly used adhesive for mosaic construction. It is a slow-drying substance and therefore allows ample time for adjusting and shifting the prints to match images. Another slow-drying adhesive that is well suited for mosaic construction is a paste made from a starch base. Both of these adhesives can be purchased ready-mixed in bulk quantities. Rubber cement and glue are other adhesives that could be used; however, the former creates difficulties because it sets too fast, and the latter presents problems with clean-up of the assembled mosaic.

Miscellaneous tools. Among the tools useful in mosaic preparation are razor blades for cutting the photos, fine sandpaper for tapering match lines to a feather edge, a plastic squeegee for applying and removing the adhesive, water and a water pail, sponges, drafting tape, and a touch-up kit for patching unavoidable image mismatches.

10-6 MOSAIC CONSTRUCTION

The definition of photogrammetry given in Sec. 1-1 implied that it was art as well as science. Mosaic construction is definitely one of those photogrammetric procedures where artistic capabilities may be utilized to advantage. Just as finger-painting is a messy procedure, mosaic preparation is also a messy task. Good-quality mosaic preparation requires a person who is not squeamish about getting his hands wet and covered with sticky adhesive. Because it is an art, a great deal of personal satisfaction can be derived from a successfully prepared mosaic.

The procedures of laying controlled and uncontrolled mosaics are essentially the same except for plotting control and matching corresponding photo images to the control. The following step-by-step procedure applies to the construction of an uncontrolled mosaic:

1. Lay out all the photographs on the mounting board so that common images overlap. Temporarily fasten them together with drafting tape to form a very crude mosaic. Shift the entire assembly so that it is centered on the mounting board and mark on the board the position of the center photo of the center strip. This photo will be laid first, and if it is laid in this marked position, the mosaic will be properly centered on the board.
2. Remove the photos from the mounting board. Trim approximately $\frac{1}{2}$ in from all four sides of the center photo of the center strip, tapering all four edges. This is done by lightly cutting through the emulsion with a razor blade, as shown in Fig. 10-3. The sides to be discarded are torn off, using a downward twisting motion, as shown in Fig. 10-4. If properly done, the edges of the central portion of the photo should have approximately a $\frac{1}{2}$-in taper from full paper thickness to a feather edge. If a suitable taper is not obtained by tearing, light sanding with fine sandpaper may be necessary to obtain the desired feather edge. Tapering to a feather edge ensures an overall smooth appearance of the mosaic even though several layers of photos are mounted one on top of the other.
3. Moisten the mounting board, using a wet sponge. Apply adhesive in liberal amounts to both the back of the photo and to the mounting board. Place the first photo on the board in its premarked position and, with the fingers, rub it flat to the board. Remove all excess adhesive from on and under the photo with a squeegee, using firm strokes outward from the center of the photo, as shown in Fig. 10-5. It is very important that all air bubbles be removed from under the photograph so that the mosaic lies flat. With a damp sponge, clean the working area in preparation for the next photo.
4. Place the second photo to be laid on the first photo with corresponding images overlapping. (The second photo can be the overlapping photo either left or right of the first photo laid.) Select the match line on the second photo along which

Figure 10-3 Making the razor cut.

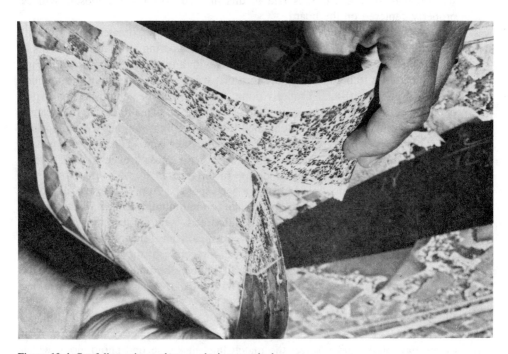

Figure 10-4 Carefully tearing a photo to obtain tapered edge.

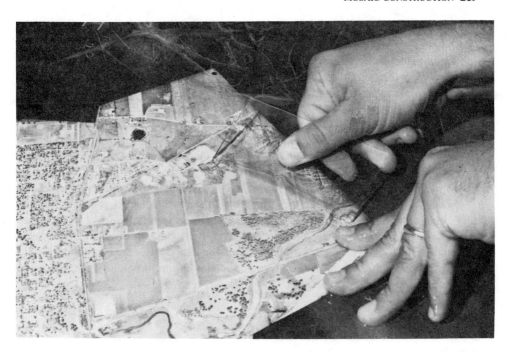

Figure 10-5 Using the squeegee.

the cut will be made. Match lines should be carefully chosen to obtain the best possible matching of images and tones. If possible, they should be selected along roads, railroads, edges of fields, or other lines of definite tone demarcation. Avoid cutting at right angles to roads, railroads, etc., as this often creates noticeable mismatches in positions and tones. Match lines in the overlap of photos in the same strip should fall about midway between principal point and conjugate principal point, and match lines in the side lap between strips should fall approximately in the middle of the side lap. If match lines are selected in these general locations, only the central portions of photos actually show on the finished mosaic, which minimizes image matching difficulties and produces a more uniform mosaic scale, since image displacements are minimal in these areas.

5. Using procedures of step 2, trim and taper the photo along the match line selected between the two photos. Also trim about $\frac{1}{2}$ in from the other three sides of the photo and taper those edges.

6. Lay the second photo using procedures of step 3. In laying this and all remaining photos, particular care must be exercised to match images. In some cases all images cannot be matched, so that a best fit compromise is selected; but major features whose mismatches would be most noticeable should be given preference in the compromise. If the photos are immersed in hot water, they may be stretched by small amounts to facilitate image matching. When the recommended adhesives

are used, the photos may be lifted and moved by small amounts for several minutes after initial laying.

7. Continue laying photographs in a systematic procedure, progressing outward from the center photograph in all directions until the entire mosaic is laid. Figure 10-6a illustrates an overlapping five-photo strip prior to trimming, and Fig. 10-6b shows the general locations of the match lines and the order of laying the mosaic.

8. When the mosaic is completed, touchup any glaring image mismatches with a touch-up paint kit, as shown in Fig. 10-7. A touch-up kit consists of a fine-pointed paintbrush and several colors of paint in shades of gray from black to white. With care in matching tones, it is possible to mask mismatches along roads, railroads, etc. Sometimes an area that should appear as one solid tone is so large that more than one photograph is required to cover it—e.g., large bodies of water. Unavoidable tonal variations from one photo to the next often cause such areas to appear patchy. In these cases the entire area may be painted a solid tone. The Pacific Ocean and San Francisco Bay in the mosaic of Fig. 10-1 were touched up in this manner.

10-7 AZIMUTH LINE METHOD

In the construction of an uncontrolled mosaic by the procedures just described, tilt of the photos and topographic relief can cause cumulative mismatch errors which result in a twist or turn of the entire mosaic in one direction. The *azimuth line* method is

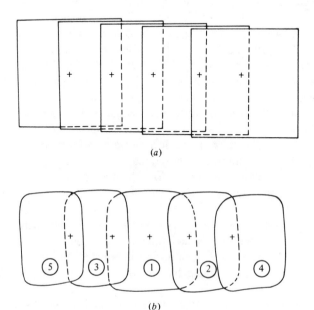

(a)

(b)

Figure 10-6 (a) Overlapping five-photo strip prior to trimming. (b) Strip mosaic illustrating match lines and order of laying photos.

Figure 10-7 Final touch-up of a mosaic.

used to prevent this objectionable twisting. In this procedure, the photographs are first laid out and taped together in strips, matching corresponding images as carefully as possible by rapidly flipping the end lap areas up and down. A long straightedge is then placed over each strip so that it passes as nearly as possible through all photo centers, and a soft pencil line called an "azimuth line" is ruled across the photos. The strips are then disassembled and the lines are prolonged across the parts of the photos that were obscured by end lap.

A long straight line representing the azimuth line of the middle strip is drawn on the mounting board. The middle strip is then laid by matching the azimuth line on the photos as closely as possible to the line on the board while at the same time matching image details as carefully as possible. If the mosaic consists of more than one strip, a straight line representing the azimuth line of the next side-lapping strip is drawn in its proper position on the mounting board. Proper position for the line is found by overlaying onto the center strip (which has been laid) the side lap area of approximately every fifth photo of the next strip. With the images of at least two widely spaced points of average elevation carefully matched in the side lap, the end of the azimuth line in each photo is transferred to the mounting board. A long straight line is then drawn on the mounting board so that it passes as nearly as possible through all points marking the azimuth line. This procedure is continued until all strips have been laid.

10-8 ORTHOPHOTOMOSAICS

An *orthophotomosaic* is an assembly of two or more *orthophotos* to form a continuous picture of the terrain. As implied by their name, orthophotos are orthographic representations of the terrain. They are derived from vertical aerial photographs using a *differential rectification* instrument. The method of producing orthophotos is discussed in detail in Chap. 13.

Orthophotos have had image displacements due to relief and tilt removed so that they show features in their true planimetric positions. Distances, angles, and areas can therefore be measured directly from orthophotos just as from maps. Figure 10-8 shows a portion of an orthophotomosaic with elevation contours superimposed. In outward appearance it looks like an ordinary vertical aerial photograph with superimposed contours. Orthophotomosaics have the pictorial advantages of aerial mosaics and the geometric correctness of maps. In addition, they have the advantage that they can usually be prepared more rapidly and economically than line and symbol maps.

In general the procedures previously described for controlled mosaic assembly are used for assembling orthophotomosaics. In orthophotomosaic preparation, however, the individual orthophotos are prepared to a common predetermined scale on stable base materials (e.g., film) rather than paper-base material. Because the orthopositives have been ratioed and because they contain no distortions, very little difficulty should be encountered in matching corresponding images during mosaic assembly. In mosaic assembly with film-base materials, featheredges must be sanded. Also, beeswax has been found to be the best adhesive.

10-9 REPRODUCTION

Before a photomap or mosaic is photographed for reproduction, a border and a title block are usually added. The lettering of the title must be large enough to be visible at the intended reproduction size. The title is usually printed on white paper and then pasted onto the original. It may also be desirable to add other information, such as grid lines or annotations of principal places or features. As shown in Fig. 10-8, contours may also be superimposed. If the average scale of the photomap or mosaic is given, it should be shown graphically so that it remains applicable for any reproduction size.

Reproduction of photomaps and mosaics may be considered as a two-step process of (1) photographing the original with a copy camera to obtain a negative and (2) production of prints from the negative. Copy cameras used in step 1 are large, sometimes having focal planes with dimensions as great as 4 ft square. Copy cameras normally contain an easel plane upon which the object to be photographed is mounted. The easel plane may be tilted with respect to the optical axis of the camera lens—a procedure necessary in rectification. For photographing a photomap or mosaic prepared from vertical photos, however, the easel plane should be perpendicular to the optical axis. Copy cameras are normally mounted on a track so that object distances can be easily varied. The focal planes of these cameras can also be moved with respect

Figure 10-8 A portion of an orthophotomosaic with elevation contours superimposed. (*Courtesy USDA Soil Conservation Service.*)

223

to the lens so that the lens formula, Eq. (2-8), can be satisfied for varying object distances.

After the mosaic has been photographed and the negative developed, reproductions may be made in one of several ways. If a small number of copies are needed, noninking reproduction methods may be most economical. One such method is photographic reproduction (exposing sensitized photo paper through the negative and then developing the positives in a darkroom process). Another noninking process in common use is the *ozalid* process. In this procedure, prints are made from halftone positives which have been prepared from the original negatives (see Sec. 3-13). If a large number of copies of the mosaic are needed, an inking process such as lithography will be most economical.

REFERENCES

American Society of Photogrammetry: "Manual of Photogrammetry," 3d ed., Falls Church, Va., 1966, chap. 17.
———: "Manual of Photogrammetry," 4th ed., Falls Church, Va., 1980, chap. 15.
Jessiman, E. G., and M. R. Walsh: An Approach to the Enhancement of Photomaps, *Canadian Surveyor,* vol. 30, no. 1, p. 11, 1976.
McNeil, G. T.: The Wet Process of Laying Mosaics, *Photogrammetric Engineering,* vol. 15, no. 2, p. 315, 1949.
Marsik, Z.: Use of Rectified Photographs and Differentially Rectified Photographs for Photo Maps, *Canadian Surveyor,* vol. 25, p. 567, 1971.
Meyer, D.: Mosaics You Can Make, *Photogrammetric Engineering,* vol. 28, no. 1, p. 167, 1962.
Moranda, P. B.: A Study of the Propagation of Errors in a Simplified Photographic Mosaic, *Photogrammetric Engineering,* vol. 26, no. 4, p. 582, 1960.
Rosenfield, G. H.: The Accuracy of Mosaics, *Photogrammetric Engineering,* vol. 21, no. 5, p. 670, 1955.

PROBLEMS

10-1 Discuss the advantages and disadvantages of photomaps and aerial mosaics as compared to conventional line and symbol maps.

10-2 Outline some of the uses of photomaps and aerial mosaics.

10-3 Name and describe the various kinds of mosaics.

10-4 What is an index mosaic and what is it used for?

10-5 Describe the "azimuth line" method of laying mosaics. Why is it useful?

10-6 A photomap is to be prepared by enlarging a 9-in-square-format aerial photo to a size of 30 in square. If a Lear jet is used to obtain the photo and a 6-in-focal-length camera is carried to an altitude of 50,000 ft above ground for the exposure, what will be the resulting scale of the photomap, and how many square miles of area will it cover?

10-7 A mosaic to serve as a map substitute for locating utilities will be prepared from contact prints. The prints will be made from negatives exposed with a 12-in-focal-length camera. If 30-in-diameter manholes must appear on the mosaic with diameters of 0.5 mm, what must be the flying height above average ground for the photography?

10-8 Repeat Prob. 10-7, except that the camera focal length is 6-in, and 24-ft-wide roadways must appear 0.10 in wide on the resulting mosaic.

10-9 A mosaic will be prepared by matching photo images to objects appearing on a $7\frac{1}{2}$-minute USGS quadrangle map. If contact prints will be obtained for the mosaic with an aerial camera having an $8\frac{1}{4}$-in focal length, what must be the flying height above average ground to obtain photos that best match the map?

10-10 A mosaic will be prepared from three sidelapping strips of aerial photos. The photos will be exposed with a 6-in-focal-length camera from an average flying height of 4,500 ft above ground, and end lap and side lap are planned to be 60 percent and 30 percent, respectively. Each strip will contain 12 photos. What will be the scale of the mosaic, and what will be the approximate ground dimensions of the area covered by the mosaic?

ELEVEN

TILTED PHOTOGRAPHS

11-1 INTRODUCTION

In spite of level vials and other stabilizing equipment, in practice it is impossible to maintain the optical axis of the camera truly vertical. Unavoidable aircraft tilts cause photographs to be exposed with the camera axis tilted slightly from vertical, and the resulting pictures are called *tilted photographs*. If vertical photography is intended, the amount by which the optical axis deviates from vertical is usually less than 1° and it rarely exceeds 3°.

Six independent parameters called the *elements of exterior orientation* express the *space position* and *angular orientation* of a tilted photograph. The space position is normally given by X_L, Y_L, and Z_L, the three-dimensional coordinates of the exposure station in a ground coordinate system. Z_L is commonly called H, flying height above datum. Angular orientation is the amount and direction of tilt in the photo. Three angles are sufficient to define angular orientation, and in this book two different systems are described: (1) the *tilt-swing-azimuth* (t-s-α) system and (2) the *omega-phi-kappa* (ω-ϕ-κ) system. The omega-phi-kappa system, possesses certain computational advantages over the tilt-swing-azimuth system, and it is therefore more widely used. The tilt-swing-azimuth system, however, is more easily understood and shall therefore be considered first.

11-2 ANGULAR ORIENTATION IN TILT, SWING, AND AZIMUTH

In Fig. 11-1, a tilted photograph is depicted showing the tilt-swing-azimuth angular orientation parameters. In the figure, L is the exposure station and o is the principal point of the photo positive. Ln is a vertical line, n being the *photographic nadir point* which occurs where the vertical line intersects the plane of the photograph. The extension of Ln intersects the ground surface at N_g, the *ground nadir point,* and it intersects the datum surface at N_d, the *datum nadir point*. Line Lo is the camera optical axis; its extension intersects the ground at P_g, the *ground principal point,* and it

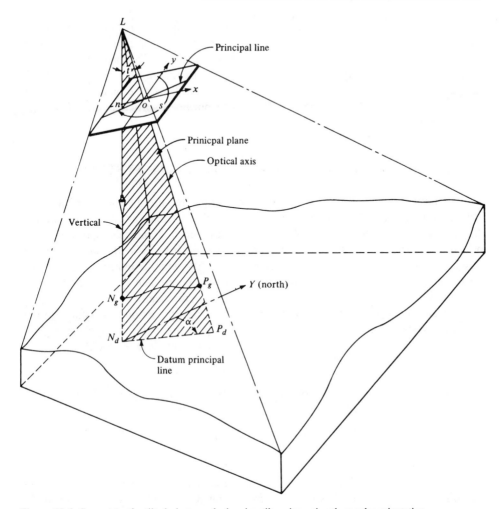

Figure 11-1 Geometry of a tilted photograph showing tilt-swing-azimuth angular orientation.

intersects the datum plane at P_d, the *datum principal point*. One of the orientation angles, *tilt,* is the angle t or nLo between the camera optical axis and the vertical line Ln. The tilt angle gives the magnitude of tilt of a photo.

Vertical plane Lno is called the *principal plane*. Its line of intersection with the plane of the photograph occurs along line *no,* which is called the *principal line*. The position of the principal line on the photo with respect to the reference fiducial axis system is given by s, the *swing* angle. "Swing" is defined as the clockwise angle measured in the plane of the photograph from the positive y axis to the downward or nadir end of the principal line, as shown in Fig. 11-1. The swing angle gives the direction of tilt on the photo. Its value can be anywhere between 0° and 360°.

The third angular orientation parameter, α or *azimuth,* gives the orientation of the principal plane with respect to the ground reference axis system. Azimuth is the clockwise angle measured from the ground Y axis (usually north) to the *datum principal line*, $N_d P_d$. It is measured in the datum plane or in a plane parallel to the datum plane, and its value can be anywhere between $0°$ and $360°$. The three angles of tilt, swing, and azimuth completely define the angular orientation of a tilted photograph in space. If the tilt angle is zero, the photo is vertical. Thus a vertical photograph is simply a special case of the general tilted photograph. For a vertical photo, swing and azimuth are undefined.

11-3 AUXILIARY TILTED PHOTOCOORDINATE SYSTEM

In the tilt-swing-azimuth system, certain computations require the use of an auxiliary $x'y'$ rectangular photographic coordinate system. This system, as shown in Fig. 11-2a, has its origin at the photographic nadir point n and its y' axis coincides with the principal line (positive in the direction from n to o). Positive x' is $90°$ clockwise from positive y'. In solving tilted photo problems in the tilt-swing-azimuth system, photocoordinates are usually first measured in the fiducial coordinate system described in Sec. 5-2 and then converted to the auxiliary system numerically.

For any point in a tilted photo, the conversion from the xy fiducial system to the $x'y'$ tilted system requires (1) a rotation about the principal point through the angle θ and (2) a translation of origin from o to n. The rotation angle θ is defined as

$$\theta = s - 180° \tag{11-1}$$

The coordinates of image point a after rotation are x'_a and y''_a, as shown in Fig. 11-2a. These are calculated using the following rotation equations:

$$x'_a = x_a \cos \theta - y_a \sin \theta$$

$$y''_a = x_a \sin \theta + y_a \cos \theta$$

The components of the above two rotation equations are illustrated in Fig. 11-2a. Auxiliary coordinate y'_a is obtained by adding the translation distance *on* to y''_a. From Fig. 11-2b, which is a side view of the principal plane, *on* is $f \tan t$. Therefore the coordinates of a point in the required auxiliary coordinate system are

$$x'_a = x_a \cos \theta - y_a \sin \theta \tag{11-2}$$

$$y'_a = x_a \sin \theta + y_a \cos \theta + f \tan t$$

11-4 SCALE OF A TILTED PHOTOGRAPH

In Chap. 6 it was shown that scale variations in a vertical photograph are the result of variations in object distance (distance from camera to ground). The shorter the object distance, the larger the scale, and vice versa. For vertical photos, variations in object distances were caused only by topographic relief. In a tilted photograph, relief variations also cause changes in scale, but scale in various parts of the photo is further

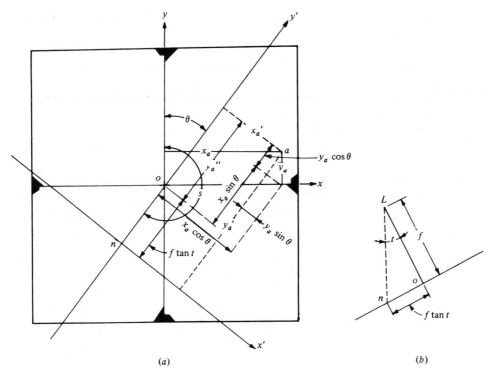

(a)   (b)

Figure 11-2 (a) Auxiliary $x'y'$ image coordinate system for a tilted photo. (b) Principal plane of a tilted photo.

affected by the magnitude and angular orientation of the tilt. Figure 11-3a illustrates the principal plane of a tilted photograph taken over a square grid on flat ground. Figure 11-3b illustrates the appearance of the grid on the resulting tilted photograph. Due to tilt, object distance LA in Fig. 11-3a is less than object distance LB, and hence a grid line near A would appear larger (at a bigger scale) than a grid line near B. This

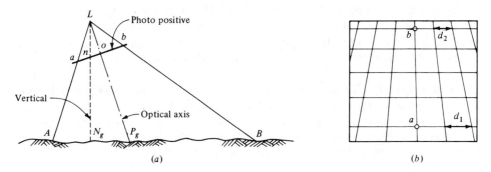

(a)   (b)

Figure 11-3 (a) Principal plane of a tilted photo taken over flat ground. (b) Image on the tilted photo of a square ground grid.

is illustrated in Fig. 11-3b, where photo distance d_1 appears longer than photo distance d_2, yet both are the same length on the ground.

The scale at any point on a tilted photograph is readily calculated if tilt and swing for the photograph and the elevation of the point are known. Figure 11-4 illustrates a tilted photo taken from a flying height H above datum. Lo is the camera focal length. The image of object point A appears at a on the tilted photo, and its photocoordinates in the auxiliary tilted photocoordinate system are x'_a and y'_a. The elevation of object point A above datum is h_A. Object plane $AA'KA''$ is a horizontal plane constructed a distance h_A above datum. Image plane $aa'kk'$ is also constructed horizontally. The scale relationship between the two parallel planes is the scale of the tilted photograph at point a because the image plane contains image point a and the object plane contains object point A. The scale relationship is the ratio of photo distance aa' to ground distance AA', and may be derived from similar triangles $La'a$ and $LA'A$, and Lka' and LKA' as follows:

$$S_a = \frac{aa'}{AA'} = \frac{La'}{LA'} = \frac{Lk}{LK} \tag{a}$$

but

$$Lk = Ln - kn = \frac{f}{\cos t} - y'_a \sin t$$

also

$$LK = H - h_A$$

Substituting Lk and LK into (a) and dropping subscripts,

$$S = \frac{f/\cos t - y' \sin t}{H - h} \tag{11-3}$$

In Eq. (11-3), S is the scale on a tilted photograph for any point whose elevation is h above datum. Flying height above datum for the photo is H, f is the camera focal length, and y' is the coordinate of the point in the auxiliary system calculated by Eqs. (11-2). If the units of f and y' are inches and if H and h are feet, then the scale ratio is obtained in inches per foot. Of course this scale can also be readily expressed as a dimensionless ratio. Examination of Eq. (11-3) shows that scale increases with increasing terrain elevation. If the photo is taken over flat ground, then h is constant but scale still varies throughout the photograph with variations in y'.

Example 11-1 A tilted photo is taken with a 6-in (152.4-mm-)-focal-length camera from a flying height of 8,200 ft above datum. Tilt and swing are 2°30' and 218°, respectively. Point A has an elevation of 1,435 ft above datum, and its image coordinates with respect to the fiducial axis system are $x_a = -2.85$ in and $y_a = 3.43$ in. What is the scale at point a?

SOLUTION By Eq. (11-1),

$$\theta = s - 180° = 218° - 180° = 38°$$

By Eq. (11-2),

$$y'_a = -2.85(0.61566) + 3.43(0.78801) + 6.00(0.04366) = 1.21 \text{ in}$$

By Eq. (11-3),

$$S_a = \frac{6.00/0.99905 - 1.21(0.04362)}{8,200 - 1,435} = \frac{5,953 \text{ in}}{6,765 \text{ ft}} = \frac{1 \text{ in}}{1,140 \text{ ft}}$$

11-5 GROUND COORDINATES FROM A TILTED PHOTOGRAPH

If tilt and swing are known for a particular photograph, ground coordinates of any points appearing in the photograph may be calculated, provided that their elevations are known. Ground coordinates are calculated in the unique $X'Y'$ rectangular coordinate system shown in Fig. 11-4. In this system the origin is at the datum nadir point, and the X' and Y' axes are in the same vertical planes and positive in the same directions as the x' and y' auxiliary photo axes, respectively. By proportionality of similar triangles LKA'' and Lkk',

$$\frac{KA''}{kk'} = \frac{LK}{Lk} \tag{b}$$

Substituting $KA'' = X'_A$, $LK = H - h_A$, $kk' = x'_a$, and $Lk = f/\cos t - y'_a \sin t$, into (b), and dropping subscripts, the following equation is obtained for the X' ground coordinate of any point in a tilted photograph:

$$X' = \left(\frac{H - h}{f/\cos t - y' \sin t} \right) x' \tag{11-4}$$

Also, by similar triangles LKA' and Lka',

$$\frac{KA'}{ka'} = \frac{LK}{Lk} \tag{c}$$

Substituting $KA' = Y'_A$, $ka' = y'_a \cos t$, $LK = H - h_A$, and $Lk = f/\cos t - y'_a \sin t$ into (c) and dropping subscripts, the following equation is obtained for the Y' ground coordinate of any point in a tilted photograph:

$$Y' = \left(\frac{H - h}{f/\cos t - y' \sin t} \right) y' \cos t \tag{11-5}$$

In Eqs. (11-4) and (11-5), X' and Y' are ground coordinates of any points appearing in a tilted photograph and all other terms are as previously defined. From X' and Y' ground coordinates, lengths of lines, angles, and areas can be calculated. Ground coordinates calculated by Eq. (11-4) and (11-5) are in the previously defined arbitrary coordinate system. However, if coordinates of two or more control points are also known in an absolute ground coordinate system (e.g., the state plane coordinate system), then coordinates of all other points can be transformed from the

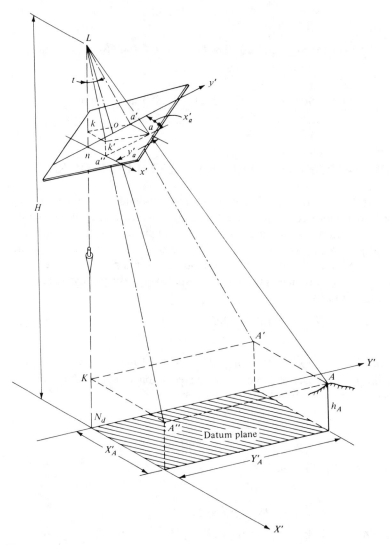

Figure 11-4 Scale of a tilted photo, and ground coordinate system.

arbitrary system into the absolute system using the coordinate transformation equations described in Sec. B-2 of Appendix B.

Example 11-2 Assume that image b of another object point B appears in the tilted photograph of Example 11-1, and that its measured photocoordinates with respect to the fiducial system are $x_b = 3.09$ in and $y_b = -1.78$ in. The elevation of B is 1,587 ft above datum. Calculate ground coordinates of A and B and the horizontal length of AB.

SOLUTION: By Eqs. (11-2),

$$x'_a = -2.85(0.78801) - 3.43(0.61566) = -4.358 \text{ in}$$

$$x'_b = 3.09(0.78801) + 1.78(0.61566) = 3.531 \text{ in}$$

$$y'_b = 3.09(0.61566) - 1.78(0.78801) + 6.00(0.04366) = 0.762 \text{ in}$$

$$y'_a = 1.210 \text{ in (from Example 11-1)}$$

By Eqs. (11-4) and (11-5),

$$X'_A = \frac{(8,200 - 1,435)(-4.358)}{6.00/0.99905 - 1.210(0.04362)} = \frac{29,481.9}{5.953} = -4,952 \text{ ft}$$

$$X'_B = \frac{(8,200 - 1,587)\, 3.531}{6.00/0.99905 - 0.762(0.04362)} = \frac{23,350.5}{5.972} = 3,910 \text{ ft}$$

$$Y'_A = \frac{(8,200 - 1,435)\, 1.210(0.99905)}{6.00/0.99905 - 1.210(0.04362)} = \frac{8,177.8}{5.953} = 1,374 \text{ ft}$$

$$Y'_B = \frac{(8,200 - 1,587)\, 0.762(0.99905)}{6.00/0.99905 - 0.762(0.04362)} = \frac{5,034.3}{5.972} = 843 \text{ ft}$$

From these coordinates length AB is found, using the pythagorean theorem, as follows:

$$AB = \sqrt{(3,910 + 4,952)^2 + (843 - 1,374)^2} = 8,878 \text{ ft}$$

11-6 RELIEF DISPLACEMENT ON A TILTED PHOTOGRAPH

Image displacements on tilted photographs caused by topographic relief occur much the same as they do on vertical photos. In Fig. 11-5, a is the image of an object point A, Point A' is in the datum plane vertically beneath A, and a' is its theoretical photographic position. Distance aa' is the image displacement due to the relief of point A. From geometry, it is known that two planes intersect along a straight line. Plane $LA''N_dA'A$ and the plane of the photograph intersect along line na. Planes $LA''N_dA'A$ and LN_dA' are coincident vertical planes because both contain the vertical line $LA''N_d$. Therefore plane LN_dA' intersects the plane of the photograph along line na'. Lines na and na' are coincident, and therefore aa' is along a radial line from n. The significance of the foregoing is that *relief displacements on tilted photographs occur along radial lines from the nadir point.* Relief displacements on a truly vertical photograph are also radial from the nadir point, but in that special case the nadir point coincides with the principal point.

On a tilted photograph, image displacements due to relief vary in magnitude depending upon flying height, height of object, amount of tilt, and location of the image in the photograph. Relief displacement is zero for images at the nadir point and increases with increased radial distances from the nadir.

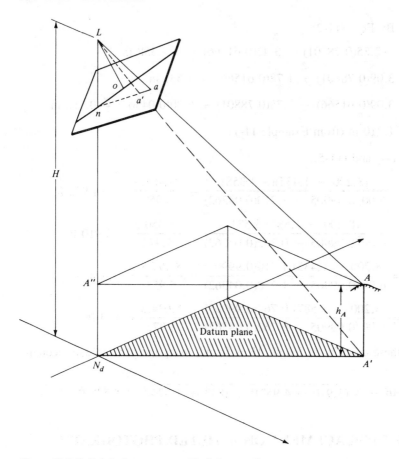

Figure 11-5 Relief displacement on a tilted photograph.

As defined in Sec. 11-1, tilted photos are those which were intended to be vertical, but which contain small unavoidable amounts of tilt. In practice, tilted photos are therefore so nearly vertical that their nadir points are generally very close to their principal points. Even for a photograph containing 3° of tilt taken with a 6-in-focal-length cameras, distance *on* is only about 0.3 in (8 mm). Relief displacements on tilted photos may therefore be calculated with satisfactory accuracy using Eq. (6-11) which applies to a vertical photograph. When this equation is used, radial distances *r* are measured from the principal point, even though theoretically, they should be measured from the nadir point.

11-7 TILT DISPLACEMENT

Figure 11-6 illustrates the principal plane of a tilted photograph taken from exposure station *L*. Shown also in the figure is the plane of an *equivalent vertical photograph*.

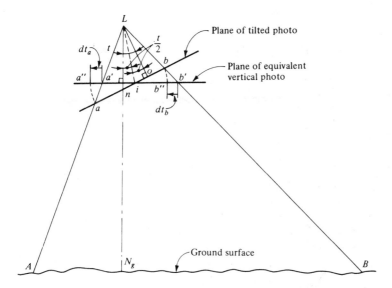

Figure 11-6 Tilt displacement in the principal plane of a tilted photograph.

An equivalent vertical photograph is an imaginary truly vertical photo taken from the same exposure station as the tilted photo with the same camera. The bisector of the tilt angle intersects the plane of the tilted photograph at point i, which is known as the *isocenter*. The line in the plane of the tilted photo, perpendicular to the principal line and passing through the isocenter, is called the *axis of tilt*. It is shown in Fig. 11-7. The axis of tilt is the line of intersection of the plane of the tilted photo and the plane of the equivalent vertical photo. Images along this line therefore have the same photographic positions in both the tilted photo and the equivalent vertical photo. At all other locations in the tilted photo, image positions differ from their corresponding positions on the equivalent vertical photo.

Tilt displacement is the radial distance from the isocenter to an image on an equivalent vertical photo minus the radial distance from the isocenter to the image on the tilted photo. In Fig. 11-6, a is the image of ground point A on a tilted photo; its image would be at a' on the equivalent vertical photo. Point a'' is located in the plane of the equivalent vertical photograph such that distance $ia = ia''$. In this case tilt displacement is of magnitude $dt_{a'}$, which is equal to ia' minus ia''. Its algebraic sign is thus negative. Similarly, images of ground point B occur at b and b' on the tilted and equivalent vertical photos, respectively, and tilt displacement is of magnitude dt_b, which equals ib' minus ib''. Its algebraic sign is positive. From this discussion it follows that images on a tilted photograph are displaced inward with respect to their positions on an equivalent vertical photo if they occur above the axis of tilt, and they are outward if they occur below the axis of tilt. Also, inward displacements are considered positive and outward displacements are negative. Tilt displacements occur even if the photo is taken over flat ground.

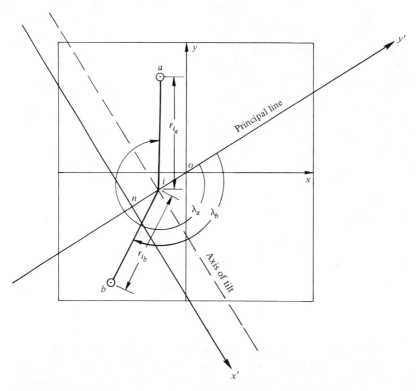

Figure 11-7 Radial distances r_i, and angular values λ, for use in Eq. (11-7).

The amount of tilt displacement for any image on a tilted photo is given by

$$dt = \frac{(r_i)^2 \sin t \cos^2 \lambda}{f - (r_i) \sin t \cos \lambda} \qquad (11\text{-}7)$$

In Eq. (11-7), dt is the amount of tilt displacement, r_i is the radial distance from the isocenter to the imaged point, f is the camera focal length, t is the tilt angle, and λ is the angle in the plane of the photograph, measured clockwise from the positive end of the principal line to the radial line from the isocenter to the imaged point. These angles can therefore have values anywhere from zero up to 360°. Figure 11-7 illustrates r_i and λ for two points, a and b. The algebraic sign of r_i is always considered positive. The units of dt will be the same as those used for r_i and f. The correct algebraic sign of dt is not automatically obtained in Eq. (11-7), but must be assigned as positive if the point lies above the axis of tilt and negative if it is below. If the value of λ is either 90° or 270°, which occurs when the point lies along the axis of tilt, the numerator of Eq. (11-7) becomes zero and dt is therefore zero. This follows logically inasmuch as a tilted photo and its equivalent vertical photo coincide along the axis of tilt.

Example 11-3 A photograph having a tilt of 3°00′ was exposed with a 152-mm-focal-length camera. Radial distance from the isocenter to a certain image point in the area above the axis of tilt measured 105 mm. The clockwise angle between the principal line and the radial line from the isocenter to the point was 40°00′. What is the tilt displacement for that image?

SOLUTION: By Eq. (11-7),

$$dt = \frac{(105)^2(0.0523)(0.7660)^2}{152 - 105(0.0523)0.7660} = +2.3 \text{ mm}$$

Note: For λ angles between 0° and 90° and between 270° and 360°, the point lies above the axis of tilt and tilt displacement is positive.

11-8 ANGULAR ORIENTATION IN OMEGA, PHI, AND KAPPA

As previously stated, besides the tilt-swing-azimuth system, angular orientation of a tilted photograph can also be expressed in terms of three rotation angles, *omega, phi,* and *kappa*. These three angles uniquely define the angular relationships between the three axes of the tilted photo (image) coordinate system and the three axes of the ground (object) coordinate system.

Figure 11-8 illustrates a tilted photo in space. The ground coordinate system is *XYZ*. The tilted photo image coordinate system is *xyz* (shown dashed), and its origin is at exposure station *L*. Consider an image coordinate system $x'y'z'$ with origin also at *L* and with its respective axes mutually parallel to the axes of the ground coordinate system, as shown in Fig. 11-8. As a result of three sequential rotations through the angles of omega, phi, and kappa, the $x'y'z'$ axis system can be made to coincide with the photographic *xyz* system. Each of the rotation angles omega, phi, and kappa is considered positive if counterclockwise when viewed from the positive end of the rotation axis.

The sequence of the three rotations is illustrated in Fig. 11-9. The first rotation, as illustrated in Fig. 11-9a, is about the x' axis through an angle omega. This first rotation creates a new axis system $x_1y_1z_1$. The second rotation phi is about the y_1 axis, as illustrated in Fig. 11-9b. As a result of the phi rotation a new axis system $x_2y_2z_2$ is created. As illustrated in Fig. 11-9c, the third and final rotation is about the z_2 axis through the angle kappa. This third rotation creates the *xyz* coordinate system which is the photographic image system. Equations which express these three rotations are developed in Sec. B-7 of Appendix B.

For any tilted photo there exist a unique set of angles omega, phi, and kappa which explicitly define the angular orientation of the photograph with respect to the reference ground coordinate system. These three angles are related to the previously described tilt, swing, and azimuth angles, and if either set of three orientation angles is known for any photo, the other three can be determined as described in Sec. C-8 of Appendix C. In the omega-phi-kappa system, as with the tilt-swing-azimuth system,

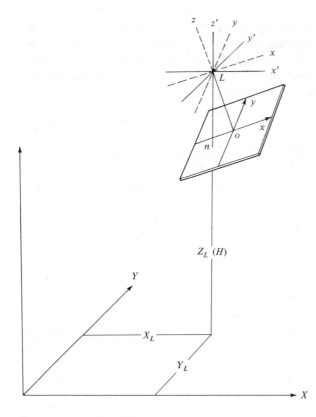

Figure 11-8 Orientation of a tilted photo in the omega-phi-kappa system.

the space position of any photo is given by the exposure station coordinates X_L, Y_L, and Z_L (or H).

11-9 DETERMINING THE ELEMENTS OF EXTERIOR ORIENTATION

Many different methods, both graphical and numerical, are available for determining the six elements of exterior orientation of a tilted photograph. In general all methods require photographic images of at least three control points whose X, Y, and Z ground coordinates are known. Also the calibrated focal length of the camera must be known. Most procedures are iterative; i.e., a solution is obtained by making successive corrections to some initially assumed values for the unknown elements. All methods for determining the elements of exterior orientation are rather lengthy, but if one of the numerical procedures is programmed for a computer, a solution is easily obtained.

In this text three methods for determining the elements of exterior orientation are considered: (1) the *Anderson scale-point* method, (2) the *Church* method, and (3) the method of *space resection by collinearity*. Students interested in studying other methods or in obtaining more details of the three methods presented herein are directed to references cited at the end of this chapter.

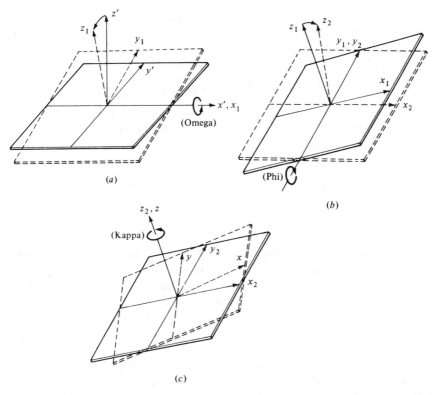

Figure 11-9 (*a*) Rotation about the *x* axis through angle omega. (*b*) Rotation about the y_1 axis through angle phi. (*c*) Rotation about the z_2 axis through angle kappa.

11-10 ANDERSON SCALE-POINT METHOD

The Anderson scale-point method is mainly a graphical procedure for determining tilt, swing, and flying height of a tilted photograph which requires only simple mathematics. When tilt, swing, and flying height have been determined, the remaining three elements of exterior orientation are readily calculated. The Anderson scale-point method cannot be applied to photos which are severely tilted. Also, because it is graphical, accuracy attainable with the method is limited. Since the emergence of the computer, the Anderson scale-point method has had little practical value, and hence the method is only briefly discussed herein.

In the Anderson scale-point method three control lines must appear in the photograph whose tilt is being determined. An economical control configuration meeting that criterion is a large, nearly equilateral triangle. Relief displacements with respect to a selected "photo datum" (arbitrary datum chosen at the elevation of the lowest of the control points) are calculated for all control points. For these calculations, the photos are assumed to be vertical, relief displacement Eq. (6-11) is used, and radial distances *r* are measured from the principal point. *Photo datum positions* of the control images are then plotted by laying off the relief displacements along radial lines *toward* the principal point. The *photo datum scales* of the three lines are then calculated from the ratios of their measured *photo datum lengths* to their corresponding ground lengths. Since the image points have all been reduced to a common datum, the photo is truly vertical if the three scales are equal; otherwise tilt exists in the photo.

The amount and direction of tilt is a function of the magnitudes of the scale differences of the three lines and of the locations of the lines in the photograph. Based upon the scale differences, a graphical technique is applied to determine tilt, swing, and flying height. The procedure is described in references cited at the end of this chapter.

11-11 CHURCH METHOD

The late Professor Earl Church of Syracuse University developed various techniques for calculating the elements of exterior orientation of a tilted photograph. The method discussed herein is his two-part "space resection and space orientation" procedure. This method is applicable to oblique photographs as well as near verticals. The Church method is included within this text to give the student one method of the direct calculation of the tilt, swing, and azimuth angular orientation elements of a tilted photograph.

The two parts of the Church method consist of (1) determining the ground space coordinates of the exposure station and (2) determining the angular elements of orientation. In Fig. 11-10, images a, b, and c of ground control points A, B, and C appear in a tilted photo positive. Angles γ, δ, and β as shown in Fig. 11-10 have apexes at L and are the included angles between rays LA, LB, and LC. According to analytic geometry, the cosines of these angles are

$$\cos \gamma = \frac{(X_L - X_A)(X_L - X_B) + (Y_L - Y_A)(Y_L - Y_B) + (Z_L - Z_A)(Z_L - Z_B)}{(LA)(LB)}$$

$$\cos \delta = \frac{(X_L - X_B)(X_L - X_C) + (Y_L - Y_B)(Y_L - Y_C) + (Z_L - Z_B)(Z_L - Z_C)}{(LB)(LC)}$$

$$\cos \beta = \frac{(X_L - X_C)(X_L - X_A) + (Y_L - Y_C)(Y_L - Y_A) + (Z_L - Z_C)(Z_L - Z_A)}{(LC)(LA)}$$

$$(11\text{-}8)$$

In Eqs. (11-8), the X's, Y's, and Z's are ground coordinates of control points A, B, and C in feet (or meters) and LA, LB, and LC are the lengths of rays from the exposure station to the respective control points in feet (or meters). Ray lengths are calculated as $LA = \sqrt{(X_L - X_A)^2 + (Y_L - Y_A)^2 + (Z_L - Z_A)^2}$, etc.

Explicit values for $\cos \gamma$, $\cos \delta$, and $\cos \beta$, regardless of the tilt of the photograph, may be found with respect to focal length and photo image coordinates using the following similar analytic geometry equations:

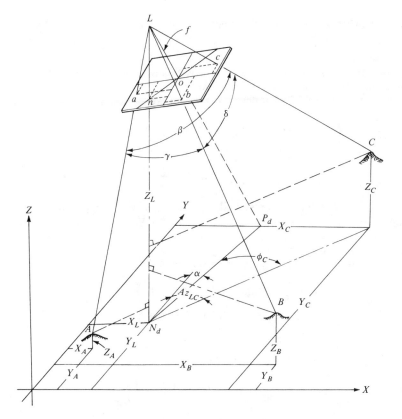

Figure 11-10 Control configuration for determining the elements of exterior orientation.

$$\cos \gamma = \frac{x_a x_b + y_a y_b + f^2}{(La)(Lb)}$$

$$\cos \delta = \frac{x_b x_c + y_b y_c + f^2}{(Lb)(Lc)} \qquad (11\text{-}9)$$

$$\cos \beta = \frac{x_c x_a + y_c y_a + f^2}{(Lc)(La)}$$

In Eqs. (11-9), the x's and y's are measured coordinates of control point images a, b, and c, f is focal length, and La, Lb, and Lc are lengths from the camera incident nodal point (exposure station) to the respective image points. The lengths are calculated as $La = \sqrt{x_a^2 + y_a^2 + f^2}$, etc., and are in the same units as photocoordinates and focal length.

In the three Eqs. (11-8), having calculated $\cos \gamma$, $\cos \delta$, and $\cos \beta$ by Eqs. (11-9), there are only three remaining unknowns: X_L, Y_L, and Z_L. The equations may therefore be solved simultaneously to obtain the space position of the tilted photograph. Equations (11-8) are nonlinear, however, and therefore must be linearized. In this presentation Taylor's theorem is used for linearization. In using Taylor's theorem, initial estimates X_{L_o}, Y_{L_o}, and Z_{L_o} are made for the unknown exposure station coordinates, and corrections dX_L, dY_L, and dZ_L to be added to these initial estimates are calculated. (Taylor's theorem for linearizing nonlinear equations may be found in any standard textbook on differential calculus.) Applying Taylor's theorem to Eqs. (11-8), and rearranging, the following three linearized equations are obtained:

$$K_1 = a_{11}dX_L + a_{12}dY_L + a_{13}dZ_L$$

$$K_2 = a_{21}dX_L + a_{22}dY_L + a_{23}dZ_L \qquad (11\text{-}10)$$

$$K_3 = a_{31}dX_L + a_{32}dY_L + a_{33}dZ_L$$

In Eqs. (11-10), the a's (coefficients) and K's (constant terms) are defined as follows (zero subscripts signify that the estimated values for X_L, Y_L, and Z_L are used in the computation):

$$a_{11} = A_1(X_L - X_B)_o + B_1(X_L - X_A)_o$$

$$a_{12} = A_1(Y_L - Y_B)_o + B_1(Y_L - Y_A)_o$$

$$a_{13} = A_1(Z_L - Z_B)_o + B_1(Z_L - Z_A)_o$$

where

$$A_1 = \left[1 - \left(\frac{LA}{LB}\right)\cos\gamma\right]_o \quad \text{and} \quad B_1 = \left[1 - \left(\frac{LB}{LA}\right)\cos\gamma\right]_o$$

$$a_{21} = A_2(X_L - X_C)_o + B_2(X_L - X_B)_o$$

$$a_{22} = A_2(Y_L - Y_C)_o + B_2(Y_L - Y_B)_o$$

$$a_{23} = A_2(Z_L - Z_C)_o + B_2(Z_L - Z_B)_o$$

where

$$A_2 = \left[1 - \left(\frac{LB}{LC}\right)\cos\delta\right]_o \quad \text{and} \quad B_2 = \left[1 - \left(\frac{LC}{LB}\right)\cos\delta\right]_o$$

$$a_{31} = A_3(X_L - X_A)_o + B_3(X_L - X_C)_o$$

$$a_{32} = A_3(Y_L - Y_A)_o + B_3(Y_L - Y_C)_o$$

$$a_{33} = A_3(Z_L - Z_A)_o + B_3(Z_L - Z_C)_o$$

where

$$A_3 = \left[1 - \left(\frac{LC}{LA}\right)\cos\beta\right]_o \quad \text{and} \quad B_3 = \left[1 - \left(\frac{LA}{LC}\right)\cos\beta\right]_o$$

and

$$K_1 = [(LA)(LB) \cos \gamma - (X_L - X_A)(X_L - X_B) - (Y_L - Y_A)(Y_L - Y_B)$$
$$- (Z_L - Z_A)(Z_L - Z_B)]_o$$

$$K_2 = [(LB)(LC) \cos \delta - (X_L - X_B)(X_L - X_C) - (Y_L - Y_B)(Y_L - Y_C)$$
$$- (Z_L - Z_B)(Z_L - Z_C)]_o$$

$$K_3 = [(LC)(LA) \cos \beta - (X_L - X_C)(X_L - X_A) - (Y_L - Y_C)(Y_L - Y_A)$$
$$- (Z_L - Z_C)(Z_L - Z_A)]_o$$

Equations (11-10) are solved to obtain corrections dX_L, dY_L, and dZ_L. These corrections are added to X_{L_o}, Y_{L_o}, and Z_{L_o} to obtain improved estimates, and the procedure is repeated to determine new corrections to be added to the improved estimates. Iterations are continued until the corrections become negligible. The closer the initial estimates are to their true values, the faster a solution will be achieved. Usually if the estimates are carefully chosen, only two or three iterations will be necessary. If the photo is nearly vertical, initial estimates for X_L, Y_L, and Z_L are readily obtained as explained in Sec. 11-12.

After the first part of the Church method has been completed, the remaining angular elements are readily determined. Expressions for the cosines of angles nLa, nLb, and nLc of Fig. 11-10, in terms of photographic nadir point coordinates x_n and y_n, photocoordinates of control points, and the camera focal length are

$$(a)\ \cos nLa = \frac{x_a x_n + y_a y_n + f^2}{(La)(Ln)}$$

$$(b)\ \cos nLb = \frac{x_b x_n + y_b y_n + f^2}{(Lb)(Ln)} \qquad (11\text{-}11)$$

$$(c)\ \cos nLc = \frac{x_c x_n + y_c y_n + f^2}{(Lc)(Ln)}$$

[Note that Eqs. (11-8) (11-9), and (11-11) are similar and are derived from the same analytic geometry equation for the cosine of an angle between two lines in space.] With X_L, Y_L, and Z_L known, explicit values for the cosines of angles nLa, nLb, and nLc may be calculated as follows (see Fig. 11-10):

$$\cos nLa = \frac{Z_L - Z_A}{LA}$$

$$\cos nLb = \frac{Z_L - Z_B}{LB} \qquad (11\text{-}12)$$

$$\cos nLc = \frac{Z_L - Z_C}{LC}$$

Rearranging Eqs. (11-11), eliminating Ln by equating (11-11a) to (11-11b), and equating (11-11b) to (11-11c), the following two equations result:

$$\left[\frac{x_a}{(La) \cos nLa} - \frac{x_b}{(Lb) \cos nLb} \right] x_n$$

$$+ \left[\frac{y_a}{(La) \cos nLa} - \frac{y_b}{(Lb) \cos nLb} \right] y_n$$

$$= \frac{f^2}{(Lb) \cos nLb} - \frac{f^2}{(La) \cos nLa} \tag{11-13}$$

$$\left[\frac{x_b}{(Lb) \cos nLb} - \frac{x_c}{(Lc) \cos nLc} \right] x_n$$

$$+ \left[\frac{y_b}{(Lb) \cos nLb} - \frac{y_c}{(Lc) \cos nLc} \right] y_n$$

$$= \frac{f^2}{(Lc) \cos nLc} - \frac{f^2}{(Lb) \cos nLb}$$

In Eqs. (11-13) there are only two unknowns, x_n and y_n, and they may be determined by solving the equations simultaneously. Then, as can be seen from Fig. 11-11, the tilt angle is

$$t = \tan^{-1} \left(\frac{\sqrt{x_n^2 + y_n^2}}{f} \right) \tag{11-14}$$

Also, with the nadir point located in the quadrant shown in Fig. 11-11 the swing angle may be calculated as

$$s = 180° + \tan^{-1} \left(\frac{x_n}{y_n} \right) \tag{11-15}$$

It should be noted that Eq. (11-15) is not general, and the student must consider the quadrant in which the nadir point falls to determine swing properly. The final element of angular orientation is the azimuth of the principal line. In Fig. 11-11, plane kwc is constructed horizontal, therefore line cw is perpendicular to the principal line and is equal to x'_c. Also lines wk and ck are perpendicular to vertical line Ln. Horizontal angle ϕ_c is calculated as

$$\phi_c = \tan^{-1} \left(\frac{x'_c}{y'_c \cos t} \right) \tag{11-16}$$

In Eq. (11-16), x'_c and y'_c are photocoordinates in the auxiliary tilted photo-coordinate system as computed by Eqs. (11-2). From ground coordinates of the exposure station and of control point C, the azimuth of the ground line LC (see Fig. 11-10) is

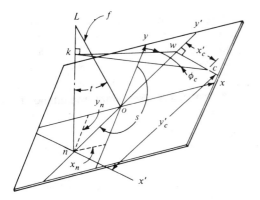

Figure 11-11 Tilt, swing, and azimuth by the Church method.

$$Az_{LC} = \tan^{-1} \left(\frac{X_C - X_L}{Y_C - Y_L} \right) \tag{11-17}$$

Finally the azimuth of the principal line, α, is calculated as:

$$\alpha = Az_{LC} - \phi_c \tag{11-18}$$

(Note horizontal angle ϕ_c in the datum plane of Fig. 11-10.)

By using another control point, a check can be obtained for the azimuth of the principal line. As a final note on the Church method, correct algebraic signs must be applied to all photocoordinates and ground coordinates used in the equations.

11-12 SPACE RESECTION BY COLLINEARITY

The method of space resection by collinearity is a purely numerical method that simultaneously yields all six elements of exterior orientation. Normally the angular values of omega, phi, and kappa are obtained in the solution, although the method is versatile and tilt, swing, and azimuth could also be obtained. Space resection by collinearity permits the use of redundant amounts of ground control; hence least squares computational techniques can be used to determine most probable values for the six elements. The procedure, although computationally lengthy, is readily programmed for computer solution so that the calculations are routinely performed. Space resection by collinearity is the preferred method of determining the elements of exterior orientation.

Space resection by collinearity involves formulating the so-called *collinearity equations* for a number of control points whose X, Y, and Z ground coordinates are known and whose images appear in the tilted photo. The equations are then solved for the six unknown elements of exterior orientation which appear in them. The collinearity equations express the condition that for any given photograph the exposure station, any object point and its corresponding image all lie on a straight line. Figure 11-10 illustrates the collinearity condition that exists for each of the three ground control points A, B, and C.

The mathematical development of the collinearity condition equations is given in detail in Appendix C. The basic equations are nonlinear and are linearized using Taylor's theorem. Collinearity equations of the linearized form [Eqs. (C-11) and (C-12) of Appendix C] can be written for each ground control point. Note, however, that since the object point coordinates of the control points are known, the unknowns dX_A, dY_A, and dZ_A drop out of Eqs. (C-11) and (C-12); hence the number of unknowns reduces to six, namely, corrections $d\omega$, $d\phi$, $d\kappa$, dX_L, dY_L, and dZ_L. These are corrections to be applied to initial approximations for omega, phi, kappa, X_L, Y_L, and Z_L respectively. The linearized forms of the space resection collinearity equations for a point A are

$$v_{x_a} = b_{11}d\omega + b_{12}d\phi + b_{13}d\kappa - b_{14}dX_L - b_{15}dY_L - b_{16}dZ_L + J$$

$$(11\text{-}19)$$

$$v_{y_a} = b_{21}d\omega + b_{22}d\phi + b_{23}d\kappa - b_{24}dX_L - b_{25}dY_L - b_{26}dZ_L + K$$

$$(11\text{-}20)$$

In Eqs. (11-19) and (11-20), the terms are as defined in Sec. C-4 of Appendix C. With two equations possible for each control point, a total of six equations are obtained from three control points. This system of equations is solved simultaneously for the six unknown corrections, and these corrections are then added to the initial values to obtain revised values. The solution is made again using the revised values as initial estimates, and new corrections are calculated. This procedure is iterated until the magnitudes of the corrections become negligible.

If more than three control points are available, then more than six equations can be formulated and their solution can be obtained by the method of least squares. If only three control points are available, the six equations that result afford a unique solution and therefore the residuals on the left-hand side of the equations, v_{x_a} and v_{y_a}, are zero.

After solving the collinearity equations iteratively and obtaining values for omega, phi, and kappa, values for tilt, swing, and azimuth (if they are desired) may be obtained. The procedure for this conversion from omega, phi, and kappa to tilt, swing, and azimuth (or vice versa) is outlined in Sec. C-8 of Appendix C.

The use of Taylor's theorem in solving the collinearity equations requires initial approximations for all unknown elements of exterior orientation. These are easily obtained if the photos are nearly vertical, because initial values of zero may be taken for omega and phi. By plotting the direction of ground north on the photo, kappa may be estimated with sufficient accuracy by simply scaling the angle from ground north to the positive y photo axis. A counterclockwise angle is considered positive. In Fig. 11-12, for example, on the basis of ground coordinates of control points A and C, the direction of north is established and plotted from the image a of control point A. A line parallel to the y photo axis is also drawn from point a and angle kappa (counterclockwise and therefore positive) is measured. Usually a solution can be obtained if the initial approximations for omega, phi, and kappa are within about 10°. It should be noted that corrections $d\omega$, $d\phi$, and $d\kappa$ in Eqs. (11-19) and (11-20) are in radian

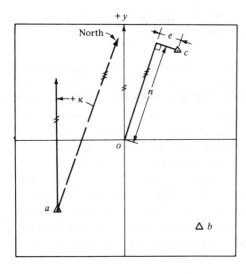

Figure 11-12 Graphical determination of intitial approximations for determining the elements of exterior orientation.

units and that initial values for omega, phi, and kappa must therefore also be in radians.

Initial values may be obtained for X_L and Y_L by scaling photo distances e (perpendicular to north) and n (parallel to north) from the principal point to a control point image such as c of Fig. 11-12. Then by multiplying e and n by average photo scale, incremental ground distances ΔX and ΔY are obtained as

$$\Delta X = eS_{avg} \tag{11-21}$$

$$\Delta Y = nS_{avg}$$

In Eqs. (11-21), if e and n are in inch units and S_{avg} is feet per inch, then ΔX and ΔY are in feet. By subtracting ΔX and ΔY from the X and Y ground coordinates of control point C, satisfactory values are obtained for X_L and Y_L, or

$$X_L = X_C - \Delta X \tag{11-22}$$

$$Y_L = Y_C - \Delta Y$$

An initial value for Z_L can be obtained by assuming a vertical photograph and solving the following modification of the scale equation:

$$H = \frac{AC}{ac} f + h_{AC} \tag{11-23}$$

In Eq. (11-23), H (or Z_L) is flying height above datum, AC is the horizontal length of the ground control line, ac is its corresponding photo length, f is camera focal length, and h_{AC} is the average elevation of control points A and C.

11-13 POSITIONS OF NEW POINTS FROM MEASUREMENTS ON STEREOPAIRS OF TILTED PHOTOS

The collinearity equations may also be used to determine X, Y, and Z ground coordinates of new points whose images appear in the overlap area of a stereopair of tilted photos. The procedure is known as *space intersection,* so called because corresponding rays to the same object point from two overlapping photos must intersect at the point, as shown in Fig. 11-13. Space intersection requires that the six elements of exterior orientation for the two overlapping tilted photos be known. These elements may be calculated by methods described in Sec. 11-12. Collinearity equations of the linearized form [Eqs. (C-11) and (C-12) of Appendix C] can be written for each new point, such as point E of Fig. 11-13. Note, however, that since the six elements of exterior orientation are known, the only remaining unknowns in Eqs. (C-11) and (C-12) are dX_E, dY_E, and dZ_E. These are corrections to be applied to initial approximations for object space coordinates X_E, Y_E, and Z_E respectively for ground point E. The linearized forms of the space intersection equations for point E are

$$v_{x_e} = b_{14}dX_E + b_{15}dY_E + b_{16}dZ_E + J \qquad (11\text{-}24)$$

$$v_{y_e} = b_{24}dX_E + b_{25}dY_E + b_{26}dZ_E + K \qquad (11\text{-}25)$$

In Eqs. (11-24) and (11-25), the terms are as defined in Sec. C-4 of Appendix C. Two equations [of the form of (11-24) and (11-25)] can be written for point e_1 of the left photo and two more for point e_2 of the right photo; hence four equations result, and the values of dX_E, dY_E, and dZ_E can be computed in a least squares solution. The corrections are applied to the initial approximations to obtain revised values for X_E, Y_E, and Z_E. The solution is then repeated until the magnitudes of the corrections become negligible.

Again because the equations are linearized using Taylor's theorem, initial approximations are required for each new point. Initial X and Y coordinates can be found

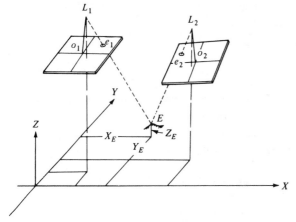

Figure 11-13 Space intersection with a stereopair of tilted photographs.

satisfactorily using Eqs. (11-21) and (11-22) together with the procedures outlined in Sec. 11-12 for finding initial values for X_L and Y_L. An estimate for Z coordinates of new points can be made by assuming truly vertical photos and using the parallax equations described in Chap. 8.

The solution of nonlinear equations, and thus the use of Taylor's theorem which requires initial approximations for X_E, Y_E, and Z_E, can be avoided in an alternate method of computation. To develop this procedure, Eqs. (a), (b), and (c) of Sec. C-3 of Appendix C may be rewritten as follows:

$$X_A - X_L = \frac{Z_A - Z_L}{z_a'} x_a'$$

$$Y_A - Y_L = \frac{Z_A - Z_L}{z_a'} y_a' \qquad (11\text{-}26)$$

$$Z_A - Z_L = \frac{Z_A - Z_L}{z_a'} z_a'$$

Substituting λ_a for the term $(Z_A - Z_L)/z_a'$, which is common to each of Eqs. (11-26), and changing subscripts from A to E to represent the unknown point in the overlap area of photos 1 and 2 of Fig. 11-13, the following expressions can be written for photo 1:

$$X_E = \lambda_{e_1} x_{e_1}' + X_{L_1} \qquad (11\text{-}27)$$

$$Y_E = \lambda_{e_1} y_{e_1}' + Y_{L_1} \qquad (11\text{-}28)$$

$$Z_E = \lambda_{e_1} z_{e_1}' + Z_{L_1} \qquad (11\text{-}29)$$

Similar expressions can be written for photo 2 as

$$X_E = \lambda_{e_2} x_{e_2}' + X_{L_2} \qquad (11\text{-}30)$$

$$Y_E = \lambda_{e_2} y_{e_2}' + Y_{L_2} \qquad (11\text{-}31)$$

$$Z_E = \lambda_{e_2} z_{e_2}' + Z_{L_2} \qquad (11\text{-}32)$$

Setting Eq. (11-27) equal to Eq. (11-30), and Eq. (11-28) equal to Eq. (11-31), and solving for λ_{e_2} gives

$$\lambda_{e_2} = \frac{y_{e_1}' (X_{L_2} - X_{L_1}) - x_{e_1}' (Y_{L_2} - Y_{L_1})}{x_{e_1}' y_{e_2}' - x_{e_2}' y_{e_1}'} \qquad (11\text{-}33)$$

Since the six elements of exterior orientation for both photos 1 and 2 have already been determined by methods described in the preceding sections, the values of X_{L_1}, Y_{L_1}, Z_{L_1}, X_{L_2}, Y_{L_2}, and Z_{L_2} are known, and x_{e_1}', y_{e_1}', z_{e_1}', x_{e_2}', y_{e_2}', and z_{e_2}' can be calculated from the angular orientation elements and measured photo coordinates of point e using Eqs. (B-26) of Appendix B. Also, λ_{e_2} can be calculated from Eq.

(11-33). Finally, Eqs. (11-30) through (11-32) can be solved directly for the unknowns X_E, Y_E, and Z_E. As a check, λ_{e_1} could be calculated in a manner similar to that used to obtain λ_{e_2}, and then Eqs. (11-27) through (11-29) could be solved directly for X_E, Y_E, and Z_E. In this check solution, λ_{e_1} has the following value:

$$\lambda_{e_1} = \frac{y'_{e_2}(X_{L_2} - X_{L_1}) - x'_{e_2}(Y_{L_2} - Y_{L_1})}{x'_{e_1}y'_{e_2} - x'_{e_2}y'_{e_1}} \tag{11-34}$$

The disadvantage of using this alternative procedure rather than Eqs. (11-24) and (11-25) is that the redundancy which exists is not utilized in a least squares solution. This alternate procedure will, however, provide very good initial approximations for solving Eqs. (11-24) and (11-25).

11-14 RECTIFICATION OF TILTED PHOTOGRAPHS

Rectification is the process of making equivalent vertical photographs (see Sec. 11-7) from tilted photo negatives. The resulting equivalent vertical photos are called *rectified photos*. Rectified photos theoretically are truly vertical photos, and as such they are free from tilt displacements. They do, however, still contain image displacements due to topographic relief. These relief displacements can also be removed in a process called *differential rectification* or *orthorectification,* and the resulting products are then called *orthophotos*. Orthophotos are often preferred over rectified photos for this reason. Nevertheless, rectified photos are still quite popular because they do make very good map substitutes where terrain variations are moderate. This chapter discusses only the methods of rectification for the removal of tilt displacements. Orthophoto production is described in Chap. 13.

Rectification can be performed in any of four basic ways: (1) graphically, (2) analytically, (3) optically-mechanically, and (4) electro-optically. The first two procedures have the disadvantage that they can be applied only to individual discrete points, i.e., points which can be specifically identified so that their locations within the tilted photo can be measured. The resulting rectified photos produced by these methods are not really photos at all since they are not composed of photo images. Rather, they are plots of individual points in their rectified locations. The optical-mechanical and electro-optical methods produce an actual continuous tone photograph wherein the images of the tilted photo have been transformed to their rectified locations. Thus, the products of these two methods can be used in the production of photomaps and mosaics. In any of these rectification procedures, the rectified photos can be simultaneously *ratioed*; i.e., their average scales can be brought to some desired value different from that of the original photo. This is particularly advantageous if rectified photos are being made for the purpose of constructing a controlled mosaic, since all photos in the strip or block can be brought to a common scale. Thus, the resulting mosaic will have a more uniform scale throughout.

11-15 GEOMETRY OF RECTIFICATION

The fundamental geometry of rectification is illustrated in Fig. 11-14. This figure shows a side view of the principal plane of a tilted photo. When the exposure was made, the negative plane made an angle t with the datum plane. Rays from A and B were imaged at a' and b', respectively, on the negative, and their corresponding locations on the tilted photo are at a and b. The plane of an equivalent vertical photo is shown parallel to the datum plane and passing through i, the isocenter of the tilted photo. The plane of a ratioed rectified photo is also shown. It is likewise parallel to the datum plane but exists at a level other than that of the equivalent vertical photo plane.

Methods of projecting points such as a and b to either a'' and b'' or a''' and b''' are subjects of the remaining sections of this chapter. Figure 11-14 also illustrates, by virtue of lines LA' and LB', that although tilt displacements are removed in rectification, displacements due to relief are still present.

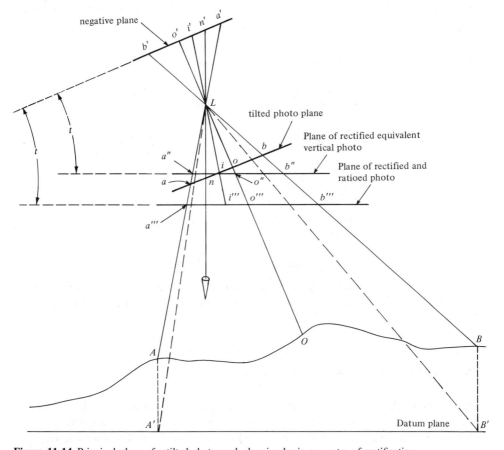

Figure 11-14 Principal plane of a tilted photograph showing basic geometry of rectification.

11-16 THEORY OF THE ANHARMONIC OR CROSS RATIO

Graphical rectification procedures are based upon the *anharmonic* or *cross ratio* of perspective geometry. This ratio applies to photographs because they are perspective projections; i.e., all light rays entering the camera at the instant of exposure pass through a common point—the emergent nodal point of the lens. This point is the perspective center of the projection.

Figure 11-15 illustrates the principle of the cross ratio. In the figure, rays passing through plane 1 at *A, B, C,* and *D* intersect at *L*, the perspective center. These same rays pass through plane 2 at *a, b, c,* and *d,* respectively. By definition, the cross ratio *R* for this perspective projection is

$$R = \frac{AC}{BC} \times \frac{BD}{AD} = \frac{ac}{bc} \times \frac{bd}{ad} \tag{11-35}$$

From Eq. (11-35) it is seen that the ratios of distances in one plane of a perspective projection are exactly the same as the corresponding distance ratios in another plane of the projection. This can be proven by applying the law of sines to Fig. 11-15 as follows:

$$AC = CL \frac{\sin \alpha}{\sin \theta} \quad \text{and} \quad ac = cL \frac{\sin \alpha}{\sin \psi}$$

$$AD = DL \frac{\sin \beta}{\sin \theta} \quad \text{and} \quad ad = dL \frac{\sin \beta}{\sin \psi} \tag{11-36}$$

$$BD = DL \frac{\sin \gamma}{\sin \phi} \quad \text{and} \quad bd = dL \frac{\sin \gamma}{\sin \xi}$$

$$BC = CL \frac{\sin \delta}{\sin \phi} \quad \text{and} \quad bc = cL \frac{\sin \delta}{\sin \xi}$$

Substituting Eqs. (11-36) into Eq. (11-35) and reducing,

$$R = \frac{\sin \alpha \sin \gamma}{\sin \delta \sin \beta} = \frac{\sin \alpha \sin \gamma}{\sin \delta \sin \beta} \tag{11-37}$$

From Eq. (11-37) not only has it been shown that the cross-ratio expression of Eq. (11-35) is true but also it is seen that it is based strictly upon the angles between the rays.

11-17 GRAPHICAL RECTIFICATION

The theory of the cross ratio is the basis for graphical rectification. In this instance, consider plane 1 of Fig. 11-15 to be the plane of the rectified or equivalent vertical photograph and plane 2 as that of the tilted photo. Figure 11-16*a* is a plan view of the plane of the tilted photograph whereupon the images of four control points, *a, b,*

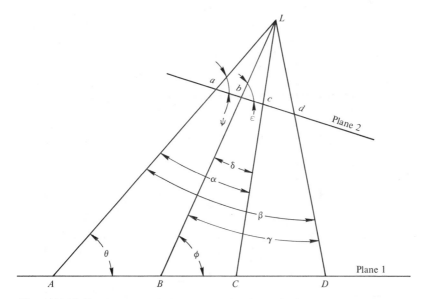

Figure 11-15 The cross ratio of planes in a perspective projection.

c, and d, appear. Point p, whose rectified location is desired, is also shown. Figure 11-16b shows a plan view of the plane of the rectified and ratioed photograph upon which the four control points A, B, C, and D have been plotted to the desired final scale of the rectified photo based upon their ground coordinates.

In Fig. 11-16a, a perspective center is chosen at point a, and rays are drawn from this point to b, c, d, and p. A paper strip (labeled paper strip 1) is laid across the rays, as shown in Fig. 11-16a, and marks are made on the strip at the locations where the various rays cross it. (The placement of the paper strip is quite arbitrary.) Because the tilted and rectified photos are both perspective projections, the cross ratio exists between the rays from point a of Fig. 11-16a and the corresponding rays from point A of Fig. 11-16b. Thus, the paper strip is laid over Fig. 11-16b and adjusted so that the various ray marks from a to points b, c, and d coincide with their corresponding rays from point A. In this location the ray indicating the direction to unknown point P is drawn (shown dashed). Next a perspective center is chosen at point b of the tilted photo (see Fig. 11-16c). A paper strip (labeled paper strip 2) is laid across the various rays from this point and the locations where the rays cross it are marked. This strip is also laid on the rectified photo (see Fig. 11-16d) such that the paper-strip marks coincide with the corresponding locations of rays from point B. In this location the ray from B to point P can be drawn (also shown dashed). The rectified location of point P exists at the intersection of its two rays from A and B. Its location could be checked by repeating the procedure using point d, for example, as a perspective center. This procedure of graphical rectification is commonly referred to as the *paper-strip* method.

The obvious drawback of this procedure is that it is painstaking, and only discrete points can be rectified, one by one. This latter problem can be somewhat overcome

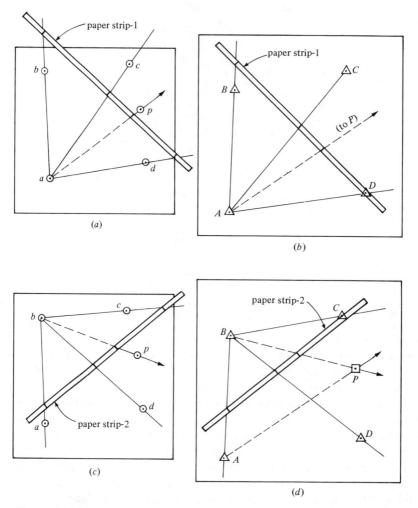

Figure 11-16 Graphical rectification by paper strip method. (*a*) Preparing paper strip for rays from *a* on tilted photo. (*b*) Placement of paper strip on rectified photo in accordance with rays from *A*. (*c*) Preparing paper strip for rays from *b* on tilted photo. (*d*) Placement of paper strip on rectified photo in accordance with rays from *B*.

by superimposing a grid onto the tilted photo and then, using the above outlined procedures, determining the rectified locations of the grid points. With these located, detail within each square can be manually traced onto the rectified photo, maintaining the general relationships to the respective grids of detail traced. With this method, the denser the grid, the greater the accuracy that can be achieved.

For the above described graphical rectification procedure to be completely rigorous, the plotted control points on the rectified and ratioed photo plane must be corrected for their relief displacements. Figure 11-17 illustrates the manner of making these corrections if the photo is not extremely tilted. The control points are first plotted

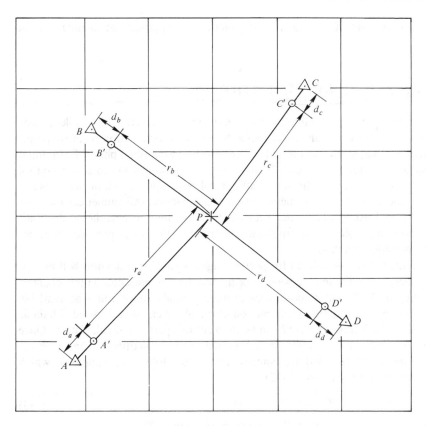

Figure 11-17 Plot of control points for rectification showing corrections made for relief displacements.

at A', B', C', and D', based upon their ground coordinates, at the desired final scale of the rectified photo. The ground principal point of the photo must also be plotted, and this can be done by three-point resection as described in Sec. 9-8. (Theoretically the ground nadir point should be located, but its position is unknown unless a computation for tilt and swing is made.) Radial distances r_a, r_b, r_c, and r_d from the ground principal point to each plotted control point are measured, and based upon their elevations h_A, h_B, h_C, and h_D their radial relief displacements d_a, d_b, d_c, and d_d are calculated using the following equation:

$$d = \frac{rh}{H - h} \qquad (11\text{-}38)$$

In Eq. (11-38), H is flying height above datum and d is the radial relief displacement. Displacements calculated by Eq. (11-38) are in inches or millimeters (the same units as the r values) and are laid off outward from the principal point (nadir point) to locate points A, B, C, and D. If control points are not plotted in this manner, a quasi-rectification will result which generally produces a very good approximation, especially if the elevation differences of control points are not severe and if flying height is relatively high. Consequently, in practice this approximate method is most often performed.

11-18 ANALYTICAL RECTIFICATION

There are several methods available for performing analytical rectification. Like graphical techniques, each of the numerical methods performs rectification point by point, and each requires that sufficient ground control appear in the tilted photo. Basic input required for all the numerical methods, in addition to ground coordinates of control points, are x and y photo coordinates of all control points plus those of the points to be rectified. These are normally measured on a comparator. All numerical methods are best performed using a computer because of the volume of calculation entailed. Owing to errors in graphical construction, numerical methods can provide better accuracy than graphical methods.

In this section two procedures for performing analytical rectification will be discussed. The first is simply an extension of principles already presented in this chapter. It requires that the X, Y, and Z coordinates of three ground control points be available so that the six elements of exterior orientation of the photo can be computed. Methods outlined in Secs. 11-9 through 11-12 can be used for this part of the calculation. Once the elements of exterior orientation have been determined, auxiliary x' and y' coordinates of points to be rectified are computed using the following equations, which are a slight modification of Eqs. (11-2):

$$x' = x \cos \theta - y \sin \theta \tag{11-39}$$

$$y' = x \sin \theta + y \cos \theta + f \tan (t/2)$$

In Eq. (11-39) the angle θ is related to the swing angle according to Eq. (11-1). Note that the only difference between Eqs. (11-39) and Eqs. (11-2) is the translation term $f \tan t/2$, which is the distance oi from the principal point to the isocenter. Thus coordinates calculated by Eqs. (11-39) have their origin at the isocenter, with the y' axis coinciding with the principal line and the x' axis coinciding with the axis of tilt.

From the $x'y'$ coordinates, r_i distances and λ angles are calculated (see Sec. 11-7), followed by the computation of tilt displacements in accordance with Eq. (11-7). The displacements, once calculated, are applied radially outward from the isocenter for points above the axis of tilt (those having positive y' coordinates), and radially inward for points below the axis of tilt (those having negative y' coordinates). This locates the rectified positions of the points.

A second method of analytical rectification utilizes the two-dimensional projective transformation equations. These are developed as Eqs. (B-36) in Appendix B. They are repeated here for convenience, as follows:

$$X = \frac{a_1 x + b_1 y + c_1}{a_3 x + b_3 y + 1}$$

$$Y = \frac{a_2 x + b_2 y + c_2}{a_3 x + b_3 y + 1} \tag{11-40}$$

In Eqs. (11-40), X and Y are ground coordinates, x and y are photo coordinates (in the fiducial axis system), and the a's, b's, and c's are eight parameters of the transformation. The use of these equations to perform analytical rectification is a two-step process. First, a pair of Eqs. (11-40) are written for each ground control point. Four control points will produce eight equations, so that a unique solution can be made for the eight unknown parameters. If more than four control points are available, more than eight equations will be possible, and an improved solution can be made using least squares.

Once the eight parameters have been determined, the second step of the solution can be performed—that of solving Eqs. (11-40) for each point whose X and Y rectified coordinates are desired. After rectified coordinates have been computed in the ground coordinates system, they can be plotted at the scale desired for the rectified and ratioed photo.

Like the graphical solution, this second analytical method is not completely rigorous unless ground coordinates of control points are also modified for relief displacements in accordance with Eq. (11-38). Otherwise, a quasi-rectification results.

11-19 OPTICAL-MECHANICAL RECTIFICATION

In practice, the optical-mechanical method of rectification is the most widely used. This method relies on instruments called *rectifiers*. They produce rectified and ratioed photos through the photographic process of projection printing (see Sec. 3-12); thus, they must be operated in a darkroom.

As illustrated in Fig. 11-18, the basic components of a rectifier consist of a lens, a light source with reflector, a stage for mounting the tilted photo negative, and an easel which holds the photographic emulsion upon which the rectified photo is exposed. The instrument is constructed with controls so that the *easel plane, lens plane* (plane perpendicular to the optical axis of the rectifier lens), and *negative plane* can be tilted with respect to each other. Thus provision is made for varying angles α and β of Fig. 11-18. With some rectifiers the lens and easel planes can be tilted to vary these angles; however, most of these instruments do not have the capability of tilting the lens plane. Rather, the α and β angles are varied by tilting the negative and easel planes. This type of instrument is referred to as a *nontilting lens* rectifier.

So that rectified photos can be ratioed to varying scales, the rectifier must also have a magnification capability, and this is achieved by varying the projection distance (distance LE of Fig. 11-18 from the lens to the easel plane). To do this, however, and still maintain proper focus in accordance with Eq. (2-8), it is necessary to simultaneously vary the image distance (distance Le of Fig. 11-18 from the lens to the negative plane). The actual magnification that results along the axis of the rectifier lens is the ratio LE/Le, but it varies elsewhere in the photo due to the variable scale of the tilted photograph.

From Fig. 11-18 it can be see that in rectification, projection distances vary depending upon the locations of points in the photo. To achieve sharp focus for all images in spite of this, the *Scheimpflug condition* (see Sec. 2-8 and Fig. 2-16) must be satisfied. The Scheimpflug condition states that, in projecting images through a

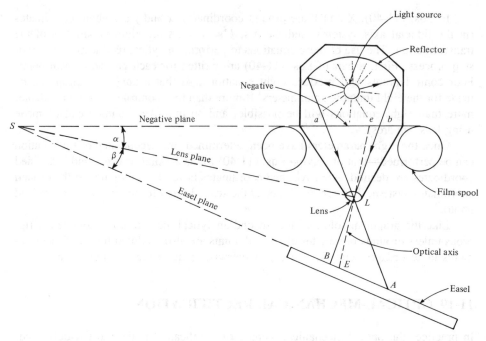

Figure 11-18 Schematic diagram of an optical-mechanical rectifier showing a side view of the principal plane.

lens, if the negative and easel planes are not parallel, the negative plane, lens plane, and easel plane must all intersect along a common line to satisfy the lens formula and achieve sharp focus for all images. Note that this condition is satisfied in Fig. 11-18 where these planes intersect at S.

In addition to being able to vary α, β, LE, and Le, rectifiers must also allow for a rotation and shift of the negative, or a provision to tilt the easel plane in any direction and shift the negative. The rotation of the negative or tilt of the easel plane is necessary to place the principal plane of the photograph perpendicular to the easel plane of the rectifier, and the shift moves the principal point of the negative a specified distance away from the optical axis of the rectifier lens along the principal line. The amount of shift necessary as well as the values of α, β, LE, and Le are functions of the tilt angle, the magnification ratio, and the focal lengths of the rectifier lens and lens of the camera used to obtain the negative.

In practice, different techniques are employed to arrive at the proper rectifier settings. One method is to calculate numerically the elements of exterior orientation for each photo, and from these values compute the required settings. Another possible method is to orient the photos in a stereoplotter, read the elements of exterior orientation, and from those values calculate the rectifier settings. This procedure requires a stereoplotter equipped with dials for reading the orientation elements, however.

The equations for calculating the settings for a nontilting lens rectifier are

$$\alpha = \sin^{-1}\left(\frac{F}{Mf}\sin t\right) \tag{11-41}$$

$$\beta = \sin^{-1}\left(\frac{F}{f}\sin t\right) \tag{11-42}$$

$$LE = \frac{F\sin(\alpha + \beta)}{\sin\alpha\cos\beta} \tag{11-43}$$

$$Le = \frac{F\sin(\alpha + \beta)}{\cos\alpha\sin\beta} \tag{11-44}$$

$$d = F\left[\sqrt{1 - \left(\frac{f}{F}\right)^2 + \cot^2\beta} - \sqrt{\left(\frac{1}{M}\right)^2 + \cot^2\beta}\right] \tag{11-45}$$

In the above equations, F is the focal length of the rectifier lens, f is the focal length of the camera used to expose the tilted negative, t is the tilt angle, and M is the desired magnification from tilted photo scale to rectified ratioed photo scale. The value d in Eq. (11-45) is the required shift of the principal point of the negative along the principal line away from the rectifier lens axis. If the algebraic sign of d is positive the shift is upward along the principal line; if negative, it is downward.

Rectifier settings that are actually needed depend upon the particular rectifier. If an *automatic rectifier* is used, for example, image and object distances are automatically held in the proper ratio to satisfy Eq. (2-8), and at the same time a special mechanical device automatically satisfies the Scheimpflug condition regardless of the tilt angle or magnification ratio. With these instruments fewer computations and settings are needed. An automatic rectifier is shown in Fig. 11-19.

Rather than computing rectifier settings, for near vertical photography a trial and error method of orienting a rectifier which yields satisfactory results may be used. This is the procedure most often applied in practice. In this method four or more control points are plotted according to their ground control coordinates at the scale desired for the rectifier photo. The plot is then placed on the easel and, by trial and error, the rectifier controls are adjusted until the projected control point images coincide with their respective plotted points. This automatically creates proper orientation of the rectifier. The plot is then removed from the easel, a sheet of photographic paper is placed there, and the rectified photo is exposed. Again, for a completely rigorous rectification to be achieved using this method, the plotted control points must be corrected for their relief displacements in accordance with Eq. (11-38); otherwise, a quasi-rectification will result.

11-20 ELECTRO-OPTICAL RECTIFICATION

Rectified photos can be produced by electro-optical methods in a process which incorporates a scanning microdensitometer and an electronic computer. This procedure

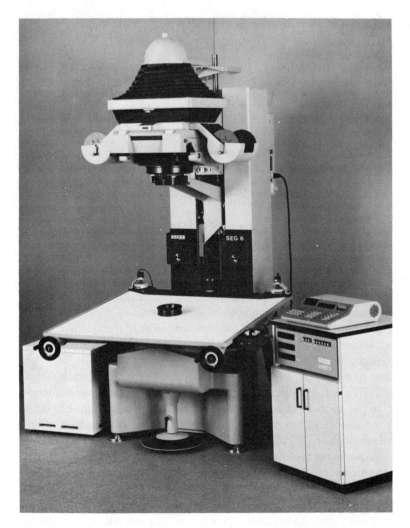

Figure 11-19 The SEG 6 automatic rectifier. (*Courtesy Carl Zeiss, Oberkochen.*)

is not widely used to produce rectified photos, but is becoming commonly used in the production of orthorectified photos (those having both tilt and relief displacements removed).

As described in Sec. 5-15, a scanning microdensitometer can record on magnetic tape, the entire contents of a photo in terms of the densities of thousands of tiny contiguous pixels. With each pixel, the scanner also records photo location. For the given photo, the elements of exterior orientation can be calculated by methods outlined in Secs. 11-9 through 11-12. With these elements known, analytical methods (see Sec. 11-18) can be applied to obtain the rectified location of each pixel. These can then be printed back onto film in their rectified positions using a *photowriter,* an instrument whose mode of operation is essentially the reverse of the scanning micro-densitometer. This procedure is described in greater detail in relation to orthophoto production in Chap. 13.

REFERENCES

American Society of Photogrammetry: "Manual of Photogrammetry," 3d ed., Falls Church, Va., 1966, chaps. 2 and 16.

―――: "Manual of Photogrammetry," 4th ed., Falls Church, Va., 1980, chaps. 2, 9, and 14.

Anderson, R. O.: Scale-Point Method of Tilt Determination, *Photogrammetric Engineering,* vol. 15, no. 2, p. 311, 1949.

Berlin, L.: The Absolute Orientation of Near Verticals, *Photogrammetric Engineering,* vol. 30, no. 6, p. 1000, 1964.

Boge, W. E.: Resection Using Iterative Least Squares, *Photogrammetric Engineering,* vol. 31, no. 4, p. 701, 1965.

Church, E., and A. Quinn: "Elements of Photogrammetry," Syracuse University Press, Syracuse, N.Y., 1944.

Clark, H. W.: The Geometry of Photorectification, *Photogrammetric Engineering,* vol. 15, no. 2, p. 288, 1949.

Estes, J. M.: The Anharmonic Method of Rectification, *Photogrammetric Engineering,* vol. 33, no. 10, p. 1171, 1967.

Gruner, H. E.: A Two Stage Rectification System, *Photogrammetric Engineering,* vol. 27, no. 4, p. 600, 1961.

Hallert, B.: Quality of Exterior Orientation, *Photogrammetric Engineering,* vol. 32, no. 3, p. 464, 1966.

Jones, A. D.: The Development of the Wild Rectifiers, *Photogrammetric Record,* vol. 2, p. 181, 1966.

Keller, M., and G. C. Tewinkel: *Space Resection in Photogrammetry,* ESSA Technical Report C&GS 32, U.S. Coast and Geodetic Survey, Washington, D.C., 1966.

Lehman, E. H., Jr.: Determining Exposure Point, Tilt, and Direction of Photograph From Three Known Ground Positions and Focal Length, *Photogrammetric Engineering,* vol. 29, no. 4, p. 702, 1963.

Levine, S. W.: A Slit-Scan Electro Optical Rectifier, *Photogrammetric Engineering,* vol. 27, no. 5, p. 740, 1961.

Mugnier, C. J.: Analytical Rectification Using Artificial Points, *Photogrammetric Engineering and Remote Sensing,* vol. 44, no. 5, p. 579, 1978.

Rosenfeld, G. H.: The Problem of Exterior Orientation in Photogrammetry, *Photogrammetric Engineering,* vol. 25, no. 4, p. 536, 1959.

Trachsel, A. F.: Electro-Optical Rectifier, *Photogrammetric Engineering,* vol. 33, no. 5, p. 513, 1967.

Wilson K. R., and J. Vlcek: Analytical Rectification, *Photogrammetric Engineering,* vol. 36, no. 6, p. 570, 1970.

PROBLEMS

11-1 A particular tilted aerial photograph exposed with a 152-mm-focal-length camera has a tilt angle of 2°45′ and a swing angle of 140°00′. On this photograph, what are the auxiliary x' and y' photocoordinates for points a and b, whose photocoordinates measured with respect to the fiducial axes are $x_a = 69.27$ mm, $y_a = -41.80$ mm, $x_b = -54.72$ mm, and $y_b = 106.38$ mm?

11-2 Repeat Prob. 11-1, except that the camera focal length is 88 mm, tilt angle is 1°55′, swing angle is 249°, $x_a = -62.41$ mm, $y_a = 76.80$ mm, $x_b = 98.55$ mm, and $y_b = -12.06$ mm.

11-3 Calculate photographic scale for image points a and b of Prob. 11-1 if flying height above datum was 7,200 ft and if elevations of points A and B were 865 ft and 1,232 ft above datum, respectively.

11-4 Calculate photographic scale for image points a and b of Prob. 11-2 if flying height above datum was 8,800 ft and if elevations of points A and B were 1,784 ft and 1,876 ft above datum, respectively.

11-5 Calculate ground length of line AB for image points a and b of Probs. 11-1 and 11-3.

11-6 Calculate ground length of line AB for image points a and b of Probs. 11-2 and 11-4.

11-7 On a tilted photo, images a, b, and c of the three corners of a triangular tract of land have the following photocoordinates measured with respect to the fiducial axes: $x_a = 42.33$ mm, $y_a = 69.85$ mm, $x_b = -50.75$ mm, $y_b = 21.73$ mm, $x_c = 16.19$ mm, and $y_c = -78.64$ mm. The tilt and swing angles of the photograph are 2°25′ and 198°, respectively. Camera focal length is 152 mm and flying height above datum is 9,650 ft. Elevations above datum of points A, B, and C are 1,265 ft, 1,047 ft, and 1,378 ft, respectively. What is the area of the tract of land in acres?

11-8 Compute the tilt displacements of the two points in Prob. 11-1.

11-9 Compute the tilt displacements of the two points in Prob. 11-2.

11-10 The images of three ground control points appear on a tilted aerial photograph taken with a camera having a focal length of 150.0 mm. Measured photocoordinates and corresponding ground control coordinates for the three points are

Point	Photocoordinates		Ground coordinates		
	x, mm	y, mm	X, ft	Y, ft	Elevation, ft
A	−100.78	71.11	1,531,367.3	500,413.2	611.7
B	106.19	76.02	1,528,225.0	501,830.2	934.4
C	2.23	−91.47	1,530,737.5	503,649.0	799.1

Using the Church method, determine the six elements of exterior orientation for the photo.

11-11 Repeat Prob. 11-10, except that the method of space resection by collinearity is to be used.

11-12 Repeat Prob. 11-10, except that the camera focal length was 152.4 mm and measured photocoordinates and corresponding ground control coordinates for the three points are

Point	Photocoordinates		Ground coordinates		
	x, mm	y, mm	X, ft	Y ft	Elevation, ft
D	61.84	6.12	1,532,353.6	501,404.4	641.1
E	−6.06	76.69	1,532,969.5	499,979.9	943.1
F	1.36	−89.87	1,533,887.8	502,412.1	860.1

11-13 Solve Prob. 11-12 using the method of space resection by collinearity.

11-14 Name and briefly describe four different methods of performing rectification. Discuss some advantages and disadvantages of the methods.

11-15 For the tilted photo coordinates (measured with respect to the fiducial axis system) and the ground coordinates given below, perform a graphical rectification to determine the rectified and ratioed coordinates of points 5 and 6. (Use a ratioed scale of 500 ft/in, and ignore elevation differences.)

Point	Tilted photo coordinates		Ground coordinates	
	x, mm	y, mm	X, ft	Y, ft
1	− 50.42	76.20	118,917.3	78,860.3
2	76.45	50.81	118,635.5	81,398.9
3	51.18	− 75.95	121,152.4	81,747.9
4	− 63.24	− 63.26	121,543.0	79,475.9
5	38.23	51.05		
6	25.53	− 35.56		

11-16 For the data of Prob. 11-15, perform an analytical rectification using Eqs. (11-40) to determine rectified coordinates of points 5 and 6.

11-17 Repeat Prob. 11-15, except that the following data applies and ratioed scale is 1,000 ft/in.

Point	Tilted photo coordinates		Ground coordinates	
	x, mm	y, mm	X, ft	Y, ft
1	− 63.48	− 50.29	879,239.4	138,346.0
2	− 38.05	89.03	877,096.9	142,679.2
3	101.85	38.10	882,328.5	144,005.4
4	50.93	− 76.45	882,809.1	139,602.3
5	50.88	− 25.41		
6	− 12.69	25.27		

11-18 For the data of Prob. 11-17, perform an analytical rectification using Eqs. (11-40) to determine the rectified coordinates of points 5 and 6.

11-19 Using Eqs. (11-41) through (11-45) calculate the settings for a nontilting lens rectifier for a photo having 10°00′ of tilt. The required magnification ratio is 4.0, and the focal lengths of the rectifier lens and camera lens were 180.00 mm and 152.40 mm, respectively.

11-20 Repeat Prob. 11-19, except that the photo tilt is 5°00′, the magnification ratio is 2.0, and the rectifier-lens focal length is 150.00 mm.

TWELVE

STEREOSCOPIC PLOTTING INSTRUMENTS

12-1 INTRODUCTION

Stereoscopic plotting instruments (commonly called stereoplotters or simply plotters) are instruments designed to provide rigorously accurate analogue solutions for object point positions from their corresponding image positions on overlapping pairs of photos. In general the optical and mechanical components of stereoplotters are manufactured to a high degree of precision, and if these instruments are properly calibrated, accurate results may be obtained from them. The fact that the photos may contain varying amounts of tilt is of no consequence in the resulting accuracy; in fact, many stereoplotters are capable of handling oblique or horizontal (terrestrial) photos. The primary use of stereoplotters is in compilation of topographic maps, and because this is currently the most widely practiced of photogrammetric applications, the subject of stereoplotters is one of the most important in the study of photogrammetry.

The basic concept underlying the design of a common type of stereoscopic plotting instruments is illustrated in Fig. 12-1. In Fig. 12-1a, an overlapping pair of aerial photos is exposed. Transparencies or *diapositives,* as they are called, carefully prepared to exacting standards from the negatives, are placed in two stereoplotter projectors as shown in Fig. 12-1b. This process is called *interior orientation.* With the diapositives in place, light rays are projected through them, and when rays from corresponding images on the left and right diapositives intersect below, they create a stereomodel (often simply called a *model*). In creating the intersections of corresponding light rays, the two projectors are oriented so that the diapositives bear the exact relative angular orientation to one another in the projectors that the negatives had in the camera at the instant they were exposed. The process is called *relative orientation* and creates, in miniature, a true three-dimensional stereomodel of the overlap area.

After relative orientation is completed, *absolute orientation* is performed. In this process the model is brought to the desired scale and leveled with respect to a reference datum. Figure 12-2 illustrates a stereomodel created by interior, relative, and absolute

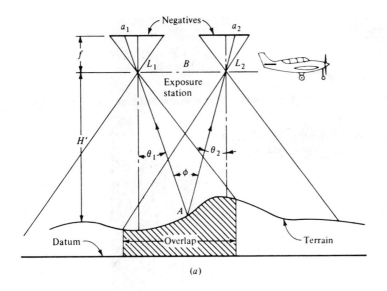

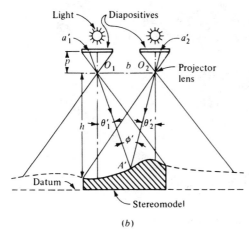

Figure 12-1 Basic concept of stereoscopic plotting instrument design. (*a*) Aerial photography; (*b*) stereoscopic plotting instrument.

orientation of a stereopair in the projectors of a Balplex plotter. To retain simplicity of the figure, the framework of the instrument has been omitted.

When orientation is completed, measurements of the model may be made and recorded either graphically or digitally, depending upon the particular plotter and project. In either case, the position of any point is determined by bringing a reference mark into contact with the model point. For the plotter of Fig. 12-2, a reference mark in the center of the *platen* (white disk mounted on the tracing table) is brought into coincidence with model points. Planimetric positions of points are plotted by means

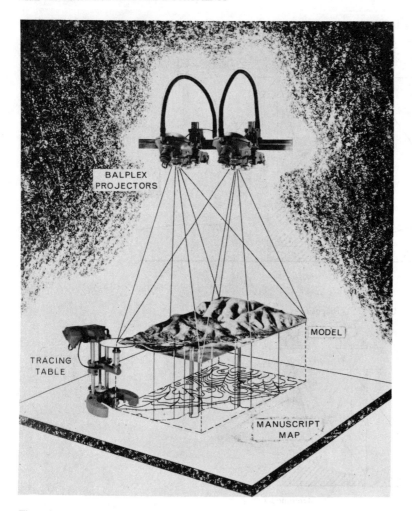

Figure 12-2 Stereomodel created with a pair of Balplex projectors.

of a pencil located vertically beneath the reference mark, and elevations are read directly from a dial which records up and down motion of the platen; the dial having been indexed to ground control during absolute orientation. Contours and other details may be traced directly, as shown on the manuscript map of Fig. 12-2, by moving the tracing table about while keeping the reference mark in contact with the model.

From an analysis of the preceding discussion, it should be apparent that stereo-plotters combine three distinct systems: (1) a *projection* system, which creates the true three-dimensional stereomodel; (2) a *viewing* system, which makes it possible for an operator to see that model; and (3) a *measuring* (or tracing) system, which enables measurements of the stereomodel to be made and recorded.

12-2 CLASSIFICATION OF STEREOSCOPIC PLOTTERS

A variety of stereoscopic plotting instruments are currently available, each designed slightly differently. As an aid to understanding stereoplotters, it is helpful to classify them into groups having common characteristics. One common method of classification is by projection. This classification shall be divided into two broad categories: (1) *direct optical projection instruments* and (2) instruments with *mechanical* or *optical-mechanical projection*. Direct optical projection instruments create a true three-dimensional stereomodel by projecting transparency images through projector lenses. This is the type of projection illustrated in Figs. 12-1 and 12-2. The model is formed by intersections of light rays from corresponding images of the left and right diapositives. An operator is able to view the model directly and make measurements on it by intercepting projected rays on a viewing screen (platen).

Instruments of mechanical projection or optical-mechanical projection also create a true three-dimensional model from which measurements are taken. Their method of projection, however, is a simulation of direct projection of light rays by mechanical or optical-mechanical means. An operator views the diapositives stereoscopically directly through a binocular optical train.

Other methods of classifying stereoplotters are (1) classification by accuracy capability (i.e., first-, second-, or third-order plotters) and (2) classification according to whether an "approximate" analogue solution or a "theoretically correct" analogue solution is obtained. The first of these classifications is unsatisfactory because of difficulties in assessing true accuracy capabilities of various instruments. Plotting accuracy is not solely a function of the instrument but also depends upon other variables such as quality of photography, operator ability, etc.

In the second of these classifications, instruments of the "approximate" category assume truly vertical photos and employ a parallax bar for measurement. These low-order instruments enable direct stereoscopic plotting, but the resulting map is a perspective projection, not an orthographic projection. For certain work, however, their accuracy is adequate. Some "approximate" plotters were discussed in Chap. 8. Plotters in the "theoretically correct" category are capable of creating a true stereomodel by means of interior, relative, and absolute orientation. This is the type of plotter discussed in this chapter.

Analytical plotters and *automated stereoplotters* are two additional classes of instruments that have emerged in recent years. Part I of this chapter discusses direct optical projection stereoplotters. These instruments are simple in design and easily understood. Part II describes instruments of mechanical and optical-mechanical projection, Part III describes analytical plotters, and Part IV discusses stereoplotters with automatic image correlators.

It would be difficult if not impossible to describe each available instrument in detail in this chapter. For the most part, therefore, descriptions are general without reference to specific plotters and without comparisons of available instruments. In order to emphasize and clarify basic principles, however, examples, are made of certain instruments and in some cases pictures are given. Omission of other comparable stereoplotters is not intended to imply any inferiority of these instruments. Operator's manuals which outline the details of each of the different instruments are provided by the manufacturers. Comprehension of the principles presented in this chapter should provide the background necessary for understanding these manuals.

PART I   DIRECT OPTICAL PROJECTION STEREOPLOTTERS

12-3 COMPONENTS

The principal components of a typical direct optical projection stereoplotter are illustrated in the schematic diagram of Fig. 12-3. The numbered parts are the (1) *main frame,* which supports the projectors rigidly in place, thereby maintaining orientation of a stereomodel over long periods of time; (2) *reference table,* a large smooth surface which serves as the vertical datum to which model elevations are referenced and which also provides the surface upon which the manuscript map is compiled; (3) *tracing table,* to which the platen and tracing pencil are attached; (4) *platen,* the viewing screen which also contains the reference mark; (5) *guide rods,* which drive the illumination lamps causing projected rays to be illuminated on the platen regardless of the area of the stereomodel being viewed; (6) *projectors;* (7) *illumination lamps;* (8) *diapositives;* (9) *leveling screws,* which may be used to tilt the projectors in absolute orientation; and (10) *projector bar,* to which the projectors are attached.

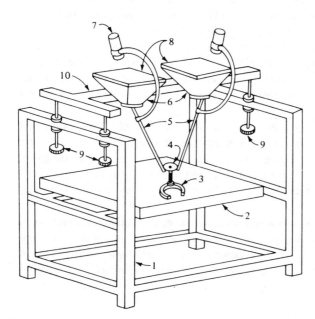

Figure 12-3 Principal components of a typical direct optical projection stereoplotter.

Although direct optical projection plotters of different manufacture vary somewhat in individual design and appearance, all are composed basically of these above parts. Reference to these parts will frequently be made in subsequent discussions in this chapter.

12-4 PROJECTION SYSTEMS

In the projection systems of direct optical projection stereoplotters, diapositives of a stereopair are placed in projectors and illuminated from above. Light rays are projected through the projector objective lenses and intercepted below on the reflecting surface of the platen. The projection systems of this type of stereoplotter require that the instruments be operated in a dark room.

Stereoplotter projectors are similar to ordinary slide projectors, differing primarily in their optical precision, physical size, and capability of adjustment in attitude relative to one another. Since projection takes place through an objective lens, the lens formula, Eq. (2-8), must be satisfied in order to obtain a sharply focused stereomodel. In terms of the stereoplotter symbols of Fig. 12-1, the lens formula is expressed as

$$\frac{1}{f'} = \frac{1}{p} + \frac{1}{h} \tag{12-1}$$

In Eq. (12-1), p is the *principal distance* of the projectors (distance from diapositive image plane to upper nodal point of the projector lens), h is the *projection distance* (distance from lower nodal point of the objective lens to the plane of optimum focus), and f' is the focal length of the projector objective lens. To obtain a clear stereomodel, intersections of projected corresponding rays must occur at a projection distance within the range of the *depth of field* of the projector lens (see Sec. 2-13).

To re-create the relative angular relationship of two photographs exactly as they were at the instant of their exposure (a process described in detail in Sec. 12-8), it is necessary that the projectors have rotational and translational movement capabilities. These motions, six in number for each projector, are illustrated in Fig. 12-4. Three of the movements are angular rotations about each of three mutually perpendicular axes: x rotation, called *omega* or *tilt*; y rotation, called *phi* or *tip*; and z rotation, called *kappa* or *swing*. The origin of the axis system about which the rotations take place is at the upper nodal point of the projector lens, with the x axis being parallel to the projector bar. The other three movements are linear translations along each of the three axes. In general, projectors of direct optical projection stereoplotters have all three angular rotations; however, they do not necessarily have all three linear translations. As a minimum, though, they must have the x translation for changing the spacing between projectors.

Two different types of systems are used in illuminating diapositives; (1) those which illuminate the entire diapositive simultaneously, and (2) those which illuminate only a small area of the diapositive at one time. Multiplex and Balplex projectors illuminate the entire diapositive. This projection system does not require guide rods

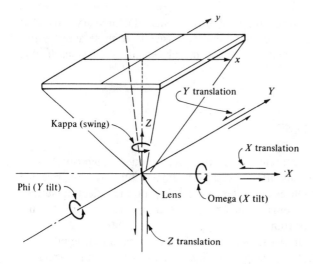

Figure 12-4 The six basic projector motions.

(item no. 5 of Fig. 12-3). To prevent excessive heat, both use reduced-sized diapositives and a system of forced air for cooling the projector housings.

The projection system of the Multiplex is illustrated in Fig. 12-5. The function of the condenser lenses is to distribute the light over the entire diapositive format and to condense all light rays through the small aperture of the objective lens. The principal distance of the commercial model of the Multiplex is 30.00 mm and its optimum projection distance for satisfying Eq. (12-1) is 360 mm. Multiplex projectors have a depth of field of 90 mm up and down from optimum projection distance.

Figure 12-6 illustrates the projection system of the Balplex (ER-55) plotter. The light source is centered at one focal point of an ellipsoidal mirror, while the objective lens is placed at the other focal point. The ellipsoidal mirror acts as a condenser by virtue of its property that any light ray emanating from a light source at one focal point is reflected through the other focal point. The principal distance of Balplex projectors is 55.00 mm. Three different models of Balplex projectors are available which provide choices in optimum projection distance of 760 mm, 525 mm, or 360 mm. Each of these different models uses a different objective lens to satisfy the lens formula. A 525-mm Balplex is shown in Fig. 12-7.

Projectors which illuminate only a small area of the diapositive at a time have the advantage that they can use full-sized diapositives, and they do not require a cooling system. This type of illumination system consists of a small, narrow-angle light source which illuminates an area on the diapositive about the size of a half dollar. When projected through the objective lens, an area slightly larger than the platen is illuminated in the model area. Driven by means of guide rods, the lamps swing above the diapositives, following the platen and illuminating it as it moves about the stereomodel. The Kelsh plotter shown in Fig. 12-8 utilizes this type of partial illumination. This instrument combines a nominal principal distance of 152 mm with an optimum projection distance of 760 mm; a combination which provides an enlargement ratio of five from diapositive scale to model scale.

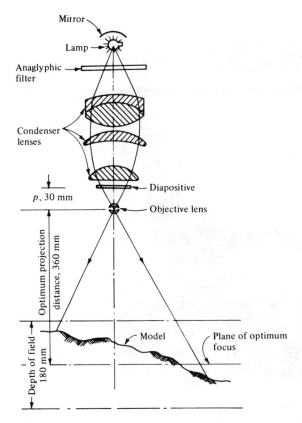

Figure 12-5 Projection system of the Multiplex instrument.

Labels in figure:
- Mirror
- Lamp
- Anaglyphic filter
- Condenser lenses
- Diapositive
- p, 30 mm
- Objective lens
- Optimum projection distance, 360 mm
- Depth of field 180 mm
- Model
- Plane of optimum focus

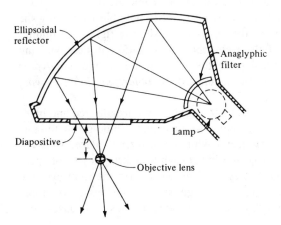

Figure 12-6 Cross section of the ellipsoidal reflector projection system of the Balplex instrument.

Labels in figure:
- Ellipsoidal reflector
- Anaglyphic filter
- Diapositive
- p
- Lamp
- Objective lens

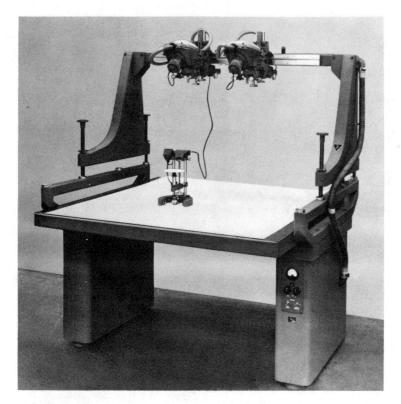

Figure 12-7 Balplex stereoplotting instrument with 525-millimeter optimum projection distance. (*Courtesy of Bausch and Lomb.*)

12-5 VIEWING SYSTEMS

The function of the viewing system of a stereoplotter is to enable the operator to view the stereomodel three-dimensionally. Stereoviewing is made possible by forcing the left eye to view only the overlap area of the left photo while the right eye simultaneously sees only the overlap area of the right photo. The different stereoviewing systems commonly used in direct optical projection plotters are (1) the *anaglyphic* system, (2) the *stereo-image alternator* (SIA), and (3) the *polarized-platen viewing* (PPV) system.

An anaglyphic system uses filters of complementary colors, usually red and blue-green, to separate the left and right projections. Assume that a blue-green filter is placed over the light source of the left projector while a red filter is placed over the right. Then, if the operator views the projected images while wearing a pair of spectacles having blue-green glass over the left eye and red glass over the right eye, the stereomodel can be seen in three dimensions. The anaglyphic viewing system is simple and inexpensive; however, it precludes the use of color diapositives and causes con-

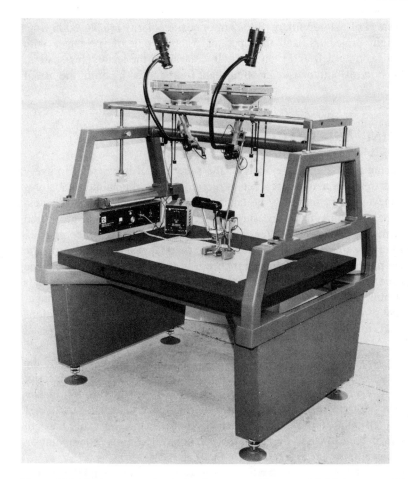

Figure 12-8 Kelsh stereoscopic plotting instrument. (*Courtesy of Kelsh Instrument Division, Danko Arlington, Inc.*)

siderable light loss, so that the model is not as bright as it could be if filters were not used.

The SIA system uses synchronized shutters to achieve stereoviewing. A shutter is placed in front of each projector lens. Also, a pair of eyepiece shutters through which the operator must look, are situated in front of the platen. The shutters are synchronized so that the left projector and left eyepiece shutters are open simultaneously while the right projector and right eyepiece shutters are closed, and vice versa. An operator therefore sees only left projector images with the left eye and right projector images with the right eye. The shutters rotate at a rapid rate so that the operator is unaware of any discontinuity in the projection. An SIA system is shown attached to the Kelsh plotter pictured in Fig. 12-8.

The PPV system operates similarly to the anaglyphic system except that polarizing filters are used instead of colored filters. Filters of opposite polarity are placed in front of the left and right projectors, and the operator wears a pair of spectacles with corresponding filters on the left and right sides. In contrast to the anaglyphic system, the SIA and PPV systems both cause very little light loss, and both permit the use of color diapositives.

12-6 MEASURING AND TRACING SYSTEMS

A system for making precise measurements of the stereomodel is essential to every stereoplotter. Measurements may be recorded as direct tracings of planimetric features and contours of elevation, or they may be taken as X, Y, and Z model coordinates. One of the principal elements of the measuring system of a direct optical projection stereoplotter is a tracing table. The platen (white disk) contains a reference mark in its center, usually a tiny speck of light. The reference mark appears to float above the stereomodel if the platen is above the terrain, hence it is called the *floating mark*. The platen can be raised or lowered by turning a screw; the total vertical run of the threads being about 120 mm. Extension shafts can be added to increase this vertical range. Vertical movement of the platen is geared to a dial, and by varying gear combinations the dial can be made to record elevations directly in feet (or meters) for varying model scales.

A manuscript map, preferably of stable base material, is placed on top of the reference table, as illustrated in Fig. 12-2. The tracing table rests on the manuscript and is moved about manually in the X and Y directions. To plot the position of any point, the platen is adjusted in X, Y, and Z until the reference mark appears to rest exactly on the desired point in the model. A pencil point which is vertically beneath the reference mark is then lowered to record the planimetric position of the point on the map, and its elevation is read directly from the dial.

To trace a feature such as a creek, the pencil is lowered to the map and the tracing table is moved in the XY plane while the platen is moved up or down to keep the floating mark in contact with the stream. The pencil thereby records a continuous trace of the feature. Contours of elevation may also be traced by locking the dial at the elevation of the desired contour and then moving the tracing table about, keeping the floating mark in contact with the terrain.

For some projects it is desirable to obtain the record of measurements in digital form rather than graphic form. This is possible if the plotter is equipped with an *XY coordinatograph* such as that pictured in Fig. 12-9. The coordinatograph is mounted on the reference table, and the tracing table is connected to it. As the tracing table is moved about the model, X and Y coordinates can be read directly from the precisely graduated scales on the two rails of the coordinatograph. Elevations (Z coordinates) may be read directly from the tracing table dial as before. Special electronic equipment enables the three coordinates at any point to be displayed visually and to be recorded automatically on magnetic tape or punched cards. In this era of computers, significant computational advantages can be realized by systematically recording X, Y, Z coordinates of a dense network of points throughout the stereomodel, thereby obtaining a so-called *digital terrain model* (DTM).

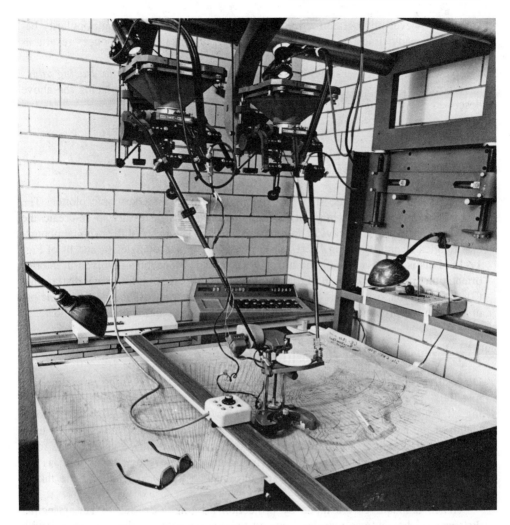

Figure 12-9 *XY* coordinatograph installed on a direct optical projection stereoplotter. (*Courtesy Auto-trol Corp.*)

Special "cross-sectioning" coordinatographs have been designed for taking terrain profiles at right angles to a center line; a procedure called *cross sectioning*. Cross sections are needed for computing earthwork volumes in planning and design of highways, railroads, canals, etc. When photogrammetric cross sections are taken, the adopted center line is plotted on the manuscript map. With the cross-sectioning coordinatograph oriented with respect to the center line, the tracing table is constrained to move perpendicular to the center line at each desired station. Elevations and distances from center line are automatically recorded on tape or cards for each point on the terrain at which an operator sets the floating mark. These recorded cross sections may then be fed directly to a computer for processing earthwork volumes, etc. Prior to the emergence of photogrammetry, cross sectioning was done exclusively by field surveying. However photogrammetric cross sectioning is rapidly gaining popularity because it is many times faster, more economical, and in general produces results as accurate as those produced by field surveying.

New systems for measuring and tracing called *digitized automatic tracing tables* have now been perfected (see Sec. 12-16), and they are rapidly replacing the above described coordinatographs.

12-7 INTERIOR ORIENTATION

As mentioned earlier, three steps are required to orient a stereoscopic plotter. The first of these, *interior orientation*, includes preparations necessary to re-create the geometry of the projected rays to duplicate exactly the geometry of the original photos, e.g., angles θ_1' and θ_2' of Fig. 12-1*b* must be exactly equal to angles θ_1 and θ_2 of Fig. 12-1*a*. This is necessary to obtain a true stereomodel. Procedures involved in interior orientation are (1) preparation of diapositives, (2) compensation for image distortions, (3) centering diapositives in the projectors, and (4) setting off the proper principal distance in the projectors. These procedures are described individually as follows:

12-7.1 Preparation of Diapositives

Diapositives are transparencies prepared either on optically flat glass or clear film base materials. They may be made either by *direct contact* printing or by *projection* printing. If contact-printed, their principal distances will be exactly equal to the focal length of the taking camera. For this reason contact-printed diapositives can be used only in plotters whose accommodation range in principal distance includes the focal length of the taking camera. Contact printing creates true geometry as long as the principal distances of the projectors are set equal to the focal length of the taking camera.

When diapositives are made by projection printing, (see Sec. 3-12), principal distances of diapositives may be prepared to some value other than the focal length of the taking camera. Projection printing is necessary when diapositives for stereoplotting are being prepared from photography having a focal length outside of the plotter's range of principal distance accommodation. An example is preparation of Multiplex and Balplex diapositives (principal distances of 30 mm and 55 mm, respectively) from 6-in-focal-length photography. If diapositives are produced having principal distances other than the focal length of the taking camera, the dimensions of the diapositives will be reduced or enlarged in the ratio of p/f.

Geometric relationships that must exist in diapositive making by projection printing are illustrated in Fig. 12-10. By similar triangles of Fig. 12-10*a*, which shows the aerial photography,

$$\frac{f}{H'} = \frac{d}{D} \qquad (a)$$

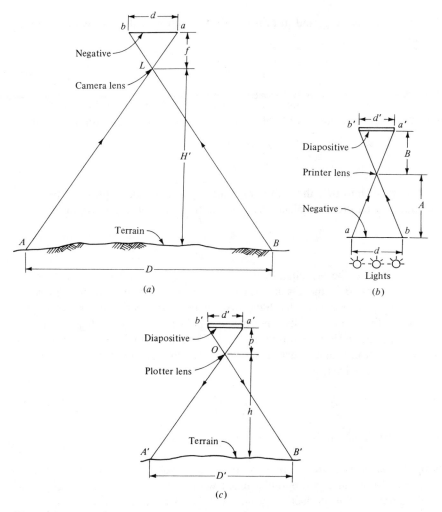

Figure 12-10 Geometric relationships required in making diapositives by projection printing. (*a*) Photography; (*b*) projection printing; (*c*) stereoplotter projection.

By similar triangles of Fig. 12-10*b*, which illustates the projection printer,

$$\frac{d'}{d} = \frac{B}{A} \tag{b}$$

And by similar triangles of Fig. 12-10*c*, which shows the stereoplotter,

$$\frac{d'}{D'} = \frac{p}{h} \tag{c}$$

Substituting (*a*) into (*b*), and in turn substituting into (*c*), the following relationship is obtained:

$$\frac{p}{f} = \frac{B}{A}\frac{hD}{H'D'} \tag{d}$$

Now since true geometry must be retained from photography to stereoplotter projection, triangles *ALB* and *A'OB'* of Figs. 12-10*a* and *c* are also similar, so that

$$\frac{H'}{D} = \frac{h}{D'} \tag{e}$$

Substituting (*e*) into (*d*), the following equation is obtained expressing the condition that must be enforced in diapositive preparation by projection printing:

$$\frac{p}{f} = \frac{B}{A} \tag{12-2}$$

In Eq. (12-2), *p* is the principal distance of the diapositive, *f* is the focal length of the taking camera, *B* is the distance in the projection printer from the emergent nodal point of the printer lens to the diapositive plane, and *A* is the distance from the negative plane to the incident nodal point of the printer lens. To get a diapositive that is in sharp focus, one further condition must be enforced, and this is that distances *A* and *B* in the projection printer must be set off so that the lens formula, Eq. (2-8), is satisfied for the printer lens, or

$$\frac{1}{f''} = \frac{1}{A} + \frac{1}{B} \tag{12-3}$$

where *f''* is the focal length of the projection printer lens.

Example 12-1 It is desired to prepare diapositives having a principal distance of 153.00 mm from lunar photography taken with an 80.00-mm-focal-length Hasselblad camera. If the projection printer lens has a focal length of 100.00 mm, what distances *A* and *B* must be set off?

SOLUTION By Eq. (12-2),

$$\frac{153.00}{80.00} = \frac{B}{A} \qquad B = \left(\frac{153.00}{80.00}\right)A$$

Substituting the above expression for *B* into Eq. (12-3),

$$\frac{1}{100.00} = \frac{1}{A} + \frac{\frac{1}{153.00}}{80}A$$

Solving the above, $A = 152.29$ mm.

Substituting and solving for B,

$$B = \frac{153.00}{80.00} \, 152.29 = 291.25 \text{ mm}$$

A projection printer manufactured especially for preparing enlarged or reduced diapositives having a wide range in versatility in distances A and B is pictured in Fig. 12-11.

If diapositives are contact-printed emulsion to emulsion, then the diapositives must be oriented emulsion up in the stereoplotter in order to re-create a geometrically correct model. This is a disadvantage with direct optical projection plotters because light rays carrying images are distorted in direction as they pass through the glass or film base (see Sec. 2-2). This condition can be avoided if the diapositives are prepared by projection printing *through the film base of the negative,* as shown in Fig. 12-12. Diapositives prepared in this manner are correctly oriented in the projectors *emulsion down,* thereby eliminating distortions due to refraction of rays passing through glass or film. (In Fig. 12-12 a dashed line for the letter F denotes that it is on the underside.)

As noted earlier, diapositives may be printed either on glass or on film. Some advantages can be realized by using film diapositives, e.g., lower cost, freedom from breakage, and less storage space required. A disadvantage in using film diapositives with many direct optical projection plotters is that for rigidity they must be placed in the projectors in a sandwich arrangement between two pieces of optically flat glass. This again creates distortions in ray paths which requires compensation. Also shrink-

Figure 12-11 Wild U-4 reduction printer for making diapositives. (*Courtesy Wild Heerbrugg Instruments, Inc.*)

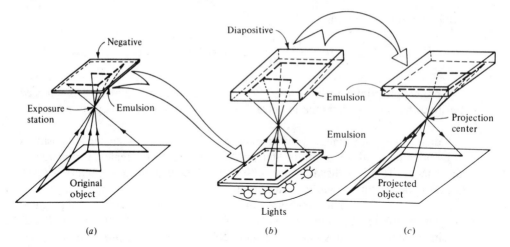

Figure 12-12 Printing diapositives by projection through the film base (note that correct model geometry is obtained with emulsion of diapositive facing down in the projector). (*a*) Original photography; (*b*) projection printing of diapositives; (*c*) stereoplotter projection.

age or expansion, which is practically nonexistent on glass, can be significant on film diapositives.

12-7.2 Compensation for Image Distortions

Compensation for radial distortion of the lens of the taking camera may be accomplished in one of the following three ways: (1) elimination of the distortion with a "correction plate" in projection printing of the diapositives, followed by use of a distortion-free projector lens; (2) varying the projector principal distance by means of a cam, thereby reconstructing true geometry; and (3) use of a projector lens whose distortion characteristics negate the camera's distortion.

In the first method, a glass correction plate of varying thickness is placed in the projection printer in the ray path between negative and diapositive. Projected light rays passing through the correction plate are deflected along radial lines from the principal point by amounts necessary to remove radial distortion. Rather than placing the correction plate in the projection printer, it can be placed in the stereoplotter projector to accomplish the same effect. Different correction plates can be obtained which correct for lens distortions of different cameras. In addition to compensating for lens distortions, correction plates are also available for removing distortions caused by atmospheric refraction and earth curvature.

In the second method of lens distortion compensation, an aspheric cam mechanically raises or lowers the projector lens (or the diapositive) so that, in spite of distortions, projected rays make the same angle with the projector optical axis as they made with the camera optical axis when they entered the camera. As a footnote to this discussion it should be mentioned that most modern aerial camera lenses are so nearly distortion-free that distortion compensation is often overlooked completely.

The Zeiss Pleogon lens is an example. Its near distortion-free characteristics are shown in the radial-lens distortion curve of Fig. 4-15.

12-7.3 Centering the Diapositives in the Projectors

Diapositives must be centered in the projectors so that the principal point is on the optical axis of the projector lens. Although this problem is solved slightly differently for each instrument, it is basically done by aligning fiducial marks of the diapositives with four calibrated collimation marks whose intersection locates the optical axis of the projectors. Before diapositives are placed in the projectors, they are laid out so that their common area overlaps. They are then separated, rotated 180° about the Z axis, placed on the plate holders of the projectors, and centered. In projection, the imagery is rotated 180°, which once again makes common areas of projected images overlap.

12-7.4 Setting Off the Proper Principal Distance in the Projectors

The final step in interior orientation is setting the diapositive principal distance on the projectors. This is unnecessary for plotters such as the Multiplex and Balplex, whose principal distances are permanently fixed and whose diapositives are prepared to conform with these principal distances. For other plotters the principal distance may be varied by either adjusting graduated screws or a graduated ring to raise or lower the diapositive image plane. These projectors are designed to accommodate a certain nominal principal distance and their range from that value is small; e.g., the Kelsh plotter pictured in Fig. 12-8 is designed for 6-in (152-mm) photography, and its range in principal distance accommodation is from only 150 to 156 mm.

12-8 RELATIVE ORIENTATION

Imagine the camera frozen in space at the instants of exposure of two photographs of a stereopair. The two negatives in the camera would then bear a definite position and attitude relationship relative to each other. In relative orientation, this relative position and attitude relationship is re-created for the two diapositives by means of movements imparted to the projectors.

The condition that is fulfilled in relative orientation is that each model point and the two projection centers form a plane in miniature just like the plane that existed for the corresponding ground point and the two exposure stations. This condition is illustrated by corresponding planes $A'O_1O_2$ and AL_1L_2 of Fig. 12-1. Also, parallactic angle ϕ' for any point of the stereomodel of Fig. 12-1 must equal the point's original parallactic angle ϕ. The implication of the foregoing condition of relative orientation is that projected rays of all corresponding points on the left and right diapositives must intersect at a point, and this is the basis of the systematic relative orientation procedures described below.

Since relative orientation is unknown at the start, the two projectors are first positioned relative to one another by estimation. Usually, if vertical photography is being used, they are set so that the diapositives are nearly level and so their x axes lie along a common line. Also, the projectors are adjusted so that their Y and Z settings (distances from the projector bar in the Y and Z directions) are equal. When the projector lamps are first turned on, corresponding light rays will not intersect, and their projected images may appear on the platen, as shown in Fig. 12-13a. Since the X component p_x of the mismatch of images is a function of the elevation of the point, it can be removed by raising or lowering the platen. The remaining Y component p_y (see Fig. 12-13b) is called y parallax, and it must be removed or *cleared* for all points in the stereomodel in order to obtain a relatively oriented model.

Rather than attempting to clear y parallax at all points in the model, conventional procedure clears five standard points (plus a sixth for a check) located in the model, as shown in Fig. 12-14. If these five points are cleared of y parallax, then the entire model should be cleared. Points 1 and 2 are vertically beneath projector I (left projector) and projector II (right projector) respectively. Points 3 and 5 are on the Y axis through projector I and points 4 and 6 are on the Y axis through projector II. Points 1, 2, 3, and 4 roughly form a square, as do points 1, 2, 5, and 6.

Before relative orientation procedures are described, it will be helpful to consider the movement of projected images in the model area caused by each of the six projector motions that were illustrated in Fig. 12-4. In Fig. 12-15a through f, these image movements are illustrated, Figure 12-15a shows that an X translation imparts only X movement to all projected images, and therefore no y parallax could be cleared using this motion. In Fig. 12-15b, Y translation causes equal Y movement of all projected images; hence any point in the stereomodel could be cleared of y parallax using this motion. Figure 12-15c shows that all points move radially outward from the projection center if a projector is translated upward in Z. Image movement would be radially inward for a downward Z translation. With the use of Z translation, Y parallax could be cleared for any points except those along the X axis. In Fig. 12-15d omega rotation (tilt) causes Y movement of all projected images. This motion could therefore be used to clear y parallax anywhere in the model. As shown in Fig. 12-15e, a phi rotation

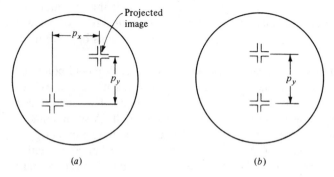

(a)   (b)

Figure 12-13 (a) Observing both y parallax and x parallax on the platen. (b) After removing x parallax, only y parallax remains.

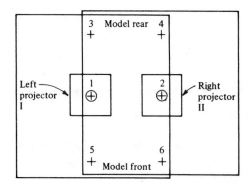

Figure 12-14 Stereomodel locations of six points conventionally used in relative orientation.

(tip) causes an X movement for all projected images, but it also imparts a Y component for points not on the X and Y axes. And finally, in Fig. 12-15f, a kappa rotation (swing) imparts Y components to points everywhere in the model except along the Y axis. A clear grasp of these image movements will help to explain why y parallax can be removed and orientation achieved using the following procedures.

Two different systematic procedures for relative orientation will be discussed in

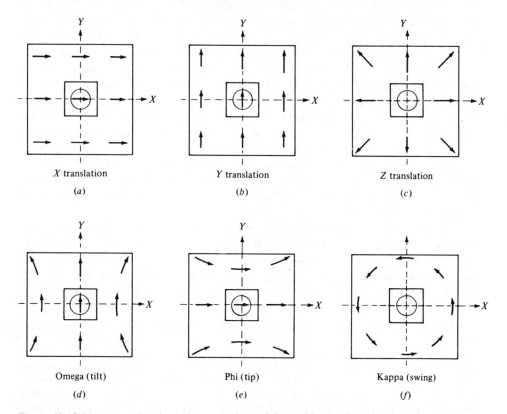

Figure 12-15 Movement of projected images in the model caused by the six projector motions.

this text: (1) the *two-projector* method, or what is commonly called the *swing-swing* method, and (2) the *one-projector* method. The steps involved in the two projector method are as follows (refer to Figs. 12-14 and 12-15):

a. Clear p_y at 1 with kappa (swing) of projector II.
b. Clear p_y at 2 with kappa (swing) of projector I.
c. Clear p_y at 3 with phi (tip) of projector II.
d. Clear p_y at 4 with phi (tip) of projector I.
e. Observe p_y at 5, remove this y parallax, and overcorrect by introducing $\frac{1}{2}$ of the original p_y in the opposite direction with omega (tilt) of either projector.†
f. Repeat steps *a* through *e* until no p_y exists at 5.
g. Check 6 for p_y.

In analyzing the above steps, it should be noted that step *b* does not introduce *y* parallax at point 1 so that both points 1 and 2 are clear following this step. Also step *c* introduces no *y* parallax at either points 1 or 2, so that points 1, 2, and 3 are clear following this step. Furthermore, step *d* introduces no *y* parallax at points 1, 2, or 3, so that points 1 through 4 are clear following this step. Step *e* introduces *y* parallax at all points, but if steps *a* through *d* are repeated, the *y* parallax at point 5 will disappear. Several repetitions of steps *a* through *e* may be required to finally clear point 5, however. If points 1 through 5 are cleared, point 6 should also be clear, and if it is not, this is generally an indication that the other five points are not truly clear.

If the anaglyphic viewing system is used, *y* parallax for the first one or two repetitions of steps *a* through *e* may be cleared by matching corresponding red and blue-green images. Final repetitions are performed with the operator wearing the spectacles. When the spectacles are worn, the left and right eyes focus on the left and right projections, respectively. Any *y* parallax in the projected images is unconsciously corrected for by a deviation of the optical axes of the two eyes. However, this causes an apparent *y* separation of the floating mark, which enables the operator to make a precise observation of the amount of *y* parallax present. Using the appropriate projector motions, *y* parallax is removed until the *y* separation of the two marks disappears and a single clear discrete floating mark is obtained.

The two-projector method described above requires rotations of both projectors. There are situations where it is necessary to retain the orientation of one projector, say, the left, and to perform relative orientation with motions of the right projector only. An example of this is in orienting a strip of three or more photos on a projector bar holding three or more projectors. An eight-projector stereoplotter is shown in Fig.

---

†The actual amount of *y*-parallax overcorrection that must be introduced into point 5 is a function of camera focal length f and distance y (measured on the original photo) from model point 1 to model point 5. It can by computed by the formula $0.5 \ (1 \ + \ f^2/y^2) \ - \ 1$. For 6-in-focal-length photography and a *y* distance of $4\frac{1}{4}$ in, the value is $\frac{1}{2}$; hence steps *e* above apply specifically to those very commonly encountered conditions. Amounts of overcorrection for other situations can be calculated, however, and appropriate *y* parallax introduced accordingly. It should be noted that overcorrection values calculated by the above formula are exact only for level terrain, but good approximations result for variable terrain. The inexactness of this step helps to explain the need to perform several iterations in relative orientation to finally clear all *y* parallax.

14-1. After the first two photos are oriented, the third must be oriented to the second without upsetting the orientation of the second, etc. The one-projector method outlined in the following steps may be used in this situation. All motions are imparted to the right-hand projector. (Refer to Fig. 12-14 for point locations.)

a. Clear p_y at 2 with Y translation.
b. Clear p_y at 1 with kappa (swing).
c. Clear p_y at 4 with Z translation.
d. Clear p_y at 3 with phi (tip).
e. Observe p_y at 6, remove this y parallax, and overcorrect by introducing $\frac{1}{2} p_y$ in the opposite direction with omega (tilt).†
f. Repeat steps *a* through *e* until no p_y exists at 6.
g. Check 5 for p_y.

12-9 ABSOLUTE ORIENTATION

After relative orientation is completed, a true three-dimensional model of the terrain exists. Although the horizontal and vertical scales of the model are equal, that scale is unknown and must be fixed at the desired value. Also the model is not yet level with respect to datum. Selecting model scale and fixing the model at that scale, and leveling the model are the purposes of absolute orientation.

12-9.1 Selecting Model Scale

Model scale is fixed within certain limits by the scale of the photography and by the characteristics of the particular stereoplotter. Comparing the geometry of Fig. 12-1*a* and Fig. 12-1*b*, model scale is seen to be the ratio of the sizes of triangles AL_1L_2 and $A'O_1O_2$. Equating these similar triangles, model scale may be expressed as

$$S_m = \frac{b}{B} = \frac{h}{H'} \tag{12-4}$$

In Eq. (12-4), S_m is model scale, b is model air base, B is photographic air base, h is plotter projection distance, and H' is flying height above ground. From Chap. 6 it will be recalled that flying height above ground and camera focal length fix photo scale according to the relationship

$$S_p = \frac{f}{H'} \tag{12-5}$$

In Eq. (12-5), S_p is photo scale and f is camera focal length. Substituting Eq. (12-5) into Eq. (12-4), the following convenient equation results for computing optimum model scale (given photo scale):

$$S_m = \frac{h}{f} S_p \tag{12-6}$$

†See footnote on p. 284.

In Eq. (12-6), the term h/f is the ratio of projection distance to camera focal length, and it is also the enlargement ratio from photo scale to model scale. From the foregoing it is apparent, then, that for a given photo scale, *optimum model scale* is fixed for a particular stereoplotter by its *optimum projection distance*. A small aperture is used in objective lenses of projectors of direct optical projection stereoplotters so that a large depth of field results. A satisfactory model can therefore be obtained for a rather wide range in projection distance, and accordingly some flexibility is afforded in selecting model scale. The range in projection distance also makes it possible to accommodate topographic relief in the stereomodel. The actual model scale that is adopted should be chosen near optimum model scale, but it should be rounded to one of the commonly used scales such as 1 in/100 ft. 1:1,000, etc. If the plotter is not equipped with a coordinatograph or pantograph to change scale from model to map, then plotting scale is equal to model scale.

Example 12-2 Mapping photography is taken with a 6-in- (152-mm-) focal-length camera from a flying height of 3,000 ft above ground. What is optimum model scale, and what actual model scale would be adopted for (*a*) a Balplex (525) plotter and (*b*) a Kelsh (760) plotter?

SOLUTION By Eq. (12-5),

$$S_p = \frac{6 \text{ in}}{3{,}000 \text{ ft}} = \frac{1 \text{ in}}{500 \text{ ft}}$$

(*a*) Balplex (optimum $h = 525$ mm):
By Eq. (12-6),

$$S_m = \frac{525 \text{ mm}}{152 \text{ mm}} \frac{1 \text{ in}}{500 \text{ ft}} = \frac{1 \text{ in}}{145 \text{ ft}}$$

Optimum model scale is 1 in/145 ft; adopt 1 in/150 ft or 1:1,800

(*b*) Kelsh (optimum $h = 760$ mm):
By Eq. (12-6),

$$S_m = \frac{760 \text{ mm}}{152 \text{ mm}} \frac{1}{500 \text{ ft}} = \frac{1 \text{ in}}{100 \text{ ft}}$$

Optimum model scale is 1 in/100 ft or 1:1,200; adopt that value.

When model scale has been adopted, an initial model air base (spacing between projectors) is set off. This is most conveniently done prior to relative orientation, so that model scale after relative orientation is close to required model scale. An initial model base can be obtained by multiplying the photo base (distance between principal point and conjugate principal point) by the actual enlargement ratio S_m/S_p. The photo base can be measured directly from the photos, or it can be computed on the basis of percent end lap.

Example 12-3 Assuming 9-in format and 60 percent end lap for the photography of Example 12-2, what should the initial model base be for (*a*) the Balplex (525) plotter and (*b*) the Kelsh (760) plotter?

SOLUTION At 60 percent end lap, the photo base is 40 percent of the 9-in format, or $0.4 \times 9 = 3.6$ in.
(*a*) Balplex:

$$b = 3.6 \frac{S_m}{S_p} = 3.6 \frac{1 \text{ in}/150 \text{ ft}}{1 \text{ in}/500 \text{ ft}} = 12.0 \text{ in}$$

(*b*) Kelsh:

$$b = 3.6 \frac{1 \text{ in}/100 \text{ ft}}{1 \text{ in}/500 \text{ ft}} = 18 \text{ in}$$

12-9.2 Scaling the Model

If a preliminary model base is calculated as described above and the projectors are set accordingly, then after relative orientation the stereomodel will be near required scale. As shown in Fig. 12-16, model scale is changed by varying the model base. If the Y and Z settings of the two projectors are equal, then model base is composed only of an X component called b_x, and model scale is varied by simply changing the model base by Δb_x, as shown in Fig. 12-16*a*.

A minimum of two horizontal control points are required to scale a stereomodel. These points are plotted at adopted model scale on the manuscript map as points A and B of Fig. 12-16*b*. The manuscript is then positioned under the model, and with

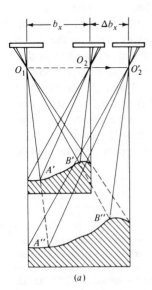
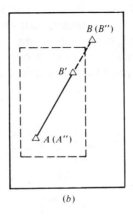

Figure 12-16 Changing model scale by adjusting model base. (*a*) Cross-sectional view; (*b*) plan view.

the floating mark set on one model point, such as A', the manuscript is moved until map point A is directly under the plotting pencil. The floating mark is then set on model point B'. Holding the manuscript firmly with a fingertip at point A, it is rotated until map line AB is collinear with model line $A'B'$. If model line $A'B'$ is shorter than map line AB, as is the case in Fig. 12-16b, model scale is too small and must be increased by increasing the model base until new model line $A''B''$ is equal in length to map line AB. The model base may be set to the required value by trial and error, or Δb_x may be calculated directly using the following formula:

$$\Delta b_x = b_x\left(\frac{AB}{A'B'} - 1\right) \tag{12-7}$$

In Eq. (12-7), AB and $A'B'$ are scaled from the manuscript in any convenient units. If the algebraic sign of Δb_x is negative, model scale is too large and b_x must be reduced by Δb_x. Once the model is scaled, it is recommended that a third horizontal control point be checked to guard against possible mistakes.

Example 12-4 Assume that a map line AB scales 17.60 in and its corresponding model length is 16.94 in. The initial model base was set at 18.00 in, and the Y and Z base settings were equal. What model base change is required to achieve the desired model scale?

SOLUTION By Eq. (12-7),

$$\Delta b_x = 18.00\left(\frac{17.60}{16.94} - 1\right) = 0.70 \text{ in}$$

Model base must be increased by 0.70 in to bring the model to correct scale.

If the Y and Z settings of the projectors are not equal, in addition to base component b_x there will be Y and Z base components b_y and b_z, respectively, in the model air base as shown in Fig. 12-17. This causes added difficulties in scaling the model. If it were necessary to increase the model base in scaling, for example, projector II would have to be moved to II′ along the base line joining I and II; otherwise, y parallax would be introduced. This necessitates introducing Δb_x, Δb_y, and Δb_z components in scaling. After the Δb_x component is introduced, the necessary Δb_y component can be introduced by clearing y parallax of a point near 2 (see Fig. 12-14) with a b_y motion of projector II, and then the required Δb_z component can be applied

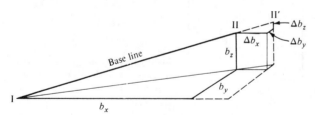

Figure 12-17 Model base containing components b_x, b_y, and b_z. (Note that an increase of b_x by Δb_x necessitates introducing Δb_y and Δb_z components as well.)

by clearing y parallax of a point near 4 or 6 with a b_z motion of projector II. These difficulties are avoided, however, by making the two Y projector settings equal and the two Z projector settings equal prior to starting orientation.

12-9.3 Leveling the Model

The final step in absolute orientation is leveling the model. This procedure requires a minimum of three vertical control points distributed in the model so that they form a large triangle. As a practical matter, four points, one near each corner of the neat model, should be used. A fifth point near the center of the model is also desirable. Before proceeding with leveling, it is imperative that the proper gears be inserted into the tracing table so that elevation changes recorded on the dial that occur with up and down motions of the platen are consistent with the required model scale.

A model with a vertical control point near each corner that has not yet been leveled is shown in Fig. 12-18. Note that there are two components of tilt in the model, an X component (also called Ω) and a Y component (also called Φ). The amount by which the model is out of level in each of these components is determined by reading model elevations of the vertical control points and comparing them with their known values. Assuming four vertical control points in a model, as ponts A through D of Fig. 12-18, the following systematic procedure would be applied in leveling the model:

(a) Set the floating mark on model point A and index the tracing table dial to read the control elevation of that point.

(b) Read model elevation of control point D.

(c) From the difference between model elevation and control elevation, determine whether the model is X-tilted up or down toward the rear. (If model elevation is higher than control elevation, the model is tilted up in the rear.)

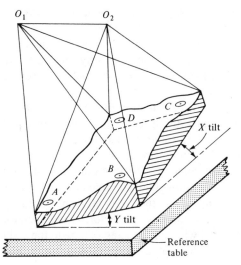

Figure 12-18 Stereomodel that is not level (note X and Y components of tilt).

(*d*) By one of the tilting methods to be discussed, introduce a corrective *X* tilt.
(*e*) Repeat steps *a* through *d* until the model is level in the direction from *A* to *D*.
(*f*) Re-index the tracing table dial to read the control elevation of point *A* with the floating mark set on model point *A*.
(*g*) Read model elevation of control point *B*.
(*h*) From the difference between model elevation and control elevation, determine whether the model is *Y*-tilted up or down to the right.
(*i*) By one of the tilting methods to be discussed, introduce a corrective *Y* tilt.
(*j*) Repeat steps *f* through *i* until the model is level in that direction.
(*k*) Check point *D* to see if its model elevation still conforms to its control elevation. If points *A* and *D* are not on a line parallel to the stereoplotter *Y* axis, it will likely not conform. If it does not conform, the above steps should be repeated until the model elevations of points *A*, *D*, and *B* all agree with their control elevations.
(*l*) Check point *C* to see if its model elevation conforms to its control elevation. If it does not, this could be an indication of model deformation caused by inaccuracies of relative orientation, or it could disclose a blunder in one or more of the vertical control points.

There are different methods available for introducing the corrective *X* and *Y* tilts, and the choice will depend partly upon the particular stereoplotter and partly upon the amount of tilt necessary. A few plotters are designed so that their reference tables may be tilted in both the *X* and *Y* directions to make them parallel with the model datum. This method is easily visualized and is convenient because relative orientation is not altered since the projectors are not touched. The method may not be practical if a severe amount of tilt is necessary, however.

A second method for introducing corrective tilts in the model is tilting the projector bar by means of the leveling screws, as shown in Fig. 12-19. Either *X* or *Y* tilts can be introduced in this way, but if the instrument has four leveling screws, only one or the other should be done. If both are performed, all four leveling screws will not rest firmly on their supports afterward. Introducing tilts by means of leveling screws is convenient because it moves both projectors simultaneously and therefore does not upset relative orientation. Again, this method may not be practical if severe amounts of tilt exist.

A third method of introducing corrective tilts consists of *X* tilting each projector with equal amounts of omega rotation (tilt) and *Y*-tilting each projector with equal amounts of phi rotation (tip), followed by a *Z* translation to remove the b_z base component introduced by the phi rotations. This is illustrated in Fig. 12-20. In applying corrective *X* tilt, a trial amount of omega rotation is first introduced to the left projector. This, of course, introduces *y* parallax throughout the model, but if the *Y* and *Z* settings of the projectors are equal, this *y* parallax is entirely removed by introducing an equal amount of omega rotation to the right projector. If the operator clears *y* parallax for any point in the stereomodel when introducing omega to the right projector, the entire model will again be clear. In applying corrective *Y* tilt, a trial

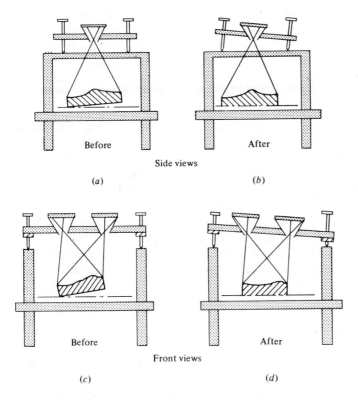

Side views

(a)   (b)

Front views

(c)   (d)

Figure 12-19 (a) and (b) Correcting X tilt of a model by X tilt of projector bar. (c) and (d) Correcting Y tilt of a model by Y tilt of projector bar.

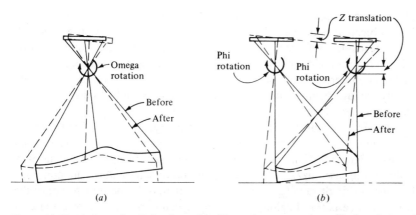

(a)   (b)

Figure 12-20 (a) Correcting X tilt of a model with equal omega rotations of both projectors. (b) Correcting Y tilt of a model with equal phi rotations of both projectors followed by a Z translation.

amount of phi rotation is introduced equally into both projectors. This has the effect of creating a b_z base component which is removed by raising or lowering either projector, the exact amount being just enough to clear y parallax for a point in one model corner. This third method has the advantage that large amounts of tilt are readily corrected, but its disadvantage is that small amounts of y parallax may remain in the model as a result of moving the projectors individually. Also, it is difficult to introduce equal amounts of phi rotation to each projector because usually there are no dials with which to read the values. In general, therefore, the best overall method of leveling consists of applying corrective X tilt using omega rotations, followed by a correction for Y tilt with the leveling screws (if the particular instrument has these capabilities).

Regardless of the method of leveling, the previously established model scale will be upset, especially if large amounts of corrective tilts are required. Also, it is likely that absolute orientation will slightly upset relative orientation. Therefore it is not practical to labor at great lengths with either relative or absolute orientation the first time through. Rather, quick orientations should be performed at first, followed by careful refinements the second time through. When orientation is completed, the manuscript map should be firmly secured to the reference table in preparation for map compilation.

12-10 MAP COMPILATION

When the model has been completely oriented and the manuscript secured to the reference table, map compilation may be started. As a general rule in compilation, planimetric details and cultural features should be traced first, followed by contouring. This is because planimetry has a very significant effect on the location and appearance of contours (e.g., the v's of contours crossing streams must peak in the stream, contours crossing roads in cut cross the road at right angles and have their ditches on either side of the road, etc.). An example of carelessly compiled contours or planimetry is shown in Fig. 12-21a, while its corresponding correct rendition is shown in Fig. 12-21b.

While planimetric features are traced, the floating mark must be kept in constant contact with the object. It is advisable to plot all features of a kind at once (e.g., all roads), before proceeding to another feature, as this reduces the likelihood of omissions. Generally, features are compiled in descending order of prominence. Any notes or labels that the compiler feels are necessary to avoid later misidentification should be made directly on the manuscript. A set of paper prints and a stereoscope are essential to the compiler as an aid in identifying features.

Contours are more difficult to compile than planimetry, especially for the beginner. But, like learning to ride a bicycle, beginning frustrations are soon forgotten by those who persist. Generally, contouring can be approached in much the same way as the assembly of a jigsaw puzzle. Easier areas and prominent features are compiled first, and the more difficult detail is filled in later. It may be helpful to study the overall model for a time before beginning to compile contours. This gives a sense of familiarity much the same as one would get in making a field visit. Studying the

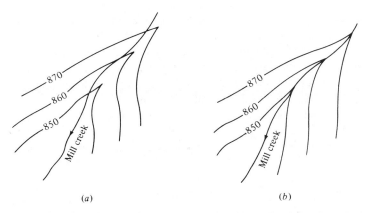

(a)   *(b)*

Figure 12-21 (*a*) Inconsistencies between planimetry and contour locations. (*b*) Consistent renditions of planimetry and contours.

model using paper prints and stereoscope can also be helpful. In more difficult areas or flat terrain, it is sometimes necessary to drop "spot elevations" and interpolate contours from them. This is helpful because spot elevations can be read to significantly greater accuracy than direct contour tracing. In areas covered with trees, it may be impossible to plot continuous contours. In these areas broken contours (contour lines plotted in open areas only) may be drawn; otherwise, spot elevations can be plotted and contours interpolated. Densely vegetated areas occasionally must be field surveyed. When compiling contours, it is advisable to lock the tracing table dial at the desired contour elevation to prevent accidental movement of the platen.

Whether compiling planimetry or contours, the operator should occasionally check the orientation of the model, including scale and level. This is important especially when returning to the plotter after being away for a time.

When the manuscript is finished, it should be carefully checked for omissions and mistakes. It is then field-checked where any missing features are noted. Also, features are identified and local names are determined. An important part of any field check is an accuracy determination. In this procedure, photogrammetrically plotted distances and elevations are checked against field measurements.

With field inspection completed, the map is ready for drafting or scribing. If the map is drafted in ink, manuscript features are traced on stable base overlay material. Lettering is usually inked first (stick-up letters can be pasted onto the manuscript as an alternative to inking them), followed by planimetric features and finally contours. Features having straight lines or known geometric shapes are drawn with the aid of straightedges or templates. A certain amount of generalization is required, especially in smoothing contours. When inking is completed, the map may be reproduced.

Scribing is another convenient method of preparing negatives for map reproduction. This procedure requires sheets of transparent stable base material that have been coated with an opaque emulsion material. In a laboratory process, the manuscript map lines are transferred to the emulsion coating. Then, using special scribing tools, lines representing contours and features are made by cutting and scraping to remove the coating material. Special tools are also available for varying line weights and for scribing cartographic symbols used for features such as roads, schools, churches, mines, etc. Scribing is often faster and more convenient than ink drafting, and unlike inking it yields a negative directly; i.e., light passes through the clear areas where lines and symbols have been scribed. Proficiency at scribing can generally be attained much faster than it can at drafting with ink. Experienced compilers can scribe planimetry and contours directly onto the map sheet while stereoplotting. In this instance, the manuscript is compiled on scribe material, and the plotting pencil is replaced with scribing tools of appropriate line widths. This can result in a significant time savings.

PART II STEREOPLOTTERS WITH MECHANICAL OR OPTICAL-MECHANICAL PROJECTION

12-11 MECHANICAL PROJECTION INSTRUMENTS

Mechanical projection stereoscopic-plotting instruments simulate direct optical projection of light rays by means of two precisely manufactured metal *space rods*. These instruments are manufactured chiefly in Europe, but their use in the United States is rapidly increasing because of their versatility, convenience of operation, accuracy capabilities, and overall stability, as well as the fact that they need not be operated in a darkroom. As with direct optical projection instruments, the many slightly different designs of mechanical projection stereoplotters cannot be described in detail in this text. Rather, basic principles of mechanical projection are explained and a few instruments which typify these basic concepts are discussed briefly. Manuals provided by manufacturers outline the details of operation for their specific instruments.

The basic principles of mechanical projection are illustrated in the simplified diagram of Fig. 12-22. Diapositives are placed in *carriers* and illuminated from above, often with neon lights. The carriers are analogous to projectors of direct optical projection instruments. Two space rods are free to rotate about *gimbal joints O' and O''*, and they can also slide up and down through these joints. The space rods represent corresponding projected light rays and the gimbal joints are mechanical projection centers, analogous to the objective lenses of projectors of direct optical projection stereoplotters. The model exposure stations are therefore represented by O' and O'', and the distance $O'O''$ is the model air base. Joints O' and O'' are fixed in position except that their spacing can be changed, either physically or theoretically, during orientation to obtain correct model scale.

The viewing system consists of two individual optical trains of lenses, mirrors, and prisms. The two optical paths are illustrated by dashed lines in Fig. 12-22. An operator looking through binocular eyepieces along the optical paths sees the diapositives directly and perceives the stereomodel. Objective lenses V' and V'' are situated in the optical trains directly beneath the diapositives. The lenses are oriented

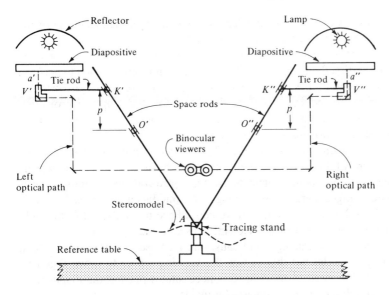

Figure 12-22 Basic principles of mechanical projection.

so that viewing is orthogonal to the diapositives; consequently the diapositive image planes (emulsion surfaces) can lie on the upper side of the diapositive glass with no refraction error being introduced because no rays pass obliquely through the glass. A reference half mark is superimposed on the optical axis of each of the lenses V' and V''. Movement is imparted to the lenses from the space rods by means of tie rods connected at another set of gimbal joints K' and K''. These joints are fixed in vertical position, and the vertical distance from lower gimbal joints O' and O'' to their corresponding upper gimbal joints K' and K'' is the principal distance p. During interior orientation, this distance is set equal to the principal distance of the diapositives.

The space rods intersect at the *tracing stand*. Manually moving the tracing stand imparts movement to the space rods, which in turn impels the viewing system and makes it possible to scan the diapositives. By manipulating the tracing stand in the X, Y, and Z directions, the optical axes of lenses V' and V'' can be placed on corresponding images such as a' and a''. This will occur when the reference half marks fuse into a single mark that appears to rest exactly on the model point. If orientation of the instrument has been carefully performed, with the floating mark fused on point a, the space rods have the same orientation that the incoming light rays from terrain point A had at the time of photography, and the intersection of space rods locates that model point. Each additional model point is located in this same manner.

By geometric comparison, the mechanical projection system illustrated in Fig. 12-22 is exactly the same as direct optical projection. Diapositives are placed in the carriers with their overlapping areas toward the outside. In scanning the diapositives, if the tracing stand is moved right, the viewing lenses move left, and vice versa. Also, if the tracing stand is pushed backward, the viewing lenses move forward, and vice versa. When the tracing stand screw is turned to raise the space rod intersection, the parallactic angle increases and the viewing lenses move apart, a manifestation of the increased x parallax which exists for higher model points.

The carriers of most mechanical projection stereoplotters are capable of rotations and translations, either directly or indirectly. These carrier motions are used to perform relative and absolute orientation. Many mechanical projection instruments are oriented using exactly the same procedures that were described for direct optical projection stereoplotters. Others which do not possess all six rotations and translations use slight variations from these basic orientation procedures.

The Wild *A-8 Autograph* shown in Fig. 12-23 is very similar in design to the hypothetical instrument of Fig. 12-22. There are some basic differences, however. Instead of the viewing lenses being connected to the space rods by tie rods, each is connected by a three-hinged parallelogram tong. The middle hinge of each tong is fixed in position. With this mechanism, a movement of the tracing stand to the right causes a corresponding movement of the viewing lenses to the right, and vice versa. Also an upward motion of the tracing stand causes an increase in parallactic angle at the intersection of space rods, but this causes the viewing lenses to move together. Unlike the previously described instrument, this necessitates placing the diapositives in the carriers with overlapping areas toward the inside.

A second basic difference is that the tracing stand is impelled in the X and Y directions by means of two handwheels in front of the instrument, while Z motion is introduced by pedaling a foot disk. Pictured with the A-8 Autograph is its standard coordinatograph, upon which map compilation is done. X and Y movements of the tracing stand are transmitted mechanically to the tracing pencil on the coordinatograph. By changing gear ratios, various enlargements or reductions from model scale to map scale may be achieved. Model coordinates can also be read directly from graduated X and Y scales on the coordinatograph. Z model coordinates are read directly from a scale graduated to record up and down motion of the tracing stand. The optional

Figure 12-23 Wild Autograph A-8. (*Courtesy Wild Heerbrugg Instruments, Inc.*)

electronic equipment shown with the A-8 Autograph in Fig. 12-23 enables coordinates to be recorded automatically.

Interior orientation of the A-8 Autograph consists of preparing the diapositives by either contact printing or projection printing, centering the diapositives in the carriers by means of collimation marks, setting off the proper principal distance, and if necessary, inserting distortion correction plates. With mechanical projection systems, a wide range of principal distances and model projection distances are readily achieved; the A-8, for example, will accommodate any principal distance between 98 and 215 mm, and its model Z range is from 175 to 350 mm.

Each carrier of the A-8 is capable of the three rotations, but its only translation is an X translation for scaling models. Relative orientation of this instrument is by the two-projector method outlined in Sec. 12-8. The model is brought to scale using the same procedure described in Sec. 12-9, except that the base is increased or decreased as required by increasing or decreasing the spacing of gimbal joints O' and O''. As described in Sec. 12-9, the model is checked for level by reading elevations of at least three vertical control points. Corrective X tilts are introduced with equal amounts of omega rotation (tilt) to each carrier, and corrective Y tilts are applied using a common phi rotation (tip) which simultaneously rotates both carriers. Graduated dials which record omega, phi, and kappa rotations facilitate both relative orientation and absolute orientation.

Another stereoplotter utilizing mechanical projection is the Galileo *Stereosimplex G-7* pictured in Fig. 12-24. The operation of this plotter is similar in principle to the hypothetical instrument illustrated in Fig. 12-22, except that the optical system is fixed and the space rods impel the diapositive carriers for scanning the model in the X and Y directions. X and Y scanning motions are introduced manually using a pantograph handle with one hand, while Z motions are introduced by turning a foot disk. Movements in X and Y are transmitted to the plotting table at the desired enlargement or reduction ratio via the pantograph. Each carrier of the instrument has omega, phi, and kappa rotations, and its common phi and common omega (which tilt both projectors simultaneously by equal amounts) facilitate leveling models. Its range in principal distance accommodation is continuous from 85 to 310 mm, and its projection distance range is from 155 to 460 mm.

The Wild *B-8 Aviograph* shown in Fig. 12-25 is a mechanical projection instrument designed principally for mapping. A schematic diagram of its projection system is shown in Fig. 12-26. Gimbal joints O' and O'' are the projection centers. The viewing system is driven by tie rods connected to the space rods at gimbal joints K' and K''. The unusual feature of this instrument is that joints K' and K'' are in a *negative* position below projection centers O' and O''. This enables the instrument to be more compact with shorter space rods. The vertical distance between projection centers O' and O'' and their corresponding gimbal joints K' and K'' is the principal distance. The B-8 does not have a continuous range of principal distance accommodation. Rather, a separate set of carriers must be obtained for each nominal principal distance desired. For a principal distance of 152 mm, its projection distance range is from 212 to 350 mm. The model is scanned in X and Y by manually pushing the tracing stand about while at the same time raising or lowering the floating mark with a finger screw on

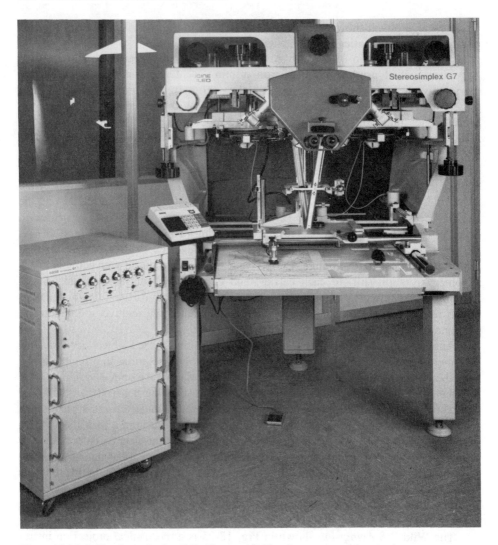

Figure 12-24 Galileo Sterersimplex G-7. (*Courtesy Galileo Corporation of America.*)

the tracing stand. If the tracing stand is pushed to the right, the viewing lenses V' and V'' also move to the right, and vice versa. Therefore, the diapositives are centered in the carriers with their overlap areas toward the inside, just as with the A-8 Autograph. A pantograph transmits X and Y motion to the tracing table. The B-8 has the same motions as the A-8 and is oriented in the same manner. There are a large number of B-8 stereoplotters in use in the United States. This instrument, however, is no longer manufactured, having been replaced by a new family of Wild *Aviomap* Stereoplotters.

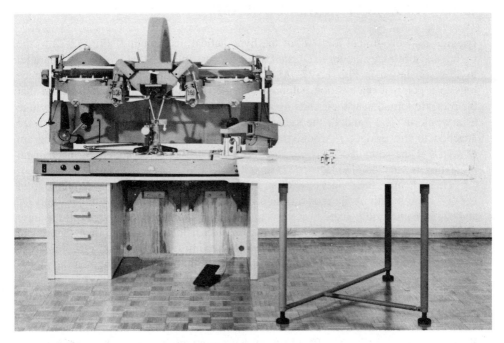

Figure 12-25 Wild B-8 Aviograph. (*Courtesy Wild Heerbrugg Instruments, Inc.*)

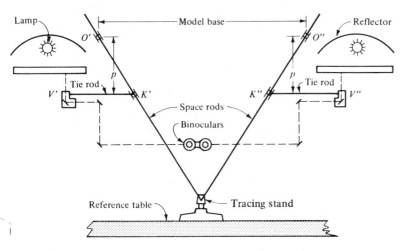

Figure 12-26 Schematic diagram of projection system of Wild B-8 Aviograph.

12-12 THE ZEISS PARALLELOGRAM

Imagine moving gimbal joint O'' of the hypothetical instrument of Fig. 12-22 to the right some arbitrary amount to O_1'', and in the process imagine that $O_1''A_1$ is held parallel to $O''A$, as shown in Fig. 12-27. Also, instead of extending the length of the space rods to a point of intersection, suppose that their original length is retained but that the rods are joined near their ends by a *base bridge*. Then, by moving the base bridge as a whole in X, Y, and Z, the stereomodel could be scanned, and if a pencil were attached to the bridge, it would trace the scanning movement. The parallelogram $O''AA_1O_1''$ is the so-called *Zeiss parallelogram*. Its use in stereoplotters of mechanical projection is very common.

Theoretically, the model base of an instrument employing the Zeiss parallelogram is $O'O''$. The length of the model base can be changed, even though gimbal joints O' and O_1'' remain rigidly fixed, by varying the length of the base bridge. If the base bridge is lengthened from S to, say, S_L, as shown in Fig. 12-28, then theoretical position O'' moves to O_L'' and the model base becomes shorter. This, of course, decreases model scale accordingly. If S is shortened, the effect is to lengthen the model base and increase model scale.

From the foregoing discussion it is seen that because of the Zeiss parallelogram, variations in the x component of the model base are introduced at the base bridge, and an increase in base bridge length causes a decrease in model base, and vice versa. Figure 12-29a and Fig. 12-29b illustrate that variations in y and z base components are also introduced at the base bridge. In Fig. 12-29a, imagine introducing a positive Δb_y at the right side of the base bridge. Since gimbal joint O_1'' is fixed in position, the space rod tilts. To complete the Zeiss parallelogram holding $O''A$ parallel to $O_1''A_1$, negative Δb_y must exist at O''. Hence a positive Δb_y introduced at the right side of the base-bridge causes negative Δb_y at the right carrier, and vice versa. Similar analysis of Fig. 12-29b illustrates that introducing positive Δb_z at the right end of the base bridge results in negative Δb_z at the right carrier, and vice versa. Translations in Y and Z to the left carrier would be introduced at the left side of the base bridge.

Some advantages realized from using the Zeiss parallelogram design are stability

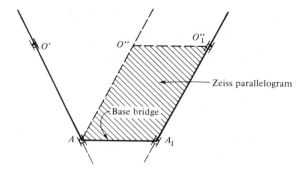

Figure 12-27 Zeiss parallelogram.

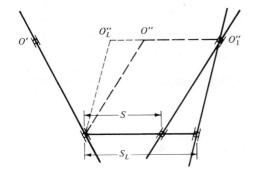

Figure 12-28 Decrease in model base is achieved by increase of base-bridge length.

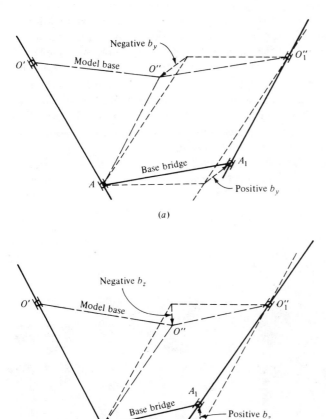

(a)

(b)

Figure 12-29 Introduction of b_y and b_z translations at base-bridge. Note that a positive b_y at base-bridge imparts a negative b_y to the right projector in (a) and positive b_z at base-bridge imparts negative b_z to the right projector in (b).

because of the rigidly fixed positions of the perspective centers, a wide range in possible model bases (including zero base), and the capability of extending the base bridge to the so-called *base-out* position. As discussed in Chap. 14, this makes possible continuous orientation of all photos in a strip; a useful feature in extending control photogrammetrically. The significance of this base-out feature has been diminished in recent years, however, by semianalytical aerotriangulation (also discussed in Chap. 14).

12-13 INSTRUMENTS USING PARALLELOGRAM PROJECTION SYSTEMS

The Wild *A-7 Universal Autograph* of Fig. 12-30 is a mechanical projection instrument which uses the Zeiss parallelogram and has the capability of being operated in the base-in or base-out position (see Sec. 14-4). This plotter can accept photos with a principal distance of anywhere from 98 to 215 mm, and its projection distance range is from 140 to 490 mm. Its viewing and measuring systems are similar to those of the A-8 Autograph. Its diapositive carriers each have the three rotations, and all three translations can be introduced at either end of the base bridge. The instrument may be relatively oriented using either the two-projector method or the one-projector method. The model base is adjusted for model scaling by translations at the base bridge. In leveling, if base components b_y and b_z are zero, corrective X tilts are applied by imparting equal omega rotations to both carriers, and corrective Y tilts are introduced with equal amounts of phi rotation to each carrier, followed by a b_z translation. The A-7 is no longer manufactured, largely because semianalytical aerotriangulation has eliminated the need for base-in and base-out instruments.

The Kern *PG-2* stereoplotter shown in Fig. 12-31 also employs the Zeiss parallelogram. This instrument has the unique feature that the diapositive carriers remain

Figure 12-30 Wild A-7 Universal Autograph. (*Courtesy Wild Heerbrugg Instruments, Inc.*)

Figure 12-31 Kern PG-2. (*Courtesy Kern and Co., Inc.*)

horizontal throughout orientation. The instrument has a continuous range in principal distance of from 85 to 172 mm, and its range in projection distance is from 102 to 172 mm by increments of 10 mm. It has the usual kappa rotations, but omega for the left carrier only, and phi for the right carrier only, are introduced by means of tilts applied to the space rods. The viewing system of the PG-2 is orthogonal to the diapositives. The procedure of relative orientation deviates slightly from the two-projector method described in Sec. 12-8 because the necessary rotations are not all available. The model is scaled by varying the length of the base bridge, and it is leveled by tilting the table. Scanning in X and Y is done by manually pushing the tracing stand—these motions being transmitted to the plotting table by means of a pantograph. Z motion is imparted at the base bridge by means of a finger screw, and model elevations can be read directly from a dial geared to the base bridge. A substantial number of PG-2 stereoplotters are in current use in the United States.

Figure 12-32 shows the *Planicart E-3,* one of a family of three basic stereoplotteers recently introduced by Carl Zeiss, Inc. The Planicart uses the Zeiss parallelogram method of projection. Its principal distance can be set at each of the nominal values of 88, 115, 152, 210, and 305 mm to accommodate the most commonly used aerial cameras. At each of these nominal settings the principal distance can be varied through a range of 6 mm. The projection distance range varies, depending upon principal

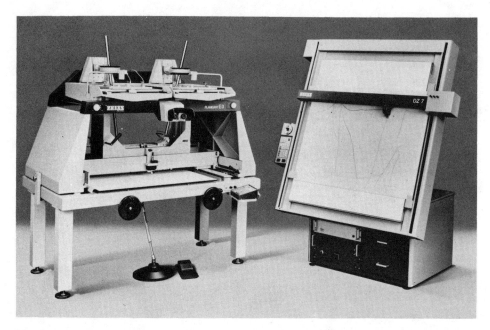

Figure 12-32 Zeiss Plaincart E-3 with digital tracing table. (*Courtesy Carl Zeiss, Oberkochen.*)

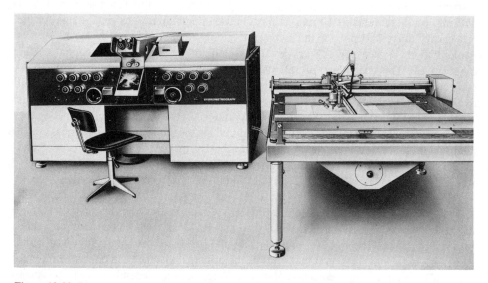

Figure 12-33 Stereometrograph with coordinatograph. (*Courtesy Jenoptik, Jena.*)

distance, and at a p of 152, Z can range from 193 to 543 mm. Pictured with the Planicart is a digitized automatic tracing table. These are described in Sec. 12-16.

The Jenoptik *Stereometrograph* shown in Fig. 12-33 also uses the Zeiss parallelogram method of projection. The instrument is completely encased to protect diapositives and internal parts from dust. The ranges of principal distance and projection distance of the Stereometrograph are continuous from 85 to 310 mm, and from 135 to 350 mm, respectively. This instrument will accommodate terrestrial as well as aerial photos.

Figure 12-34 shows the Wild *A-10 Autograph*. This instrument uses a parallelogram projection system that differs somewhat from the standard Zeiss parallelogram in that the two space rods exist in two different XZ planes. As with the B-8 Aviograph, gimbals K' and K'' are in a negative position below the projection centers. The ranges of principal distance and projection distance for the A-10 are continuous from 85 to 308 mm, and from 130 to 450 mm, respectively. This instrument will also accommodate both aerial and terrestrial photos.

Each of the instruments shown in Figs. 12-32 through 12-34 is driven in the X and Y directions by means of handwheels, while their Z motions are controlled by a foot disk.

12-14 INSTRUMENTS WITH OPTICAL-MECHANICAL PROJECTION

A few stereoplotters have projection systems which are partly optical and partly mechanical, but these instruments are not as common as those of purely mechanical

Figure 12-34 Wild A-10 Autograph. (*Courtesy Wild Heerbrugg Instruments, Inc.*)

projection. The Zeiss *C-8 Stereoplanigraph* pictured in Fig. 12-35 is an instrument using a combination of optical and mechanical projection. With this instrument, diapositives are centered in the two projectors and illuminated from above. Corresponding images are optically projected through the projector objective lenses and come to focus on a pair of reference mirrors, as shown in the diagram of Fig. 12-36. The rays are reflected by the mirrors into two optical trains and are viewed by an observer through binoculars. A special mechanical linkage aligns the mirrors so that the reflected rays are received in the optical trains regardless of the area of the diapositives being viewed. Reference half marks superimposed at each reference mirror fuse into a floating mark that appears to rest exactly on a model point when the half marks are set on corresponding images.

An auxiliary *Bauersfeld* lens system attached beneath each projector objective lens, as illustrated in Fig. 12-37, assures perfect focus of projected rays on the reference mirrors regardless of projection distance. The focal length f' of the projector lenses is equal to the nominal principal distance, and therefore projected rays emerge from the projector lenses parallel. These parallel rays enter the positive and negative Bauersfeld lens combination—the two lenses being separated by a spacing e. Distance e is mechanically varied to maintain focus on the mirrors according to variations in the projection distance h.

Each projector of the C-8 Stereoplanigraph has the customary three angular rotations, but translations are introduced as movements of the reference mirrors. This is the case because the instrument utilizes the Zeiss parallelogram, as illustrated in Fig. 12-36. Base components b_x, b_y, and b_z are varied in essentially the same manner as described for mechanical projection instruments that use the Zeiss parallelogram, except that with the C-8 the reference half marks are translated instead of the ends of the base bridge.

Model scanning in the X and Y directions is introduced with the C-8 by means of handwheels, while Z motion is imparted by pedaling the foot disk. A unique feature

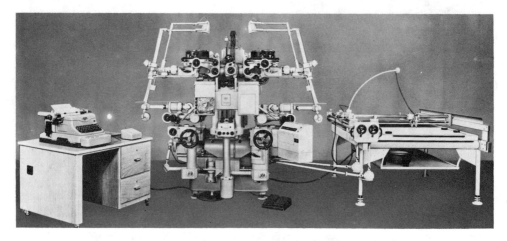

Figure 12-35 Zeiss C-8 Stereoplanigraph. (*Courtesy Carl Zeiss, Oberkochen.*)

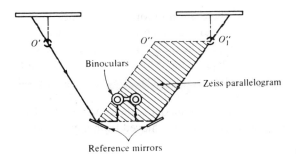

Figure 12-36 Schematic diagram of projection system of C-8 Stereoplanigraph.

of the Z motion of the C-8 is that the two projectors are raised or lowered to accommodate variations in terrain. The Y and Z motions of the C-8 can be interchanged for plotting from terrestrial photographs. It can accommodate a wide range of model scales, having a projection distance variable from 170 to 605 mm. This instrument is a universal stereoplotter capable of the base-in and base-out positions and is therefore suitable for continuous strip control extension as well as for map compilation.

12-15 MODEL AND MAP SCALES OF MECHANICAL PROJECTION INSTRUMENTS

As noted in the preceding sections, with mechanical projection stereoplotting instruments, images are not projected through objective lenses, but rather projection is

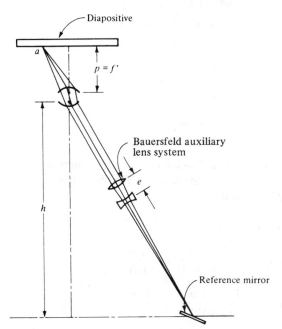

Figure 12-37 Auxiliary lens system which maintains sharp focus for all parts of the stereomodel.

accomplished by means of space rods. Thus, there are no projection distance limitations imposed by the depth of field of a lens, and the range in projection distance, which is simply a matter of physical limitations, can be quite large. As a result, model scales calculated by Eq. (12-6) can vary considerably more for mechanical projection stereoplotters than they can for optical projection instruments. Also, for mechanical projection instruments that use coordinatographs and pantographs, a wide range in enlargement or reduction ratios can be achieved from model scale to map scale. This means that great versatility in map plotting scales can be achieved from a given photo scale.

Example 12-5 What range in map scales is possible for a Wild A-8 Autograph for photography taken with a 6-in- (152-mm-) focal-length camera at a flying height of 6,000 ft above mean terrain? Assume that the A-8 is equipped with a coordinatograph capable of $4\times$ reduction or enlargement from model scale. (The A-8 has a projection distance range of from 175 to 350 mm.)

SOLUTION (a) By Eq. (12-5), photo scale is

$$S_p = \frac{6 \text{ in}}{6,000 \text{ ft}} = \frac{1 \text{ in}}{1,000 \text{ ft}}$$

(b) By Eq. (12-6), model scale range is

$$\text{Minimum } S_m = \frac{175}{152}\frac{1 \text{ in}}{1,000 \text{ ft}} = \frac{1 \text{ in}}{870 \text{ ft}}$$

$$\text{Maximum } S_m = \frac{350}{152}\frac{1 \text{ in}}{1,000 \text{ ft}} = \frac{1 \text{ in}}{435 \text{ ft}}$$

(c) Map scale range is

$$\text{Minimum} = \frac{1 \text{ in}}{870 \text{ ft}}\frac{1}{4} = \frac{1 \text{ in}}{3480 \text{ ft}}$$

$$\text{Maximum} = \frac{1 \text{ in}}{435 \text{ ft}}4 = \frac{1 \text{ in}}{110 \text{ ft}}$$

Example 12-6 Solve Example 12-5 for a PG-2 Stereoplotter equipped with a pantograph of $2\times$ reduction and $3.333\times$ enlargement. (The PG-2 has a projection distance range of from 102 to 172 mm.)

SOLUTION (a) By Eq. (12-6), model scale range is

$$\text{Minimum } S_m = \frac{102}{152}\frac{1 \text{ in}}{1000 \text{ ft}} = \frac{1 \text{ in}}{1490 \text{ ft}}$$

$$\text{Maximum } S_m = \frac{172}{152}\frac{1 \text{ in}}{1000 \text{ ft}} = \frac{1 \text{ in}}{885 \text{ ft}}$$

(*b*) Map scale range is

$$\text{Minimum} = \frac{1 \text{ in}}{1490 \text{ ft}} \frac{1}{2} = \frac{1 \text{ in}}{2980 \text{ ft}}$$

$$\text{Maximum} = \frac{1 \text{ in}}{885 \text{ ft}} \, 3.333 = \frac{1 \text{ in}}{265 \text{ ft}}$$

12-16 DIGITIZED AUTOMATIC COORDINATOGRAPHS

Digitized automatic coordinatographs which can be directly connected to stereoplotting instruments have recently been introduced into the photogrammetric industry by several manufacturers. One of these is shown in Fig. 12-32. These systems have greatly increased the overall speed and efficiency of map compilation and drafting.

Automatic tracing systems consist basically of precision coordinatographs driven by servosystems upon commands from built-in microprocessors or external computers. The operator can interact with the system and control the various modes of operation by means of a keyboard. While the individual capabilities of the various models differ somewhat, in general each of them can be operated in three basic modes. In one of these modes, the systems can be used as automatic coordinatographs to plot positions of points from their input coordinates. Rapid and accurate plotting of photo control points and coordinate grid lines for the preparation of manuscripts is one of the useful applications of this mode of operation.

In a second mode, the instruments can be operated in a manner similar to conventional stereoplotting. In this mode the systems will simultaneously follow the X and Y movements of the plotter operator throughout the stereomodel and plot them on the map in real time. Extreme flexibility in the selection of scale factors is possible, and a variety of line types can be drawn, e.g., solid lines, dashed lines, dotted lines, or dot-dashed lines.

In the third mode of operation, the systems will draft straight lines defined by two points. In plotting planimetry in this mode, an operator need not trace entire straight-line features. Rather, the reference mark is set in the stereomodel on each end of the straight-line feature, whereupon the chosen line type is automatically drawn between them. This is particularly useful for plotting roads, railroads, property boundaries, buildings, etc. Special software has been developed so that in plotting rectangular features such as buildings, only three points need be located and the system will automatically plot a closed rectangle. Similar programming enables automatic plotting of circles if three points on the arc are located in the stereomodel. Frequently used symbols for feature representation can also be automatically plotted on command. Crosses, squares, triangles, circles, tree symbols, etc., are some examples, and their size can be readily varied. Lettering can also be automatically performed by these devices.

Plotting can be accomplished with lead pencils, ball point pens, ink pens, or scribing tools. Thus, maps in near final cartographic form can be directly produced.

The tables can be tilted for ease of viewing by the operator. The technology of digitized automatic coordinatographs has recently become so advanced that it is now possible for an operator to give voice commands to control the plotting and recording of information. This has circumvented the need to make keyboard entries and made compilation efficiency even greater.

PART III   ANALYTICAL PLOTTERS

12-17 INTRODUCTION

In photogrammetry, a number of operations have been partially automated for many years. One of the earliest applications of partial automation was in digitizing data from the measurement systems of various instruments through the use of encoders and digital recording apparatuses such as card punchers and magnetic tape units. One example of an early application of this type, discussed briefly in Sec. 5-7, has been in connection with comparators for automatic recording of photocoordinates. This type of equipment has also been used for several years to digitize X, Y, Z model coordinates from stereoplotters, as noted in Sec. 12-6. Another more recent example of the application of automatic recording equipment is in digitizing data obtained from scanning microdensitometers (see Sec. 5-15).

Automatic coordinatographs, described in Sec. 12-16, are now used in conjunction with stereoplotters. This equipment exploits the concept of automatically driving an interconnected system component (in this case a coordinatograph connected electronically to a stereoplotter) by means of servomotors which act on commands from a computer. The computer is able to comprehend stereoplotter model coordinates on the basis of information supplied to it from encoders on the stereoplotter, and is able to drive the coordinatograph to its required map locations also on the basis of encoders mounted on the coordinatograph. Of course some rather sophisticated software is needed to control the operations of the computer.

The basic components mentioned above—encoders, servosystems, and computers—have been linked with other conventional photogrammetric instruments and procedures to develop some truly ingenious automated equipment. Analytical plotters, described in the following sections, are an example.

12-18 SYSTEM COMPONENTS AND METHOD OF OPERATION

Analytical plotters consist basically of a precision stereocomparator (see Sec. 14-11) and a coordinatograph interfaced with a computer. These components are illustrated in the schematic diagram of Fig. 12-38. Servomotors and encoders are integral parts of the system enabling the computer to drive the various components of the equipment

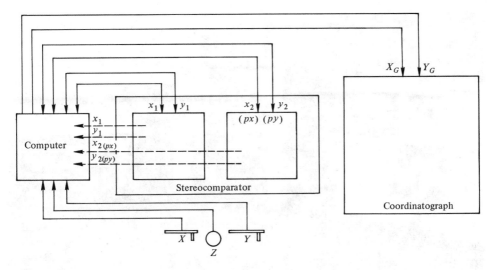

Figure 12-38 Schematic diagram of components and operation of an analytical plotter.

so that it can perform a wide variety of photogrammetric operations in a highly automated fashion. Again, sophisticated software controls the operation of the system.

Analytical plotters form neither an *optical* nor a *mechanical* model, as do all the stereoplotters described in the preceding sections of this chapter. Rather they compute a *mathematical* model using the collinearity equations (see Appendix C). External input for the solution of the equations consists of camera interior orientation parameters and ground coordinates of control points, and internal input consists of image coordinates measured by the instrument itself. From these data the computer calculates, in real time, model coordinates and other forms of useful output data and then displays the information on a screen, prints it in hard copy form, or transmits it to a coordinatograph for plotting.

The concept of the analytical plotter was first patented in 1957 by Dr. U. V. Helava, and these instruments have been undergoing constant research and development since that time. The AP/C, shown in Fig. 12-39, was the first analytical plotter marketed for commercial use, and this occurred in 1964. Because of their many advantages, acceptance of the analytical plotters has been tremendous, and since 1976 approximately 15 new models have been introduced. Figures 12-40 and 12-41 show two of these: the Zeiss Planicomp C-100, and the Matra Traster.

12-19 ADVANTAGES OF ANALYTICAL PLOTTERS

Because they have no optical or mechanical limitations in the formation of their mathematical models, analytical plotters have great versatility. They can handle any type of photography, including vertical, tilted, low oblique, convergent, high oblique, panoramic, and terrestrial photos. They can also be used with radar imagery. In

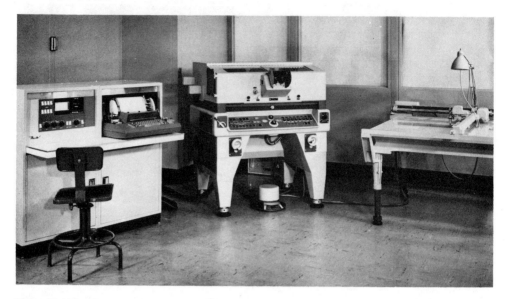

Figure 12-39 Analytical plotter AP/C. (*Courtesy O.M.I. Corp. of America.*)

addition, they can accommodate photography from any focal length camera, and in fact can simultaneously use two photos of different focal lengths to form a model.

In comparison with analog plotters, analytical plotters can provide results of superior accuracy for basically three reasons. First of all, because they do not form model points by intersecting projected light rays or mechanical space rods, optical and mechanical errors from these sources are not introduced. Second, they can effec-

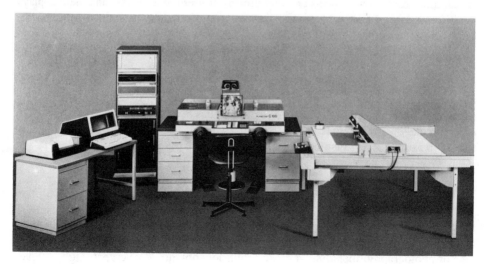

Figure 12-40 Zeiss Planicomp C-100 analytical plotter. (*Courtesy Carl Zeiss, Oberkochen.*)

Figure 12-41 Matra Traster analytical plotter. (*Courtesy Matra Technology, Inc.*)

tively correct for any combination of systematic errors caused by camera-lens distortions, film shrinkage or expansion, atmospheric refraction, and earth curvature, Third, in almost every phase of their operation, they can take advantage of redundant observations and incorporate the method of least squares into the solution of the equations.

12-20 ANALYTICAL PLOTTER ORIENTATION

As is necessary with all analogue stereoplotters, interior, relative, and absolute orientation are also required for analytical plotters prior to going into most modes of operation. In all phases of orientation, and operation for that matter, a dialogue is maintained between operator and instrument. The system's computer directs the operator through each of the various steps of operation, calling for data when needed and giving the operator opportunities to make certain decisions. Messages from the computer are displayed on a TV screen, and data is input by the operator either via a keyboard or by way of measurements using the instrument.

The orientation and operation of all analytical plotters is quite similar. Figure 12-42, which illustrates the various reference coordinate systems pertinent to analytical plotters, will facilitate a discussion of the orientation of the instruments.

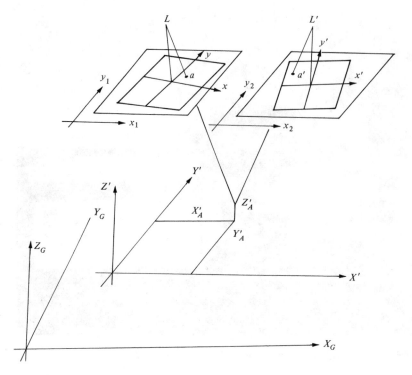

Figure 12-42 Reference coordinate systems of an analytical plotter.

12-20.1 Interior Orientation

In interior orientation, a stereopair of diapositives with xy and $x'y'$ fiducial coordinate systems is placed on the measuring stages as illustrated in Fig. 12-42. Precise centering precautions need not be taken. The principal distances of the diapositives and fiducial coordinates are input to the computer. Machine image coordinates x_1y_1 and x_2y_2 of the diapositive fiducials are then read. This phase of the operation can be aided by computer activated servomotors which automatically drive the measuring mark to the vicinity of the fiducials. A fine pointing is then made by the operator. Hand wheels and a foot disk are the usual means provided to move the reference measuring marks. As few as two fiducials can be measured, but more are recommended and up to eight should be measured if they are available to increase redundancy. From this information the computer solves a coordinate transformation, using least squares if sufficient measurements were taken, to locate the principal points of the diapositives and determine the relationships of the two photo coordinate systems with respect to the instrument's image coordinate measurement systems. Corrections for shrinkage or expansion are included in the transformation. A choice of coordinate transformations is available but usually the affine or projective types are used (see Secs. B-6 and B-8 of Appendix B). Residuals for this solution will be displayed so that the operator can either accept it or remeasure the fiducials. When the solution is accepted, the interior orientation parameters are stored in the computer. From information entered into the computer, corrections for lens distortions, atmospheric refraction, and earth curvature can be made.

12-20.2 Relative Orientation

For relative orientation, xy and $x'y'$ machine image coordinates are measured at a minimum of five points (at least six are recommended) located in the approximate positions shown in Fig. 12-14. Again the computer will drive the measuring mark to the approximate locations, whereupon the operator makes a precise pointing. More than six points are recommended to increase redundancy, and most analytical plotters can accommodate up to 20 or more points for this phase of orientation. Based upon these measurements the computer calculates the elements of relative orientation using basically the methods described in Sec. 14-13. The computations are performed using least squares if more than five points are involved in the solution. Again the residuals will be displayed, and the operator has the option of discarding certain points, adding others, or accepting the solution. When relative orientation is accepted, the operator notifies the computer, and the orientation parameters are stored for future use. Based upon these interior and relative orientation parameters now in storage, given any set of machine image coordinates for conjugate points such as a and a', the computer can calculate the corresponding $X_A'Y_A'Z_A'$ model coordinates of that point, as shown in Fig. 12-42.

12-20.3 Absolute Orientation

In absolute orientation, the ground coordinates of all control points must be first input to the computer. The operator then places the reference mark stereoscopically on corresponding images of the ground control points. For absolute orientation a minimum of two horizontal and three well-distributed vertical control points are required, just as for absolutely orienting analogue plotters. More than the minimum is recommended, however, so that a least squares solution can be made. When the measurements have been taken the computer solves a three-dimensional coordinate transformation (see Sec. B-7 of Appendix B) to determine the parameters that relate the $X'Y'Z'$ model coordinate system of Fig. 12-42 to the $X_GY_GZ_G$ ground coordinate system. As before, a display of residuals is produced and the operator can delete or add points, or accept the solution. When the solution is accepted, the computer stores the parameters of absolute orientation.

An analytical plotter can be oriented considerably faster than an analog plotter and usually it can be accomplished in 15 min or less. With the three steps of orientation completed, and the associated parameters all in memory, an operator merely need set the floating mark on any unknown point, whereupon the computer can instantaneously calculate ground coordinates of the point. Conversely, given a set of ground coordinates, the computer can immediately determine the corresponding machine image coordinates and drive the measuring mark to that location. These capabilities set up a multitude of uses for analytical plotters.

12-21 MODES OF USE OF ANALYTICAL PLOTTERS

The modes of use of analytical plotters vary only slightly with the different instruments. Generally all of them can perform planimetric and topographic mapping. In this case the operator simply traces out features and contours by keeping the floating mark in contact with the feature. The computer continually, and extremely rapidly, solves the equations to transform machine image coordinates into the ground system. The information can be transmitted on-line to an automatic plotting table to provide a graphic rendition (map), or the data can be stored in digital form for subsequent off-line plotting or the development of digital terrain models.

Analytical plotters can also be operated in a profiling mode. In this operation X, Y, and Z coordinates can be measured along predetermined lines in the model. The locations and directions of desired profile lines can be input to the computer in terms of XY ground coordinates and the machine will automatically drive the measuring mark to the corresponding locations while an operator monitors the floating mark to keep it in contact with the ground. The output is normally recorded in digital form. Profile data can be used directly for various engineering applications such as investigating earthwork quantities along proposed centerlines, or it can be used to drive off-line orthophoto printers (see Chap. 13). In addition, digital terrain models can be developed from the data.

Another common application of analytical plotters is in aerotriangulation. Here various modes of operation are possible, from independent model triangulation to simultaneous block adjustment (see Chap. 14). When the aerotriangulation is completed, the diapositives can be used in other stereoplotters or orthophoto instruments, or they can be reset in the analytical plotter based upon the now available ground control information. As a result of aerotriangulation, the parameters of relative and absolute orientation become known. Analytical plotters have a "reset" capability which enables reading in these known parameters, whereupon an instantaneous relative and absolute orientation can be achieved automatically, thus bypassing the measurements otherwise needed to determine those two steps. In fact, if for any reason a model that was once set should need to be reset to obtain additional data, the orientation parameters from the first setting can be recalled, if they were held in storage, and the system will automatically perform relative and absolute orientation.

Analytical plotters can also be used simply as monocomparators or stereocomparators, where image coordinates are recorded for use in aerotriangulation using other external computers.

The methods of operation and capabilities of the Planicomp C-100 Analytical Plotter shown in Fig. 12-40 are closely representative of the general descriptions given herein. This instrument, which has been kept relatively compact, has a conventional binocular stereoviewing system. Motions in the X and Y directions are imparted by means of the two handwheels shown on the front of the instrument, while Z motion is controlled by means of a foot disk.

The Traster shown in Fig. 12-41 has a unique and convenient viewing system which significantly reduces operator fatigue. With this instrument, corresponding imagery from the left and right diapositives is projected onto a large screen at the front of the instrument. Light of opposite polarity is used in the projection, and an operator wearing spectacles of the same opposite polarity is able to view the stereoimage in three dimensions and set the floating mark. Motions in X, Y, and Z are introduced by hand controls with the Traster—X and Y being regulated by a "track ball" on the right side of the console, and Z being controlled by a "rotating drum" on the left side of the console.

PART IV   STEREOPLOTTERS WITH AUTOMATIC IMAGE CORRELATORS

12-22 INTRODUCTION

Whereas the analytical plotters described in the preceding sections have been automated to a very high degree, they are still dependent upon human operators who must make measurements stereoscopically. Instruments have been developed, however, in which automation has been advanced to the point where humans can be almost totally replaced. As with analytical plotters, these instruments use computers, encoders, and servomotors, but in addition they also employ *automatic image correlators* to perform stereoscopic measurements.

Image correlation is the process of selecting images from within the overlap areas of the left and right photos of a stereopair that correspond. It is done on the basis of comparisons of the shapes, sizes, and densities of the images. The human eye/brain combination can make these comparisons, almost unconsciously, in a very rapid and accurate manner. Although they cannot perform the task as well as humans, automatic image correlators, which are products of recent advances in electronic and computer technology, can also perform these comparisons to select corresponding imagery.

Image correlation is necessary in many photogrammetric operations and is especially essential in the use of stereoplotting instruments. As described in earlier sections of this chapter, image correlation is performed in relative orientation as x and y parallax are cleared in the stereomodel to bring corresponding imagery into coincidence. Image correlation is also performed in absolute orientation, and in map compilation and other phases of stereoplotter use, as x parallax is removed to raise or lower the floating mark and keep it in contact with model points.

12-23 AUTOMATIC IMAGE CORRELATORS

The operation of automatic image correlators involves basically two distinct functions: (1) *scanning* the imagery to record the spatial positions and density differences of the various picture elements of the two overlapping photos, and (2) *correlating* the data obtained in scanning to identify corresponding imagery. In general, the scanning operation is performed electronically, while correlation may be done either numerically or electronically. Numerical image correlation is described in Sec. 14-20 in connection with automated systems for analytical aerotriangulation, while an electronic system, used in conjunction with a standard stereoplotter, is described in the following section.

12-24 ELECTRONIC IMAGE CORRELATORS AND THE WILD B-8 STEREOMAT

An electronic image correlation system has been combined with the standard Wild B-8 Aviograph stereoplotter (see Fig. 12-25) to produce an instrument known as the *B-8 Stereomat*. The system can operate almost totally without human intervention.

Figure 12-43 illustrates the principle of automatic operation of the electronic image correlator of the B-8 Stereomat. The instrument contains two identical electronic scanning systems—one for each of the diapositives of a stereopair. A "flying spot" of light systematically scans back and forth across the face of each of the cathode ray tubes above the diapositives. The lens between the cathode ray tubes and the diapositives focus the flying spots of light onto the diapositives, and the lenses below the diapositives project the spots onto the photomultiplier tubes below. Electronic currents (I) generated by the photomultiplier tubes alternate according to the intensity of light received.

The flying spots above the two diapositives move synchronously with respect to the left and right diapositives. As the spots scan, if they cross corresponding light-dark borders of the two diapositives at different times, the resulting alternating currents generated by the photomultiplier tubes will have different phases. A difference in phase such as T_L and T_R of Fig. 12-43 indicates the presence of parallax, in which case the space rod intersection occurs at an incorrect model point such as A'. If any difference in phase is detected, a servomotor is automatically activated which moves the space rods to such position that the phase difference becomes zero. Then the flying spots will be simultaneously illuminating corresponding images and the space rod intersection will occur at correct model point A. This operation is exactly analogous to a human operator clearing parallax by manually moving the space rods until corresponding images are seen simultaneously by the operator's two eyes. The Stereomat does have an optical viewing system so that an operator can manipulate the instrument manually.

Since the flying spots have both x and y components, it is possible to automatically detect both x parallax and y parallax in a stereomodel. The instrument can relatively orient itself through automatic detection of y parallax. The presence of any y parallax activates servomotors that introduce projector motions to clear it. Automatic clearance of y parallax at the five standard points relatively orients a model. After a model has been oriented, automatic detection and elimination of x parallax enables profiles to be read, contours to be traced, and models to be digitized, i.e., recording X, Y, and Z model coordinates of a dense network of points throughout the model. Profiles can be scanned in either the X or Y direction.

Contours can be traced automatically by the B-8 Stereomat, but an operator must

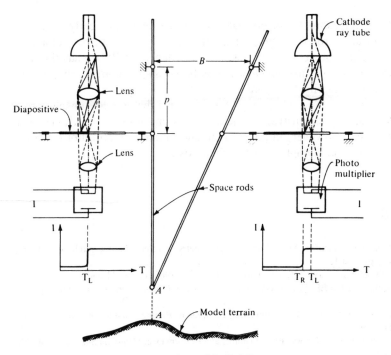

Figure 12-43 Automatic image correlator of the B-8 Stereomat.

first bring the measuring mark in contact with the terrain at the desired starting contour elevation. The instrument will automatically compile the contour until it leaves the model, whereupon it will step to the next contour and begin tracing it. If a contour closes upon itself, the instrument would endlessly retrace that contour unless set to the next contour by an operator. In the automatic contouring mode, therefore, an operator is needed to monitor compilation.

It should be noted that an image correlator such as that used on the B-8 Stereomat compares corresponding images and measures their respective positions on a stereopair of diapositives. It does not recognize or memorize forms or objects. Therefore if drastically differing corresponding imagery is encountered, as sometimes occurs in steep terrain, for example, the instrument may not be able to function. This, of course, would pose no problem for human operators, who can associate common images even though they appear different. When contouring, the automatic image correlator may occasionally come to a point where it cannot find corresponding images. While attempting to find them, it may stray so far away that it cannot correct itself. An alarm alerts an operator to reset the instrument when this occurs.

12-25 OTHER INSTRUMENTS THAT UTILIZE AUTOMATIC IMAGE CORRELATORS

Automatic image correlators have been used successfully for several years in conjunction with analytical plotters. The Universal Automatic Map Compilation Equipment (UNAMACE), for example, incorporated image correlators as early as 1965. This instrument can also produce orthophotos automatically, and utilizes automatic image correlation for this process. The Gestalt Photo Mapper (GPM) is another automated orthophoto-producing instrument which is based upon automatic image correlation. This system is described in more detail in Sec. 13-8.3.

At the present time, automated stereoplotting systems that employ image correlation devices are relatively expensive, and hence the extent of their use is somewhat limited. They are found, however, in government and military organizations, whose volume of mapping is great enough to justify their cost.

REFERENCES

American Society of Photogrammetry: "Manual of Photogrammetry," 3d ed., Falls Church, Va., 1966, chaps. 3, 12, 13, 14, and 15.

————: "Manual of Photogrammetry," 4th ed., Falls Church, Va., 1980, chaps. 11 and 12.

Bean, R. K.: Development of the ER-55 Projector, *Photogrammetric Engineering*, vol. 19, no. 1, p. 71, 1953.

Bertram, S.: The Universal Automatic Map Compilation Equipment, *Photogrammetric Engineering*, vol. 31, no. 2, p. 244, 1965.

Blachut, T. J., et al.: ANAPLOT—History, Basic Features and Performance, *Canadian Surveyor*, vol. 33, no. 2, p. 89, 1979.

Case, J. B.: ASP-DTM Symposium, *Photogrammetric Engineering and Remote Sensing*, vol. 44, no. 12, p. 1477, 1978.

Collins, S. H.: Terrain Parameters Directly from a Digital Terrain Model, *Canadian Surveyor*, vol. 29, no. 5, p. 507, 1975.

Crabtree, J. S., and J. D. McLaurin: The BAI Image Correlator, *Photogrammetric Engineering*, vol. 36, no. 1, p. 70, 1970.

Danko, J. O., Jr.: Color, the Kelsh, and the PPV, *Photogrammetric Engineering*, vol. 38, no. 1, p. 83, 1972.

DeGraaf, R. M.: Automation Characteristics of the Stereomat B8, *Photogrammetric Engineering*, vol. 30, no. 5, p. 818, 1964.

Dorrer, E.: Software Aspects in Desk-Top Computer Assisted Stereoplotting, *Photogrammetria*, vol. 33, no. 1, p. 1, 1977.

Dowman, I. J.: A Working Method for the Calibration of Plotting Instruments Using Computers, *Photogrammetric Record*, vol. VII, no. 42, p. 662, 1973.

————: Model Deformation—An Interactive Demonstration, *Photogrammetric Engineering and Remote Sensing*, vol. 43, no. 3, p. 303, 1977.

Esten, R. S. : Automatic Photogrammetric Instruments, *Photogrammetric Engineering*, vol. 30, no. 4, p. 544, 1964.

Forest, R. B.: AP-C Plotter Orientation, *Photogrammetric Engineering*, vol. 31, no. 6, p. 1024, 1966.

Ghosh, S. K.: *Theory of Stereophotogrammetry*, The Ohio State University Book Stores, Columbus, Ohio, 1968.

————: Photo/Model/Map Scales, *Photogrammetric Engineering*, vol. 32, no. 11, p. 1154, 1971.

Helava, U. V.: The Analytical Plotter—Its Future, *Photogrammetric Engineering and Remote Sensing*, vol. 43, no. 11, p. 1361, 1977.

Hobrough, G., and T. Hobrough: Image Correlator Speed Limits, *Photogrammetric Engineering*, vol. 37, no. 8, p. 1045, 1971.

Katibah, G. P.: Model Flatness—A Guide for Stereo-Operators, *Photogrammetric Engineering*, vol. 30, no. 2, p. 299, 1964.

Kasper, H.: The Wild B-8 Aviograph—A Simple Photogrammetric Plotter, *Photogrammetric Engineering,* vol. 27, no. 4, p. 590, 1961.

Knauf, J. W.: The Stereoimage Alternator, *Photogrammetric Engineering,* vol. 33, no. 10, p. 1113, 1967.

Lawrence, C. H.: Stereomat IV, Automatic Plotter, *Photogrammetric Engineering,* vol. 33, no. 4, p. 394, 1967.

Newton, I.: The CP1—A New Photogrammetric Plotter, *Survey Review,* vol. 22, no. 167, p. 43, 1973.

Noonan, R. P.: A Systematic Procedure for Stereocompilers, *Photogrammetric Engineering,* vol. 36, no. 1, p. 56, 1970.

Petrie, G., and M. O. Adam: The Design and Development of a Software Based Photogrammetric Digitising System, *Photogrammetric Record,* vol. X, no. 55, p. 39, 1980.

Struck, L.: The Multiplex, Kelsh Plotter and Wild Autographs, *Photogrammetric Engineering,* vol. 18, no. 1, p. 84, 1952.

Tewinkel, G. C.: Stereoscopic Plotting Instruments, *Photogrammetric Engineering,* vol. 17, no. 4, p. 635, 1951.

Theis, J. B.: Automation and Photogrammetry, *Photogrammetric Engineering,* vol. 31, no. 2, p. 281, 1965.

Thompson, M. M., and J. G. Lewis: Practical Improvements in Stereoplotting Instruments, *Photogrammetric Engineering,* vol. 30, no. 5, p. 802, 1964.

Trow, S. W., and M. Keller: Transfer of Absolute Orientation from One Type of Stereoscopic Plotting Instrument to Another, *Photogrammetric Engineering,* vol. 19, no. 5, p. 831, 1953.

Veres, S. A.: The Use and Adoption of Conventional Stereoplotting Instruments for Bridging and Plotting of Super Wide Angle Photography, *Canadian Surveyor,* vol. 23, no. 4, p. 359, 1969.

Whiteside, A. E., and C. W. Matherly: Recent Analytical Stereoplotter Developments, *Photogrammetric Engineering,* vol. 38, no. 4, p. 373, 1972.

Zarzycki, J. M.: An Integrated Digital Mapping System, *Canadian Surveyor,* vol. 32, no. 4, p. 443, 1978.

PROBLEMS

12-1 Describe the basic differences between stereoplotters with direct optical projection and instruments with mechanical or optical-mechanical projection.

12-2 Three basic orientation steps are necessary prior to using a stereoplotter. Name them and give the objective of each.

12-3 What is the focal length of the lens in a Balplex projector whose principal distance and optimum projection distance are 55 mm and 525 mm, respectively?

12-4 Repeat Prob. 12-3, except that a Kelsh plotter is being considered whose principal distance and optimum projection distance are 210 mm and 760 mm, respectively.

12-5 Discuss the different viewing systems used in direct optical projection stereoplotters.

12-6 Outline the steps of interior orientation.

12-7 Photography is taken at 60 percent end lap and 30 percent sidelap with a 152-mm-focal-length 9-in-square format camera from a flying height of 8,600 ft above average ground. If diapositives of this photography are to be prepared for use in a Balplex plotter, what values of A and B should be set off in the reduction printer if the reduction printer has a lens with a focal length of 115 mm?

12-8 For the photography of Prob. 12-7, what will be optimum model scale for a Balplex stereoplotter having an optimum projection distance of 525 mm?

12-9 If a plotting scale to the nearest 100 ft/in is adopted for the stereomodel of Prob. 12-8, what will be the nominal dimensions of the neat model? What will be the projection distance to average ground?

12-10 If terrain varies 500 ft up and down from average ground in the photography of Prob. 12-7, and the model scale of Prob. 12-9 is selected, how many millimeters up and down will the terrain vary at model scale? If the depth of field of the Balplex projection system is 120 mm up and down from optimum projection distance, can this range in terrain be accommodated in the instrument? If the vertical run on the tracing table is 100 mm, will the tracing table's vertical range accommodate the terrain variation?

12-11 Photography is taken at 65 percent end lap with a 152-mm-focal-length 9-in-format camera from a flying height of 14,500 ft above average ground. What is optimum model scale for a Kelsh plotter having a 760-mm optimum projection distance and projectors which will accommodate contact printed diapositives made from these photos?

12-12 If a plotting scale to the nearest 100 ft/in is adopted for the stereomodel of Prob. 12-11, what will be the nominal dimensions of the neat model? (Assume 25 percent side lap.) What will be the model dimensions of the entire overlap area? What will be the projection distance to average ground?

12-13 If the photography of Prob. 12-11 is to be used in a Multiplex plotter, what plotting scale will be adopted assuming a scale to the nearest 100 ft/in is selected? What approximate base (bx) between projectors should be set off prior to relative orientation?

12-14 What are the advantages and disadvantages of using film diapositives in stereoplotters as opposed to glass diapositives?

12-15 Draw a sketch of the stereoscopic neat model and label the six points where y parallax is conventionally removed in relative orientation. For each of the right projector rotations and translations, list the point numbers at which y parallax may be introduced.

12-16 Outline the steps of the two-projector "swing-swing" method of relative orientation.

12-17 Outline the steps of the one-projector method of relative orientation.

12-18 Assume that a map line AB scales of 26.38 in, and its corresponding model length $A'B'$, prior to absolute orientation, measures 25.95 in. If the initial model base was set at 18.0 in, to what length should the model base be adjusted to scale the model? (Assume that Y and Z components of the model base are zero.)

12-19 Discuss the different methods of leveling a model in a stereoplotter.

12-20 Explain how a model can be leveled without using the leveling screws on the frame of the Kelsh instrument. Under what conditions would this procedure be necessary?

12-21 Explain why it is convenient to set base components b_y and b_z equal to zero prior to plotter orientation.

12-22 A Kelsh plotter having a usable depth of field of 12 in (6 in up and down from optimum projection distance) is oriented with average ground at optimum projection distance of 760 mm. What range in relief can be accommodated up and down from average ground if flying height was 1,500 ft above average ground? (Assume that the diapositive principal distance is equal to the 152-mm focal length of the taking camera.)

12-23 Repeat Prob. 12-22, except that a Multiplex plotter is used. The plotter has a usable depth of field of 90 mm up and down from its optimum projection distance of 360 mm.

12-24 The Kern PG-2 stereoplotter has a range of principal distance of from 85 to 172 mm, a nominal range in projection distance of from 102 to 172 mm, and a maximum pantograph enlargement ratio from model to map of 1:3.333. If the diapositive principal distance is equal to camera focal length, what maximum manuscript scale is possible for the photographs of Prob. 12-7?

12-25 For Prob. 12-24, if map scale of 1:4,800 is selected and a pantograph ratio of 1:3.333 is used, what will actual projection distance be for terrain at average ground? If a nominal projection distance of 172 is used and the tracing table has an up and down range of ± 30 mm from this nominal distance, can a range in terrain variation of 500 ft up and down from average ground be accommodated with the instrument at these settings?

12-26 Repeat prob. 12-24, except that a Wild A-8 instrument is used. It has a principal distance range of from 98 to 215 mm, a projection distance range of from 175 to 350 mm, and a coordinatograph enlargement ratio up to 1:4.

12-27 For Prob. 12-26, if map scale of 1:2,400 is selected and a coordinatograph enlargement ratio of 1:4 is used, what will model projection distance be for average terrain? Assuming terrain varies 500 ft up and down from average, what will model projection distance be for high terrain? Low terrain?

12-28 Repeat Prob. 12-24, except that it applies to the photography of Prob. 12-11.

12-29 Repeat Prob. 12-26, except it applies to the photography of Prob. 12-11.

12-30 Draw a sketch illustrating the Zeiss parallelogram. How are b_x, b_y, and b_z base components varied for instruments incorporating the Zeiss parallelogram?

12-31 Discuss the capabilities and advantages of digitized automatic tracing tables.

12-32 Explain how analytical plotters differ from analog stereoplotters.

12-33 Discuss the advantages of analytical plotters.

12-34 Describe the method of operation of the electronic image correlator of the B-8 Stereomat.

THIRTEEN

ORTHOPHOTOGRAPHY

13-1 INTRODUCTION

An orthophoto is a photograph showing images of objects in their true orthographic positions. Orthophotos are therefore geometrically equivalent to conventional line and symbol planimetric maps which also show true orthographic positions of objects. The major difference between an orthophoto and a map is that an orthophoto is composed of images of features, whereas maps utilize lines and symbols plotted to scale to depict features. Because they are planimetrically correct, orthophotos can be used as maps for making direct measurements of distances, angles, positions, and areas without making corrections for image displacements. This, of course, cannot be done with perspective photos.

Orthophotos are produced from perspective photos (usually aerial photos) through a process called *differential rectification,* which eliminates image displacements due to photographic tilt and relief. Tilt displacements, as described in Sec 11-7, exist in any photo if at the instant of exposure the photo plane is tilted with respect to the datum plane. Rectification eliminates the effects of tilt and yields an equivalent vertical photo. Unless the terrain is perfectly flat, however, a rectified equivalent vertical photo will still contain scale variations as a result of image displacements due to changes in relief. In the process of removing relief displacements from any photo, scale variations are also removed and scale becomes constant throughout the photo. Any photo which has a constant scale throughout is an orthophoto having the same planimetric correctness as a map. It should be mentioned that although relief displacements due to variable terrain are removed, a shortcoming of orthophotos is that relief displacements of vertical surfaces such as walls of buildings cannot be removed.

At first glance, an orthophoto looks the same as a perspective photo. But upon comparison of an orthophoto and a perspective photo of the same area, differences can usually be detected. Figure 13-1a is a portion of a perspective photo taken in Oklahoma. In Fig. 13-1b an orthophoto produced from that portion of the perspective photo is shown. Note in particular how relief displacement has made the power line on the perspective photo appear crooked, whereas it appears straight on the orthophoto because relief displacements have been removed.

(a)   (b)

Figure 13-1 (a) Portion of a perspective photo taken in Oklahoma. (Note the apparent crookedness of the power line caused by relief displacement.) (b) Orthophoto of same portion of perspective photo shown in (a). (Note straightness of the power line after relief displacement is eliminated.) (*Courtesy U.S. Geological Survey.*)

The desirability of orthophotos over perspective photos has been recognized for many years. As early as 1903, Scheimpflug had conceived the idea of directly producing orthophotos from perspective photos. In the early 1930s, the Gallus-Ferber restitution machine was introduced in France. In addition to its use for stereoscopic plotting, this instrument could produce orthophotos directly from perspective photos. (Its operating principles were much the same as those incorporated into present-day orthophoto-producing equipment.) Even though the instrument was operational, inefficiencies of the system together with a general lack of enthusiasm for the finished product caused orthophoto production to lie dormant until the 1950s.

In 1950 Russell Bean of the U.S. Geological Survey began experimenting with equipment for producing orthophotos, and his work led to the development of an instrument known as the *orthophotoscope*. The first orthophotoscope was introduced in 1953, and it was followed by several improved generations. Further developments led to the model T-64 orthophotoscope, pictured in Fig. 13-2, which was one of the earliest instruments used exclusively by the Geological Survey in producing orthophotos. Continued research and development in recent years has resulted in the introduction of many modern and ingenious instruments for producing orthophotos.

13-2 ADVANTAGES AND USES OF ORTHOPHOTOS

Orthophotomaps prepared from orthophotos offer significant advantages over aerial photos and line maps because they possess the advantages of both. On the one hand, orthophotos have the pictorial qualities of air photos because the images of an infinite number of ground objects can be recognized and identified. Furthermore, because of the planimetric correctness with which images are shown, measurements may be taken directly from orthophotos just as from line maps.

The capability of being able to correlate images on an orthophotomap with what is observed on the ground is an asset in many fields of endeavor. Engineers, planners, surveyors, foresters, geologists, agronomists, etc., can use orthophotos advantageously as base maps for plotting field observations. Foresters, for example, can classify and delineate different timber types in their true positions directly on the orthophotomap by correlating images with what they observe in the field. Also, soil scientists can plot locations where soil samples were taken and delineate soil-type boundaries directly on orthophotomaps by correlating images with their corresponding objects on the ground. Property surveyors can utilize orthophotomaps to advantage because fence lines and other key items of evidence that can be field-identified are shown in true plan view on the orthophotomap.

Because images can be correlated with their corresponding objects on the ground, orthophotos make excellent base maps for preparing flight plan maps. Orthophotos are also very useful communication tools. Property owners and lay people can understand an orthophotomap, whereas they are frequently awestruck by a line and symbol map. Using orthophotos, engineers are better able to communicate with property owners to discuss right-of-way purchases, access to property, etc.

Orthophotos themselves contain no information regarding elevations. They can,

Dark   Projector
curtain   bar

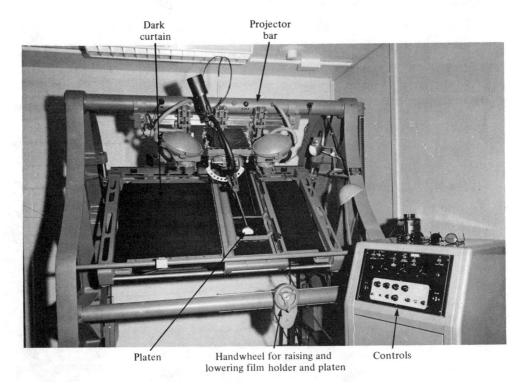

Platen   Handwheel for raising and   Controls
                  lowering film holder and platen

Figure 13-2 Model T-64 Orthophotoscope. (*Courtesy U.S. Geological Survey.*)

however, serve as planimetric base maps upon which elevation contours are super-imposed, and the resulting product is termed a *topographic orthophotomap*. Contours for topographic orthophotomaps are normally derived in a separate stereoplotting oper-ation. The use of orthophotomaps as base maps for elevation contours eliminates the need for mapping planimetric data in stereoplotter compilation and can result in sub-stantial time savings. An example of a topographic orthophotomap produced by the U.S. Geological Survey is shown in Fig. 13-3. In 1980 the U.S. Geological Survey produced over 3,000 different such orthophotomaps.

Orthophotos can be used as maps without cartographic treatment, or they can be enhanced and supplemented with lines, symbols, names, etc. It can be seen in Fig. 13-3, for example, that lines and names have been added. Colors may also be added to enhance the quality of the orthophotomap.

Compared to conventional line maps prepared through stereoplotter compilation, orthophotomaps can generally be prepared faster and cheaper. This is especially true for small-scale maps, maps of urban areas, or maps of areas with dense planimetric and cultural features. Added reproduction costs of orthophotomaps, however, offset savings in labor somewhat. Orthophotography has made it possible to map areas that would otherwise have gone unmapped, and it has made more frequent map revision plausible.

Orthophotomap prepared by the Geological Survey
Control by USC&GS and USCE

Orthophotomosaic by photogrammetric methods from aerial photographs
taken Sept. 3, 1969. Topography from Army Map Service
1:50,000 series maps, surveyed 1955

Selected hydrographic data compiled from USC&GS Chart 9472 (1968)
This information is not intended for navigational purposes.

Universal Transverse Mercator projection. 1927 North American datum
10,000-foot grid based on Alaska coordinate system, zone 4
1000-meter Universal Tranverse Mercator grid ticks
zone 6, shown in blue

Township exteriors are surveyed. Township subdivisions are
predetermined by the Bureau of Land Management.
Folio U-3, Umiat Meridian

Lake elevations are unchecked

SCALE 1:24 000

CONTOUR INTERVAL 25 FEET
DATUM IS MEAN SEA LEVEL
DEPTH CURVES AND SOUNDINGS IN FEET-DATUM IS MEAN LOWER LOW WATER
SHORELINE SHOWN REPRESENTS THE APPROXIMATE LINE OF MEAN HIGH WATER
THE MEAN RANGE OF TIDE IS APPROXIMATELY 0.5 OF A FOOT

UTM GRID AND 1969 MAGNETIC NORTH
DECLINATION AT CENTER OF SHEET

Figure 13-3 Topographic orthophotomap of Prudhoe Bay, Alaska. (*Courtesy U.S. Geological Survey.*)

328

13-3 CLASSIFICATION OF INSTRUMENTS
FOR PRODUCING ORTHOPHOTOS

Instruments used in the production of orthophotos operate by exposing images from original perspective photo negatives onto a piece of unexposed photographic film. This exposed film, when photographically processed, yields an *orthonegative*. The orthonegative is in turn processed photographically to derive an orthophoto.

Conventional orthophoto instruments can generally be classified into one of two broad categories: (1) those which produce images by direct optical projection, and (2) those which produce images electronically. Equipment of the first category exposes orthonegatives by directly projecting, through optical elements, imagery from the original negatives. Many of these types of instruments are simply standard optical or mechanical projection stereoplotters, of the kinds described in Chap. 12, which incorporate some added components. Equipment of the second category utilizes a *printing cathode ray tube* to expose the orthonegative. In addition to these two conventional systems, a third method of producing orthophotos by digital image processing is described in Sec. 13-9.

Within the direct optical projection category, a distinction may be made between instruments which operate *on-line* and those which operate *off-line*. This on-line or off-line designation stems from the fact that two distinct operational phases are entailed in orthophoto production: (1) deriving elevation information of stereomodels (usually scanning profiles of the terrain of a stereomodel at uniformly spaced increments), and (2) exposing the orthonegative (usually in a series of narrow strips, each having a width exactly equal to the spacing between adjacent profile scans). On-line instruments are those in which profile scanning and orthonegative exposure are performed simultaneously. A record of profile elevations is normally not retained, and the only product of on-line operation is usually the orthonegative. Off-line instruments, on the other hand, produce orthophotos in two separate operations. Profile information is first obtained and recorded in either graphical or digital form. Subsequently these profiles are read into an instrument which automatically exposes the orthonegative on the basis of the profiles. Some instruments are capable of operating either on-line or off-line. In discussing the details of operation of orthophoto-producing instruments in the remaining sections of this chapter, it is assumed that the student has studied Chap. 12 or has an understanding of the principles of stereoscopic plotting instruments.

13-4 ON-LINE OPTICAL PROJECTION
ORTHOPHOTO INSTRUMENTS

To introduce the subject of on-line optical projection orthophoto instruments, the U.S. Geological Survey's model T-64 Orthophotoscope will be discussed. This instrument, although no longer used extensively, is still representative of the fundamental method of operation of several instruments in the on-line optical projection category. Following the discussion of the model T-64, other similar instruments are described. Available space limits the discussion to only a few of the many instruments now available for orthophoto production. Exclusion of any instrument in this text is no reflection of the author's opinion of the quality of that instrument.

13-4.1 Model T-64 Orthophotoscope

The model T-64 Orthophotoscope of Fig. 13-2 contains three stereoplotter projectors supported on a projector bar. The center projector is a standard Kelsh projector which uses full-sized 9-in diapositives. The other two are Balplex projectors which require diapositives reduced in size to achieve a 55-mm principal distance. Figure 13-4a illustrates a front view of the three projectors, and Fig. 13-4b shows a side view of the instrument. It, unlike an ordinary optical projection stereoplotter, has a flat *film holder* instead of a reference table. So that the operator can more conveniently see the rear of the model, the film holder (and projectors) is inclined upward at the rear of the model, making an angle of 40° with the horizontal.

Diapositives of three consecutive overlapping photographs of a flight strip, say, photos 1, 2, and 3, are placed in the projectors and oriented using methods described in Chap. 12. The three diapositives produce two stereomodels, models 1-2 and 2-3. The *two-projector* (swing-swing) method is first used to relatively orient one of the models say, model 1-2. That model is then leveled to ground control. The accuracy of the vertical ground control and the accuracy of the leveling operation need not be extremely precise, inasmuch as horizontal accuracy of the orthophoto is not appreciably affected by small leveling errors. When the first model is relatively oriented and leveled, model 2-3 is then relatively oriented to model 1-2 using the *one-projector* method. All motions are imparted to projector 3 so as not to upset previously oriented

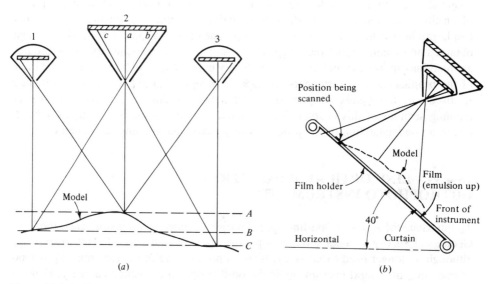

(a) (b)

Figure 13-4 (a) Front view of the projectors of the Model T-64 Orthophotoscope. (b) Side view of the Model T-64 Orthophotoscope.

model 1-2. The second model is automatically leveled in the process of relatively orienting it to model 1-2. Finally, the scale of model 2-3 is made equal to the scale of model 1-2. This is achieved by adjusting model base 2-3 until the elevation of a *carry-over* point (point near the center of diapositive 2 and common to both models) is the same in model 2-3 as it is in model 1-2. It is unnecessary to scale the models absolutely to ground control, because scaling is readily done later by photographic reduction or enlargement. For reasons which will be explained later, model scale is initially fixed by selecting a model base that produces an average projection distance of about 430 mm. When orientation is completed, a true three-dimensional stereomodel exists beneath the projectors, and as shown in Fig. 13-4a, the stereomodel includes the entire area of the center diapositive.

The model T-64 Orthophotoscope uses the anaglyphic viewing system. Red filters are placed in front of the two outside projectors and a blue filter is placed over the center projector. An operator wearing spectacles with red and blue glasses over the left and right eye, respectively, can view model 1-2 stereoscopically. By reversing the spectacles, model 2-3 can be seen in three dimensions. When model 1-2 is viewed, the illumination lamp of projector 3 is turned off. Likewise, while viewing model 2-3 the lamp from projector 1 is off. At any one time an operator is able to see only the small portion of the stereomodel that is projected onto the reflecting surface of the circular platen (see Fig. 13-2).

The model T-64 Orthophotoscope must be operated under darkroom conditions. A large piece of unexposed film which will become the orthonegative is placed emulsion up on the film holder. Because the film emulsion is sensitive only to blue light, a yellow or red "safe light" may be used to light the work area during operation. The unexposed film is covered by a dark curtain to protect it from undesired exposure (see Fig. 13-2). An operator systematically scans the stereomodel back and forth, and in the process the orthonegative is exposed in narrow strips through a thin slit in the center of the platen. This slit also serves as the operator's reference floating mark.

During the scanning process the film remains stationary. The platen is automatically motor-driven back and forth across the model in the Y direction. A guide rod connecting the platen and illumination lamp of the center projector causes the lamp to continually follow and illuminate the platen as it scans. When a Y scan is completed, the platen automatically "steps over" (moves sideways in the X direction) an amount equal to the width of the platen's scanning slit and begins scanning again in the opposite Y direction. Each Y scan exposes the negative to a narrow strip of terrain.

As illustrated in Fig. 13-5, the dark curtain protecting the unexposed film consists of two pieces which move with respect to one another during scanning. The platen is fixed to a narrow section (unshaded area) which is able to move back and forth in the Y direction by means of rollers. This narrow Y-scanning section of curtain is attached to the large section of the curtain (shaded), which covers the remainder of the film holder. This large curtain passes over two large rollers, thereby enabling the scanning platen to step over in the X direction at the completion of each Y scan.

The scanning process is begun in a corner of one of the two stereomodels and is continued without interruption until that model is completed. The operator then switches the lighting to the other model and continues. The total time required to scan the

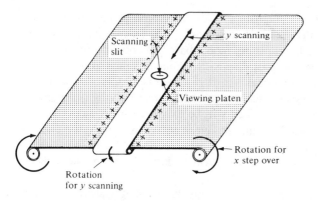

Figure 13-5 Scanning mechanism of the Model T-64 Orthophotoscope.

entire area of the center diapositive is about 2 hr. Continuous scanning at constant speed is important to ensure uniform exposure throughout the orthonegative.

Since the emulsion of the film is sensitive only to blue light, the orthonegative is exposed only to images projected through the blue filter of the center projector. The center diapositive is full-sized and therefore no loss of detail occurs from reduction printing of diapositives, as would be the case if the center projector, like the side projectors, also used reduced-size diapositives. (An earlier generation of the model T-64 also used a Balplex projector in the center position.) Optimum projection distance for blue light through the lens of the center projector is 430 mm. This is the basis for an earlier statement about scale being arbitrarily fixed during orientation so as to achieve an average projection distance of 430 mm. With a nominal principal distance of 152 mm for the Kelsh projector, this 430 mm projection distance produces an orthonegative enlarged approximately $2.8\times$ from photo scale. An orthophoto is then produced at the desired final scale by careful photographic enlargement or reduction of the orthonegative.

Turning the handwheel on the front of the instrument (see Fig. 13-2) permits the film holder and platen to be raised and lowered. During the scanning process an operator views the stereomodel on the platen and adjusts its height so as to keep the scanning slit continually in contact with the model terrain. This is illustrated with points A, B, and C of Fig. 13-4a. The scale of the stereomodel is uniform throughout, and if the platen is kept continually in contact with the stereomodel, the entire orthonegative is exposed at model scale. The result is a negative of uniform scale throughout, which is, of course, an orthonegative.

To achieve a theoretically perfect orthophoto, each model point of varying elevation would have to be exposed at the proper height. Orthophotos obtained with the model T-64 are therefore approximations of theoretically perfect orthophotos because each individual point is not exposed separately. Rather, the width of a narrow strip is exposed simultaneously, a procedure which assumes equal model elevation for the width of the strip. The strips are sufficiently narrow so that, for practical purposes, the assumption is nearly met and satisfactory results are achieved. Normally the width of the scanning slit is set at 5 mm. The slit width as well as the scan speed can be varied, however, depending upon the type of terrain in any model. Best results are achieved in rugged terrain by using a narrow strip. In flat terrain a wider strip may be scanned, thereby increasing the speed of the operation yet maintaining satisfactory accuracy.

The procedure described above is used for producing black-and-white ortho-negatives with the T-64 Orthophotoscope. If color orthophotos are desired, a some-what different technique is necessary. Only a two-photo stereomodel can be printed at one time. Three diapositives are used, two of them being of the same photo of the stereopair. The Kelsh projector, which must contain a color diapositive, is moved to one side on the projector bar, say, the left side. It is completely oriented to one of the Balplex projectors. This establishes the position and orientation of the Kelsh projector. The first Balplex projector is then moved along the projector bar to the far right. The other Balplex projector, which contains a diapositive of the same photo as that contained in the Kelsh projector, is placed on the projector bar between the other two projectors. The model is oriented again using the two Balplex projectors. The one-projector method is used and all motions are imparted to the center projector. This technique establishes the center Balplex projector in the same relative orientation as the Kelsh projector.

With orientation completed, scanning is performed by an operator who raises and lowers the platen and film holder following terrain profiles of the stereomodel beneath the Balplex projectors. Color film on the film holder beneath the Kelsh projector is exposed through an exposing slit which moves back and forth in unison with the viewing platen. In this way the orthonegative is simultaneously exposed to the same area viewed by the operator. The diapositives in the Balplex projectors can be black and white, and the operator can view the model using the anaglyphic system. How-ever, white light must be projected through the color diapositive in the Kelsh projector to expose the color film.

13-4.2 SFOM 693

The SFOM 693 orthophoto instrument is designed so that it can be attached to most optical projection stereoplotters. Its method of operation is very similar to that of the model T-64 orthophotoscope. The film holder, which lies in a horizontal plane rather than being tilted, is raised and lowered during scanning. Scanning proceeds back and forth in the X direction with step over in the Y direction. Scanning speed is varied electronically to provide uniform light intensity over the entire orthonegative in spite of light intensity variations within the stereomodel. The SFOM 693 is capable of producing color orthonegatives through the use of the stereo-image alternator (SIA) viewing system.

13-4.3 Kelsh K-320 Orthoscan

The Kelsh K-320 Orthoscan pictured in Fig. 13-6 can also be operated in a manner similar to the model T-64 Orthophotoscope. This instrument contains three standard Kelsh projectors and is capable of scanning the entire area of the center diapositive in a single setup. An operator varies the height of the platen to keep it in contact with

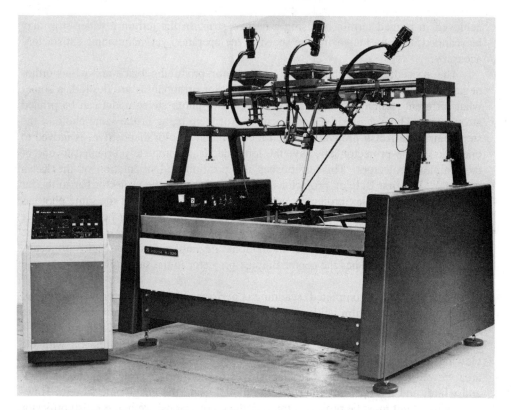

Figure 13-6 Kelsh K-320 Orthoscan. (*Courtesy Kelsh Instrument Div., Danko Arlington, Inc.*)

the terrain as the instrument scans back and forth in the Y direction. What is unique about the Kelsh K-320 orthoscan is that the film plane remains stationary, and only the platen (which contains the exposure slit) is raised and lowered by the operator during scanning. Images are transmitted through a fiber optic coil from the exposure slit to the orthonegative below. The K-320 provides an optimum enlargement ratio of five from photo scale to orthonegative scale. This is convenient since no photographic enlargement or reduction is needed to make the scale of the resulting orthonegatives compatible with contour maps produced at a $5 \times$ enlargement ratio from the same photography. As discussed in Sec. 13-5, this instrument can also be operated off-line.

13-4.4 Zeiss Ortho-3 Projector

The Zeiss Ortho-3 Projector pictured in Fig. 13-7 combines a two-projector stereoplotter and a third *orthoprojector* into one instrument. The stereomodel of the Ortho-3 instrument is formed in the two front projectors, and the orthonegative is exposed through the rear projector. The rear projector contains a duplicate of one of the two diapositives within the front projectors, and it is placed in the same orientation as the

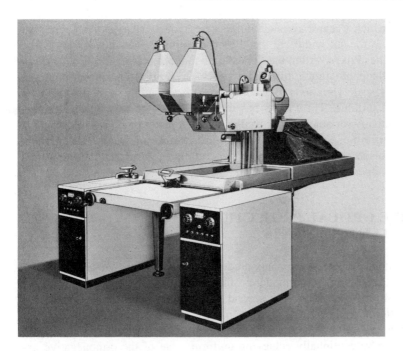

Figure 13-7 Zeiss Ortho-3 projector. (*Courtesy Carl Zeiss, Oberkochen.*)

front projector which contains the common diapositive. An operator follows stereomodel profiles on the viewing platen as it moves back and forth in the Y direction. This is done by turning the handwheel in front of the instrument, which causes all three projectors to be raised or lowered synchronously. An exposure slit beneath the rear projector moves back and forth in unison with the viewing platen and simultaneously exposes the orthonegative (which is enclosed and thereby protected from undesired light) to the same area viewed by the operator. Color orthonegatives can be produced with this instrument.

13-4.5 Wild PPO-8

The Wild PPO-8 Orthophoto instrument is operated in conjunction with the Wild A-8 Autograph stereoscopic plotter (see Fig. 12-23). Even though this stereoplotter uses mechanical projection to form the stereomodel, orthonegatives are produced by optical projection of images. With the PPO-8 an operator turns a handwheel and maintains the floating mark in contact with the terrain as the stereoplotting instrument automatically scans back and forth in the Y direction. Imagery from the scans is projected from the left diapositive of the stereopair via an optical train of lenses and prisms to a drum at the rear of the instrument. The unexposed film is attached on the circumference of the drum. Exposure is made onto the film through a slit in the optical projection system. The drum rotates about its longitudinal axis at a speed synchronized with the speed at which y scans are made in the stereomodel. Steps in the X direction which correspond to the slit width are made by translating the drum sideways at the end of each scan line. Prisms within the optical train that projects imagery to the drum rotate to account for photographic tilt, and a variable magnification optical unit constantly corrects for scale variations in accordance with variations in model elevation along the scanned profiles. The PPO-8 has the capabilities of varying its scanning speed and slit width. A variable density wedge operated in conjunction with scanning speed provides uniform exposure intensity. The instrument can produce color orthophotos as well as black and white, and since the drum is enclosed within a light-tight cylinder, it can be operated in a lighted room.

13-5 OFF-LINE OPTICAL PROJECTION ORTHOPHOTO INSTRUMENTS

As previously stated, off-line instruments produce orthophotos in two separate operations. In the first operation, stereomodel profiles are obtained, recorded, and stored in either graphical or digital form. The second operation, which may take place at a later date, consists of reading these profiles into an instrument, which, in turn, automatically drives an orthoprojector and regulates projection distance according to these profiles, thereby continually achieving uniform scale of the orthonegative exposure.

Following are descriptions of some optical projection instruments capable of producing orthophotos off-line.

13-5.1 Kelsh K-320 Orthoscan

In addition to its capability of being operated on-line, the Kelsh K-320 Orthoscan can be operated off-line. With this instrument a stereomodel can be scanned without placing film in the holder. Each scan profile is recorded on punched tape. When the entire model has been scanned and all profiles recorded, film may be placed in the holder and the instrument switched to automatic scanning. In this mode only the center projector lamp is turned on. The exposure platen automatically follows the model terrain according to the information contained on the punched tape. Either color or black-and-white orthonegatives can be exposed.

13-5.2 Gigas-Zeiss GZ-1 Orthoprojector

The Gigas-Zeiss GZ-1 Orthoprojector shown in Fig. 13-8 contains only a single projector. It was designed expressly for orthophoto production. This projector is the same as those of the C-8 Stereoplanigraph (see Sec. 12-14) and therefore incorporates the Bauersfeld auxiliary lens system for maintaining sharp focus regardless of variations in projection distance. The GZ-1 can either be operated on-line with a C-8 Stereoplanigraph, Planimat, or other stereoplotter, or it can be operated off-line from stored

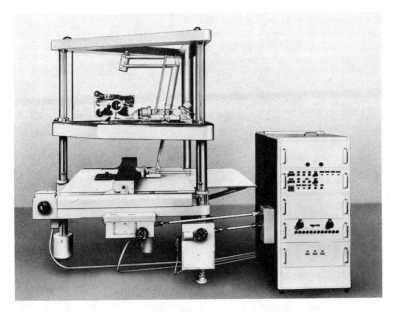

Figure 13-8 Gigas-Zeiss GZ-1 Orthoprojector. (*Courtesy Carl Zeiss, Oberkochen.*)

profiles. As with the Zeiss Ortho-3 Projector, the film holder remains stationary and the scale of imagery projected onto the orthonegative is kept constant by raising and lowering the projector to vary the projection distance.

An accessory called the Optical Interpolation System is available with the GZ-1 Orthoprojector. This accessory is used when the GZ-1 is operating in the off-line mode. The Optical Interpolation System reads two adjacent profiles simultaneously and interpolates across them to determine terrain slopes at right angles to the scan lines. With this instrument the scanning slit exposes the width between adjacent profiles instead of centering the scanning slit over a single profile. By means of coherent fiber optics, imagery is transmitted to the orthonegative.

Figure 13-9 is a section taken at right angles to the scan lines which illustrates the concept of the Optical Interpolation System. Instead of imaging horizontal scans S_1, S_2, S_3, and S_4 which are centered on profiles P_1, P_2, P_3, and P_4, exposures are made onto the orthonegative of the sloping scans, *ab, bc, cd,* and *de*. An affine scale change is produced by the fiber optics in conducting images so that the sloping scans are imaged orthogonally onto the orthonegative film below. Advantages of this system are an orthophoto with greater planimetric accuracy, the possibility of increased scan width, and better matching of images on the orthophoto between adjacent scan lines. This last advantage is particularly important in rugged terrain where great differences in terrain elevation from one side of the scan slit to the other cause noticeable mismatching of images. Note on Fig. 13-9 for example, that scan lines, *ab, bc, cd,* and *de* fit the terrain much better than do scans S_1, S_2, S_3, and S_4.

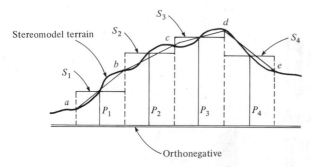

Figure 13-9 Concept of the Zeiss Optical Interpolation System.

13-5.3 Wild Avioplan OR-1 Orthophoto System

The Wild Avioplan OR-1 illustrated in Fig. 13-10 is another orthophoto instrument utilizing direct optical projection that can be operated off-line. Like the GZ-1 it also simultaneously uses two adjacent profiles to correct for sloping terrain. The profile information may be derived by scanning stereomodels, or it may be generated from digital terrain models or contour maps. The operation of the OR-1 is controlled by a computer.

The basic operating principle of the OR-1 is illustrated in Fig. 13-11. In the figure, P_1, P_2, P_3, etc., are profile lines along the terrain imaged in the photograph. The terrain profile information is in an XYZ coordinate system which is parallel to the axis system of the instrument, the Y axis being defined by the direction of scan. The original information from which these profiles are derived will not normally exist in this parallel coordinate system; thus it is placed in the required configuration by means of a coordinate transformation. Besides being in a parallel coordinate system, profiles must be equally spaced in the X direction by an amount S which is consistent with the width of the exposure slit that will be used.

In Fig. 13-11, T_1 and T_2 are model points in adjacent profiles P_1 and P_2, and they have equal Y coordinates. Line L is the straight segment in the model joining these two points, and if the spacing S between adjacent profiles is small, it closely approximates the terrain surface. Image points t_1 and t_2, and image line l, correspond to model points T_1 and T_2 and model line L, respectively. Once the photo has been properly oriented on the photo carriage of the instrument, and given the camera focal length and the X, Y, and Z model coordinates of T_1 and T_2, the instrument calculates the x and y image coordinates of t_1 and t_2 using relatively simple perspective geometry. (Proper orientation of the photo places it on the photo carriage, with its y axis parallel to the instrument Y axis.) Then the angular orientation between lines l and L, as well as their scale relationship, can be calculated.

Based upon the computed image coordinates, servomotors drive the photo carriage to the location necessary for image line l to enter the projection optics. Servomotors also drive a dove prism and zoom lens which rotate and scale line l by the proper

Figure 13-10 Wild Avioplan OR-1 Orthophoto System. (*Courtesy Wild Heerbrugg Instruments, Inc.*)

amounts. The result is that image line l is projected orthogonally through the exposure slit onto the orthonegative at L_x.

The OR-1 systematically repeats this process by increments in the Y direction until the scan between profiles P_1 and P_2 is completed. The almost infinite number of contiguous projected lines L_x projected onto the orthonegative uniformly exposes that scan. At the end of the scan, the instrument steps over a distance S and performs the same process along the scan between profiles P_2 and P_3. Because the same model coordinates of profile P_2 are used for this scan, gaps and overlaps between adjacent scans cannot occur, and thus a high-quality orthophoto results.

Like the PPO-8, the orthonegative is attached to the circumference of a drum which is enclosed within a cylinder. The OR-1 also incorporates a gray wedge to automatically provide uniform exposures for variations in scanning speeds. Its range in slit widths is from 3 mm to 16 mm by 1 mm increments. Scanning speed can be varied between 10 and 30 mm/sec, the slower speeds being used for very rugged terrain. The instrument can accommodate photography taken with a camera of any focal length.

In orientation of the OR-1, it has already been mentioned that the photo coordinate system must be made parallel to the XY axis of the orthophoto projection unit. This is accomplished by making photo x and y shifts and a kappa rotation. Two image

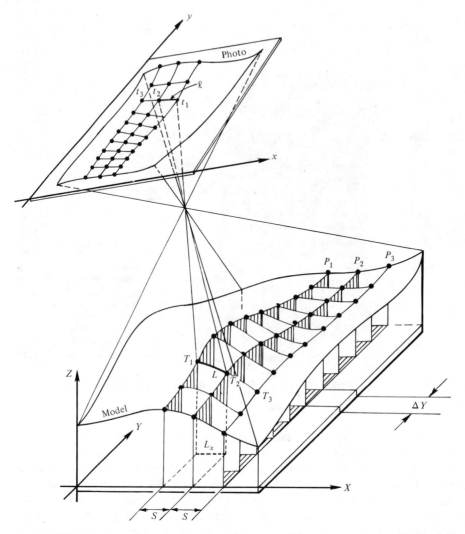

Figure 13-11 Basic operating principle of the Avioplan OR-1. (*Courtesy Wild Heerbrugg Instruments, Inc.*)

points are needed for the orientation and they need not be ground control points. A third point provides a check.

The OR-1 can be operated on-line if desired by directly connecting it to a stereoplotter which can generate model profiles.

13-5.4 Zeiss Orthocomp Z-2 Orthoprojector

Figure 13-12 shows the Zeiss Orthocomp Z-2 orthoprojector. Principally designed for off-line production of orthophotos, the system operates in a manner somewhat similar to the OR-1. A computer controls the operations of the instrument. Terrain information

Figure 13-12 Zeiss Orthocomp Z-2 Orthoprojector. (*Courtesy Carl Zeiss, Oberkochen.*)

can be read into the system in the form of scanned model profiles, digital terrain models, or contour map data. From camera focal length and exterior orientation parameters of the photo, image coordinates are computed based upon model coordinates of profile points. The computer then activates servomotors which drive the photo carriage, dove prism, and zoom optics so that the corresponding imagery is projected orthogonally through the exposure slit onto the orthonegative, which is mounted on the circumference of a drum below.

In addition to orienting the photo properly on the photo carriage, orientation also includes providing image and model coordinates of at least three ground control points. From this information, the system's computer determines the elements of exterior orientation needed for the subsequent operation of the system. The Z-2 can accommodate any camera focal length and it will produce orthophotos from oblique and panoramic photos as well as verticals. It also has the capability of projecting alphanumeric characters onto the orthophoto so that name places, feature identifications, etc., can be added. This instrument can be operated on-line as well as off-line.

13-6 ADVANTAGES OF OFF-LINE ORTHOPHOTO PRODUCTION

Several significant advantages can be realized through the use of off-line orthophoto production techniques. First of all, greater accuracy results in deriving the terrain profiles for the following reasons: (1) the stereoplotter reference mark is the tiny floating mark rather than the large exposure slit; (2) an operator can occasionally stop to rest so that the tedium of prolonged uninterrupted scanning is removed, thereby reducing errors arising from operator fatigue; (3) scanning speed can be reduced to very slow speeds to accommodate profiling difficult areas; (4) errors made on any profile can be immediately corrected; and (5) spot checks can be made and profiles corrected before removing the model from the stereoplotter. (For the on-line instrument described in Sec. 13-4.5, advantages 1, 2, and 3 apply.)

In addition to increased accuracy of the orthophoto, several other advantages result from off-line production techniques. Efficiency of orthophoto production can be increased because the rate at which an orthonegative can be exposed off-line is increased. Profiles can be derived from several stereoplotters and fed to one off-line orthoprojector. Double models (the entire area of the center diapositive of a triplet of photos) can be printed nonstop without the necessity of a three-projector stereoplotter. In this procedure, profiles from two adjacent models can be read independently and fed to the orthoprojector as a single set of data. Nonstop printing of the entire diapositive eliminates the possibility of a visible match line occurring if the two independent models were printed separately and pieced together later as an *orthophotomosaic*. Also two adjacent profiles can be read simultaneously and corrections for sloping terrain made by projecting orthogonally onto the orthonegatives.

Off-line orthophoto production enables profiles of an area to be stored for long periods of time and then reused for map revisions. Planimetric details, especially in heavily populated areas, may change a great deal over relatively short periods of time, but these changes rarely affect terrain profiles. Orthophotos for planimetric map revision can therefore generally be prepared from new photography, but using the old stored profiles of the same area. Off-line orthophoto production also gives rise to the possibility of obtaining the necessary model profile information from digital terrain models, or of digitizing contours from existing maps and then numerically deriving the profiles. Also, contour locations can automatically be computed and plotted from the stored digitized profiles using a computer driven flat-bed *XY* plotter. One final advantage of off-line orthophoto production is the ease with which the procedure enables color orthophotos to be produced.

13-7 AUTOMATIC CONTOURING DURING ORTHOPHOTO PRODUCTION

During the process of exposing orthonegatives, locations of elevation contours can be marked by a technique called *line-dropping*. Imagine that while scanning profiles, a pen which moves with the scanning device is automatically lowered to a map sheet when the platen passes through the elevation of a desired contour. Suppose that the pen stays in contact with the map, marking a line until another contour elevation is crossed by the platen, at which time the pen is automatically raised. Alternate raising and lowering of the pen at contour elevation crossings permits the spaces between adjacent contours to be marked. Upon completion of scanning, the ends of the lines marking corresponding points of equal elevation can be joined to form continuous contours. Contour maps produced in this manner are satisfactory for some work, although they will not generally be as accurate as maps compiled in the conventional manner in a stereoplotter.

It is sometimes difficult to decipher which lines correspond to equal elevations in drawing contours from the line-dropped plots. This problem may be alleviated somewhat through a photographic line-dropping technique. In this procedure, lines of varying width are recorded photographically on a separate film as contour elevations are crossed during profile scanning. The intensity of the light signal which exposes the line-drop film is varied each time a new contour elevation is crossed. This causes a variation in thickness of dropped lines which greatly reduces the difficulty of drawing the contours. Figure 13-13*a* illustrates a portion of a line-dropped contour map produced in the GZ-1 Orthoprojector simultaneously with the production of the corresponding orthophoto pictured in Fig. 13-13*b*. Note that the contours on this map have already been manually drawn by connecting the ends of lines of common thickness.

The *Automatic Contourliner* is an accessory operated in conjunction with the GZ-1 Orthoprojector which affords another possibility in achieving automatic contours while exposing orthonegatives. The instrument simultaneously reads two adjacent profiles that have been read off-line, and it makes extremely high-speed interpolations between them. During interpolation, small dots marking positions where contour-line elevations exist are imaged photographically on a film by pulses of light. The density of imaged points is so great that the impression of continuous contour lines is achieved. Figure 13-13*c* shows a contour map of the area of Fig. 13-13*a* and *b* that was compiled by the Automatic Contourliner.

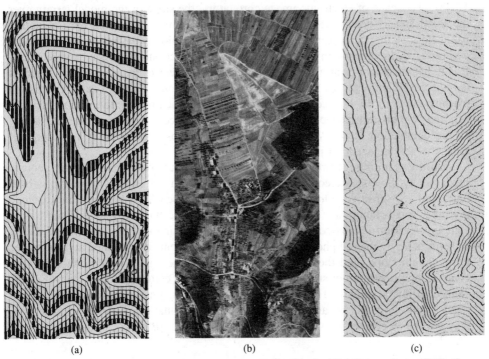

| (a) | (b) | (c) |

Figure 13-13 (*a*) Portion of a line-dropped contour map produced in the GZ-1 Orthoprojector. (*b*) Portion of an orthophoto (same area) produced by the GZ-1 Orthoprojector. (*c*) Portion of a contour map (same area) produced by the Zeiss Automatic Contourliner in conjunction with the GZ-1. (Courtesy Carl Zeiss, Oberkochen.)

13-8 INSTRUMENTS FOR IMAGING ORTHOPHOTOS ELECTRONICALLY

Instruments which produce orthophotos by electronic imaging were developed more or less as by-products of efforts toward automating stereocompilation. These instruments utilize image correlation devices (see Secs. 12-22 through 12-25) which automatically scan model profiles and expose orthonegatives through cathode ray tubes. Electronic imaging instruments are capable of producing high-quality orthophotos very rapidly. In spite of these advantages, their high cost has prevented them from becoming as abundant as optical imaging instruments. Following is a brief description of some electronic imaging instruments.

13-8.1 B-8 Stereomat

The B-8 Stereomat, as described in Sec. 12-24, is an automated stereoplotting instrument. When operated in its automatic profiling mode, the instrument is capable of producing orthophotos. In this process, the stereomodel is scanned in the Y direction. The space rods drive two cathode ray tubes which project the imagery of corresponding small rectangular areas of the diapositives onto photomultiplier tubes below (see Fig. 12-43). By means of electronic image correlators, the slope of the model terrain over the small area is determined. Based upon the slope information and the orientation parameters of the stereopair that were determined analogically when the instrument was oriented, imagery from the right-hand photo is electronically scaled and rectified. It is transmitted to a printing cathode ray tube at the rear of the instrument, where it is redisplayed and exposes the small rectangular area onto the orthonegative. As the machine scans in the Y direction the process is performed repeatedly. The result is an exposed strip on the orthonegative composed of the successive contiguous small areas. At the completion of one Y scan, the instrument steps over in the X direction and scans the next adjacent strip.

During scanning for orthophoto production, contours can also be simultaneously produced by the B-8 Stereomat. For this procedure an emulsion is used that is both blue- and red-sensitive. Orthophoto imagery is exposed by a blue-printing cathode ray tube, and contours are imaged by a red-sensitive cathode ray tube. Each time a contour elevation is crossed during scanning, a short line segment is exposed on the orthonegative by the red phosphor tube. The line segments are oriented at right angles to maximum slope, and they are equal in length to the width of the scan strip. The contours are therefore not smooth lines, but rather, are composed of a series of short, connected line segments. The orthophoto imagery and contours can be printed separately by filtering, and because they were exposed simultaneously onto the same film, they will be in perfect register.

13-8.2 Analytical Plotter AS11-C

The Analytical Plotter AS11-C is also equipped with an electronic imaging orthoprinter. Therefore, besides being able to perform its varied battery of photogrammetric tasks, which include aerotriangulation, automatic contouring, automatic profiling, and automatic model digitizing, the AS11-C is also capable of automatically exposing orthonegatives. Initial setup procedures and orientation for printing orthophotos is the same as for operating in any of its modes.

The AS11-C orthoprinter has an *XY* carriage which enables the orthonegative to move back and forth in unison with the scanner. When operating in the automatic orthophoto mode, the image correlator scans in the *Y* direction until it has advanced a preset distance. The computer then interrupts the scanner and transmits the scanned imagery to the printing cathode ray tube, which makes the exposure onto the ortho-negative. Prior to exposure, the imagery of each area is electronically rectified and scaled according to camera parameters, geometry of the stereopair, and terrain in that particular area. After exposure, the scanner resumes its scanning in the *Y* direction until it has again advanced the preset distance, whereupon another exposure is made. Scanning is therefore also performed in a series of parallel strips, with exposures being made onto the orthonegative film as a series of small, individual, contiguous rectan-gular areas or patches.

The AS11-C is also able to plot contours as a by-product of printing orthophotos. Each time a contour elevation is crossed by the scanner, a pulse is transmitted to the contour generator, which exposes a short line segment onto a separate sheet of film located on a separate stage in the orthoprinter. The scanner senses terrain slopes as well as heights, and consequently it is able to orient the line segments perpendicular to the direction of maximum slope. The resulting contours are therefore not smooth lines, but instead they are composed of a series of short connected line segments like those produced in the B-8 Stereomat orthoprinter.

13-8.3 Gestalt Photomapper GPM

The Gestalt Photomapper GPM produces orthophotos in a manner similar to that of the AS11-C described above. It exposes orthonegatives in small hexagonal-shaped contiguous patches. The use of hexagonal patches eliminates the lineations that some-times occur in orthophotos produced by imaging contiguous strips or rectangular patches. The size of the hexagonal patch is precisely controlled through the use of a mask and is 48 mm^2, which is large in comparison to areas imaged by other orthophoto instruments. When each patch is scanned, it is electronically transformed into a scaled orthogonal projection and transmitted to a cathode ray tube, which makes the exposure onto the orthonegative. A unique feature about the GPM is that the *x* parallaxes measured by the image correlators are used to determine the *X*, *Y*, and *Z* model coordinates of a matrix of 3,000 points within each individual area. These are then used to perform a rectification which accounts for curvature of the terrain within the area. This not only results in high-quality orthophotos, but as noted previously, it enables the use of relatively large patches which speeds the orthophoto production process. As an adjunct to orthophoto printing, the GPM can produce line-drop con-tours, digital terrain profiles, and it can annotate the orthophotos by superimposing alphanumeric characters onto them.

13-9 ORTHOPHOTOS BY DIGITAL IMAGE PROCESSING

It is possible to produce orthophotos using data in digital form. In this procedure, data from aerial photos are converted to digital form and recorded on magnetic tape using scanning microdensitometers (see Sec. 5-15). The density value of each sampled picture element (pixel) is stored, together with its location, in terms of the row and column that it occupied during the scanning process. The row/column locations of fiducial marks and ground control points are also obtained during scanning.

Based upon fiducial coordinates obtained from camera calibration, a coordinate transformation is applied to convert row/column locations of pixels to a standard fiducial system. From camera focal length, photo coordinates of ground control points, and ground coordinates of the control points, the space resection problem can be solved to determine the exterior orientation parameters of the photo.

Given the X, Y, and Z ground coordinates of points along terrain profiles, the computer can calculate the differentially rectified xy photo coordinates that correspond to these points based upon their elevations and the interior and exterior orientation parameters. Pixels that correspond to these rectified locations of points can then be stored on magnetic tape, whereupon a computer can drive a *photowriter* which will print them orthogonally onto an unexposed film to create an orthonegative. The photowriter operates in a manner which is essentially a reversal of the scanning process. In fact, instruments have been developed which can perform both the initial density scanning and the photowriting processes.

The X, Y, and Z ground coordinates along terrain profiles could be obtained in a separate operation by scanning stereomodels, or they could be derived from digital terrain models or contour maps. Another possibility would be to derive these coordinates using numerical image correlation and scanned density data from a stereopair of photos. (This procedure is described in Sec. 14-20.) Thus, the determination of model profile coordinates could become an integral part of the software which controls the system.

The system described above theoretically offers great potential. It could be applied to photography taken with any focal length camera, and would be applicable to oblique, panoramic, and terrestrial photos, as well as to verticals. Contours or digital terrain models could be developed as an adjunct to the system. The procedure is not limited to the production of conventional orthophotos which show a vertical orthogonal view. It could be used to produce any projection of images that can be modeled with a computer. As an example, it is possible to develop a horizontal perspective from aerial views. The system can also operate with imagery already in digital form such as LANDSAT multispectral scanner data. In this case the need for initial scanning is eliminated. The procedure can utilize an off-line digital computer on a time-sharing basis so that the only additional equipment needed would be a scanning microdensitometer and photowriter. Research and development are currently active in this area to perfect the methodology.

13-10 FLIGHT PLANNING FOR ORTHOPHOTOGRAPHY

Many of the principles discussed in Chap. 16 apply directly in planning aerial photography for orthophotos, and therefore it is recommended that the student review that chapter. There are some additional factors that must also be considered in planning aerial photography for othophotos, however.

Aerial photography taken for orthophotography must be planned in accordance with the equipment to be used in exposing the orthonegatives. If an optical projection instrument will be used, diapositives must be prepared with the proper principal distance, the same as is necessary in preparing diapositives for stereocompilation. A 6-in-(152-mm-) focal-length camera should be used, for example, if the projectors are designed to accept nominal 6-in-focal-length photography. Some orthophoto instruments such as those described in Secs. 13-5.3, 13-5.4, 13-8.2, and 13-8.3 do not restrict the choice of camera.

Photographic scale is, of course, of paramount consideration in planning any aerial photographic mission. Specific project requirements normally fix the scale of the orthophotomap, and unless photographic enlargement or reduction of the orthonegatives is planned, this fixes photo scale to within certain limits according to the enlargement ratio from photo scale to orthonegative scale that will occur with a given orthophoto instrument. Orthophotomaps may be produced at scales larger than orthonegative scale by projection printing, and this affords some latitude in planning aerial photo scale. This, of course, is an added expense, but in addition image quality is somewhat degraded with enlargement and therefore the limits of possible enlargement are restricted. In general approximately a $5 \times$ enlargement from original photo scale to final orthophoto is recommended as maximum; however, with some instruments, up to $8 \times$ enlargement may still produce good-quality results. Once camera focal length and photo scale are selected, flying height is also fixed.

In certain special situations, the ground coverage of each orthophoto may be a consideration that overrides other factors in planning the photography. With a 6-in-(152-mm-) focal-length camera and a flight height of approximately 40,000 ft, for example, an orthophoto printed from two adjacent neat models (using only approximately the center 7-in square of the middle photo of a triplet) will cover an entire $7\frac{1}{2}$-min quadrangle. In executing flights for *orthophotoquads*, as they are called by the U.S. Geological Survey, desired locations for ground principal points are marked on the flight map to ensure that the exposure station of the center photo of a triplet of photos occurs in the center of the quadrangle. By placing the center photo of the triplet in the center projector of the model T-64 Orthophotoscope or Kelsh K-320 Orthoscan, an orthonegative of the entire quadrangle is obtained by scanning the double model. This arrangement is not only economical but also eliminates the possibility of a visible splice occuring if a mosaic of two or more orthonegatives was necessary to cover the quad. In cases where individual orthonegatives must be pieced together in mosaic form, it is most economical to plan the coverage of each orthonegative so as to minimize the total number of stereomodels that must be scanned.

End lap and side lap for orthophotography should generally be about the same as that for mapping photography. Sixty percent end lap and approximately 30 percent side lap allows for complete stereocoverage in spite of aircraft tilts, flying height variations, terrain variation, crab, and drift. Although the entire stereo overlap area can be scanned, deterioration of image quality along the edges of the picture format also add to the justification for using the end lap and side lap suggested above. In areas of extremely rugged terrain, or in urban areas, these may be increased to produce better results.

Although best results are usually obtained if aerial photography is planned specifically for orthophotography, the possibility of other uses should be considered. It is possible, for example, that aerial photos can serve both for orthophoto production and for stereoplotter compilation of contours.

After all factors have been considered, a flight plan including a flight map and specifications should be prepared to guide the aerial flight crew.

REFERENCES

Ahrend, M., W. Bruchlacher, and H. Meier: The Gigas-Zeiss Orthoprojector, *Photogrammetric Engineering,* vol. 31, no. 6, p. 1039, 1965.

American Society of Photogrammetry: "Manual of Photogrammetry," 3d. ed., Falls Church, Va., 1966, chap. 15.

———: "Manual of Photogrammetry," 4th ed., Falls Church, Va., 1980, chaps. 13 and 15.

Bean, R.: Development of the Orthophotoscope, *Photogrammetric Engineering,* vol. 21, no. 4, p. 529, 1955.

Blachut, T., and M. C. Van Wijk: 3-D Information From Orthophotos, *Photogrammetric Engineering,* vol. 36, no. 4, p. 365, 1970.

——— and ———: Results of the International Orthophoto Experiment 1972–1976, *Photogrammetric Engineering and Remote Sensing,* vol. 42, no. 12, p. 1483, 1976.

Brum, M. G., and J. G. Waters: Photodensity Control System for Orthophoto Products, *Photogrammetric Engineering and Remote Sensing,* vol. 43, no. 9, p. 1177, 1977.

Chapelle, W., and J. Edmond: The AS-11C Automatic System, *Photogrammetric Engineering,* vol. 35, no. 10, p. 1059, 1969.

Collins, S. H.: The Accuracy of Optically Projected Orthophotos and Stereo Orthophotos, *Canadian Surveyor,* vol. 23, no. 5, p. 450, 1969.

———: Image Quality in Orthophotography, *Canadian Surveyor,* vol. 30, no. 3, p. 171, 1976.

Crawley, B. G.: Gestalt Contours, *Canadian Surveyor,* vol. 28, no. 3, p. 237, 1974.

Danko, J. O.: Quick Orientation of the Kelsh K-320 Orthoscan, *Photogrammetric Engineering,* vol. 40, no. 9, p. 1071, 1974.

Esten, R.: Automatic Photogrammetric Instruments, *Photogrammetric Engineering,* vol. 30, no. 4, p. 544, 1964.

Fleming, E. A.: Photo Maps as Part of a Map Series, *Canadian Surveyor,* vol. 24, no. 2, p. 173, 1970.

———: Quality of Production Orthophotos, *Photogrammetric Engineering,* vol. 39, no. 11, p. 1151, 1973.

Hobrough, G., and T. Hobrough: Image Correlator Speed Limits, *Photogrammetric Engineering,* vol. 37, no. 8, p. 1045, 1971.

Hughes, T., A. Shope, and F. Baxter: U.S.G.S. Automatic Orthophoto System, *Photogrammetric Engineering,* vol. 37, no. 10, p. 1055, 1971.

Jaksic, Z.: Man-Machine Photogrammetric Systems and System Components, *Canadian Surveyor,* vol. 27, no. 4, p. 308, 1973.

Keating, T. J., and D. R. Boston: Digital Orthophoto Production Using Scanning Microdensitometers, *Photogrammetric Engineering and Remote Sensing,* vol. 45, no. 6, p. 735, 1979.

Konecny, G.: Methods and Possibilities for Digital Differential Rectification, *Photogrammetric Engineering and Remote Sensing,* vol. 45, no. 6, p. 727, 1979.

Marckwardt, W.: The Accuracy of Orthophotos and Simultaneously Collected Terrain Height Data, *Photogrammetric Engineering and Remote Sensing*, vol. 44, no. 5, p. 575, 1978.

Marsik, Z.: Use of Rectified Photographs and Differentially Rectified Photographs for Photo Maps, *Canadian Surveyor*, vol. 25, no. 5, p. 567, 1971.

O'Brien, T. J.: Orthophotomapping for Prudhoe Bay Development, *ASCE Journal of the Surveying and Mapping Division*, vol. 97, no. Su2, p. 199, 1971.

Parenti, G.: Orthophoto Printing with the Analytical Plotter, *Photogrammetric Engineering*, vol. 33, no. 4, p. 411, 1967.

Pumpelly, J. W.: Cartographic Treatments in the Production of Orthophotomaps, *Surveying and Mapping*, vol. 24, no. 4, 1964.

————: Color-Separation and Printing Techniques for Photomaps, *Surveying and Mapping*, vol. 27, no. 2, p. 277, 1967.

Radlinski, W.: Orthophotomaps Versus Conventional Maps, *Canadian Surveyor*, vol. 22, no. 1, p. 118, 1967.

Scarano, F., and A. Jeric: Off-Line Orthophoto Printer, *Photogrammetric Engineering and Remote Sensing*, vol. 41, no. 8, p. 977, 1975.

Scher, M. B.: Research in Orthophotography, *Photogrammetric Engineering*, vol. 30, no. 5, p. 756, 1964.

————: Orthophotomaps for Urban Areas, *Surveying and Mapping*, vol. 29, no. 3, p. 413, 1969.

Seeger, E.: Orthophotography in Architectural Photogrammetry, *Photogrammetric Engineering and Remote Sensing*, vol. 42, no. 5, p. 625, 1976.

Tanden, D.: Progress in Orthophotography, *Photogrammetric Engineering*, vol. 40, no. 3, p. 265, 1974.

Thompson, M. M., and E. M. Mikhail: Automation in Photogrammetry—Recent Developments and Applications, *Photogrammetria*, vol. 32, no. 4, p. 111, 1976.

Winikka, C. C., and S. A. Morse, Jr.: Orthophoto Quads for Arizona Land-Use Mapping and Planning, *ASCE Journal of the Surveying and Mapping Division*, vol. 100, no. SU1, p. 1, 1974.

PROBLEMS

13-1 Explain the basic differences between an orthophoto and a perspective photo.

13-2 Discuss the advantages and uses of orthophotos.

13-3 Explain the differences between on-line and off-line procedures for producing orthophotos.

13-4 What are the advantages of off-line orthophoto production as opposed to on-line production?

13-5 Describe the method of orienting black-and-white diapositives in the model T-64 Orthophotoscope.

13-6 How are color orthophotos produced on the model T-64 Orthophotoscope?

13-7 Aerial photography taken with 60 percent end lap with a 6-in-focal-length 9-in-square-format camera is to be scanned in the Kelsh K-320 Orthoscan. How many scanning strips are necessary to cover the model of the 7-in-square central portion of the center diapositive if a 10-mm scanning slit width is used and if the model is oriented with a projection distance of 760 mm for average ground?

13-8 For Prob. 13-7, if movement of the scanning slit is 10 mm/sec, how long would it take to expose the orthophoto derived from the central 7-in-square portion of the center diapositive? (Neglect step-over time.)

13-9 If it is required to produce an orthonegative at a scale of 1:15,000 with the Kelsh K-320 Orthoscan from 6-in-focal-length photography, what must be the flying height of the photography? (The optimum projection distance of the K-320 is 760 mm.)

13-10 Repeat Prob. 13-9, except that the required orthonegative scale is 1:4,800.

13-11 It is desired to cover a 6-mi^2 township with a single orthophoto made from the center diapositive of a triplet of overlapping photos. What must be flying height above ground for the overlapping aerial photography if a 6-in-focal-length 9-in-square-format camera is used? (Use only the center 7-in-square area of diapositive.)

13-12 For the photography of Prob. 13-11, what will be the dimensions of the exposed orthonegative if a Kelsh K-320 Orthoscan is used? (Assume the instrument is oriented so that its optimum projection distance of 760 mm applies at average ground elevation.)

13-13 If a final orthophotomap at a scale of 1,000 ft/in will be produced from the orthonegative of Prob. 13-12, what will be the required enlargement (or reduction) ratio from orthonegative to final orthophotomap? What will be the dimensions of the final orthophotomap?

13-14 What will be the dimensions of the orthonegative of the center diapositive of Prob. 13-11 if it is scanned in a GZ-1 Orthoprojector using a projection distance of 550 mm?

13-15 Describe the "line-dropping" technique of automatic contouring.

13-16 Consult the references listed at the end of this chapter. Write a brief report on one use of orthophotos in the solution of a practical problem.

FOURTEEN

AEROTRIANGULATION

14-1 INTRODUCTION

Aerotriangulation is the term most frequently applied to the process of determining X, Y, and Z ground coordinates of individual points based on measurements from photographs. *Phototriangulation* is perhaps a more general term, however, because the procedure can be applied to terrestrial photos as well as aerial photos. The principles involved are extensions of the material presented in Chaps. 11 and 12. In recent years, with the advent of computers and improved photogrammetric equipment and techniques, accuracies to which ground coordinates can be determined by these procedures have become very high.

Aerotriangulation is used extensively for many purposes. One of the principal applications is in extending or densifying ground control through strips or blocks of photos for use in subsequent photogrammetric operations. When used for this purpose it is often called *bridging,* because in essence a "bridge" of intermediate control points is developed between field surveyed control that exists in only a limited number of photos in a strip or block. Establishment of the needed control for compilation of topographic maps with stereoplotters is an excellent example to illustrate the value of aerotriangulation. In this application, as described in Chap. 12, the practical minimum number of control points necessary in each stereomodel is three horizontal and four vertical points. For large mapping projects, therefore, the number of control points needed is extensive, and the cost of establishing them can be extremely high if done exclusively by field survey methods. Much of this needed control is now routinely being established by aerotriangulation from only a sparse network of field surveyed ground control and at a substantial cost savings.

Besides having an economic advantage over field surveying, aerotriangulation has other benefits, such as (1) most of the work is done under laboratory conditions, thus minimizing delays and hardships due to adverse weather conditions; (2) access to much of the property within a project area is not required; (3) field surveying in difficult areas, such as marshes, extreme slopes, hazardous rock formations, etc., can be minimized; and (4) the accuracy of the field-surveyed control necessary for bridging is verified during the aerotriangulation process, and as a consequence, chances of finding erroneous control values after compilation has begun are minimized and usually eliminated. This latter advantage is so meaningful that some organizations perform bridging even though adequate field-surveyed control exists for stereomodel control. It is for this reason also that some specifications for mapping projects are written which require that aerotriangulation be used to establish photo control.

Apart from bridging for subsequent photogrammetric operations, aerotriangulation is also used in a variety of other applications where precise ground coordinates are needed. In property surveying it is used to locate section corners and property corners or to locate evidence that will assist in finding these corners. In topographic mapping, aerotriangulation can be used to develop Digital Terrain Models (DTM's) by simply computing X, Y, and Z ground coordinates of a systematic network of points in an area. Other applications include determining ground coordinates of points at various time intervals to monitor movements of dams, retaining walls, and other structures or to measure ground subsidence due to mining activity or water pumping. Special applications include the densification of geodetic control networks and the precise determination of the relative positions of large machine parts during fabrication. It has been found especially useful in such industries as ship building and aircraft manufacture. Many other applications of aerotriangulation are also being pursued.

Methods of performing aerotriangulation may be classified into one of three categories: *analogue, semianalytical,* or *analytical.* Analogue procedures involve manual relative and absolute orientation of models using stereoscopic plotting instruments, followed by measurement of model coordinates. Aerotriangulation performed in this manner is called *stereotriangulation.* Although still occasionally performed, this procedure is now principally of historical interest, having given way to the other two methods.

Semianalytical aerotriangulation involves manual relative orientation of stereomodels within a stereoplotter, followed by measurement of model coordinates. Absolute orientation is performed numerically—hence the term "semianalytical" aerotriangulation. Analytical methods consist of photocoordinate measurements followed by numerical relative and absolute orientation from which model coordinates are determined. As a practical matter, this procedure requires the use of a computer. Various specialized techniques have been developed within each of the three aerotriangulation categories. This chapter briefly describes some of these techniques. Part I covers analogue aerotriangulation, Part II describes semianalytical procedures, and Part III covers analytical methods. The discussion predominantly relates to bridging for subsequent photogrammetric operations because this is the principal use of aerotriangulation. Extension of these basic principles can readily be translated into the other areas of application, however.

PART I ANALOGUE AEROTRIANGULATION

14-2 AEROTRIANGULATION WITH MULTIPROJECTOR INSTRUMENTS

Perhaps the most easily visualized (although now seldom practiced) method of photogrammetric aerotriangulation utilizes a multiprojector stereoplotter of the type shown in Fig. 14-1. In extending control by this method, adjacent stereomodels of a strip are successively oriented to each other to form a continuous "strip model." Coordinates of all pass points and control points are read from the strip model and then all points are adjusted to a limited amount of field-surveyed ground control to arrive at their final coordinates.

The first step in aerotriangulation with a multiprojector instrument is preparation of a long manuscript map which will accommodate the strip model. All available field-surveyed horizontal and vertical ground control is plotted thereon. The manuscript is placed on the plotting table, and assuming enough control exists in the first model, it is completely oriented. It is convenient, although not necessary, to have enough field-surveyed ground control in the beginning model of the strip to enable that model to be absolutely oriented.

Figure 14-1 Multiprojector Balplex Stereoplotting instrument. (*Courtesy Bausch and Lomb.*)

After completion of the orientation of the first model, the third diapositive of the strip is placed in projector III, as illustrated in Fig. 14-2. An initial model base $b'_{2\text{-}3}$ for model 2-3 is set approximately equal to the model base for model 1-2. The third photo is then relatively oriented to photo 2, using the one-projector method (see Sec. 12-8) to avoid upsetting previously oriented model 1-2. All motions during relative orientation of model 2-3 are imparted to projector III only.

Relative orientation of model 2-3 by the one-projector method automatically levels the model (it has the same datum as model 1-2), but model 2-3 is not yet at proper scale. In Fig. 14-2 point B', approximately vertically beneath projector II, has a lower elevation in model 2-3 than corresponding point B in model 1-2. This indicates that model base $b'_{2\text{-}3}$ is too large, causing the scale of model 2-3 to be too large. To give model 2-3 the same scale as model 1-2, the platen is raised until the tracing table dial reads the elevation of point B as determined in model 1-2. Model base $b'_{2\text{-}3}$ is then decreased by $\Delta b_{2\text{-}3}$ until the floating mark appears to rest exactly on point B. Points such as B are known as *carry-over* points because they make it possible to carry scale throughout the strip.

Because relative orientation was done by one-projector method, the model base contains b_y and b_z components. Therefore, besides introducing a Δb_x correction, small Δb_y and Δb_z components will also be needed to clear y parallax in model 2-3. This procedure is explained in Sec. 12-9. The process is repeated for each successive model of the strip until all photos have been oriented or until all projectors have been utilized.

Suppose that a control extension of the seven-model strip shown in Fig. 14-3a has been performed using the instrument pictured in Fig. 14-1. Positions of all horizontal and vertical control points have been plotted on the manuscript as they were measured from the model, as shown on the figure. Also, model locations of all pass points have been plotted, and model elevations of all control points and pass points

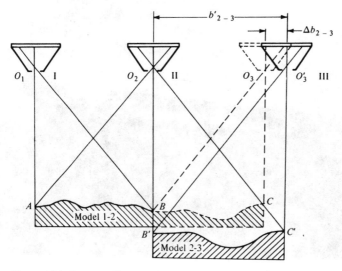

Figure 14-2 Scaling adjacent model 2-3 to model 1-2 by adjusting model base.

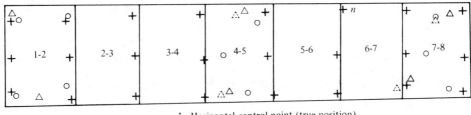

△. Horizontal control point (true position)
o Vertical control point
+ Pass point
△ Horizontal control point (model position)

(a)

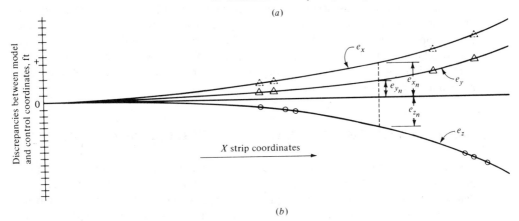

(b)

Figure 14-3 (a) Plan view of control extension of seven-model strip. (b) Smooth curves indicating accumulation of errors in X, Y, and Z coordinates during control extension of a strip.

have been read and recorded. Note that six pass points have been placed in each model, a standard procedure for stereotriangulation.

Except for the ground control in the first model, which was used to absolutely orient that model, discrepancies exist between plotted model positions of horizontal control points and their corresponding plotted field-surveyed positions. This is the result of various systematic and random errors that occur in the process of stereotriangulation. On the manuscript map of Fig. 14-3a, solid triangles represent model positions of horizontal control points and dashed triangles represent their corresponding "true" positions plotted from field-survey information. Note the discrepancies between model and true positions. If the X-coordinate discrepancies (e_x) and Y-coordinate discrepancies (e_y) for all horizontal control points are plotted versus their X-strip coordinates, smooth curves such as those of Fig. 14-3b should result. Assuming that discrepancies along the entire strip correspond to these curves, corrections to pass points for X and Y discrepancies can be taken directly from the curves on the basis of their respective X-strip coordinates. Pass point n, for example, would receive corrections of magnitude e_{x_n} and e_{y_n}, as shown on Fig. 14-3b.

In similar fashion, by comparing measured model elevations of vertical control points with their field surveyed values, discrepancies will be noted as the strip progresses. These discrepancies (e_z) may also be plotted versus X-strip coordinates and corrections made to pass point elevations on the basis of the curve. As shown on Fig. 14-3b, for example, pass point n would receive a correction of e_{z_n}. When the positions and elevations of the pass points have been corrected, they are ready for use as control points for map compilation or other photogrammetric operations.

Instead of obtaining the corrections graphically, as illustrated in Fig. 14-3b, they may be determined numerically by approximating the graphical curves with numerical polynomials. Strip adjustment by polynomials is readily performed using a computer. The procedure is described further in Sec. 14-15.

The amount of field-surveyed control needed in a strip depends upon the length of the strip. As a minimum, about two horizontal and three vertical points should exist in approximately every fifth model of the strip. Accuracy requirements, terrain and vegetation conditions, and access to the area for surveys will govern the actual configuration somewhat. A satisfactory control scheme for photogrammetric control extension of a strip is shown in Fig. 14-3.

14-3 PASS POINTS FOR ANALOGUE AEROTRIANGULATION

Pass points for analogue aerotriangulation are normally selected in the general photographic locations shown in Fig. 14-4a. The points may be images of natural well-defined objects that appear in the required photo areas, but if such points are unavailable, pass points may be artificially marked using a special point-marking device such as that shown in Fig. 14-12. Point-marking devices make small holes in the emulsion which become the pass points.

Even though satisfactory natural points may exist in the required general locations on the photographs, many photogrammetrists prefer to mark pass points artificially for two reasons. First, a more discrete point is obtained so that more accurate measurements of its position can be obtained. Second, the likelihood of misidentifying pass points is greatly reduced. In analogue control extension only three pass points near the y axis of each photo are marked, as shown in Fig. 14-4a. When stereopairs of photos with pass points marked in this manner are oriented in a plotter, six points appear in each stereomodel, as shown in Fig. 14-4b.

14-4 AEROTRIANGULATION WITH UNIVERSAL INSTRUMENTS

Continuous strip aerotriangulation by the previously described multiprojector technique requires a rather large instrument equipped with many projectors. This same procedure can be accomplished with a stereoplotter having only two projectors if it is a *universal instrument* which possesses *base-in* and *base-out* capabilities. The Wild A-7 Autograph and Zeiss C-8 Stereoplanigraph described in Chap. 12 are two universal instruments capable of performing continuous strip triangulation.

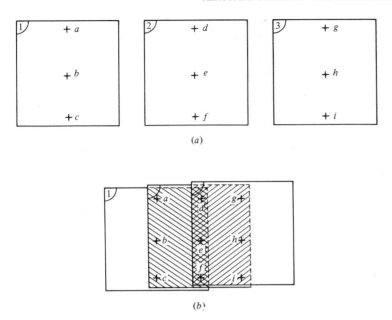

Figure 14-4 (*a*) Idealized pass point locations for analogue control extension. (*b*) Locations of pass points in two adjacent stereomodels.

The principles of strip formation using a universal instrument are illustrated in Fig. 14-5. Model 1-2 of the strip is relatively and absolutely oriented in the conventional manner with the instrument in the base-in mode. The geometry of this situation is illustrated in Fig. 14-5*a*. The base-in mode is said to exist when the space rods are pointing toward each other, and the two model base components b_L and b_R are to the inside of perspective centers O' and O''. With model 1-2 oriented, X, Y, and Z coordinates of all control points and pass points are read and recorded. With universal instruments these coordinates may be read directly from graduated scales so that errors due to a graphical plot are removed. A preferred data recording arrangement uses electronic equipment to automatically record model coordinates on punched cards, punched paper tape, or magnetic tape. This equipment avoids mistakes in reading and recording and saves a great deal of time.

After model 1-2 has been completed, photo 1 is replaced by photo 3 in the left projector and the base bridge is extended to the base-out mode shown in Fig. 14-5*b*. The base-out position is so called because the two base components b_L and b_R are to the outside of the perspective centers. In the base-out mode an optical switch enables switching the view of the left eye to the right diapositive and the right eye to the left diapositive. The net result of the base-out mode, in geometrical equivalence, is that the right projector is moved to the left, as shown dashed in Fig. 14-5*b*.

In the base-out mode, model 3-2 is relatively oriented using the one-projector method and imparting all motions to the left projector only. The model is then scaled as described in the previous section by changing the model base until the elevation

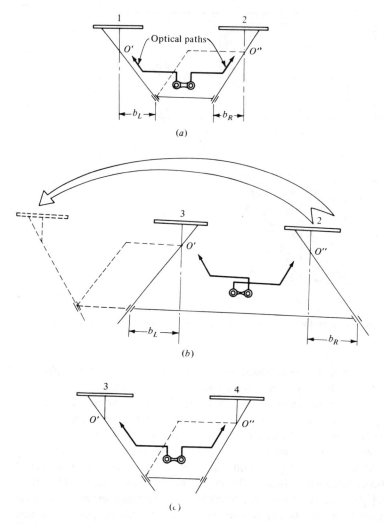

Figure 14-5 Principles of strip formation with a universal stereotriangulation instrument. (a) Model 1-2 oriented with base in; (b) Model 2-3 oriented with base out; (c) model 3-4 oriented with base in. In (b) note that the equivalent of the base-in projector configuration (shown dashed) is achieved.

of a carry-over point in model 3-2 reads the same as its elevation was in model 1-2. Coordinates of model points are then read and recorded. In model 3-2, Y and Z coordinates will be in the same coordinate system as model 1-2. Depending upon the instrument, there may be a discontinuity of X coordinates, however, and a numerical translation may be necessary.

When model 3-2 is completed, photo 2 is replaced by photo 4 in the right projector, the optical switch is flipped back, and the instrument is returned to the base-in mode, as shown in Fig. 14-5c. Photo 4 is relatively oriented to photo 3 by the one-projector method and the model is scaled using a carry-over point as before. Model coordinates for model 3-4 are then read and recorded. This procedure of alternating base-in and base-out is continued until the strip has been completed. Strip coordinates are in the arbitrary coordinate system of model 1-2, which is an approximate and unadjusted coordinate system. When the strip is completed, an adjustment is made to all pass point coordinates according to discrepancies between field-surveyed coordinates of ground control points and their model coordinates. Because the recorded data is in digital form, this adjustment is most logically performed numerically using polynomials, as described in Sec. 14-15, which approximates the graphical curves. After adjustment is made, the pass points are ready for use as control for other photogrammetric work.

Some advantages of the universal instrument method over the multiprojector approach are that theoretically there is no limit on the length of strip that can be handled, and accuracy is improved because digital recording is more accurate than plotting. In recent years both of these procedures of analogue aerotriangulation have largely been replaced by the method of independent models, described in the next section.

PART II   SEMIANALYTICAL AEROTRIANGULATION

14-5   GENERAL DESCRIPTION

Semianalytical aerotriangulation, often referred to as *independent model* aerotriangulation, is a partly analogue and partly analytical procedure that emerged with the development of computers. It involves manual relative orientation in a stereoplotter of each stereomodel of a strip or block of photos. Contiguous models are then joined analytically to form a continuous strip or block model, and then absolute orientation is performed numerically to adjust the strip or block model to ground control. The most significant advantage of the method over analogue aerotriangulation is that any two-projector stereoplotter can be used, provided that it is equipped with a coordinatograph for reading model coordinates. Other advantages are that absolute orientation is performed analytically and is therefore unnecessary in the plotter; thus time is saved, and least squares can be used in numerical strip formation which increases precision.

In semianalytical aerotriangulation, each stereopair of a strip is relatively oriented in the plotter, the coordinate system of each model being independent from the others. Model coordinates of all control points and pass points are then read and recorded for each stereomodel in turn. Figure 14-6a and Fig. 14-6b illustrate the first three relatively oriented stereomodels of a strip and show plan views of their respective independent coordinate systems. By means of pass points common to adjacent models, a three-dimensional coordinate transformation (see Sec. B-7 of Appendix B) is used to tie each successive model to the previous one. To gain needed geometrical strength in the transformations, the coordinates of the perspective centers (model exposure sta-

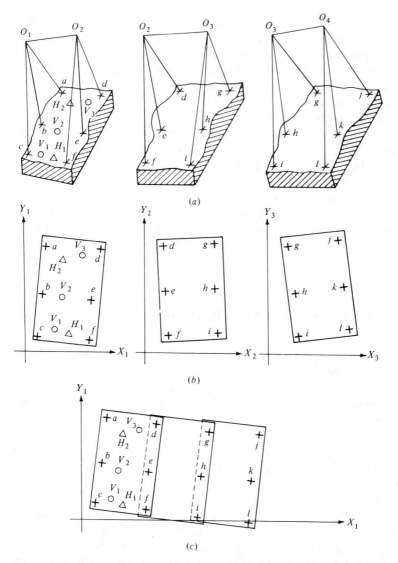

Figure 14-6 Independent model or semianalytical stereotriangulation. (*a*) Three adjacent relatively oriented stereomodels. (*b*) Individual arbitrary coordinate systems of three adjacent stereomodels. (*c*) Continuous strip of stereomodels formed by numerically joining the individual arbitrary coordinate systems into one system.

tions) are also measured in each independent model and included as common points in the transformation. The right exposure station of model 1-2, for example, is the same point as the left exposure station of model 2-3. To transform model 2-3 into model 1-2, therefore, coordinates of common points *d, e, f,* and O_2 of model 2-3 are made to coincide with their corresponding model 1-2 coordinates.

By applying successive coordinate transformations, a continuous strip of stereo-models may be formed, as illustrated in Fig. 14-6c. This strip may then be numerically adjusted to ground control using polynomial methods (see Sec. 14-15) to obtain corrected coordinates for all pass points. An alternative procedure for adjusting model coordinates to ground control has been found to yield superior results. In this method all independent stereomodels of the strip or block are simultaneously adjusted to each other by virtue of their common pass points and perspective centers, while at the same time the models are also adjusted to all ground control. The procedure is described in references cited at the end of this chapter.

14-6 PERSPECTIVE CENTER COORDINATES

Since model perspective centers are used in the strip formation coordinate transformations, their coordinates must be determined. Different techniques are applied for determining these coordinates. The most convenient method is their direct readout in the model coordinate system, but this requires a stereoplotter with equipment designed specifically for that purpose. This equipment is standard for a few mechanical projection plotters, such as the Kern PG-2, and consists basically of a collimating telescope and mirror for aligning each space rod vertically by the autocollimation principle. With the space rods vertical, X and Y coordinates of the perspective centers can be read. Also, by adjusting the Z motion of the tracing stand, a reference mark on the space rod can be brought into coincidence with a reference mark on the plotter frame. By addition of a calibrated constant to the measured Z value at this setting, the Z coordinate of the perspective center is obtained.

A second method of perspective center coordinate determination, sometimes referred to as the *two-point* method, is depicted in Fig. 14-7. In this approach discrete points from any diapositive are read monoscopically in the model coordinate system at the extreme high and low levels of the stereoplotter's Z range. The following two-point form of the equation for a straight line is then written for each point:

$$\frac{X_0 - X_{A_1}}{X_{A_2} - X_{A_1}} = \frac{Y_0 - Y_{A_1}}{Y_{A_2} - Y_{A_1}} = \frac{Z_0 - Z_{A_1}}{Z_{A_2} - Z_{A_1}} \tag{14-1}$$

In this equation X_0, Y_0, and Z_0 are perspective center coordinates and X_{A_1}, Y_{A_1}, Z_{A_1}, X_{A_2}, Y_{A_2}, and Z_{A_2} are model coordinates for point a read at the two different Z levels, as shown in Fig. 14-7. A minimum of two points must be read at two Z levels to obtain a solution, but, as a practical minimum, four corner points should be read and the solution obtained using least squares. The advantage of this method is that no special equipment is needed.

A third method, commonly called the *grid-plate* method, is illustrated in Fig. 14-8. A precisely manufactured grid plate is carefully centered in each projector, and X, Y, and Z model coordinates of the number of grid intersection points (preferably those in the corners of the model space) are read monoscopically in the model coordinate system. Treating the projector principal distance as analogous to the camera focal

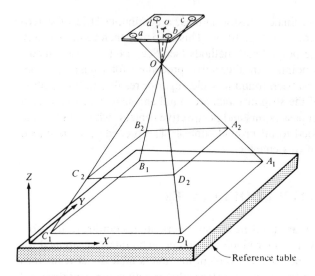

Figure 14-7 *Two point* method of determining model coordinates of perspective centers of stereoplotter projectors.

length, the coordinates of grid intersection points on the grid plate as analogous to photocoordinates, and model coordinates of grid intersections as ground coordinates, the space resection problem described in Sec. 11-12 is solved to obtain the X, Y, and Z coordinates of the perspective center. A minimum of three grid intersection points must be read for a solution but four corner points are recommended as a practical minimum. If four or more points are read, a least squares solution is obtained. Tests show that the accuracy with which perspective center coordinates can be determined increases by increasing the number of grid points read; however, no appreciable accuracy is gained by reading more than about 10 points.

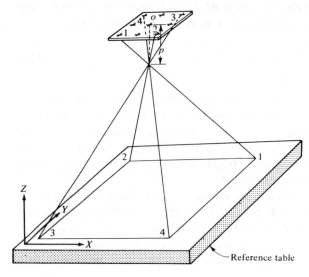

Figure 14-8 *Grid plate* method of determining model coordinates of perspective centers of stereoplotter projectors.

Each of the above-described methods is capable of yielding perspective center coordinates with good accuracy; however, the most precise method for the determination of perspective center coordinates involves the measurement of a stereomodel formed by relatively orienting a pair of precise grid plates that have been carefully centered in the projectors. Ideally, the resultant model should be a plane with a uniform grid size (the actual size will be dependent on the base set in the stereoplotter). After an operator performs relative orientation and measures X, Y, and Z model coordinates of a number of grid intersections, transformation constants as expressed by Eqs. (B-28) of Appendix B between model X, Y, and Z coordinates and an $X'Y'Z'$ "control" coordinate system are analytically determined. The control coordinate system in $X'Y'$ consists of coordinate values of the actual grid intersections, and Z' is zero for all points. The X, Y, and Z coordinates of the left perspective center are then determined by applying the transformation constants according to Eqs. (B-28) to a point whose X', Y' coordinates correspond to the principal point of the left grid plate and whose Z' coordinate is equal to the principal distance set in the two projectors. Likewise, X, Y, and Z coordinates of the right perspective center are determined by applying the transformation constants to a point whose X', Y' coordinates correspond to the principal point of the right grid plate and whose Z' coordinate is also equal to the principal distance set in the projector.

As previously mentioned, the stereomodel formed by the grid plates should be a uniform grid that exists in a plane. Discrepancies between the measured X, Y, and Z model coordinates of grid intersections and their X', Y', and Z' coordinates are an indication of any departure from the ideal condition. These discrepancies may be due to instrument maladjustment, scale differences between the X annd Y projections in the stereoplotter, nonorthogonality between the X and Y axes of the stereoplotter coordinatograph, etc. These discrepancies (or errors) may be used to mathematically express stereomodel deformations through the use of observation equations such as

$$v_x = a_{11}X + a_{12}Y + a_{13}XY + a_{14}X^2 + a_{15}Y^2$$
$$+ a_{16}X^2Y + a_{17}XY^2 + a_{18}X^2Y^2$$

$$v_y = a_{21}X + a_{22}Y + a_{23}XY + a_{24}X^2 + a_{25}Y^2 \qquad (14\text{-}2)$$
$$+ a_{26}X^2Y + a_{27}XY^2 + a_{28}X^2Y^2$$

$$v_z = a_{31}X + a_{32}Y + a_{33}XY + a_{34}X^2 + a_{35}Y^2$$
$$+ a_{36}X^2Y + a_{37}XY^2 + a_{38}X^2Y^2$$

In Eqs. (14-2), the v's are discrepancies in X, Y, and Z, and the a's are coefficients which describe the model deformation. A set of three equations of the type of Eqs. (14-2) can be written for each grid model point read. If 8 points are read, a unique solution for the 24 unknown a's can be made; if more than 8 are available, then least squares can be applied in their calculation. After these coefficients have been determined, Eqs. (14-2) can then be used to correct for model deformations for every model point subsequently read in semianalytical aerotriangulation. Thus, this grid model procedure not only yields perspective center coordinates but also enables corrections to be made for systematic errors of the stereoplotter, thereby increasing the overall accuracy of the aerotriangulation.

PART III   ANALYTICAL AEROTRIANGULATION

14-7 INTRODUCTION

The most elementary approaches to analytical aerotriangulation consist of the same basic steps as those of analogue methods and include (1) relative orientation of each stereomodel, (2) connection of adjacent models to form a continuous strip, and (3) adjustment of the strip to field-surveyed ground control. What is different about analytical methods is that the basic input consists of precisely measured photocoordinates of control points and pass points. Relative orientation is then performed analytically based upon the measured coordinates and known camera constants.

Analytical aerotriangulation tends to be more accurate than analogue or semi-analytical analogue, largely because analytic techniques can more effectively eliminate systematic errors such as film shrinkage, atmospheric refraction distortions, camera-lens distortions, etc. It is not uncommon, for example, for X and Y coordinates of pass points to be located analytically to an accuracy of within $\frac{1}{15,000}$ of the flying height, and for Z coordinates to be located to an accuracy of $\frac{1}{10,000}$ of the flying height. Even higher accuracies are possible with specialized procedures. Another advantage of analytical methods is freedom from the mechanical or optical limitations imposed by stereoplotters. Photography of any focal length, tilt, and flying height can be handled with the same efficiency. A disadvantage of analytical methods is that the computations are somewhat complicated and difficult to comprehend. Also, a computer with relatively large storage capacity and computing capability is necessary to contain and compute the volume of data involved in extensive problems with economy and speed.

Several different variations in analytical aerotriangulation techniques have evolved. Basically, however, all methods consist of writing condition equations which express the unknown elements of exterior orientation of each photo in terms of camera constants, measured photocoordinates, and ground coordinates. The equations are solved to determine the unknown orientation parameters, and either simultaneously or subsequently, coordinates of pass points are calculated. The most commonly used methods enforce one of two conditions: *collinearity* or *coplanarity*. Analytical procedures have been developed which can simultaneously enforce collinearity or coplanarity conditions onto units which consist of stereopairs, triplets, small blocks, and even large blocks of photos.

14-8 COLLINEARITY CONDITION

Collinearity, as described in Appendix C, is the condition that the exposure station, any object point, and its photo image all lie along a straight line. The collinearity condition is illustrated in Fig. 14-9, where L, a, and A lie along a straight line. Two equations express the collinearity condition for any point on a photo: one equation for the x photocoordinate and another for the y photocoordinate. The basic equations are Eqs. (C-5) and (C-6) of Appendix C, which are repeated here for convenience:

$$x = -f\left[\frac{m_{11}(X - X_L) + m_{12}(Y - Y_L) + m_{13}(Z - Z_L)}{m_{31}(X - X_L) + m_{32}(Y - Y_L) + m_{33}(Z - Z_L)}\right] \qquad (14\text{-}3)$$

$$y = -f\left[\frac{m_{21}(X - X_L) + m_{22}(Y - Y_L) + m_{23}(Z - Z_L)}{m_{31}(X - X_L) + m_{32}(Y - Y_L) + m_{33}(Z - Z_L)}\right] \qquad (14\text{-}4)$$

In Eqs. (14-3) and (14-4), x and y are photocoordinates of an image point; X, Y, and Z are ground coordinates of the object point; X_L, Y_L, and Z_L are ground coordinates of the exposure station; f is the camera focal length; and the m's (as described in Sec. B-7 of Appendix B) are functions of rotation angles omega, phi, and kappa.

The collinearity equations are nonlinear and are linearized using Taylor's theorem as described in Appendix C. The linearized forms are Eqs. (C-11) and (C-12) of Appendix C, and they are also repeated here for convenience.

$$v_x = b_{11}(d\omega) + b_{12}(d\phi) + b_{13}(d\kappa) - b_{14}(dX_L) - b_{15}(dY_L) - b_{16}(dZ_L)$$
$$+ \; b_{14}(dX) + b_{15}(dY) + b_{16}(dZ) + J \qquad (14\text{-}5)$$

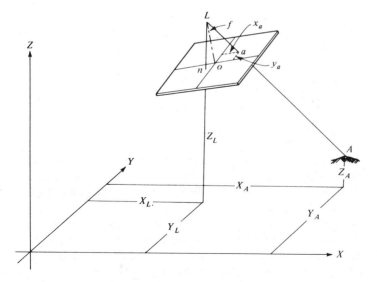

Figure 14-9 The *collinearity* condition.

$$v_y = b_{21}(d\omega) + b_{22}(d\phi) + b_{23}(d\kappa) - b_{24}(dX_L) - b_{25}(dY_L) - b_{26}(dZ_L)$$

$$+ b_{24}(dX) + b_{25}(dY) + b_{26}(dZ) + K \qquad (14\text{-}6)$$

In Eqs. (14-5) and (14-6), v_x and v_y are residual errors in measured x and y image coordinates; $d\omega$, $d\phi$, and $d\kappa$ are corrections to initial approximations for the orientation angles of the photo; dX_L, dY_L, and dZ_L are corrections to initial approximations for the exposure station coordinates; and dX, dY, and dZ are corrections to initial values for the object space coordinates of the point. The b's are coefficients which are described in Appendix C. Because nonlinear terms are ignored in linearization by Taylor's theorem, the linearized forms of the equations are approximations. They are therefore solved iteratively, as described in Appendix C, until the magnitudes of corrections to initial approximations become negligible.

14-9 COPLANARITY CONDITION

Coplanarity, as illustrated in Fig. 14-10, is the condition that the two exposure stations of a stereopair, any object point, and its corresponding image points on the two photos all lie in a common plane. In the figure, for example, points L_1, L_2, a_1, a_2, and A all lie in a common plane. The coplanarity condition equation is

$$0 = B_X(D_1F_2 - D_2F_1) + B_Y(E_2F_1 - E_1F_2) + B_Z(E_1D_2 - E_2D_1) \qquad (14\text{-}7)$$

where

$$B_X = X_{L_2} - X_{L_1}$$

$$B_Y = Y_{L_2} - Y_{L_1}$$

$$B_Z = Z_{L_2} - Z_{L_1}$$

$$D = (m_{12})x + (m_{22})y - (m_{32})f$$

$$E = (m_{11})x + (m_{21})y - (m_{31})f$$

$$F = (m_{13})x + (m_{23})y - (m_{33})f$$

In Eq. (14-7), subscripts 1 and 2 affixed to terms D, E, and F indicate that the terms apply to either photo 1 or photo 2. The m's again are functions of the three rotation angles omega, phi, and kappa. They are defined in Sec. B-7 of Appendix B. One coplanarity equation may be written for each object point whose images appear on both photos of the stereopair. The coplanarity equations do not contain object space coordinates as unknowns; rather, they contain only the elements of exterior orientation of the two photos of the stereopair. Therefore, after solving for the elements of exterior orientation, object point coordinates are calculated, by solving the space resection problem as described in Sec. 11-13.

Like collinearity equations, the coplanarity equation is nonlinear and must be

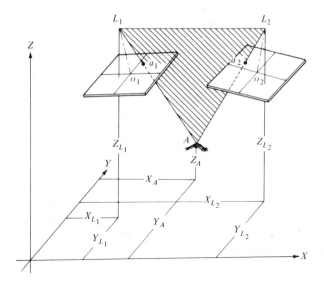

Figure 14-10 The *coplanarity* condition.

linearized using Taylor's theorem. Linearization of coplanarity equations is somewhat more difficult than collinearity equations and is beyond the scope of this text. References cited at the end of this chapter explain these procedures, however. Coplanarity is not used nearly as extensively as collinearity in analytical photogrammetry.

14-10 PASS POINTS FOR ANALYTICAL AEROTRIANGULATION

In analytical aerotriangulation, pass points are located in the standard positions illustrated in Fig. 14-11. The pass points may be natural features, but artificially marked points are preferred for the same reasons cited in analogue aerotriangulation. If a stereocomparator (see Sec. 14-11) is used for taking the photocoordinate measurements, only the three pass points along the y axis through the center of each photo need to be marked, the same as for analogue or semianalytical aerotriangulation (see Fig. 14-4a). If a monocomparator is used, however, all nine points (six points on end photos of a strip) must be marked on each plate, as shown in Fig. 14-11.

In artificially marking pass points for monocomparator measurement, some points such as d, e, and f of Fig. 14-11, for example, appear on three successive photos. Each of these points may be located quite arbitrarily on one photograph, which is usually the center photo on which they appear; but once marked they must be *carefully* transferred to their corresponding locations on the two adjacent plates. The Wild PUG-4 point-transfer instrument of Fig. 14-12 is a device designed specifically for marking corresponding pass points. With this instrument a stereopair of diapositives is placed on the illuminated stage plate. The diapositives are viewed stereoscopically through a binocular viewing system which has a reference half mark superimposed on each of its optical paths. By means of slow-motion screws, the diapositives can be adjusted slightly in the x and y directions so that the floating mark appears to rest exactly on

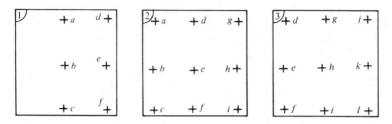

Figure 14-11 Pass point locations for analytical control extension using a monocomparator.

the point to be marked. Tiny drills which are collinear with the half marks are then lowered, and holes drilled into the emulsion to mark corresponding points. With any point-marking and point-transfer instrument it must be recognized that, upon making the holes, the emulsion in those spots is destroyed. Thus no further refinement in the identification of the image position can be made, and any error in positioning is fixed. It must therefore be done with extreme caution.

14-11 PHOTOCOORDINATE MEASUREMENT

Once pass points have been marked on the diapositives, they are ready for comparator measurement. Image coordinates of all field-surveyed control points are also measured. Either *monocomparators* or *stereocomparators* may be used. Measurements with a monocomparator are taken monoscopically using one plate at a time, as discussed in Sec. 5-7.

Stereocomparators simultaneously measure photocoordinates of corresponding points on a stereopair of diapositives. These instruments have two separate measuring systems, one for each photo. While viewing through binoculars along optical paths, the positions of the diapositives are adjusted until a reference floating mark appears to rest exactly on the desired point. In this position measurements are recorded for both

Figure 14-12 Wild PUG-4 stereoscopic point transfer instrument. (*Courtesy Wild Heerbrugg Instruments, Inc.*)

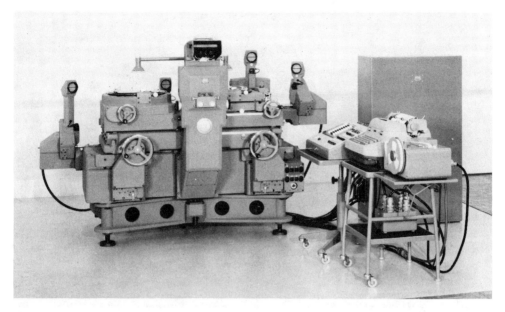

Figure 14-13 Wild STK-1 Stereocomparator. (*Courtesy Wild Heerbrugg Instruments, Inc.*)

photos. The basic measuring system of the Wild STK-1 stereocomparator shown in Fig. 14-13 consists of lead screws. Two measuring stages are utilized in the instrument: (1) a lower stage which carries both plates and records x_1 and y_1, the coordinates of a point on the left plate; and (2) an upper stage which records p_x and p_y, the small x and y translations necessary to place the floating mark in contact with the desired point. Coordinates x_2 and y_2 for the right photo are then

$$x_2 = x_1 + p_x$$

$$y_2 = y_1 + p_y$$

(14-8)

The Zeiss PSK Stereocomparator of Fig. 14-14 uses a precisely graduated glass grid as its basic measuring device. Diapositives are mounted firmly against the grid and with the floating mark set, x and y increments from the point being measured to the nearest calibrated grid lines are measured by a micrometer. Both the STK-1 and PSK stereocomparators are capable of readings to the nearest micrometer (micron).

The Kern CPM-1 shown in Fig. 14-15 serves as both a point-marking/point-transfer instrument and as a comparator. Its stereoviewing system enables precise placement of the floating mark on points to be marked and measured. When the mark has been placed, drill holes collinear with the half marks can be made to identify the point on both plates, and the photocoordinates of the point on the left plate can be measured. By combining these two operations, efficiency as well as accuracy can be increased.

The O.M.I.-Nistri TA-3/P uses three plates simultaneously. An operator can view

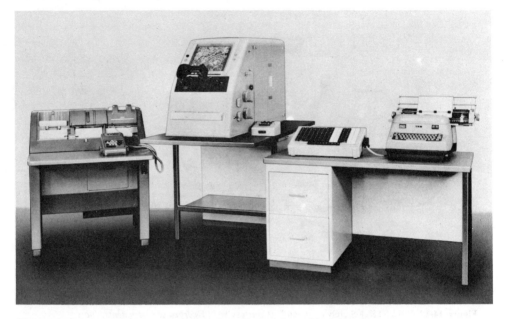

Figure 14-14 Zeiss PSK Stereocomparator. (*Courtesy Carl Zeiss, Oberkochen.*)

either the left stereopair or the right stereopair. Three separate measuring stages, each operating on the lead-screw principle, enable the simultaneous measurement of photocoordinates of common points on all three photos. This technique increases measurement efficiency, but perhaps its greatest advantage is the elimination of mistakes in misidentification of points from one stereopair to the next.

The measurement systems of all comparators are subject to small systematic errors, but these can be corrected for by reading coordinates of precise grid plates, comparing the readings with the precisely known grid coordinates, and then modeling the errors numerically.

14-12 PHOTOCOORDINATE REFINEMENT

To obtain the highest possible degree of accuracy in analytical solutions, measured photocoordinates must be corrected for several systematic errors which cause distortions in image positions. This is true whether the measurements were taken by monocomparator or stereocomparator. The systematic image distortion corrections normally applied in analytical photogrammetry are (1) reduction of measured coordinates to an axis system with origin at the principal point, (2) correction for film shrinkage or expansion, (3) compensation for camera-lens distortions, and (4) corrections for atmospheric refraction. If object point positions are to be determined in a plane coordinate system from high-altitude photography, it may also be necessary to apply a

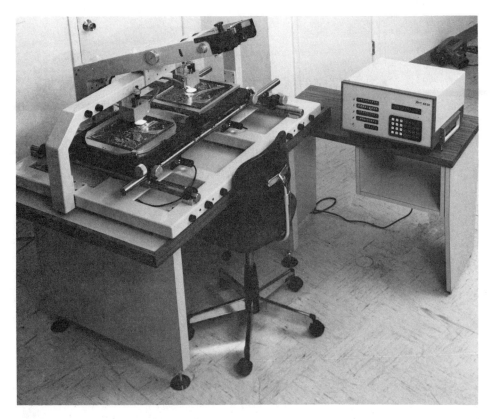

Figure 14-15 Kern CPM-1 Comparator/Point Marker. (*Courtesy Kern Instruments, Inc.*)

correction for earth curvature. All these corrections are described in Secs. 5-8 through 5-14.

14-13 RELATIVE ORIENTATION OF STEREOPAIRS BY COLLINEARITY

Analytical relative orientation of a stereopair is essentially a numerical duplication of the one-projector method of stereoplotter relative orientation. The left photo is fixed in position, its orientation is also fixed, and model scale is initially set by assigning an arbitrary model base. The right photo is then adjusted by applying rotations and translations until all y parallax is cleared.

Analytical relative orientation is depicted in Fig. 14-16. As illustrated in the figure, the left photo is arbitrarily fixed in position and orientation by setting ω_1, ϕ_1, κ_1, X_{L_1} and Y_{L_1} equal to zero. Z_{L_1} is also arbitrarily fixed at some rounded value, for example, approximately equal to the actual flying height of the photos. In addition, X_{L_2} is fixed at some arbitrary value, for example, approximately equal to the actual

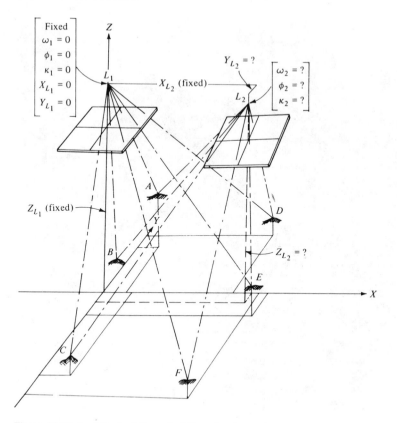

Figure 14-16 Analytical relative orientation of a stereopair.

air base. This fixes the scale of the mathematical stereomodel approximately equal to true scale. An alternative procedure is convenient in which Z_{L_1} is also fixed at zero and X_{L_2} is fixed at a value approximately equal to the photo base (photo distance between principal point and conjugate principal point). This establishes a model scale which is approximately equal to photo scale, and is advantageous for analyzing the results of the solution since standard deviations in computed X, Y, and Z model coordinates are then approximately in the units of the measured photo coordinates.

As with stereoplotter relative orientation, analytical relative orientation is achieved by enforcing the condition that corresponding rays intersect at a point. In Fig. 14-16, corresponding rays for six points A through F are shown satisfying this condition. The six points are located in basically the same areas as they were for stereoplotter relative orientation and they coincide with the pass point locations shown in Fig. 14-11. The simultaneous enforcement of collinearity for intersecting corresponding rays is achieved by writing collinearity equations for both photos for at least five object points. Intersection of corresponding rays is automatically enforced because the equations from both photos for a given object point contain the same object space coordinates. The system of collinearity equations that is formulated contains five unknown elements of exterior orientation for photo 2 (ω_2, ϕ_2, κ_2, Y_{L_2}, and Z_{L_2}) plus three unknown object space coordinates, (X, Y, and Z) for each point used in the solution.

For each point used in relative orientation, four collinearity equations can be written: an x and y equation of the form of Eqs. (14-5) and (14-6) for each photo of the stereopair. Using 5 object points, 20 equations can be written and a unique solution results because the number of unknowns is also 20, that is, 5 unknown exterior orientation parameters for photo 2 plus 15 unknown object point coordinates. (Recall that five model points was also the minimum number required for relative orientation with a stereoplotter.)

More than five points may be used in analytical relative orientation, and in that case each additional point adds four equations and only three new unknowns. Therefore each additional point adds one redundant equation. If redundancy exists in the system of equations, least squares techniques may be used to obtain most probable values. If 6 points were used for relative orientation, a system of 24 equations and 23 unknowns would result, and if 10 points were used, 40 equations involving 35 unknowns would result. In the solution, matrix methods which employ partitioning can considerably reduce the computational burden of solving these large systems of equations.

Prior to solving the collinearity equations, initial approximations for all unknown values must be determined. For photography that was intended to be vertical, values of zero are commonly used for initial estimates of ω_2, ϕ_2, κ_2, and Y_{L_2}. An initial value for Z_{L_2} may be selected equal to the value used for Z_{L_1}. Approximations for object space coordinates may be scaled from the photographs, or more conveniently, they may be calculated using parallax Eqs. (8-5) through (8-7) assuming a vertical stereopair.

Suppose that the six points of Fig. 14-16 were used in analytical relative orientation. In matrix form, the system of 24 equations involving 23 unknowns could be expressed as follows:

$$_{24}A_{23}\ _{23}X_1 = _{24}L_1 + _{24}V_1 \qquad (14\text{-}9)$$

Note that Eq. (14-9) conforms to Eq. (A-10) of Appendix A. The individual elements of the A matrix are given on pages 374 and 375, and those of X, L and V are shown on page 376.

The terms of these matrices are from Eqs. (14-5) and (14-6), and the method of calculating each is explained in Sec. C-4 of Appendix C. The subscripts a, b, c, . . . , f refer to corresponding points a, b, c, . . . , f; subscript 1 refers to the left photo, and subscript 2 refers to the right photo.

Upon studying these matrices, particularly the A matrix, their systematic nature becomes apparent. The fact that many submatrices of zeros exist also indicates the relative ease with which partitioning could be applied in the solution. Equation (14-9) can be solved by least squares using Eq. (A-12), which will yield most probable values for corrections to be applied to the initial approximations. By adding these corrections to the initial values, new improved estimates for the unknowns are obtained, and these are used to formulate the matrices again and re-solve the problem. This second iteration will yield a new set of smaller corrections, which are added to

$$A = \begin{bmatrix}
0 & 0 & 0 & 0 & 0 & (ba_{14})_1 & (ba_{15})_1 & (ba_{16})_1 & 0 & 0 & 0 \\
0 & 0 & 0 & 0 & 0 & (ba_{24})_1 & (ba_{25})_1 & (ba_{26})_1 & 0 & 0 & 0 \\
0 & 0 & 0 & 0 & 0 & 0 & 0 & 0 & (bb_{14})_1 & (bb_{15})_1 & (bb_{16})_1 \\
0 & 0 & 0 & 0 & 0 & 0 & 0 & 0 & (bb_{24})_1 & (bb_{25})_1 & (bb_{26})_1 \\
0 & 0 & 0 & 0 & 0 & 0 & 0 & 0 & 0 & 0 & 0 \\
0 & 0 & 0 & 0 & 0 & 0 & 0 & 0 & 0 & 0 & 0 \\
0 & 0 & 0 & 0 & 0 & 0 & 0 & 0 & 0 & 0 & 0 \\
0 & 0 & 0 & 0 & 0 & 0 & 0 & 0 & 0 & 0 & 0 \\
0 & 0 & 0 & 0 & 0 & 0 & 0 & 0 & 0 & 0 & 0 \\
0 & 0 & 0 & 0 & 0 & 0 & 0 & 0 & 0 & 0 & 0 \\
0 & 0 & 0 & 0 & 0 & 0 & 0 & 0 & 0 & 0 & 0 \\
0 & 0 & 0 & 0 & 0 & 0 & 0 & 0 & 0 & 0 & 0 \\
(ba_{11})_2 & (ba_{12})_2 & (ba_{13})_2 & (-ba_{15})_2 & (-ba_{16})_2 & (ba_{14})_2 & (ba_{15})_2 & (ba_{16})_2 & 0 & 0 & 0 \\
(ba_{21})_2 & (ba_{22})_2 & (ba_{23})_2 & (-ba_{25})_2 & (-ba_{26})_2 & (ba_{24})_2 & (ba_{25})_2 & (ba_{26})_2 & 0 & 0 & 0 \\
(bb_{11})_2 & (bb_{12})_2 & (bb_{13})_2 & (-bb_{15})_2 & (-bb_{16})_2 & 0 & 0 & 0 & (bb_{14})_2 & (bb_{15})_2 & (bb_{16})_2 \\
(bb_{21})_2 & (bb_{22})_2 & (bb_{23})_2 & (-bb_{25})_2 & (-bb_{26})_2 & 0 & 0 & 0 & (bb_{24})_2 & (bb_{25})_2 & (bb_{26})_2 \\
(bc_{11})_2 & (bc_{12})_2 & (bc_{13})_2 & (-bc_{15})_2 & (-bc_{16})_2 & 0 & 0 & 0 & 0 & 0 & 0 \\
(bc_{21})_2 & (bc_{22})_2 & (bc_{23})_2 & (-bc_{25})_2 & (-bc_{26})_2 & 0 & 0 & 0 & 0 & 0 & 0 \\
(bd_{11})_2 & (bd_{12})_2 & (bd_{13})_2 & (-bd_{15})_2 & (-bd_{16})_2 & 0 & 0 & 0 & 0 & 0 & 0 \\
(bd_{21})_2 & (bd_{22})_2 & (bd_{23})_2 & (-bd_{25})_2 & (-bd_{26})_2 & 0 & 0 & 0 & 0 & 0 & 0 \\
(be_{11})_2 & (be_{12})_2 & (be_{13})_2 & (-be_{15})_2 & (-be_{16})_2 & 0 & 0 & 0 & 0 & 0 & 0 \\
(be_{21})_2 & (be_{22})_2 & (be_{23})_2 & (-be_{25})_2 & (-be_{26})_2 & 0 & 0 & 0 & 0 & 0 & 0 \\
(bf_{11})_2 & (bf_{12})_2 & (bf_{13})_2 & (-bf_{15})_2 & (-bf_{16})_2 & 0 & 0 & 0 & 0 & 0 & 0 \\
(bf_{21})_2 & (bf_{22})_2 & (bf_{23})_2 & (-bf_{25})_2 & (-bf_{26})_2 & 0 & 0 & 0 & 0 & 0 & 0
\end{bmatrix} \begin{array}{l} \\ 24 \end{array}$$

the current estimates. The solution is iterated until all corrections become negligible in size, whence the final values used as estimates thus become the solution for the unknowns.

After relative orientation, the model coordinates of the pass points are known. They are the same values that would be read from a stereoplotter if a one-projector relative orientation were performed with the left projector and the model base fixed as described for the analytical procedure. The model coordinate system is arbitrary with its axes parallel to the left photo axis system and its origin a distance Z_{L_1} vertically beneath the left exposure station.

With the elements of exterior orientation known after relative orientation, model coordinates of any field-surveyed ground control or other points that occur within the stereomodel can be obtained using the methods described in Sec. 11-13. These ad-

$$\begin{bmatrix}
0 & 0 & 0 & 0 & 0 & 0 & 0 & 0 & 0 & 0 & 0 & 0 \\
0 & 0 & 0 & 0 & 0 & 0 & 0 & 0 & 0 & 0 & 0 & 0 \\
0 & 0 & 0 & 0 & 0 & 0 & 0 & 0 & 0 & 0 & 0 & 0 \\
0 & 0 & 0 & 0 & 0 & 0 & 0 & 0 & 0 & 0 & 0 & 0 \\
(b_{c_{14}})_1 & (b_{c_{15}})_1 & (b_{c_{16}})_1 & 0 & 0 & 0 & 0 & 0 & 0 & 0 & 0 & 0 \\
(b_{c_{24}})_1 & (b_{c_{25}})_1 & (b_{c_{26}})_1 & 0 & 0 & 0 & 0 & 0 & 0 & 0 & 0 & 0 \\
0 & 0 & 0 & (b_{d_{14}})_1 & (b_{d_{15}})_1 & (b_{d_{16}})_1 & 0 & 0 & 0 & 0 & 0 & 0 \\
0 & 0 & 0 & (b_{d_{24}})_1 & (b_{d_{25}})_1 & (b_{d_{26}})_1 & 0 & 0 & 0 & 0 & 0 & 0 \\
0 & 0 & 0 & 0 & 0 & 0 & (b_{e_{14}})_1 & (b_{e_{15}})_1 & (b_{e_{16}})_1 & 0 & 0 & 0 \\
0 & 0 & 0 & 0 & 0 & 0 & (b_{e_{24}})_1 & (b_{e_{25}})_1 & (b_{e_{26}})_1 & 0 & 0 & 0 \\
0 & 0 & 0 & 0 & 0 & 0 & 0 & 0 & 0 & (b_{f_{14}})_1 & (b_{f_{15}})_1 & (b_{f_{16}})_1 \\
0 & 0 & 0 & 0 & 0 & 0 & 0 & 0 & 0 & (b_{f_{24}})_1 & (b_{f_{25}})_1 & (b_{f_{26}})_1 \\
0 & 0 & 0 & 0 & 0 & 0 & 0 & 0 & 0 & 0 & 0 & 0 \\
0 & 0 & 0 & 0 & 0 & 0 & 0 & 0 & 0 & 0 & 0 & 0 \\
0 & 0 & 0 & 0 & 0 & 0 & 0 & 0 & 0 & 0 & 0 & 0 \\
0 & 0 & 0 & 0 & 0 & 0 & 0 & 0 & 0 & 0 & 0 & 0 \\
(b_{c_{14}})_2 & (b_{c_{15}})_2 & (b_{c_{16}})_2 & 0 & 0 & 0 & 0 & 0 & 0 & 0 & 0 & 0 \\
(b_{c_{24}})_2 & (b_{c_{25}})_2 & (b_{c_{26}})_2 & 0 & 0 & 0 & 0 & 0 & 0 & 0 & 0 & 0 \\
0 & 0 & 0 & (b_{d_{14}})_2 & (b_{d_{15}})_2 & (b_{d_{16}})_2 & 0 & 0 & 0 & 0 & 0 & 0 \\
0 & 0 & 0 & (b_{d_{24}})_2 & (b_{d_{25}})_2 & (b_{d_{26}})_2 & 0 & 0 & 0 & 0 & 0 & 0 \\
0 & 0 & 0 & 0 & 0 & 0 & (b_{e_{14}})_2 & (b_{e_{15}})_2 & (b_{e_{16}})_2 & 0 & 0 & 0 \\
0 & 0 & 0 & 0 & 0 & 0 & (b_{e_{24}})_2 & (b_{e_{25}})_2 & (b_{e_{26}})_2 & 0 & 0 & 0 \\
0 & 0 & 0 & 0 & 0 & 0 & 0 & 0 & 0 & (b_{f_{14}})_2 & (b_{f_{15}})_2 & (b_{f_{16}})_2 \\
0 & 0 & 0 & 0 & 0 & 0 & 0 & 0 & 0 & (b_{f_{24}})_2 & (b_{f_{25}})_2 & (b_{f_{26}})_2
\end{bmatrix} \tag{23}$$

ditional points could also have been used in relative orientation, thereby adding redundancy to the solution.

14-14 STRIP FORMATION

There are different methods of analytically forming strips, and three of these are described here. In the first method, each stereopair in a strip of photos may be relatively oriented in the manner described in Sec. 14-13. The result is a series of independent models, each having its own arbitrary coordinate system. Just as with semianalytical aerotriangulation, adjacent models contain points that are common to both models. By use of coordinates of common points, and also by utilizing common

$$X = \begin{bmatrix} d_{\omega_2} \\ d_{\phi_2} \\ d_{\kappa_2} \\ d_{Y_{L_2}} \\ d_{Z_{L_2}} \\ d_{X_A} \\ d_{Y_A} \\ d_{Z_A} \\ d_{X_B} \\ d_{Y_B} \\ d_{Z_B} \\ d_{X_C} \\ d_{Y_C} \\ d_{Z_C} \\ d_{X_D} \\ d_{Y_D} \\ d_{Z_D} \\ d_{X_E} \\ d_{Y_E} \\ d_{Z_E} \\ d_{X_F} \\ d_{Y_F} \\ d_{Z_F} \end{bmatrix}_{23} \qquad L = \begin{bmatrix} (-J_a)_1 \\ (-K_a)_1 \\ (-J_b)_1 \\ (-K_b)_1 \\ (-J_c)_1 \\ (-K_c)_1 \\ (-J_d)_1 \\ (-K_d)_1 \\ (-J_e)_1 \\ (-K_e)_1 \\ (-J_f)_1 \\ (-K_f)_1 \\ (-J_a)_2 \\ (-K_a)_2 \\ (-J_b)_2 \\ (-K_b)_2 \\ (-J_c)_2 \\ (-K_c)_2 \\ (-J_d)_2 \\ (-K_d)_2 \\ (-J_e)_2 \\ (-K_e)_2 \\ (-J_f)_2 \\ (-K_f)_2 \end{bmatrix}_{24} \qquad V = \begin{bmatrix} (v_{x_a})_1 \\ (v_{y_a})_1 \\ (v_{x_b})_1 \\ (v_{y_b})_1 \\ (v_{x_c})_1 \\ (v_{y_c})_1 \\ (v_{x_d})_1 \\ (v_{y_d})_1 \\ (v_{x_e})_1 \\ (v_{y_e})_1 \\ (v_{x_f})_1 \\ (v_{y_f})_1 \\ (v_{x_a})_2 \\ (v_{y_a})_2 \\ (v_{x_b})_2 \\ (v_{y_b})_2 \\ (v_{x_c})_2 \\ (v_{y_c})_2 \\ (v_{x_d})_2 \\ (v_{y_d})_2 \\ (v_{x_e})_2 \\ (v_{y_e})_2 \\ (v_{x_f})_2 \\ (v_{y_f})_2 \end{bmatrix}_{24}$$

exposure station coordinates, three-dimensional coordinate transformations may be successively performed to join adjacent models and form a continuous model strip. This procedure is known as the *analytical method of independent models.*

Another approach to strip formation known as the *sequential method* is done simultaneously with relative orientation. It is more economical in terms of computational time because the need for strip formation by numerous three-dimensional coordinate transformations is eliminated. In the sequential method, the first model is relatively oriented as previously described. For stereopair 2-3, however, the elements of exterior orientation of the left photo are fixed equal to their values determined from relative orientation of model 1-2. This automatically retains the coordinate system of model 1-2 for the second model. The procedure is analogous to the multiprojector method of aerotriangulation. As with the multiprojector method of aerotriangulation, the scale of model 2-3 must initially be arbitrarily set by fixing a model base by estimation. Relative orientation is then performed and model coordinates for pass points calculated. The scale of model 2-3 is then checked against the scale of model 1-2 by computing lengths of lines common to both models. The scales are generally not equal because the initial estimate of model base is inexact. In Fig. 14-17, for example, points D and F are in model 1-2, and their corresponding locations computed in model 2-3 are at D' and F'. A scale factor relating the two models is calculated based on the two line lengths as follows:

$$S = \frac{DF}{D'F'} = \frac{\sqrt{(X_F - X_D)^2 + (Y_F - Y_D)^2 + (Z_F - Z_D)^2}}{\sqrt{(X_F' - X_D')^2 + (Y_F' - Y_D')^2 + (Z_F' - Z_D')^2}} \quad (14\text{-}10)$$

Generally the average of two or three scale factors determined from two or three different "scale lines" is used. All coordinates in model 2-3 are then multiplied by the scale factor to obtain scaled coordinates. Model base components for model 2-3 are also multiplied by the scale factor and new exposure station coordinates for photo

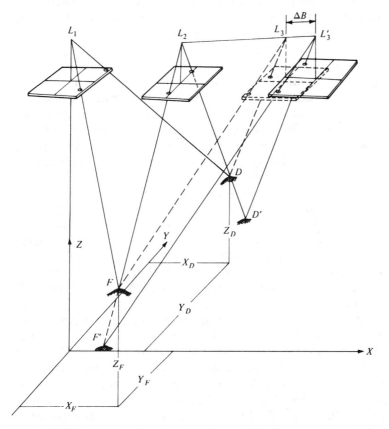

Figure 14-17 Analytical scaling of model 2-3 to model 1-2 by means of a scale line.

3 obtained by adding the respective scaled base components to the X, Y, and Z coordinates for exposure station 2. In Fig. 14-17, L_3' represents the original estimated exposure station and L_3 is the final exposure station obtained by scaling the model base. The change in model base length is Δb.

Scaled coordinates of a carry-over point in model 2-3 are compared to the co-ordinates of the carry-over point in model 1-2, and any difference represents a trans-lation that must be applied to all scaled model 2-3 coordinates to obtain final model coordinates for model 2-3.

Upon completion of model 2-3, model 3-4 is relatively oriented by holding the exterior orientation parameters of photo 3 fixed at their values determined in relative orientation of model 2-3. This procedure is continued until all models in the strip have been relatively oriented. The resulting strip will be continuous and in the coordinate system of the first stereopair.

Another approach to strip formation known as *three-photo aerotriangulation* is done simultaneously with relative orientation. In this method, three photos are con-sidered as a basic unit which is known as a "triplet." In its normal use, all parameters of the first photo of a flight line are assigned zero values. The scale of the output strip coordinate system is fixed by assigning an arbitrary value for X_L of the second photo. Again, if X_{L_2} is assigned a value approximately equal to the distance between the principal point and conjugate principal point, a scale close to photo scale results which simplifies analysis of the solution. If this is done, Z coordinates for all object points will be somewhat near the negative value of the camera focal length. Consider the three photos of Fig. 14-11 as the first triplet of a flight line. Having fixed all six parameters of the first photo and X_L of the second photo, there remain five unknown parameters for the second photo, six for the third photo, and X, Y, and Z for pass points a through i, for a total of 38 unknowns. These may be determined with the 42 measurements of x and y for points a through i. It may be observed that for points a, b, c, g, h, and i, there is no redundancy for the x measurements in this initial triplet. This can be readily visualized by considering parallax equations, Eqs. (8-5) through (8-7). Since parallax p_a exists in each equation, and because $p_a = x_a - x_a'$, x measurements on both photos are required to solve each equation. An additional equation similar to Eq. (8-7) could be developed, however, in which y_a' is simply interchanged with y_a. Thus, there is redundancy in y but not in x for points appearing on only two photos. Points d, e, and f each have one degree of freedom in the x-coordinate measurements since each appears on three photos. It will be seen pres-ently that, assuming the flight line continues toward the right, points g, h, and i will appear on three photos in the second triplet.

After the solution for the above-mentioned unknowns in the first triplet, all six parameters for the second photo are considered final along with the X, Y, and Z strip coordinates for points a through f. The next triplet would consist of photos 2, 3, and 4. The (now known) parameters for the second photo and X, Y, and Z strip coordinates for points d, e, and f serve as control for the solution. This triplet contains 12 unknown parameters for photos 3 and 4 and X, Y, and Z for points g through l, for a total of 30. These may be determined with the 42 x and y observations of points d through l. This process continues to the end of the flight line. With the completion of the computations for each triplet, the first photo of the triplet is dropped, the parameters for the center photo of the triplet are considered final (along with the three pass points for that photo) and it becomes the first photo of the next triplet. Naturally, object points other than pass points, such as control points, tie points, etc., may be included to add geometric strength to the solution. Again, it should be observed that any object point appearing on only two photos has no redundancy in x-coordinate measurements. This is particularly critical for vertical control points, and thus they should appear on as many photos as possible. A blunder in a vertical control value may go undetected if the point is measured on only two photos.

14-15 STRIP ADJUSTMENT

When a strip has been formed by one of the methods described in the previous section, model strip coordinates will be available for all pass points and ground control points. For the pass points to be useful for controlling subsequent photogrammetric operations, however, it is required to adjust the model strip coordinates for their inherent systematic and random errors and to transform them into the ground coordinate system. Analytically derived strips, like those formed by analogue methods or semianalytically could be adjusted graphically, as illustrated in Fig. 14-3b, using smooth curves to represent the errors that have accumulated along the strip. The adjustment is preferably done numerically by modeling the graphical curves with polynomials, especially in this age of computers.

Most of the polynomials in use for adjusting strips formed by aerotriangulation are variations of the following equations:

$$\Delta X = a_0 + a_1 X + a_2 X^2 + a_3 X^3$$

$$\Delta Y = b_0 + b_1 X + b_2 X^2 + b_3 X^3 \qquad (14\text{-}11)$$

$$\Delta Z = c_0 + c_1 X + c_2 X^2 + c_3 X^3$$

In Eqs. (14-11) the ΔX, ΔY, and ΔZ terms are discrepancies between strip coordinates of control points and ground control coordinates for the same points (the latter having been transformed into the same coordinate system as that of the strip), the X terms are model strip coordinates, and the a's, b's, and c's are coefficients which define the shapes of the curves. As illustrated in Fig. 14-3b, and as apparent in Eqs. (14-11), errors in X, Y, and Z at any point are principally functions of the linear distance (X coordinate) of the point along the strip. However the nature of error propagation along strips formed by aerotriangulation is such that discrepancies in X, Y, and Z coordinates are also each somewhat related to the Y positions of the points in the strip. Therefore some agencies have modified basic Eqs. (14-11) to derive polynomials which also include Y coordinates. Two of these—one devised in the United States by the National Geodetic Survey (NGS) and the other developed in Canada by the National Research Council (NRC)—have been widely used throughout the world. Discussion of these polynomials, and others in use for adjusting aerotriangulation, is beyond the scope of this text, but references cited at the end of this chapter describe them.

14-16 BLOCKS OF PHOTOS

Thus far only control extension through a strip of photos has been considered. Control extensions can also be made through blocks of photos (two or more side-lapping strips). In dealing with blocks, a number of pass points must be carefully selected so that they appear in the side lap area of adjacent strips as shown in Fig. 14-18. These points, called *tie points,* make it possible to tie the individual strips together to form the continuous block.

Whether the control extension procedure is analogue, semianalytical, or analytical, strip model coordinates are determined for all tie points as well as pass points and control points. Using numerical methods, the strips are successively joined by matching common tie points, and the block is adjusted to available horizontal and vertical field-surveyed ground control. This general procedure is known as *block adjustment.* As depicted in Fig. 14-18, control can be extended photogrammetrically through rather large areas using relatively few field-surveyed ground control points.

A technique that has been successfully used in photogrammetric control extension is bridging from high-altitude photography to establish photo control for map compilation with lower-altitude photos. Supplemental control extended in this manner can be established to satisfactory accuracy, and the use of high-altitude photos enables

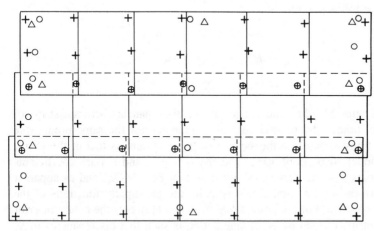

△ Horizontal control points

O Vertical control points

+ Pass points

⊕ Tie points

Figure 14-18 Block of photos prepared for analytical control extension showing horizontal and vertical ground control points, pass points, and tie points.

greater coverage from fewer photos, thereby increasing the efficiency of photogrammetric control extension. Pass points established on the high-altitude photographs can
be transferred to the lower-altitude photos using a point transfer instrument such as
that shown in Fig. 14-12, which has variable zoom optics to accommodate the different
photo scales.

14-17 SIMULTANEOUS "BUNDLE" ADJUSTMENT

The ultimate extension of the principles described in the preceding sections is to adjust
all photogrammetric measurements to ground control values in a single solution. Some
systems also allow for errors in the ground control values, and thus these are adjusted
along with the photo measurements. Such an adjustment minimizes the sum of the
squares of the weighted residuals for both photo and control measurements. The
process is relatively simple in concept since one merely forms observation equations,
Eqs. (14-5) and (14-6), for each x and y photo measurement. Field-survey values for
X, Y, and/or Z of control points either are considered absolutely known, in which
case they are eliminated as unknowns, or they are constrained to their measured (or
computed) values by weighting. All photo parameters are considered as unknowns,
and thus the total number of unknowns for a particular adjustment is six times the
number of photos plus three times the number of unknown object points (such as pass
points, tie points, etc.). Obviously, even a moderate size block of photos generates a
large number of unknowns. For example, the small block illustrated with Fig. 14-18
has 126 unknown camera parameters (6×21), plus 147 unknown pass point (includes
tie points) coordinates (3×49), plus 10 unknown Z values for the horizontal control
points, plus 34 unknown X, Y values for the vertical control points (2×17), for a
grand total of 317 unknown values. The same block would generate 506 independent
observation equations of the kind given by Eqs. (14-5) and (14-6). Fortunately, the
normal equation matrix can be broken down into smaller matrices because the X, Y,
and Z coordinates of all object points are directly related to the camera station parameters; that is, given the camera station parameters, X, Y, and Z of all object points
can be computed by intersection, and likewise given the X, Y, and Z of all object
points, the camera parameters can be computed by resection. Because of this the
largest matrix that need be inverted would be 126×126. Even this matrix is "sparse"
(contains many zero off-diagonal elements), and as a result it can be broken down
into still smaller units.

The simultaneous bundle adjustment requires rather sophisticated computer software and is the most costly of analytical methods in terms of execution time. The
results are, however, theoretically most rigorous and of the highest precision.

14-18 CONTROL REQUIREMENTS

Aerotriangulation programs generally use state plane coordinates and elevations related
to mean sea level as the ground control system, which is suitable in most cases.

However, for high altitude photography, or on projects covering extensive areas, to maintain theoretical correctness, coordinates of control points should be converted from state plane coordinates and mean sea level to a geocentric or local secant or tangent plane system. After performing the aerotriangulation in one of these systems, all coordinates are transformed back to state plane, and elevations referred to mean sea level to make them useful for mapping purposes. Methods of performing these transformations are beyond the scope of this text, but are described in references cited at the end of this chapter.

Figure 14-3*a* exhibits a reasonable control configuration for a single strip, and the configuration shown for Fig. 14-18 is satisfactory for a block of photos. Note that in Fig. 14-18 horizontal control is included around the periphery of the area only, whereas vertical control must be included in the interior to support the "bridge." Control points should appear on as many photos as possible to increase the redundancy for these important points. For this reason, the control points shown in Fig. 14-18 are located in the side-lap area between flight lines wherever possible. It is for this reason also that selection of control points after the photography has been taken is better than premarking control, although in barren country there is little choice except pre-marking.

In addition to using ground coordinates of points as control, some aerotriangulation systems can utilize airborne profile recording data as control. Other systems can take advantage of known lengths of lines or of constant elevations known to exist along shorelines and use them as control.

14-19 USE OF COMPUTERS AND DATA ANALYSIS

The magnitude of calculations involved in analytical aerotriangulation makes the use of digital computers imperative. In terms of computer execution time, the simultaneous bundle method is most expensive and anything that can be done to ensure error-free input data is advisable. It is for this reason that some systems perform relative orientation and strip adjustment even though these are not required prior to the simultaneous adjustment. Relative orientation provides a convenient means of checking most point-marking and photogrammetric measurements if the program provides for output of residuals on the measurements. These should be studied by the user, and any abnormally large values or systematic trends must be justified or reworking may be required. Point marking or transfer of common points (tie points) between flight lines, which is most difficult, cannot be checked by relative orientation since relative orientation concerns itself usually with a single flight line.

Strip adjustment provides a means of checking field-surveyed control and also serves to check tie points in the case of a block configuration. Again, the user should study the residuals or misclosures on control to verify that they are acceptable.

Besides adjusted coordinates for all pass points, etc., the simultaneous solution should also output residuals on photogrammetric measurements and ground control values. It may be necessary to edit measurements which exhibit abnormally large residuals and follow this with a rerun. Of course, in the solution it is necessary to do a series of iterations until corrections to the unknown quantities are so small that they can be considered insignificant. The number of iterations required for an adjustment varies, depending on control density and configuration and the quality of the approximate values input at the beginning. However, in the normal case, three or four iterations is usually sufficient.

14-20 ANALYTICAL AEROTRIANGULATION FROM DIGITIZED IMAGE DENSITIES

In recent years a new technique in analytical aerotriangulation has emerged which uses digitized image densities rather than comparator coordinates as basic input. (Systems for automatically digitizing photo image densities were described in Sec. 5-15.) The procedure involves numerical *image correlation* to match points on the left photo of a stereopair to their conjugate images on the right photo; thus, the process of manual point transfer is eliminated.

Figure 14-19 illustrates a stereopair of photos which have had their overlapping areas scanned by a microdensitometer. The individual pixels (although greatly exaggerated in size) are represented by small squares. For each of these pixels, the density value and row and column of its location in the photo can be stored in an electronic computer. Figure 14-19 also illustrates, in a geometric sense, how numerical image correlation is performed. A *target area* on the left photo of a certain row and column dimension (five by five in the example of the figure) is selected, and its location in the photo is identified by the row and column numbers of its corner elements. A larger corresponding *search array* (15 by 15 in the example of Fig. 14-19) is chosen in the right photo so that it will encompass the corresponding imagery of the target area. The necessary corresponding location of the search area can be determined on the basis of the left photo location of the target area, the camera focal length, and the percent end lap of the stereopair.

In the process of numerical image correlation, the computer tries to select the pixels within the search area that correspond to those of the target area. This is done by systematically moving about the search area and comparing densities of elements of the target area to those of equal-sized trial "subsearch" areas within the search array. Various algorithms have been used for making these comparisons. One of the simplest ones sums the squares of the density differences of corresponding elements, which in equation form is

$$C = \sum_{i=1}^{m} \sum_{j=1}^{n} (u_{i_j} - v_{i_j})^2 \tag{14-12}$$

In Eq. (14-12) C represents the correlation coefficient (which is a measure of the degree of correspondence between the target array and the trial subsearch array), u

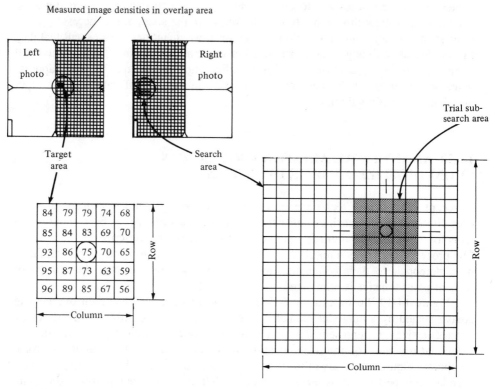

Figure 14-19 Density array configuration of a pair of overlapping aerial photographs. Shown also are target and corresponding search areas used in numerical image correlation.

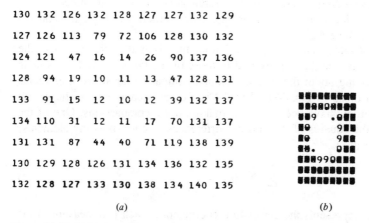

(a)

(b)

Figure 14-20 (a) Listing of density values by row and column obtained for a fiducial mark with a scanning microdensitometer. (b) Computer printout based on density values illustrating tonal variations of the fiducial mark as compared to surrounding imagery.

represents density values of target pixels, v represents densities of subsearch array elements, and i and j are row and column locations, respectively, within the two arrays. The density difference algorithm compensates somewhat for systematic tonal variations that occur from one photo to another due to differences in photoprocessing, lighting conditions, etc.

Having calculated the correlation coefficient for the first trial subsearch array, the computer systematically steps through the entire search array by moving one element to the right and/or one element up or down, calculating correlation coefficients of each trial subsearch array. One of these trial subsearch areas near the center of the search array is shown in Fig. 14-19. The subsearch array having the best correlation [for Eq. (14-12) the one producing the smallest value of C] is the area which best corresponds to the target array. The image at the row and column location of the center pixel of this subsearch array thus best corresponds to the image at the row/column position of the center element of the target array. Numerical image correlation, as described above, can be performed in each of the six conventional pass point locations to find corresponding pass point image locations.

During the original scanning process, the densities of the fiducial marks are also digitized and are located by row and column. Figure 14-20a shows the density values obtained for a fiducial mark, and Fig. 14-20b shows a printout of characters (similar to Fig. 5-21b) that shows the tonal variations of this fiducial mark. Note how strikingly different these densities are from the densities of surrounding photo images. The computer can be trained to identify the fiducial marks on the basis of these extreme density differences. Based upon the row/column locations of the fiducials, and their corresponding coordinates from camera calibration, a two-dimensional affine coordinate transformation (see Sec. B-6 of Appendix B) can be performed. Use of the resulting transformation parameters permits the row and column locations of the six pass points of each photo to be determined in their respective photocoordinate systems. With their photocoordinates known, relative orientation of the stereopair can be determined by applying the collinearity equations in accordance with procedures described in Sec. 14-13.

With the elements of relative orientation of the stereopair known, image correlation techniques can be continued to calculate X, Y, and Z coordinates of other model points. This procedure can be greatly aided by taking advantage of *epipolar* geometry, as illustrated in Fig. 14-21. This figure depicts the condition of coplanarity and shows the lines of intersection of the epipolar plane (plane L_1PL_2) with the left and right photo planes. These lines of intersection, pk and $p'k'$, are the so-called *epipolar lines*. They are important because, given the left photo location of image p, its corresponding point p' on the right photo is known to lie along epipolar line $p'k'$, and this epipolar line can be located on the basis of the known elements of relative orientation. This greatly reduces image correlation computational time because smaller target and subsearch arrays can be used, and the trial subsearch areas can be directed along the known location of the epipolar line.

By use of epipolar geometry, many points can be correlated throughout the overlap area. Their row and column locations can then be transformed into photocoordinates using the two-dimensional affine coordinate transformation parameters that were de-

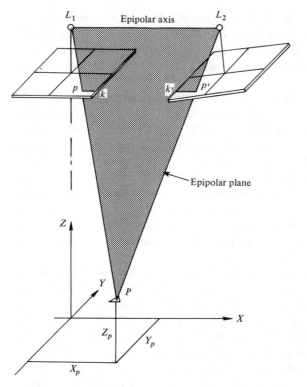

Figure 14-21 Epipolar geometry of a stereopair of photos. (*Courtesy Dr. T. J. Keating*)

termined earlier. With photocoordinates known for corresponding points, their X, Y, and Z model coordinates can be calculated by methods outlined in Sec. 11-13.

Adjustment of the model coordinates to ground is the final step, which is performed on the basis of photo control. For this process to be possible, the photo control points have to be uniquely marked prior to doing the densitometer scanning so that the computer can also identify them on the basis of their densities. Premarked panel points that appear uniquely in the photos, or drill holes made with a point marking instrument, will suffice for this purpose.

One of the uses of the above-described analytical procedure is in calculating digital terrain models (DTM's), from which contour maps and other products can be obtained. The procedure is nearly totally automated in that both manual point transfer and comparator measurement are eliminated. The system requires only a scanning microdensitometer and a computer. The procedure is not without its problems, which consist principally of difficulties in numerical image correlation in heavily vegetated areas or areas having a monotone type of texture and in urban areas having significant relief displacements of buildings. Research and development are active in this area, however, and it can be expected that these problems will be overcome in the future.

REFERENCES

Ackerman, F.: Results of Recent Tests in Aerial Triangulation, *Photogrammetric Engineering and Remote Sensing*, vol. 41, no. 1, p. 91, 1975.

Allam, M. M.: Mathematical Statistical Methods for the Analysis of Aerial Triangulation Adjustments, *Canadian Surveyor*, vol. 29, no. 2, p. 175, 1975.

American Society of Photogrammetry: "Manual of Photogrammetry," 4th ed., Falls Church, Va., 1980, chaps. 2 and 9.

Anderson, J. M., and E. H. Ramey: Analytic Block Adjustment, *Photogrammetric Engineering*, vol. 39, no. 10, p. 1087, 1973.

Berstein, R., and D. Ferneyhough, Jr.: Digital Image Processing, *Photogrammetric Engineering and Remote Sensing*, vol. 41, no. 12, p. 1465, 1975.

Brucklacher, W.: Instruments for Marking Natural Points and Producing Artificial Points in the Preparation of Aerial Photography for Aerotriangulation, *Photogrammetric Engineering*, vol. 29, no. 5, p. 800, 1963.

Colcord, J. E.: Aerial Triangulation Strip Adjustment with Independent Geodetic Control, *Photogrammetric Engineering*, vol. 27, no. 1, p. 117, 1961.

Derenyi, E. E.: Triangulation with Super-Wide Angle Photographs, *Photogrammetric Engineering*, vol. 38, no. 1, p. 71, 1972.

——— and A. Maarek: Photogrammetric Control Extension for Route Design, *ASCE Journal of the Surveying and Mapping Division*, vol. 100, no. SU1, p. 49, 1974.

Ebner, H., and R. Mayer: Numerical Accuracy of Block Adjustments, *Photogrammetria*, vol. 32, no. 3, p. 101, 1976.

Eden, J. A.: Point Transfer from One Photograph to Another, *Photogrammetric Record*, vol. VII, no. 41, p. 531, 1973.

Erio, G.: Three-Dimensional Transformations of Independent Models, *Photogrammetric Engineering and Remote Sensing*, vol. 41, no. 9, p. 1117, 1975.

Forster, B. C.: Aerotriangulation Accuracy, *Photogrammetric Engineering and Remote Sensing*, vol. 41, no. 4, p. 533, 1975.

Gauthier, J., et al: The Planimetric Adjustment of Very Large Blocks of Models—Its Application to Topographic Mapping in Canada, *Canadian Surveyor*, vol. 27, no. 2, p. 99, 1973.

Ghosh, S. K.: Strip Triangulation with Independent Geodetic Control, *Photogrammetric Engineering*, vol. 28, no. 5, p. 801, 1962.

Granshaw, S. I.: Bundle Adjustment Methods in Engineering Photogrammetry, *Photogrammetric Record*, vol. X, no. 56, p. 149, 1980.

Harley, I. A.: The Determination of XYZ Coordinates Using Numerical Photogrammetry, *Australian Surveyor*, vol. 25, no. 2, p. 89, 1973.

Harris, W. D., G. C. Tewinkel, and C. A. Whitten: Analytic Aerotriangulation, *Photogrammetric Engineering*, vol. 28, no. 1, p. 44, 1962.

Holden, G. J.: AIM-Independent Model Aerial Triangulation Desk Calculator Package, *Australian Surveyor*, vol. 26, no. 4, p. 283, 1974.

Hull, W. V.: Control Densification by Analytic Photogrammetry, *ASCE Journal of the Surveying and Mapping Division*, vol. 101, No. SU1, p. 11, 1975.

Karara, H. M.: Maximum Bridging Distance in Spatial Aerotriangulation, *Photogrammetric Engineering*, vol. 27, no. 4, p. 542, 1961.

Keating, T. J., Wolf, P. R., and F. L. Scarpace: An Improved Method of Digital Image Correlation, *Photogrammetric Engineering and Remote Sensing*, vol. 41, no. 8, p. 993, 1975.

Keller, M.: Block Adjustment Operation at C and GS, *Photogrammetric Engineering*, vol. 33, no. 11, p. 1266, 1967.

——— and G. C. Tewinkel: "Aerotriangulation Strip Adjustment," Technical Bulletin no. 23, U.S. Coast and Geodetic Survey, Washington, D.C., 1964.

——— and ———: "Three-Photo Aerotriangulation," Technical Bulletin no. 29, U.S. Coast and Geodetic Survey, Washington, D.C., 1966.

—— and ——:"Space Resection in Photogrammetry," Technical Bulletin no. 32, U.S. Coast and Geodetic Survey, Washington, D.C., 1966.

Kenefick, J. F., et al.: Bridging with Independent Horizontal Control, *Photogrammetric Engineering and Remote Sensing,* vol. 44, no. 6, p. 687, 1978.

Kratky, V.: Use of Aerotriangulation in Large Scale Mapping, *Canadian Surveyor,* vol. 25, no. 5, p. 542, 1971.

Leatherdale, J. D., and K. M. Keir: Digital Methods of Map Production, *Photogrammetric Record,* vol. IX, no. 54, p. 757, 1979.

Leupin, M. M.: Analytical Photogrammetry; an Alternative to Terrestrial Point Determination, *Australian Surveyor,* vol. 28, no. 2, p. 73, 1976.

Maarek, A.: Practical Numerical Photogrammetry, *Photogrammetric Engineering and Remote Sensing,* vol. 43, no. 10, p. 1295, 1977.

Marks, G. W., et al.: Block Triangulation by Bundles and Stereo-Units, *ASCE Journal of the Surveying and Mapping Division,* vol. 106, no. SU1, p. 1, 1980.

Merchant, D. C.: Surveying by the Aerial Photogrammetric Post-Block Adjustment Method, *Surveying and Mapping,* vol. 36, no. 1, p. 43, 1976.

Morgan, P.: Rigorous Adjustment of Strips, *Photogrammetric Engineering,* vol. 37, no. 12, p. 1271, 1971.

Parsic, Z.: Results of Aerotriangulation with Independent Models Using the Wild A-10 Autograph, *Photogrammetria,* vol. 33, no. 6, p. 209, 1977.

Saxena, K. C.: Independent Model Triangulation Using Different Transformations, *Photogrammetria,* vol. 30, no. 2, p. 67, 1975.

——: Independent Model Triangulation—An Improved Method, *Photogrammetric Engineering and Remote Sensing,* vol. 42, no. 9, p. 1187, 1976.

Schut, G. H.: Development of Programs for Strip and Block Adjustment at the National Research Council of Canada, *Photogrammetric Engineering,* vol. 30, no. 2, p. 283, 1964.

——: Selection of Additional Parameters for the Bundle Adjustment, *Photogrammetric Engineering and Remote Sensing,* vol. 45, no. 9, p. 1243, 1979.

——: Block Adjustment by Bundles, *Canadian Surveyor,* no. 34, no. 2, p. 139, 1980.

——: Block Adjustment by Polynomial Transformations, *Photogrammetric Engineering,* vol. 33, no. 9, p. 1042, 1967.

——: Formation of Strips from Independent Models, *Photogrammetric Engineering,* vol. 34, no. 7, p. 690, 1968.

Shmutter, B.: Triangulation with Independent Models, *Photogrammetric Engineering,* vol. 35, no. 6, p. 548, 1969.

Smith, G. L.: Analytical Photogrammetry Applied to Survey Point Coordination, *Australian Surveyor,* vol. 28, no. 5, p. 263, 1977.

Soliman, A. H.: Standard Error in Strip Adjustment, *Photogrammetric Engineering,* vol. 35, no. 1, p. 87, 1969.

Tewinkel, G. C.: Block Analytic Aerotriangulation, *Photogrammetric Engineering,* vol. 32, no. 6, p. 1056, 1966.

——: Aerotriangulation for Control Surveys, *Surveying and Mapping,* vol. 32, no. 1, p. 39, 1972.

Thompson, E. H.: Aerial Triangulation by Independent Models, *Photogrammetria,* vol. 19, no. 7, p. 262, 1964.

Thompson, L. G.: Determination of the Point Transfer Error, *Photogrammetric Engineering and Remote Sensing,* vol. 45, no. 4, p. 535, 1979.

Trinder, J. C.: Some Remarks on Numerical Absolute Orientation, *Australian Surveyor,* vol. 23, no. 6, p. 368, 1971.

Veres, S.: Aerial Triangulation Using Independent Photo Pairs, *American Society of Civil Engineers, Journal of the Surveying and Mapping Division,* vol. 91, no. SU2, p. 27, 1965.

Weissman, S.: Semi-Analytical Aerotriangulation, *Photogrammetric Engineering,* vol. 35, no. 8, p. 789, 1969.

Williams, H.: Analogue Aerial Triangulation, *ASCE Journal of the Surveying and Mapping Division,* vol. 90, no. SU2, p. 49, 1964.

Wolf, P. R.: Independent Model Triangulation, *Photogrammetric Engineering,* vol. 36, no. 12, p. 1262, 1970.

Wong, K.: Computer Programs for Strip Aerotriangulation, *ASCE Journal of the Surveying and Mapping Division,* vol. 95, no. SU1, p. 71, 1969.

Zarzycki, J. M.: An Integrated Digital Mapping System, *Canadian Surveyor,* vol. 32, no. 4, p. 443, 1978.

PROBLEMS

14-1 Explain the basic differences between analogue and analytical aerotriangulation.

14-2 In stereotriangulation with multiprojector instruments, explain how adjacent models are scaled after relative orientation by the one-projector method.

14-3 Briefly describe the ''graphical error curves'' method of adjusting strips of stereotriangulation to ground control.

14-4 Compare the multiprojector stereotriangulation method with stereotriangulation using universal instruments.

14-5 Discuss the advantages of independent model stereotriangulation over other methods of analogue stereotriangulation.

14-6 Describe the different methods available for determining model coordinates of perspective centers in semianalytical aerotriangulation.

14-7 Compute model coordinates of one of the perspective centers of a stereomodel using the two-point equation [Eq. (14-1)]. Model coordinates of two points A and B measured at two different elevations in the model are

Point	X and Y at $Z = 50.00$ mm		X and Y at $Z = 280.00$ mm	
	X, mm	Y, mm	X, mm	Y, mm
A	591.81	785.71	594.30	656.30
B	902.47	505.10	762.31	504.55

14-8 Repeat Prob. 14-7, except that the model coordinates of points A and B are

Point	X and Y at $Z = 100.00$ mm		X and Y at $Z = 330.00$ mm	
	X, mm	Y, mm	X, mm	Y, mm
A	420.53	217.89	409.83	347.48
B	652.77	787.13	535.43	655.30

14-9 Discuss the advantages of analytical methods of photogrammetric control extension as opposed to analogue methods.

14-10 Describe two different conditions that are commonly enforced in analytical photogrammetry.

14-11 Describe the different requirements in marking and transferring pass points for analytical photogrammetry when a monocomparator is used—as opposed to a stereocomparator—for measuring photocoordinates.

14-12 List the systematic errors in measured photocoordinates that are normally corrected for in analytical photogrammetry.

14-13 If 12 pass points are used in the analytical relative orientation of a stereopair, how many independent collinearity equations can be written?

14-14 Perform analytical relative orientation using collinearity for the following stereomodel, given the refined photocoordinates for the left and right photos. In the solution set ω_1, ϕ_1, κ_1, X_{L_1} and Y_{L_1} all equal to zero, and set Z_{L_1} = 5,750 ft and X_{L_2} = 2, 425 ft. The camera focal length is 151.992 mm. Compute ω_2, ϕ_2, κ_2, Y_{L_2} and Z_{L_2} and the X, Y, and Z model coordinates of points 1 through 6.

| | Refined photocoordinates | | | |
| | Left photo | | Right photo | |
Point	x, mm	y, mm	x, mm	y, mm
1	− 4.617	0.392	−84.078	− 1.637
2	87.296	− 0.309	7.322	− 4.153
3	1.470	98.289	−75.107	94.664
4	71.917	73.563	− 6.402	69.085
5	− 2.274	− 91.876	−84.680	− 94.334
6	83.690	− 100.003	1.120	− 104.202

14-15 Repeat Prob. 14-14, except that Z_{L_1} = 151.992 mm, X_{L_2} = 84.000 mm, the camera focal length is 151.992 mm, and the photocoordinates of pass points 1 through 6 are

| | Refined photocoordinates | | | |
| | Left photo | | Right photo | |
Point	x, mm	y, mm	x, mm	y, mm
1	4.233	7.065	−80.347	9.789
2	85.121	11.168	1.157	12.788
3	− 0.829	91.816	−86.327	95.366
4	85.163	96.363	2.201	99.231
5	14.835	−86.132	−70.288	−83.471
6	96.905	−98.899	12.888	−97.735

14-16 Briefly describe the procedure of numerical image correlation and how it can be used in analytical aerotriangulation.

14-17 Discuss some of the advantages of analytical aerotriangulation from digitized image densities.

FIFTEEN

GROUND CONTROL
FOR AERIAL PHOTOGRAMMETRY

15-1 INTRODUCTION

Photogrammetric control consists of any points whose positions are known in an object-space reference coordinate system and whose images can be positively identified in the photographs. In aerial photogrammetry the object space is the ground surface, and various reference ground coordinate systems are used to describe control point positions. Photogrammetric control, or "ground control" as it is commonly called in aerial photogrammetry, provides the means for orienting or relating aerial photographs to the ground. Almost every phase of photogrammetric work requires some ground control.

Photogrammetric control is generally classified as either *horizontal control* (the position of the point in object space is known with respect to a horizontal datum), or *vertical control* (the elevation of the point is known with respect to a vertical datum). Separate classifications of horizontal and vertical control have resulted primarily because of differences in horizontal and vertical reference datums, and because of differences in surveying techniques for establishing horizontal and vertical control. Also, horizontal and vertical control are often considered separately in photogrammetric processes. Sometimes both horizontal and vertical object-space positions of points are known, so that these points serve a dual control purpose.

Field surveying for photogrammetric control is generally a two-step process. The first step consists of establishing a network of *basic control* in the project area. This basic control consists of horizontal control monuments and bench marks of vertical control which will serve as a reference framework for subsequent photo control surveys. The second step involves establishing object-space positions of *photo control* by means of surveys originating from the basic control network. Photo control points are the actual image points appearing in the photos that are used to control photogrammetric operations. The accuracy of basic control surveys is generally higher than that of subsequent photo control surveys.

The two-step procedure of field surveying for photogrammetric control is illustrated in Fig. 15-1. In the figure a basic control survey originates from existing control stations E_1 and E_2 and establishes a network of basic control points B_1 through B_6 in the project area. With these basic stations established, the second step of conducting subordinate surveys to locate photo control can occur. This is illustrated with the surveys that run between B_5 and B_6 and locate photo control points P_1 and P_2.

The establishment of good ground control is an extremely important aspect of any overall photogrammetric mapping operation. The accuracy of a finished map can be no better than the ground control upon which it is based. Many maps that have been carefully prepared in the office to exacting standards have failed to pass field inspection simply because the ground control was of poor quality. Because of its importance, the ground control phase of any photogrammetric project should be carefully planned and executed. Depending upon conditions, the cost of establishing ground control for photogrammetric mapping can be expected to constitute between 20 and 50 percent of the total mapping cost.

15-2 SELECTING PHOTO CONTROL IMAGES

In general, images of acceptable photo control points must satisfy two requirements: (1) they must be sharp, well defined, and positively identified on all photos, and (2) they must lie in favorable locations in the photographs. (Reasons for this latter requirement are described in Sec. 15-3.) Control surveys for photogrammetry are normally conducted after the photography has been obtained. This assures that the above

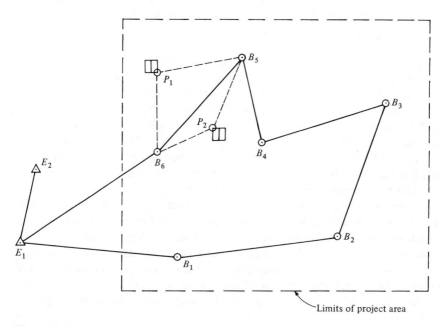

Figure 15-1 Field surveys for establishing photogrammetric control.

two requirements can be met. Photo control images are selected after careful study of the photos. The study should include the use of a stereoscope to ensure a clear stereoscopic view of all points selected. This is important because many of the subsequent photogrammetric measurements will be made stereoscopically. A preliminary selection of photo control images may be made in the office, but the final selection should be made in the field with the photos in hand. This enables positive identification of objects to be made, and it also permits making a first-hand assessment of object point accessibility, terrain conditions, and surveying convenience.

Images for horizontal control have slightly different requirements than images for vertical control. Because their horizontal positions on the photographs must be precisely measured, images of horizontal control points must be very sharp and well defined horizontally. Some objects whose images are commonly satisfactory for horizontal control are intersections of sidewalks, intersections of roads, manhole covers, small lone bushes, isolated rocks, corners of buildings, fence corners, power poles, points on bridges, intersections of small trails or watercourses, etc. Care must be exercised to ensure that control points do not fall in shadowed areas on some photos.

Images for vertical control need not be as sharp and well defined horizontally. Points selected should, however, be well defined vertically. Best vertical control points are small flat or slightly crowned areas. The small areas should have some natural features nearby, such as trees or rocks, which help to strengthen stereoscopic depth perception. Large open areas such as the tops of grassy hills or open fields should be avoided, if possible, because of the difficulties they cause in stereoscopic depth perception. Intersections of roads and sidewalks, small patches of grass, small bare spots, etc., make excellent vertical control points.

The importance of exercising extreme caution in locating and marking objects in the field that correspond to selected photo images cannot be overemphasized. Mistakes in point identification are common and costly. A power pole for example, may be located in the field, but it may not be the same pole whose image was identified on the photos. Mistakes such as this can be avoided by identifying enough other details in the immediate vicinity of each point so that verification is certain. A pocket stereoscope taken into the field can be invaluable in point identification, not only because it magnifies images but also because hills and valleys which aid in object verification can be seen both on the photos and on the ground.

15-3 NUMBER AND LOCATION OF PHOTO CONTROL

The required number of control points and their optimum location in the photos depend upon the use that will be made of them. For a very simple problem such as calculating the flying height of a photo which is assumed to be vertical (see Sec. 6-9), only the horizontal length of a line and the elevations of its end points are needed. A line of as great a length as possible should be chosen. For controlling mosaics (see Chap. 10), only a sparse network of horizontal control may be needed. The network should be uniformly distributed throughout the block of photos.

In solving the *space resection* problem for determining the position and orientation of a tilted photo (see Chap. 11), a minimum of three vertical control points and two horizontal control points are required in each photo. Images of the vertical control points should ideally form a large, nearly equilateral triangle, and the horizontal control points should be widely spaced. Although these are required minimums for space resection, redundant control is recommended to increase the accuracy of the photogrammetric solution and to prevent mistakes from going undetected.

If photo control is being established for the purpose of orienting stereomodels in a plotting instrument for topographic map compilation, the absolute minimum amount of control needed in each stereomodel is the same as the minimum needed for space resection. Again, the prudent photogrammetrist will utilize some amount of redundant control. As a practical minimum, each stereomodel oriented in a plotter should have three horizontal and four vertical control points. The horizontal points should be fairly widely spaced and the vertical control points should be near the corners of the model. A satisfactory configuration is shown in Fig. 15-2. Some organizations require a fifth vertical control point in the center of each stereomodel.

If aerotriangulation (see Chap. 14) is planned to supplement photo control, then lesser amounts of ground-surveyed photo control points are needed. The amount of ground-surveyed photo control needed for aerotriangulation will vary, depending upon the size, shape, and nature of the area to be covered, the resulting accuracy required, and the procedures, instruments, and personnel to be used. In general, the more dense the ground-surveyed network of photo control, the better the resulting accuracy in the supplemental control determined by aerotriangulation. There is an optimum amount of ground-surveyed photo control, however, which affords maximum economic benefit from aerotriangulation and at the same time maintains a satisfactory standard of accuracy. On the average, if a strip of photos is to be bridged for the purpose of obtaining control for orienting stereomodels in a stereoplotter, a minimum of about two horizontal and three or four vertical ground-surveyed photo control points should appear

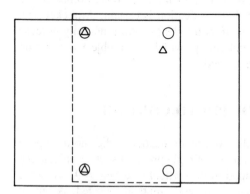

△ Horizontal control point
○ Vertical control point
◎ Horizontal and vertical control point

Figure 15-2 Control recommended for orienting stereomodels in a stereoscopic plotting instrument. (△) Horizontal control point. (○) Vertical control point. (◎) Horizontal and vertical control point.

in approximately every fifth stereomodel along the strip. This configuration is shown in Fig. 15-3. For bridging blocks of photos, the ground-surveyed control should be systematically arranged throughout the block. Best control configurations consist of horizontal control along the periphery of the block with a uniform distribution of vertical control throughout the block. Figure 14-18 illustrates one example. Experience generally dictates the best control configurations to use, and organizations involved in bridging normally develop their own standards which meet accuracy requirements for their particular combination of procedures, instruments, and personnel.

15-4 PLANNING THE CONTROL SURVEY

A great deal of planning should precede the photogrammetric control survey. A decision must be made early on required accuracy of the survey. The type of equipment needed and the field techniques to be used must also be settled. These aspects and many more are interrelated and must be considered simultaneously in reaching the best overall control survey plan. Previous experience in planning photogrammetric control surveys is a most valuable asset. A thorough understanding of surveying field techniques and a knowledge of instrument capabilities are also essential.

The accuracy required in any photogrammetric control survey will govern, to a large extent, the type of equipment and surveying techniques that can be used. Required accuracy of photo control depends primarily upon the accuracy required in the photogrammetric map or computation that it controls. This is not the only consideration, however, since accuracy may also depend upon whether the control will serve other purposes in addition to controlling the photogrammetric work. As an example, suppose a map is to be prepared by photogrammetric techniques for the purpose of planning and designing an urban transportation system. In that case the basic ground control should be accurate enough to also be used in describing property for right of

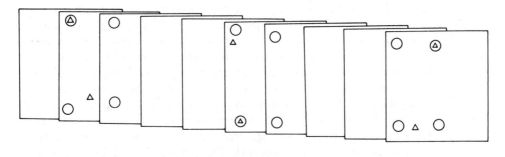

△ Horizontal control point

○ Vertical control point

⊚ Horizontal and vertical control point

Figure 15-3 Example control configuration for bridging a strip. (△) Horizontal control point. (○) Vertical control point. (⊚) Horizontal and vertical control point.

way acquisition and to serve in precisely laying out the construction alignment, structures, etc. In this example the basic ground control should at least meet second-order standards and it should be referenced to the state plane coordinate system (see Sec. 15-5). Basic horizontal control points and bench marks should be permanently monumented and well described. If, on the other hand, a map is being prepared photogrammetrically for purposes of forest inventory, not nearly as accurate a control survey is required; in fact, control taken directly from U.S. Geological Survey quadrangle maps is sometimes adequate.

The *national map standards of accuracy* govern accuracy requirements of maps and therefore indirectly also govern control surveying accuracy. If a map produced photogrammetrically is to be labeled as complying with these standards, it is required that 90 percent of the principal planimetric features be plotted to within $\frac{1}{30}$ in of their true positions for map scales of 1:20,000 or larger, and to within $\frac{1}{50}$ in for scales smaller than 1:20,000. On a map plotted at a scale of 1 in/50 ft, this represents an allowable horizontal map error of 1.6 ft on the ground. On a map at a scale of 1 in/2,000 ft, allowable horizontal map error is 40 ft. If national map accuracy standards are to be met, horizontal photo control must be located to better accuracy than the allowable horizontal map error. As a general rule of thumb, photo control should contain error no greater than one-half the horizontal map accuracy tolerance. Some organizations require stricter tolerances than this. Of course basic control must be more accurate than photo control.

National map standards of accuracy also require that 90 percent of all points tested for elevation be correct to within half the contour interval. To meet this standard, a rule of thumb in topographic mapping states that elevations of vertical photo control points should be correct to within plus or minus one-fifth of the contour interval, but as an additional safety factor, some agencies require that their accuracy be within one-tenth of the contour interval. According to this latter rule, a map being plotted with a contour interval of 2 ft requires vertical photo control accurate to within ± 0.2. Again, the basic control must be more accurate than this.

A separate set of accuracy standards have been developed for highway mapping by photogrammetric methods. These are published by the Federal Highway Administration in the *Reference Guide Outline,* a reference cited at the end of this chapter. This publication also contains a great deal of other information relative to establishing ground control for aerial photography. The American Society of Civil Engineers (ASCE) is also in the process of developing a new set of large-scale mapping standards.

In planning the control survey, maximum advantage should be taken of existing control in the area. The National Geodetic Survey has established numerous first- and second-order triangulation monuments, traverse stations, and bench marks in their work of extending the national control network. The U.S. Geological Survey has also established a network of reliable horizontal and vertical control monuments in their topographic mapping operations. In certain localities other agencies of the federal government such as the Tennessee Valley Authority, Corps of Engineers, Bureau of Land Management, etc., have established control. Also, various state, county, and municipal agencies may have established control monuments. Caution should always be exercised in using existing control if it is of unknown accuracy.

15-5 GROUND COORDINATE SYSTEMS FOR HORIZONTAL CONTROL

Monuments established throughout the United States by the National Geodetic Survey form the basis of the *National Horizontal Control Network*. The national survey organizations of Canada and Mexico have also established networks of control in their countries. These networks are all tied together in a combined system referred to as the *North American Datum of 1927,* so called because the last general simultaneous adjustment of the network occurred in 1927. As a result of this adjustment, published values for geodetic latitude and longitude are available to local surveyors for each monument. Since 1927 a multitude of new stations have been added to the network. Many of these were established by agencies other than the National Geodetic Survey. A new general adjustment is currently in progress which will incorporate all these monuments now in existence.

Surveys for establishing horizontal photo control normally originate from existing monuments of the National Network. Based upon measurements from these stations, positions of new points are determined. If the survey covers a large area, the positions may be computed in terms of geodetic latitude and longitude. This has the advantage that relative positions of widely spaced points are not in error due to earth curvature, a situation which cannot be totally avoided in using plane rectangular coordinate systems. The mathematics of computing geodetic coordinates is somewhat complicated, however, and therefore special plane rectangular coordinate systems called *state plane coordinate* systems have been developed.

State plane coordinate systems retain the simplicity of plane rectangular coordinate computation yet permit coverge of large areas without introducing errors greater than one part in 10,000 between actual mean sea level distances and their equivalent distances as represented in the state plane coordinate systems. This has been accomplished by breaking the entire United States into numerous zones and limiting zone sizes. Contiguous zones overlap, and therefore calculations for surveys of large extent are readily carried from one zone to another.

State plane coordinates of any point can be computed from the geodetic latitude and longitude of the point, and conversely, latitude and longitude of any point can be calculated given its state plane coordinates. The National Geodetic Survey has published tables containing the information necessary for computing positions of points in state plane coordinate systems. These tables are published in booklet form; a separate booklet being available for each of the 50 states.†

State plane coordinates have been computed for all horizontal control stations of the National Network. Coordinates of new points in state plane systems may therefore readily be established from surveys originating on these monuments. Because of their advantage of simplicity of computation, together with the capability of wide coverage without significant loss in accuracy from earth curvature, state plane coordinates are

---

†Available from the Superintendent of Documents, U.S. Government Printing Office, Washington, D.C., 20402.

the most commonly used reference ground coordinate systems for photogrammetric work in the United States.

For projects of a limited size, local plane rectangular coordinate systems which are related to an arbitrary or imaginary datum have sometimes been used. These arbitrary coordinate systems should be avoided, however, because they are unrelated to latitude and longitude or to any other absolute coordinate system.

Both geodetic and state plane coordinate systems are described in detail in the textbooks on surveying and geodesy cited at the end of this chapter, and students interested in further study in these areas should consult them.

15-6 VERTICAL DATUMS

Vertical positions or elevations of points in object space are given in terms of their vertical distances above or below some datum surface. The most commonly adopted vertical datum is *mean sea level,* although on some projects arbitrary datums have been used. Local vertical datums, for example, have been adopted by some United States cities. Their future use should be discouraged as much as possible, however, because they are not related to mean sea level and their use frequently causes confusion and leads to mistakes.

In the United States mean sea level has been determined on the basis of many years of observations at tidal gaging stations located along the Pacific and Atlantic Oceans and the Gulf of Mexico. A general adjustment incorporating the observed data of these stations was made in 1929, and hence the vertical datum in current use in the United States is referred to as the *National Geodetic Vertical Datum of 1929.*

A network of bench marks has been established throughout the nation, principally by the National Geodetic Survey. These bench marks were set by precise differential leveling and their elevations are based upon the sea level datum of 1929. Because leveling generally followed roads and railroads, these bench marks are readily accessible to local surveyors.

15-7 FIELD METHODS FOR ESTABLISHING HORIZONTAL CONTROL

Instruments and techniques for field surveying are numerous and varied. In this text only a very brief discussion of some basic methods is presented. The textbooks on surveying listed as references at the end of this chapter provide a much more thorough treatment of these subjects.

Horizontal control surveys, both for basic control and for photo control, may be conducted using any one of the conventional field methods: *traversing, triangulation,* or *trilateration.* Of these methods, traversing is most common. Regardless of the method used, however, the survey must originate from some existing reference control in the proximity of the project area. Existing control normally consists of a minimum of two intervisible points whose horizontal positions (e.g., state plane coordinates) are accurately known. The direction (e.g., azimuth from north) of the line connecting the two points is also normally known.

Traversing, as illustrated in Fig. 15-4*a*, consists of measuring horizontal angles and horizontal distances between consecutive stations of a closed network. The existing reference control stations are included in the network. Angles are measured with a theodolite or transit, while distances may be measured electronically or by taping. Based on the existing reference coordinates and reference direction, along with the newly measured angles and distances, coordinates of all new stations may be calculated trigonometrically in the rectangular coordinate system of the existing reference control. An adjustment is normally made to account for measurement errors, and the least squares method is best for this purpose.

Triangulation involves precise measurement of one or more "base lines" such as B_1 and B_2 of Fig. 15-4*b*. All possible horizontal angles between intervisible stations are also carefully measured. Then coordinates of new stations may be calculated trigonometrically. Trilateration is similar in concept to triangulation except that no angles are measured. Rather, all possible horizontal distances between intervisible stations in the network are measured, as illustrated in Fig. 15-4*c*. In trilateration, distances are generally measured electronically. Coordinates of all new stations may be calculated based on the measured distances. Often, a horizontal control survey will combine the techniques of triangulation and trilateration.

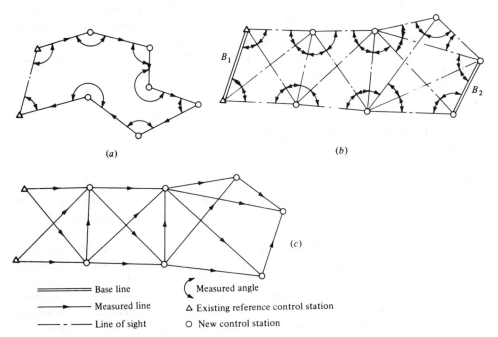

Figure 15-4 (*a*) Example of a traverse network. (*b*) Example of a triangulation network. (*c*) Example of a trilateration network. (═) Base line. (➤) Measured line. (---) Line of sight. (◖) Measured angle. (△) Existing reference control station. (○) New control station.

For any horizontal control survey, an important task which must precede taking field measurements is establishment of the network of stations in the project area whose positions are to be determined. In basic control surveys, the stations will normally be artificial monuments such as wooden stakes or iron rods driven into the ground. These are carefully referenced to permanent nearby features so that they can be recovered at a later date if lost. In photo control surveys, some of the stations will be artificial monuments and some of them will be the natural features selected for photo control points.

The particular combination of instruments and techniques that should be used for the horizontal control survey will depend upon the conditions of each project. Existing topography or the presence of certain constructed features such as roads and railroads may render one procedure more economical than others. If the project area is rugged or perhaps covers a river valley with bluffs on either side, triangulation or trilateration may prove to be the best approach. Triangulation and trilateration may also be best if the area is rather flat and densely wooded, although in this case towers may be required to raise the instruments above the trees to achieve intervisibility between stations. If a road or railroad runs along a strip of photographs in an area which provides satisfactory image point locations in the photos, or if the area is flat or gently rolling and is rather open for lines of sight, traversing may be most convenient. Accessibility of points, and surveying cost and convenience are aspects that must be kept in mind when planning the control survey.

Accuracy required of photo control is another major consideration in deciding upon equipment and procedures to use. If a high accuracy standard for horizontal control is necessary, high standards of surveying will, of course, be required. Distance measurements by electronic means or by precise taping will be necessary, and angle measurements by precise theodolites will be required. Lower-order horizontal surveys may be adequately accomplished using rough taping or even stadia in some cases.

15-8 FIELD METHODS FOR ESTABLISHING VERTICAL CONTROL

For vertical control surveys, *differential* leveling is the most common field procedure, and where highest accuracy is required it is the method used. The basic equipment for differential leveling is a leveling instrument and a graduated rod. A level vial nearly filled with fluid is mounted on a leveling instrument. When the air bubble of the vial is carefully centered, the telescopic line of sight is horizontal. The procedure of differential leveling is illustrated in Fig. 15-5. The vertical control survey originates from a *bench mark* (monument of known elevation) such as BM_x. The elevation of BM_x above datum is h_{BM_x}. To establish the elevation of new point A, the leveling instrument is set up between BM_x and point A, the bubble is centered, and readings R_1 and R_2 are taken on the graduated rod held vertically at BM_x and A, respectively. The elevation of point A is then equal to the elevation of BM_x, plus rod reading R_1, minus rod reading R_2.

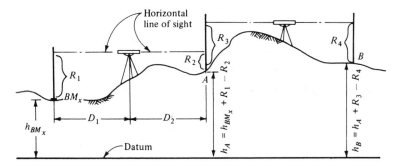

Figure 15-5 Spirit or "differential" leveling.

To compensate for earth curvature and errors of atmospheric refraction of the line of sight, lengths D_1 and D_2 of sights to BM_x and A should be approximately equal. Then the magnitude of error from these sources will be equal in readings R_1 and R_2, and since the error is added in R_1 and subtracted in R_2, it is eliminated. With the elevation of point A known, the elevation of point B may be established by setting the instrument between A and B and taking rod readings R_3 and R_4. The elevation of B is then equal to the elevation of A plus rod reading R_3, minus rod reading R_4. This process is continued until elevations have been determined for all desired points. The leveling circuit should terminate on either the initial bench mark or on some other bench mark so that adjustments can be made for errors that accumulated in the leveling process.

Another technique for determining differences in elevation is *trigonometric leveling*. It may be used where moderate accuracy is required and is especially well suited for rugged terrain. In trigonometric leveling, as illustrated in Fig. 15-6, vertical angle α and horizontal distance D or inclined distance S between two points A and B are measured. Then difference in elevation is either $\Delta h = D \tan \alpha$ or $\Delta h = S \sin \alpha$. The distance may be measured electronically or by taping, but electronic methods are generally most convenient. The vertical angle is measured with a theodolite. On long sights, to compensate for earth curvature and atmospheric refraction, the vertical angle should be measured in both directions from each station and averaged.

For small-scale mapping which requires only rough elevations, *barometric* leveling, *airborne profile recording* (APR), or *elevation metering* may provide satisfactory results. Barometric leveling is performed on the ground using two or more precise barometric altimeters. The procedure can yield elevations having average errors less than 5 ft, and best accuracy is achieved during stable atmospheric conditions. An airborne profile recorder incorporates a barometric altimeter and radar, and determines a profile of the ground beneath the path of the aircraft in which it is carried. It is capable of achieving absolute elevations correct to within 10 ft, and it is particularly well adapted for use in inaccessible terrain. The elevation meter is a device attached to a wheeled vehicle which continually measures the inclination angle and length of travel as the vehicle moves. These parameters allow for an automatic recording to be

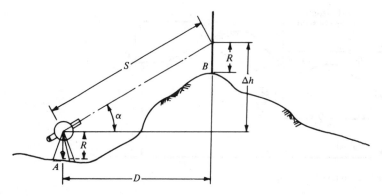

Figure 15-6 Trigonometric leveling.

made of differences in elevation from point to point. Accuracies to within plus or minus 2 ft have been obtained with this instrument.

15-9 OTHER SYSTEMS FOR PHOTO CONTROL SURVEYS

A unique system called the *Airborne Control (ABC) System* has been used by the U.S. Geological Survey for low-order horizontal and vertical control surveys. In this system, distances and horizontal and vertical angles are measured to a helicopter hovering over a station of unknown position and elevation. The measurements are taken from two or more established control points while the pilot is aligned vertically above the unknown point by means of a special alignment device called a *hoversight*. The helicopter's distance above the point is read from graduations on a plumb line lowered to the point. Based on these measurements, the position and elevation of the new point can be calculated. This system is especially useful for small-scale mapping over rugged or inaccessible terrain.

Inertial surveying systems are now being used to great advantage in obtaining photo control. These systems use precise gyroscopes to sense the earth's rotation and orient themselves with respect to north-south and east-west, as well as to the direction of gravity. As the system is moved from one point to another, the gyros maintain their orientation while accelerometers measure components of position change in the cardinal directions and elevation. The instrument, having been initialized at an existing control station, can therefore yield direct readings of positions and elevations of points visited. These systems, which can be carried in land vehicles or aircraft, are capable of yielding positional accuracies of less than 1 m. Figure 15-7 shows the *Spanmark*, an inertial surveying system, mounted in a helicopter. If it is carried in a helicopter, measurements can be taken as the aircraft hovers over new photo control points. A graduated plumb line lowered to the point provides a vertical translation that must be subtracted from the instrument elevation to obtain the elevation of the point.

Doppler systems have also been used in photo control surveying, principally to establish basic control points in large project areas. Using special receivers called

Figure 15-7 Inertial surveying system mounted in a helicopter. (*Courtesy Span International, Inc.*)

Geoceivers, these systems measure changes in frequency of a broadcast signal emitted from a passing satellite. From known satellite orbital data together with precise measurements of time, the measured frequency changes allow computation of the occupied position of the geoceiver. These systems have produced positions accurate to within 1 m or less.

15-10 ARTIFICIAL TARGETS

In some areas such as prairies, dense forests, deserts, etc., natural points suitable for photogrammetric control may not exist. In these cases artifical points called *panel points* may be placed on the ground prior to taking the aerial photography. Their positions are then determined by field survey or in some cases by aerotriangulation. This procedure is called *premarking* or *paneling*.

Artificial targets provide the best possible photographic images, and therefore they are used for controlling the most precise photogrammetric work, whether or not natural points exist. Artificial targets are also used to mark section corners and boundary lines for photogrammetric cadastral work.

Besides their advantage of excellent image quality, their unique appearance makes misidentification of artificial targets unlikely. Disadvantages of artificial targets are that (1) extra work and expense are incurred in placing the targets, (2) the targets could be moved between the time of their placement and the time of photography, and (3) the targets may not appear in favorable locations on the photographs. To guard against the second disadvantage, the photography should be obtained as near as possible to the time of placing targets. To obtain target images in favorable positions on the photographs, the coverage of each photo can be planned in relation to target locations, and the positions of ground principal points can be specified on the flight plan.

A number of different types of artificial targets have been successfully used for photogrammetric control. The main elements in target design are good color contrast, a symmetrical target that can be centered over the control point, and a target size that yields a satisfactory image on the resulting photographs. Contrast is best obtained using light-colored targets against a dark background or dark-colored targets against light backgrounds. The target shown in Fig. 15-8 provides good symmetry for centering over the control point. The center panel of the target should be centered over the control point, since this is the image point to which measurements will be taken.

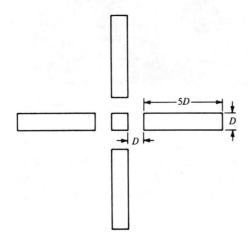

Figure 15-8 Artificial photogrammetric target.

The legs help in identifying targets on the photos, and they also help in determining the exact center of the target should the image of the center panel be unclear.

Target sizes must be designed on the basis of intended photo scale so that the target images are the desired size on the photos. An image size of about 0.03 mm to about 0.10 mm for the sides of the central panel is generally ideal. As shown in Fig. 15-8, if the ground dimension of the central panel of the target is D, then the leg width should also be D, leg length should be $5D$, and the open space between the central panel and the leg should be D. Target sizes are readily calculated once photo scale and optimum target image size are selected. If, for example, a central panel size of 0.05 mm is desired and photography at a scale of 1:12,000 is planned, then D should be 2.0 ft.

Materials used for targeting are quite variable. In some cases satisfactory targets are obtained by simply painting white crosses on blacktop roads. In other cases targets are painted on plywood, masonite, or heavy cloth, in which case they may be salvaged and reused. Satisfactory targets have also been made by placing stones against an earth background in the shape of a cross. The stones may be painted white for added contrast. Lime placed in the shape of a cross against a dark background has also produced satisfactory targets. Old tires painted white centered over the control points are also good for low-altitude large-scale photography.

A procedure known as *post marking* can be performed if panel points are needed in an area after the photography has already been obtained. In this method, targets as described above are placed in the desired positions. Supplemental vertical photographs are taken of each target and its surrounding area with a small-format camera carried in a light aircraft flying at low altitude. Flying height can be calculated for the supplemental photography so that it has the same scale as the original photography. Locations of the targets can be transferred stereoscopically from the supplemental photography to the original photography using a point transfer device. Zoom magnification of the individual viewing systems of the point transfer device eliminates the need for having the scales of the supplemental and original photography equal.

15-11 INDEXING GROUND CONTROL

For each project it is advisable to prepare a set of paper prints for indexing the ground control. This set of photos should be examined carefully, and images of all control appearing on them should be identified from their field descriptions. When positive identification is made, the control point images should be lightly pricked with a pin to avoid the possibility of later misidentification mistakes; e.g., if a control point lies at a particular sidewalk intersection corner, pricking that corner will offset the possibility of mistakenly using a different corner at some later time simply because the written description was misinterpreted. The pinprick should just penetrate the emulsion and the pin should be held at right angles to the paper. Pricking should be done with the aid of a magnifying glass.

Control point images should be further marked by surrounding them with appropriate symbols, depending upon the type of control point. Triangles are commonly used to identify horizontal control points, while circles may be used for vertical control. A triangle inside a circle indicates both horizontal and vertical control on that point. Identifying numbers or names of control points should be written on the control photos beside the points. Some organizations have adopted numbering systems which correspond to the type of control point. As an example, vertical points may be given numbers from 1 to 999, horizontal points may be numbered from 1,000 to 1,999, and points of both vertical and horizontal control may be numbered from 2,000 to 2,999. Short written descriptions should be placed on the backs of the photos near the pinpricks. A set of control index photos carefully prepared can be a valuable asset in performing subsequent photogrammetric operations.

REFERENCES

American Society of Photogrammetry: "Manual of Photogrammetry," 3d ed., Falls Church, Va., 1966, chap. 8.
———:"Manual of Photogrammetry," 4th ed., Falls Church, Va., 1980, chap. 8.
Biege, R. R., Jr.: Photogrammetric Control Methods of the State Highway Commission of Kansas, *Surveying and Mapping,* vol. 29, no. 4, p. 679, 1969.
Bomford, G.: "Geodesy," 2d ed. Oxford University Press, New York, 1962.
Brinker, R. C., and P. R. Wolf: "Elementary Surveying," 6th ed., Harper & Row, Publishers, Incorporated, New York, 1977.
Burger, T. D.: Use of the Elevation Meter in Topographic Mapping, *Surveying and Mapping,* vol. 21, no. 4, p. 481, 1961.
Carriere, R. J., et al.: Experience with the Inertial Survey System at the Geodetic Survey of Canada, *Canadian Surveyor,* vol. 32, no. 3, p. 341, 1978.
Danner, C. S.: Horizontal Control Problems in Private Practice, *Surveying and Mapping,* vol. 30, no. 2, p. 265, 1970.
Davis, R. E., et al.: "Surveying Theory and Practice", 6th ed., McGraw-Hill Book Co., Inc., New York, 1981.
Eckhardt, C. V.: Airborne Control for Topographic Mapping, *Surveying and Mapping,* vol. 26, no. 1, p. 49, 1966.
Federal Highway Administration: Reference Guide Outline, *Specifications for Aerial Surveys and Mapping by Photogrammetric Methods for Highways,* U.S. Dept. of Transportation, Washington, D.C., 1968.
Haig, M. D., et al.: A Simplified Explanation of Doppler Positioning, *Surveying and Mapping,* vol. 40, no. 1, p. 29, 1980.
Halliday, J.: The Vital Communications Link—Photoidentification of Horizontal Control, *Photogrammetric Engineering,* vol. 29, no. 5, p. 804, 1963.
Hittel, A., et al.: Doppler Satellite Applications in Manitoba, *Canadian Surveyor,* vol. 31, no. 2, p. 167, 1977.
Hothern, L. D., et al.: Doppler Satellite Surveying System, *ASCE Journal of the Surveying and Mapping Division,* vol. 104, no. SU1, p. 79, 1978.
Krakiwsky, E. J., et al.: Geodetic Control from Doppler Satellite Observations of Lines Under 200 KM, *Canadian Surveyor,* vol. 27, no. 2, p. 141, 1973.
Kratky, V.: Real Time Photogrammetric Support of Dynamic Three-Dimensional Control, *Photogrammetric Engineering and Remote Sensing,* vol. 45, no. 9, p. 1231, 1979.
Lachapelle, G.: Redefinition of National Vertical Geodetic Datum, *Canadian Surveyor,* vol. 33, no. 3, p. 273, 1979.
Lee, D. R.: Vertical Control for Mapping the Okefenokee Swamp, *Surveying and Mapping,* vol. 27, no. 1, p. 73, 1967.
Lennon, G. W.: Mean Sea Level as a Reference for Geodetic Leveling, *Canadian Surveyor,* vol. 28, no. 5, p. 524, 1974.

Lippold, H. R.: Readjustment of the National Geodetic Vertical Datum, *Surveying and Mapping,* vol. 40, no. 2, p. 155, 1980.

Loving, H. G.: Airborne Control System, *Surveying and Mapping,* vol. 23, no. 1, p. 91, 1963.

O'Leary, W. V.: A New Development Program for the Airborne Profile Recorder, *Photogrammetric Engineering,* vol. 29, no. 5, p. 872, 1963.

Schwieder, W. H.: Laser Terrain Profiler, *Photogrammetric Engineering,* vol. 34, no. 7, p. 658, 1968.

Theurer, C.: Control for Photogrammetric Mapping, *Photogrammetric Engineering,* vol. 23, no. 2, p. 318, 1957.

Thompson, M. M., and G. H. Rosenfeld: Map Accuracy Specifications, *Surveying and Mapping,* vol. 31, no. 1, p. 57, 1971.

Todd, M. S.: The Development of the Inertial Rapid Geodetic Survey, *Canadian Surveyor,* vol. 32, no. 4, p. 465, 1978.

VanWijk, M. C.: Test Areas and Targeting in the Hull Project, *Canadian Surveyor,* vol. 25, no. 5, p. 514, 1971.

Watts, R. G.: Simplicity of State Plane Coordinate System in Surveying, *Surveying and Mapping,* vol. 25, no. 4, p. 543, 1965.

Wolf, P. R.: "Adjustment Computations: Practical Least Squares for Surveyors," 2d ed., P.B.L. Publishing Co., Madison, Wis., 1980.

PROBLEMS

15-1 Explain the difference between basic control and photo control.

15-2 Describe the characteristics of good horizontal photo control points.

15-3 Describe the characteristics of good vertical photo control points.

15-4 State the national map standards of accuracy for both horizontal positions and elevations.

15-5 Discuss the advantages of using state plane coordinate systems as a reference for horizontal photo control.

15-6 If a map is being prepared photogrammetrically to a scale of 500 ft/in, and photo control must be established to an accuracy four times greater than the allowable error for plotted points as specified by national map accuracy standards, how accurately on the ground must photo control points be located?

15-7 Repeat Prob. 15-6, except that map scale is 1:12,000 and photo control must be accurate to within ± 0.005 in on the map.

15-8 What are the photo dimensions of the square at the intersection of two sidewalks of 6-ft width if photo scale is 1:9,600?

15-9 What are the photographic dimensions in millimeters of a 36-in-diameter manhole cover if photo scale is 1:6,000?

15-10 Describe three conventional field methods used in horizontal control surveys.

15-11 Discuss briefly the different field techniques used in establishing vertical control.

15-12 Describe the Airborne Control System for photo control surveying.

15-13 Explain briefly how an inertial surveying system operates, and how it can be used in photo control surveys.

15-14 Discuss briefly Doppler positioning systems and their application in photo control surveying.

15-15 Discuss the advantages and disadvantages of using artificial targets as opposed to using natural targets.

15-16 What must be the ground dimension D (see Fig. 15-8) of artificial targets if their corresponding photo dimension is to be 0.05 mm on photos exposed from 10,000 ft above ground with a 152-mm-focal-length camera?

15-17 Repeat Prob. 15-16, except that photo dimension is 0.10 mm, flying height above ground is 6,000 ft, and camera focal length is 210 mm.

SIXTEEN

PROJECT PLANNING

16-1 CONSIDERATIONS IN PROJECT PLANNING

Successful execution of any photogrammetric project requires thorough planning prior to proceeding with the work. Planning, more than any other area of photogrammetric practice, must be performed by knowledgeable and experienced persons who are familiar with all aspects of the subject.

One of the important considerations that must be addressed early in planning between the client and photogrammetrist involves the decision of exactly what products will be prepared, together with their scales and accuracies. This can be done only if the planner thoroughly understands what the client's needs are so that the best overall products can be developed to meet those needs. The client will also naturally be concerned with the anticipated costs of the items, as well as with the proposed schedule for their delivery. Therefore, successful planning will probably require several meetings with the client prior to commencing the work, and depending upon the nature and magnitude of the project, continued meetings may be needed as production progresses.

A variety of products may be developed in a given photogrammetric project, including prints of aerial photos, photomaps, mosaics, planimetric and topographic maps, cross sections, digital terrain models, orthophotos, cadastral maps, and others. In addition to the wide variation in products that could be developed for a given project, there are normally other major considerations that will have definite bearing on procedures, costs, and scheduling. These include the location of the project area, its size, shape, topography, and vegetation cover, the availability of existing ground control, etc. Thus, every project presents unique problems to be considered in the planning stages.

Assuming that the products to be developed have been agreed upon with the client, the balance of the work of project planning can generally be summarized into the following categories:

1. Planning the aerial photography†
2. Planning the ground control
3. Selecting instruments and procedures necessary to achieve the desired results
4. Estimating costs and delivery schedules

When planning has been completed for these categories, the photogrammetrist will normally prepare a detailed proposal which outlines plans, specifications, estimate of costs, and delivery schedules for the project. The proposal often forms the basis of an agreement or contract for the performance of the work.

Of the above four categories, item 2 has been discussed in detail in Chap. 15, and item 3 has also been discussed in earlier chapters where the various photogrammetric products and instruments for producing them have been described. This chapter presents the two remaining categories in separate parts: Part I discusses flight planning, and Part II covers the subject of cost estimating and scheduling.

PART I   FLIGHT PLANNING

16-2 INTRODUCTION

Because the ultimate success of any photogrammetric project probably depends more upon good-quality photography than on any other aspect, flight planning is of major concern. If the photography is to satisfactorily serve its intended purposes, the photographic mission must be carefully planned and faithfully executed according to the *flight plan*. A flight plan generally consists of two items: (1) a *flight map* which shows where the photos are to be taken, and (2) *specifications* which outline how to take them, including specific requirements such as camera and film requirements, scale, flying height, end lap, side lap, tilt and crab tolerances, etc. A flight plan which gives optimum specifications for a project can be prepared only after careful consideration of all the many variables which influence aerial photography.

An aerial photographic mission is an expensive operation involving two or more crewpersons and expensive aircraft and equipment. In addition, in many areas periods of time that are acceptable for aerial photography are quite limited by weather and ground cover conditions which are related to seasons of the year. Failure to obtain satisfactory photography on a flight mission not only necessitates costly reflights but in all probability it will also cause long and expensive delays on the project for which the photos were ordered. For these reasons flight planning is one of the most important operations in the overall photogrammetric project. The following sections present various considerations in flight planning.

---

†Terrestrial and close-range photogrammetry also comprise a growing and significant amount of photogrammetric activity. Special considerations for these types of projects are covered in Chap. 18.

16-3 PHOTOGRAPHIC END LAP AND SIDE LAP

Before discussing the many aspects which enter into consideration in planning an aerial photographic mission, it will be helpful to redefine the terms "end lap" and "side lap." As discussed in Sec. 1-4, vertical aerial photographic coverage of an area is normally taken as a series of overlapping flight strips. As illustrated in Fig. 16-1, *end lap* is the overlapping of successive photos along a flight strip. Figure 16-2 illustrates *side lap,* the overlap of adjacent flight strips.

In Fig. 16-1, *G* represents the dimension of the square of ground covered by a single photograph (assuming level ground and a square camera focal-plane format), and *B* is the air base or distance between exposure stations of a stereopair. The amount of end lap of a stereopair is commonly given in percent. Expressed in terms of *G* and *B*, it is

$$PE = \left(\frac{G - B}{G}\right) \times 100 \qquad (16\text{-}1)$$

In Eq. (16-1), *PE* is percent end lap. If stereoscopic coverage of an area is required, the absolute minimum end lap is 50 percent. However, to prevent gaps from occurring in the stereoscopic coverage due to crab, tilt, flying height variations, and terrain variations, end laps greater than 50 percent are used. Also, if the photos are to be used for photogrammetric control extension, images of some points must appear on three successive photographs—a condition requiring greater than 50 percent end lap. For these reasons aerial photography for mapping purposes is normally taken with about 60 percent end lap, plus or minus about 5 percent.

Crab, as explained in Sec. 4-7, exists when the edges of the photos in the *x* direction are not parallel with the direction of flight. It causes a reduction in stereoscopic coverage, as was indicated in Fig. 4-11*b*. Figures 16-3 through 16-5 illustrate reductions in end lap causing loss of stereoscopic coverage due to tilt, flying height variations, and relief variations, respectively.

Side lap is required in aerial photography to prevent gaps from occurring between flight strips as a result of drift, crab, tilt, flying height variations, and terrain variations. *Drift* is the term applied to a failure of the pilot to fly along planned flight lines. It is often caused by strong winds. Excessive drifts are the most common cause for gaps in photo coverage; when this occurs, reflights are necessary.

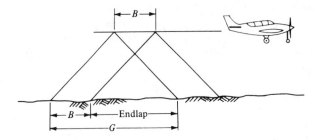

Figure 16-1 End lap, the overlapping of successive photos along a flight strip.

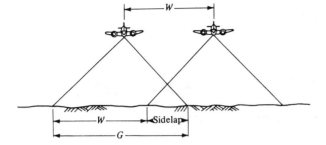

Figure 16-2 Side lap, the overlapping of adjacent flight strips.

In Fig. 16-2, G again represents the dimension of the square of ground coverage of a single photograph and W is the spacing between adjacent flight lines. An expression for PS, percent side lap in terms of G and W, is

$$PS = \left(\frac{G - W}{G}\right) \times 100 \qquad (16\text{-}2)$$

Mapping photography is normally taken with a side lap of about 30 percent. An advantage realized from using this large a percentage is elimination of the need to use the extreme edges of the photography, where the imagery is of poorer quality. Photography for mosaic work is sometimes taken with greater than 30 percent side lap since this reduces the size of the central portion of the photograph that must be used, thereby lessening distortions of images due to tilt and relief. In some cases where aerial photography is to be used for very precise photogrammetric control extension, it may be taken with 60 percent side lap as well as 60 percent end lap.

Example 16-1 The air base of a stereopair of vertical photos is 4,600 ft and flying height above average ground is 8,000 ft. The camera has a 6-in (152.4-mm) focal length and a 9-in (23-cm) format. What is the percent end lap?

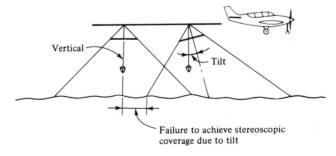

Figure 16-3 Failure to achieve stereoscopic coverage due to tilt.

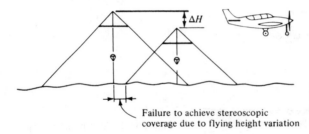

Figure 16-4 Failure to achieve stereoscopic coverage due to flying height variations.

Failure to achieve stereoscopic coverage due to flying height variation

SOLUTION

(a) Average photo scale $= \dfrac{f}{H'_{avg}} = 6 \text{ in}/8{,}000 \text{ ft} = 1 \text{ in}/1{,}333 \text{ ft}$

(b) Ground coverage dimension $G = 9 \text{ in} \times 1{,}333 \text{ ft/in} = 12{,}000 \text{ ft}$

(c) Percent end lap, by Eq. (16-1),

$$PE = \left(\frac{12{,}000 - 4{,}600}{12{,}000} \right) \times 100 = 62\%$$

Example 16-2 In Example 16-1, assume that the spacing between adjacent flight strips is 8,200 ft. What is the percent side lap?

SOLUTION By Eq. (16-2),

$$PS = \left(\frac{12{,}000 - 8{,}200}{12{,}000} \right) \times 100 = 32\%$$

16-4 PURPOSE OF THE PHOTOGRAPHY

In planning aerial photographic missions, the first and foremost consideration is the purpose for which the photography is being taken. Only with the purpose defined can optimum equipment and procedures be selected. In general, aerial photographs are

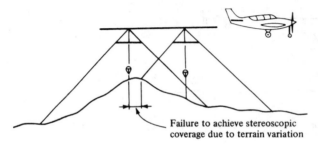

Failure to achieve stereoscopic coverage due to terrain variation

Figure 16-5 Failure to achieve stereoscopic coverage due to terrain variations.

desired which have either good *metrical* qualities or high *pictorial* qualities. Photos having good metrical qualities are needed for topographic mapping or other purposes where precise quantitative photogrammetric measurements are required. High pictorial qualities are required for qualitative analysis such as photographic interpretation or for constructing photomaps and aerial mosaics.

Photographs of good metrical quality are obtained using calibrated cameras and films having fine-grained, high-resolution emulsions. For topographic mapping, photography is preferably taken with a wide- or super-wide-angle (short-focal-length) camera so that a large *base-height (B/H')* ratio is obtained. The B/H' ratio, as described in Sec. 7-8, is the ratio of the air base of a pair of overlapping photographs to average flying height above ground. The larger the B/H' ratio, the greater the intersection angles or parallactic angles between intersecting light rays to common points. In Fig. 16-6a and Fig. 16-6b, for example, the air bases are equal but the focal length and flying height in Fig. 16-6a are half those in Fig. 16-6b. Photographic scales are therefore equal, but the B/H' ratio of Fig. 16-6a is double that of Fig. 16-6b, and parallactic angle ϕ_1 to point A in Fig. 16-6a is nearly double the corresponding angle ϕ_2 in Fig. 16-6b.

It can be shown that errors in computed positions and elevations of points in a stereopair increase with increasing flying heights and decrease with increasing x parallax. Large B/H' ratios denote low flying heights and large x parallaxes, conditions favorable to higher accuracy. The photos of Fig. 16-6a are therefore superior to those of Fig. 16-6b from a mapping or quantitative point of view.

Photography of high pictorial quality does not require a calibrated camera, but the camera must have a good-quality lens. In many cases films having fast, large-grained emulsions produce desirable effects. For some photo interpretation work, normal color films are useful. For other special applications, black-and-white infrared or color infrared films are desirable. Special effects can also be obtained using filters in combination with various types of films. Timber types, for example, can be delineated quite effectively using a red filter in combination with black-and-white infrared film.

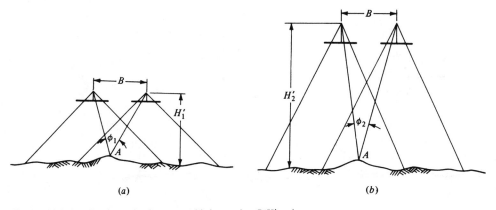

(a)   (b)

Figure 16-6 Parallactic angles increase with increasing B/H' ratios.

For mosaic work, relief displacements, tilt displacements, and scale variations produce objectionable degradations of pictorial quality. These may be minimized, however, by increasing flying height, thereby decreasing the B/H' ratio. Increased flying height, of course, reduces photo scale; but this can be compensated for by using a longer focal length camera. The photo of Fig. 16-7a was exposed at half the flying height of the photo of Fig. 16-7b. The scales of the two photos are equal, however, because the focal length f_2 of Fig. 16-7b is double f_1 of Fig. 16-7a. The photo of Fig. 16-7b is more desirable for mosaic construction because its scale variations and image distortions due to relief, tilt, and flying height variations are much less than those of the photo of Fig. 16-7a. On Fig. 16-7a, for example, relief displacement d_1 is double the corresponding relief displacement d_2 of Fig. 16-7b.

16-5 PHOTO SCALE

Average photographic scale is one of the most important variables that must be selected in planning aerial photography. It is normally fixed within certain limits by specific project requirements. For topographic mapping, photo scale is usually dictated by required map scale, required contour interval, and capabilities of the instruments that will be used in compiling the map. On the other hand, aerial photographic coverage for mosaic preparation or for photo interpretation must be planned at a scale which enables the smallest objects of importance to be resolved on the photos.

In topographic mapping with a stereoscopic plotting instrument, the enlargement ratio capabilities from photo scale to map compilation scale must be considered. With some plotters the range is quite wide; with others it is restricted within narrow limits

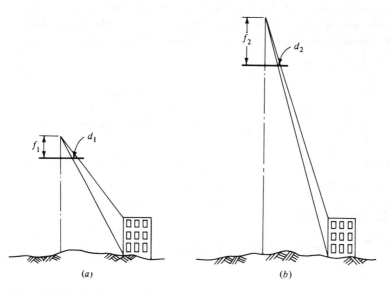

Figure 16-7 Reduction in relief displacement is achieved by increasing flying height.

(see Chap. 12). For some mechanical projection stereoplotters, enlargement ratio capabilities vary from less than 1 up to 10 or more. Although ratios as great as 10 may be possible, in practice they should normally be held to a maximum of approximately 8; otherwise, required map accuracies may not be achieved. For many double-projection direct-viewing plotters, the optimum enlargement ratio is 5, with only a slight variation possible. If one of these instruments is to be used and if map scale is fixed, optimum photo scale is automatically fixed at one-fifth map scale.

Example 16-3 A map must be compiled at a scale of 1:6,000 using a Kelsh plotter whose optimum enlargement ratio from photo scale to map scale is 5. What is optimum photo scale?

SOLUTION Photo scale is one-fifth as large as map scale, therefore

$$S_{\text{photo}} = \frac{1}{6,000} \times \frac{1}{5} = \frac{1}{30,000} \qquad \text{or } 1 \text{ in}/2,500 \text{ ft}$$

Selection of optimum map scale depends upon the purpose of the map. It should be carefully planned, because compilation at a larger scale than necessary is uneconomical, and compilation at too small a scale reduces the usefulness of the map or may even render it unsatisfactory. The accuracy to which planimetric positions of points can be measured from a map depends upon the scale of the map. Assume that map positions of planimetric features can be plotted correctly to within $\frac{1}{30}$ inch, a condition necessary to meet National Map Accuracy Standards. Then if points must be accurate to within ± 1.0 ft on a cadastral map, required map scale is 1 in/30 ft. On the other hand, if points need only be accurate to within ± 10 ft on a topographic map, then a scale of 1 in/300 ft is all that is required.

Required contour interval for a map must be considered as well as planimetric accuracy in planning aerial photography. As vertical mapping accuracy requirements increase (contour interval decreases), flying height must decrease and hence photographic scale increases. Like planimetric accuracy, contour interval also depends upon the intended use of the map. Assume that elevations can be interpolated correctly from a map to within one-half the contour interval, a condition required for meeting National Map Accuracy Standards. If elevations must be interpolated to within ± 0.5 ft on a highway design map, then a 1 ft contour interval is necessary. If elevations must be interpolated to ± 10 ft on a map prepared for studying the volume of water impounded in the reservoir of a large dam, then a 20-ft contour interval is all that is required.

Recommended contour interval depends not only on the use to be made of the map but also on the type of terrain. If the map is being prepared for planning a sewer system for a city such as Las Vegas, Nevada, which lies on very flat terrain, perhaps a 1-ft contour interval is required. On the other hand, if a topographic map of San Francisco is being prepared for the same purpose, because of the large range of relief in that city, perhaps a 5- or 10-ft contour interval would be used.

Contour interval and map scale must be selected so that they are compatible. As map scale decreases, contour interval must increase, otherwise the contours would become too congested on the map. In large-scale mapping of average types of terrain, the scale and contour interval relationships shown in Table 16-1 generally provide satisfactory compatibility.

16-6 FLYING HEIGHT

Once camera focal length and required average photo scale have been selected, required flying height above average ground is automatically fixed in accordance with scale Eq. (6-3).

Example 16-4 Aerial photography having an average scale of 1:6,000 is required to be taken with a 6-in- (152.4-mm-) focal-length camera over terrain whose average elevation is 1,400 ft above mean sea level. What is required flying height above mean sea level?

SOLUTION By Eq. (6-3) (*Note:* 6 in = 0.5 ft),

$$S = \frac{f}{H - h_{avg}} \qquad \frac{1}{6,000} = \frac{0.5}{H - 1,400}$$

$$H = 6,000(0.5) + 1,400 = 4,400 \text{ ft above sea level}$$

Flying heights above average ground may vary from a few hundred feet in the case of large-scale helicopter photography, to several hundred miles if satellites are used to carry the camera. Flying heights used in photographing for topographic mapping normally vary between about 1,500 and 30,000 ft. If a portion of the project area lies at a substantially higher or lower elevation than the other part, a different flying height above mean sea level for that portion may be necessary to maintain uniform flying height above ground.

Ground coverage per photo for high-altitude photography is greater than for low-altitude photography (see Table 16-2). Fewer high-altitude photos are therefore required to cover a given area. Very-high-altitude coverage is more expensive to obtain than low-altitude photography because of the special equipment that it requires. Some

Table 16-1 Compatible map scales and contour intervals for average terrain

English system		Metric system	
Map scale	Contour interval	Map scale	Contour interval
1 in/50 ft	1 ft	1:500	0.5 m
1 in/100 ft	2 ft	1:1,000	1 m
1 in/200 ft	5 ft	1:2,000	2 m
1 in/500 ft	10 ft	1:5,000	5 m
1 in/1,000 ft	20 ft	1:10,000	10 m

of the problems encountered at high flying heights are decreasing available oxygen, decreasing pressure, and extreme cold. When flying heights exceed about 10,000 ft, an oxygen supply system is necessary for the flight crew. At altitudes above 30,000 ft, pure oxygen under pressure is required. Also, the cabin must be pressurized and heaters are required to protect the crew against the cold. Most aerial photography is taken using single- or twin-engine aircraft. Supercharged single-engine aircraft can reach 20,000-ft altitudes, and supercharged twin-engine aircraft are capable of approaching 30,000 ft. Higher altitudes require turbocharged or jet aircraft.

During photography the pilot maintains proper flying height by means of an altimeter. Since altimeters give elevations above mean sea level, the proper reading is the sum of average ground elevation and required flying height above ground necessary to achieve proper photo scale. Altimeters are barometric instruments and consequently their readings are affected by varying atmospheric pressure. They must be checked daily and adjusted to base airport air pressure.

16-7 STEREOSCOPIC PLOTTER CONSIDERATIONS

If topographic mapping with a stereoscopic plotting instrument is to be performed, the photography must be planned in accordance with certain limiting factors of the particular plotter to be used. Important plotter considerations are (1) principal distance of the projectors, (2) optimum enlargement ratio from photo scale or diapositive scale to map compilation scale, (3) C factor of the plotter, and (4) vertical operating range of the plotter. These factors influence the choice of both camera and flying height.

In interior orientation, the principal distance of the plotter projectors must be set exactly equal to the principal distance of the diapositives. If the diapositives are prepared by contact printing or by one-to-one projection printing, then their principal distance is equal to the focal length of the taking camera (see Sec. 12-7). Some plotters are capable of accommodating a wide range of principal distances; others are limited either optically or mechanically to a very narrow range. Wide-angle Kelsh projectors, for example, accommodate a nominal principal distance of 152 mm (6 in), and diapositive principal distances can vary only a few millimeters from this value. If a Kelsh or equivalent plotter is to be used, it is both logical and convenient to use a camera whose nominal focal length is equal to the nominal principal distance of the projectors. If the focal length of the taking camera is not within the range of the projector principal distance, the diapositives can be reduced or enlarged by projection printing to obtain the proper principal distance. Reduction causes a reduced model size, however, and enlargement may cause loss of imagery on the edges of the photos.

Optimum enlargement ratio of the stereoscopic plotter from photo scale to map compilation scale must also be considered in planning aerial photography. If it is required that a topographic map be compiled to a certain scale, and if the enlargement ratio of the plotter is fixed, photo scale is also fixed, and this in turn fixes flying height. Example 16-3 illustrates this condition. With stereoplotters having a range in enlargement ratios from photo scale to map scale, more flexibility exists in planning aerial photography.

Relative vertical accuracy capabilities of various stereoscopic plotters are commonly compared on the basis of their *C factors*. The *C* factor is the ratio of the flying height above ground of the photography to the contour interval that can be reliably plotted using that photography, or in equation form.

$$C \text{ factor} = \frac{H'}{C.I.} \tag{16-3}$$

The units of H' and C.I. (contour interval) of Eq. (16-3) are the same. Based upon Eq. (16-3), if a plotter has a *C* factor of 1,000 and a map with a contour interval of 5 ft is required, then a flying height above ground of 5,000 ft *or less* is required. The more precise the plotter, the greater its *C* factor rating. Manufacturers commonly specify *C* factors for their instruments, and they may vary from 800 to about 2,000, depending upon the instrument. Contour accuracy depends not only on the plotting instrument but also upon the nature of the terrain, the camera and its calibration, the quality of the photography, the density and quality of ground control, and the capability of the plotter operator. These conditions all combine to yield a total "system *C* factor."

Example 16-5 A topographic map having a scale of 200 ft/in with 5-ft contour interval is to be compiled from contact printed diapositives using a stereoplotter having a nominal 6-in principal distance. Determine the required flying height for the photography if the stereoplotter has a *C*-factor rating of 1,200 and a fixed enlargement ratio capability of 5 from photo to map.

SOLUTION Considering *C*-factor:

$$H' = 1,200(5) = 6,000 \text{ ft}$$

Considering map scale and enlargement ratio:

$$\text{Photo scale} = 5 \times 200 = 1,000 \text{ ft/in}$$

Thus, $H' = 6(1,000) = 6,000$ ft

In this instance map scale and contour interval are compatible, and flying height by either criteria should be 6,000 ft above mean terrain.

Example 16-6 The example is the same as Example 16-5, except required map scale is 1:1,000 with 2-ft contour interval, and the stereoplotter has a *C*-factor rating of 1,500 with enlargement ratio capabilities up to $7\frac{1}{2}$.

SOLUTION Considering *C*-factor:

$$H' = 1,500(2) = 3,000 \text{ ft}$$

Considering map scale and enlargement ratio:

$$\text{At } H' = 3,000 \text{ ft, photo scale} = 1 \text{ in}/500 \text{ ft} = 1:6,000$$

Actual enlargement ratio utilized is

$$ER = \frac{\text{map scale}}{\text{photo scale}} = \frac{\dfrac{1}{1,000}}{\dfrac{1}{6,000}} = 6$$

This is within the range of the stereoplotter capability; thus a flying height of 6,000 ft will meet the criteria and should be used.

In areas where terrain elevation varies considerably, mechanical or optical limitations of the stereoscopic plotter can place a limit on the lowest possible flying height. Several optical projection plotters, for example, can accommodate maximum relief variations in the stereoscopic model of about 20 percent of the projection distance due to depth of field limitations of their projector lenses. Projection distance is analogous to flying height above ground, and therefore for these instruments flying height should be at least 5 times greater than the maximum terrain variations of any model. If, for example, an area has terrain variations of 600 ft, flying height must be at least 3,000 ft in order that the terrain variation does not exceed 20 percent of the flying height.

16-8 GROUND COVERAGE

Once average photographic scale and camera format dimensions have been selected, the ground surface area covered by a single photograph may be readily calculated. In addition, if end lap and side lap are known, the ground area covered by the stereoscopic *neat model* can also be determined. The neat model, as illustrated in Fig. 16-8, is the stereoscopic area between adjacent principal points and extending out sideways in both directions to the middle of the side lap. The neat model has a width of B and a breadth of W. Its coverage is important since it represents the approximate mapping area of each stereopair.

Example 16-7 Aerial photography is to be taken from a flying height of 6,000 ft above average ground with a camera having a 6-in (152.4-mm) focal length and a 9-in (23-cm) format. End lap will be 60 percent and side lap will be 30 percent. What is the ground area covered by a single photograph and by the stereoscopic neat model?

SOLUTION By Eq. (6-1),

(a) $S = 6 \text{ in}/6,000 \text{ ft} = 1 \text{ in}/1,000 \text{ ft}$ or 1:12,000
(b) The dimension, G of the square ground area covered by a single photo is

$$G = 1,000 \text{ ft/in} \times 9 \text{ in} = 9,000 \text{ ft}$$

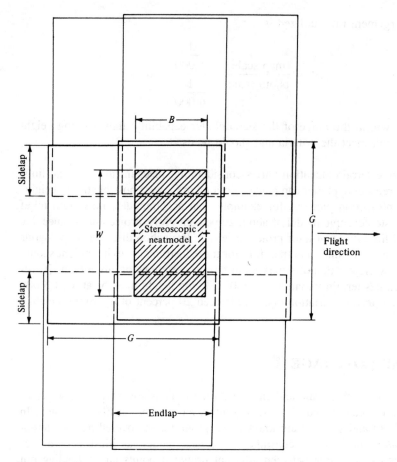

Figure 16-8 The area covered by a stereoscopic neat model.

(c) The area in acres covered on the ground by a single photo is

$$A = \frac{(9{,}000)^2}{43{,}560 \text{ ft}^2/\text{acre}} = 1{,}861 \text{ acres}$$

(d) At 60 percent, end lap B is 0.4 G and at 30 percent side lap W is 0.7 G.

Therefore the dimensions of the rectangular stereoscopic neat model are

$$B = 0.4(9{,}000) = 3{,}600 \text{ ft}$$
$$W = 0.7(9{,}000) = 6{,}300 \text{ ft}$$

The area of the neat model is

$$A_N = \frac{3{,}600 \times 6{,}300}{43{,}560} = 521 \text{ acres}$$

Table 16-2 gives ground dimensions and ground areas covered by a single photo and by the stereoscopic neat model for various commonly used photo scales. The table is based on a 6-in-(152.4-mm-)focal-length camera having a 9-in-(23-cm-) square format. End lap and side lap of 60 percent and 30 percent, respectively, are assumed.

16-9 WEATHER CONDITIONS

The weather, which in most locations is uncertain for any given day, is a very important consideration in aerial photography. In most cases, an ideal day for aerial photography is one that is free from clouds; although if the sky is less than 10 percent cloud-covered, the day may be considered satisfactory. If clouds of greater than 10 percent coverage are present but are so high that they are above planned flying height, this may still be objectionable since large cloud shadows will be cast on the ground, obscuring features. The number of satisfactory cloudless days varies with time of year and locality. There are certain situations where overcast weather can be favorable for

Table 16-2 Dimensions and areas of single photos and neat models for a 6-in focal length camera and various commonly used photo scales

Photo scale	Photo scale, in	Flying height, ft	G, ft	W (0.7 G), ft	B (0.4 G), ft	Acres per photo	Acres per neat model
1:1,800	1"/150'	900	1,350	945	540	42	12
1:2,400	1"/200'	1,200	1,800	1,260	720	74	21
1:3,000	1"/250'	1,500	2,250	1,575	900	116	33
1:3,600	1"/300'	1,800	2,700	1,890	1,080	168	47
1:4,200	1"/350'	2,100	3,150	2,205	1,260	228	64
1:4,800	1"/400'	2,400	3,600	2,520	1,440	298	83
1:5,400	1"/450'	2,700	4,050	2,835	1,620	376	105
1:6,000	1"/500'	3,000	4,500	3,150	1,800	465	130
1:6,600	1"/550'	3,300	4,950	3,465	1,980	563	158
1:7,200	1"/600'	3,600	5,400	3,780	2,160	669	187
1:7,800	1"/650'	3,900	5,850	4,095	2,340	786	220
1:8,400	1"/700'	4,200	6,300	4,410	2,520	911	255
1:9,000	1"/750'	4,500	6,750	4,725	2,700	1,046	293
1:9,600	1"/800'	4,800	7,200	5,040	2,880	1,189	333
1:10,800	1"/900'	5,400	8,100	5,670	3,240	1,507	422
1:12,000	1"/1,000'	6,000	9,000	6,300	3,600	1,861	521
1:15,000	1"/1,250'	7,500	11,250	7,875	4,500	2,907	814
1:18,000	1"/1,500'	9,000	13,500	9,450	5,400	6.5†	1.8†
1:24,000	1"/2,000'	12,000	18,000	12,600	7,200	11.6†	3.3†
1:30,000	1"/2,500'	15,000	22,500	15,750	9,000	18.2†	5.1†
1:40,000	1"/3,333'	20,000	30,000	21,000	12,000	32.3†	9.0†
1:50,000	1"/4,167'	25,000	37,500	26,250	15,000	50.4†	14.1†
1:60,000	1"/5,000'	30,000	45,000	31,500	18,000	72.6†	20.3†

† Square miles

aerial photography. This is true, for example, when large-scale photos are being taken for topographic mapping over built-up areas, forests, steep canyons, or other features which would cast troublesome shadows on clear sunny days.

A particular day can be cloudless and still be unsuitable for aerial photography due to atmospheric haze, smog, dust, smoke, high winds, or air turbulence. Atmospheric haze scatters almost entirely in the blue portion of the spectrum and it can therefore be effectively eliminated from the photographs by using a yellow filter in front of the camera lens. Smog, dust, and smoke scatter throughout the entire spectrum and cannot be filtered out satisfactorily. Best days for photographing over industrial areas which are susceptible to smog, dust, and smoke occur after heavy rains or during moving cold fronts which clear the air. Windy, turbulent days can create excessive image motion and cause difficulties in keeping the camera oriented for vertical photography, in staying on planned flight lines, and in maintaining constant flying heights.

The decision to fly or not to fly is one that must be made daily. The flight crew should be capable of interpreting weather conditions and of making sound decisions as to when satisfactory photography can be obtained. If possible, the flight crew should be based near the project so that they can observe the weather firsthand and quickly take advantage of satisfactory conditions.

16-10 SEASON OF YEAR

The season of the year is a limiting factor in aerial photography because it affects ground cover conditions and the sun's altitude. If photography is being taken for topographic mapping, the photos should be taken when the deciduous trees are bare, so that the ground is not obscured by leaves. In many places this occurs twice a year for short periods in the late fall and in early spring. Oak trees tend to hold many of their leaves until spring, when the buds swell and cause them to fall. In areas with heavy oak cover, therefore, the most satisfactory period for aerial photography is that very short period in the spring between budding and leafing out. Sometimes aerial photography is taken for special forestry interpretation purposes, in which case it may be desirable for the trees to be in full leaf. Normally aerial photography is not taken when the ground is snow-covered. Heavy snow not only obscures the ground but also causes difficulties in interpretation and in stereoviewing. Occasionally, however, a light snow cover can be helpful by making the ground surface more readily identifiable in tree-covered areas.

Another factor to be considered in planning aerial photography is the sun's altitude. Low sun angles produce long shadows which can be objectionable because they obscure detail. Generally about a 30° sun angle is the minimum acceptable for aerial photography. During the winter months of November through February, the sun never reaches a 30° altitude in some northern parts of the United States due to the sun's southerly declination. Aerial photography should therefore be avoided in those areas during these winter months if possible. Often snow cover will prevent photography during these periods anyway. For the other months, photography should be exposed during the middle portion of the day after the sun rises above 30° and before it falls below that altitude. For certain purposes, shadows may be desirable, since they aid in identifying objects. Shadows of trees, for example, help to identify the species. Shadows may also be helpful in locating photo-identifiable features such as fence posts, power poles, etc., to serve as photo control points.

16-11 FLIGHT MAP

A flight map, as shown in Fig. 16-9, gives the project boundaries and flight lines the pilot must fly to obtain the desired coverage. The flight map is prepared on some existing map which shows the project area. United States Geological Survey quadrangle maps are frequently used. The flight map may also be prepared on small-scale photographs of the area if they are available. In executing the planned photographic mission, the pilot finds two or more features on each flight line which can be identified both on the flight map and on the ground. The aircraft is flown so that lines of flight pass over the ground points.

Rectangular project areas are most conveniently covered with flight lines oriented north and south or east and west. As illustrated in Fig. 16-9, this is desirable because the pilot can take advantage of section lines and roads running in the cardinal directions and fly parallel to them.

If the project area is irregular in shape or if it is long and narrow and skewed to cardinal directions, it may not be economical to fly north and south or east and west. In planning coverage for such irregular areas, it may be most economical to align flight lines parallel to project boundaries as nearly as possible. Flight planning templates are useful for determining best and most economical photographic coverage for mapping, especially for small areas. These templates, which show blocks of neat models, are prepared on transparent plastic sheets at scales which correspond to the scales of the base maps upon which the flight plan is prepared. The templates are then simply superimposed on the map over the project area and oriented in the position which yields best coverage with the fewest number of neat models. Such a template is shown in Fig. 16-10. The crosses represent exposure stations, and these may be individually marked on the flight map. This template method of flight planning is exceptionally useful in planning exposure station locations when artificial targets are used (see Sec. 15-10).

Once camera focal length, photo scale, end lap, and side lap have been selected, the flight map can be prepared. The following example illustrates flight map preparation for a rectangular project area.

Example 16-8 A project area is 10 mi long in the east-west direction and $6\frac{1}{2}$ mi wide in the north-south direction (see Fig. 16-11). It is to be covered with vertical aerial photography having a scale of 1:12,000. End lap and side lap are to be 60 percent and 30 percent, respectively. A 6-in- (152.4-mm-) focal-length camera with a 9-in- (23-cm-) square format is to be used. Prepare the flight map on a base map whose scale is 1:24,000, and compute the total number of photographs necessary for the project.

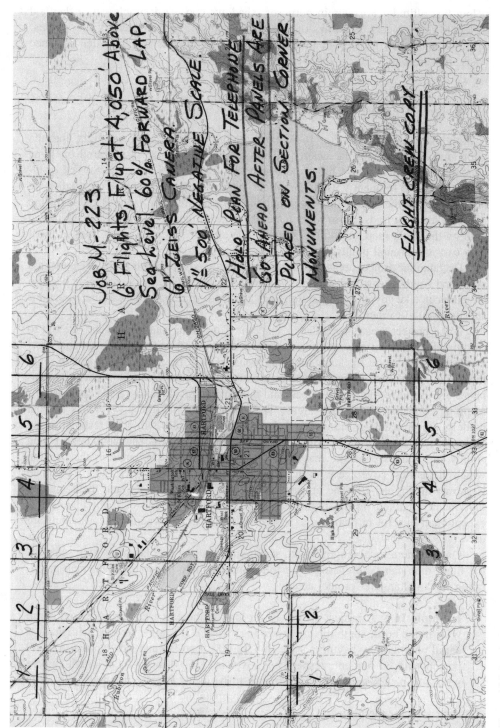

Figure 16-9 Example of a flight plan. (*Courtesy Owen Ayres & Associates, Inc.*)

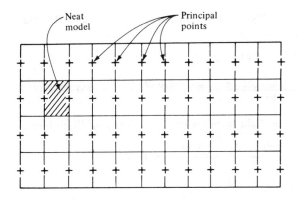

Figure 16-10 Transparent template of neat models used for planning aerial photography.

SOLUTION

(a) Fly east-west to reduce the number of flight lines
(b) Dimension of square ground coverage per photograph [Photo scale = 1:12,000 (1 in/1,000 ft).]:

$$G = 9 \text{ in} \times 1,000 \text{ ft/in} = 9,000 \text{ ft}$$

(c) Lateral advance per strip:

$$W = 0.7 \, G \text{ at 30 percent side lap}$$

$$= (0.7)9,000 = 6,300 \text{ ft}$$

(d) Number of flight lines (Align the first and last line with 0.3G coverage outside the north and south project boundary lines, as shown in Fig. 16-11. This ensures lateral coverage outside of the project area.):

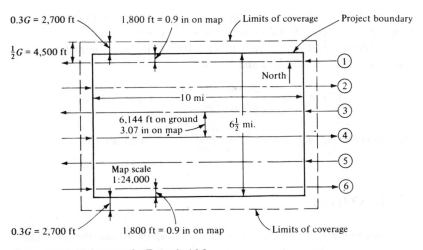

Figure 16-11 Project area for Example 16-8.

Distance of first and last flight lines inside their respective north and south project boundaries (see Fig. 16-11):

$$\tfrac{1}{2}G - 0.3G = 4,500 - 2,700 = 1,800 \text{ ft}$$

$$\text{No. flight lines} = \frac{6\tfrac{1}{2} \text{ mi} \times 5,280 \text{ ft/mi} - 2 \times 1,800}{6,300 \text{ ft per strip}} + 1 = 5.9 \text{ (use 6)}$$

Adjusted spacing between flight lines for integral number of flight lines:

$$W_a = \frac{6\tfrac{1}{2} \times 5,280 - 2 \times 1,800}{5} = 6,144 \text{ ft}$$

(e) Linear advance per photo (air base):

$$B = 0.4 \, G \text{ at 60 percent end lap}$$

$$= (0.4)9,000 = 3,600 \text{ ft}$$

(f) Number photos per strip (Take two extra photos beyond project boundary at each end of the strips to ensure coverage.):

$$\text{No. photos per strip} = \frac{10 \text{ mi} \times 5,280 \text{ ft/mi}}{3,600 \text{ ft per photo}} + 2 + 2 = 18.7 \text{ (use 19)}$$

(g) Total no. of photos:

$$19 \text{ photos per strip} \times 6 \text{ strips} = 114$$

(h) Spacing flight lines on the map:

$$\text{Map scale} = 1{:}24,000(1 \text{ in} = 2,000 \text{ ft})$$

$$\frac{6,144 \text{ ft per strip}}{2,000 \text{ ft/in}} = 3.07 \text{ in}$$

(Draw the flight lines at 3.07 in spacing on the map, with the first and last lines 0.9 in—for 1,800 ft—inside the project boundaries.)

(i) Assuming the aircraft flies at a velocity of 120 mph, the intervalometer setting necessary to obtain the desired end lap is

$$120 \text{ mph} \times \frac{5,280 \text{ ft/mi}}{3,600 \text{ sec/hr}} = 176 \text{ ft/sec}$$

$$\text{Intervalometer setting} = \frac{3,600 \text{ ft per photo}}{176 \text{ ft/sec}} = 20.45 \text{ sec (use 20)}$$

16-12 SPECIFICATIONS

Most flight plans include a set of detailed specifications which outline the materials, equipment, and procedures to be used on the project. These specifications include requirements and tolerances pertaining to photographic scale (including camera focal length and flying height), end lap, side lap, tilt, crab, and photographic quality. The following is a sample set of detailed specifications for aerial photography (courtesy Owen Ayres and Associates, Inc.).

A. *General* The Engineer shall perform the necessary flying and photography to provide photographic coverage of an area approximately 8 square miles in extent shown on the sketch map attached hereto as exhibit "A." The engineer may sublet this phase of the work to a qualified and experienced aerial photographic firm. The city, however, retains the right to approve or reject any or all such firms which the engineer may wish to engage.

B. *Scale* Flight height above average ground shall be such that the negatives will have an average scale of 1 inch equals 500 feet. Negatives having a departure from the specified scale by more than 5 percent because of tilt or abrupt changes in flying altitude must be corrected. The photographs shall be suitable for the compilation of the topographic maps specified herein, and the mapping flight height shall not vary from 3,000 feet above mean terrain, by more than 5 percent.

C. *End lap and Side lap* End lap shall be sufficient to provide full stereoscopic coverage of the area to be mapped. End lap shall average 63 percent, plus or minus 5 percent. End lap of less than 58 percent or more than 68 percent in one or more negatives shall be cause for rejection of the negatives in which such deficiency or excess occurs; unless within a stereoscopic pair, end lap exceeding 68 percent is necessary in areas of low elevation to attain the minimum 58 percent end lap in adjacent areas of high elevation. Wherever there is a change in direction of the flight lines, vertical photography on the beginning of a forward section shall end lap the photography of a back section by 100 percent. Any negatives having side lap of less than 20 percent or more than 55 percent may be rejected.

D. *Tilt* Negatives made with the optical axis of the aerial camera in a vertical position are desired. Tilt of any negative by more than three degrees, an average tilt of more than one degree for the entire project, or tilt between any two successive negatives exceeding four degrees may be cause of rejection.

E. *Crab* Crab in excess of three degrees may be cause of rejection of the flight line of negative or portions thereof in which such crab occurs.

F. *Quality* The photographs shall be clear and sharp in detail and of average uniform density. They shall be free from clouds, cloud shadows, light streaks, static marks, or other blemishes which would interfere with their intended use. All photography shall be taken when the area to be mapped is free of snow, before foliation, and at such time as to insure a minimum solar angle of 30 degrees, except upon written authorization to the contrary by the city.

G. *Camera* For topographic and contour mapping, photographs shall be exposed with a distortion free 6-inch focal length precision aerial mapping camera equipped with a between-the-lens element shutter to produce negatives 9 inches × 9 inches in size. The engineer shall furnish the city with a precision camera calibration report from the National Bureau of Standards for the camera to be used.

H. *Contact Prints* The contact prints from the vertical negatives shall be printed on double weight semimatte paper of suitable contrast.

I. *Photo Index* Photo indices shall be prepared by directly photographing on safety base film at a convenient scale, the assembly of contact prints from all indexed and evaluated prints used. One photo index map shall be delivered on cronapaque or equal. The photo index shall carry a suitable title, scale and north point.

J. *Ownership of Negatives* All negatives shall become the property of the city and shall be delivered to the city upon completion of this contract, or may be stored indefinitely in the film library of the engineer at no added charge.

PART II   COST ESTIMATING AND SCHEDULING

16-13 COST ESTIMATING

Cost estimating is an area of critical concern in the operation of any photogrammetric business, for projects let to contract that are underestimated can have devastating financial results. In general the items that must be considered in a cost analysis include material, labor, and overhead. In addition, a reasonable allowance for profit must be included.

Material costs are directly related to the quantity of each photogrammetric product to be prepared, and the procedures for calculating these quantities are quite straightforward. Thus, material costs can usually be estimated with fair accuracy. Overhead costs, which consist of salaries of administrative personnel, office and laboratory rental, electricity, water, heat, telephone, miscellaneous office supplies, etc., are also rather straightforward to determine. On the other hand, labor costs are considerably more difficult to estimate accurately, and this presents the greatest challenge to the estimator.

The most successful cost estimators base their computations largely on their past experiences with projects of a similar nature. Obviously, therefore, it is important to keep detailed records of the actual costs incurred on the individual items of all projects. Because of significant variations of project complexities, however, and rapidly changing cost factors, past records alone cannot be relied upon completely, and a good deal of intuition and subjective judgment is also necessary. This can normally be obtained only through years of experience.

In the following subsections, various factors that must be considered in estimating costs for some of the major photogrammetric operations are discussed.

16-13.1 Aerial Photography

Costs for aerial photography are generally estimated on the basis of an hourly equipment ownership charge, plus actual operating expenses. An hourly ownership cost for the aircraft and camera can be calculated by dividing the cost per year for ownership by the average yearly hours of its use. Operating expenses include aircraft maintenance and storage, landing fees, gasoline, oil, oxygen, cost of pilot and camera operator, and the cost for film, photo processing, etc.

The number of hours of operation required on a given project is a critical item to be estimated. It is related to aircraft speed and the travel distance to and from the project area. It is also related to the total number of flight line miles involved in photographing the area. In addition, extra time and expense must be figured for attaining higher altitudes, so that required flying height is a factor to be considered, as is the number of flight lines because of the additional time needed to make turns between lines. Costs for film and photo processing can be estimated on the basis of total number of photos to be exposed. The various quantities needed for estimating these costs can be calculated by methods illustrated in Example 16-8, or they can be scaled from the flight map. Difficult elements to estimate in aerial photography are the extra costs incurred in photographic missions that have to be aborted due to unfavorable weather and the costs for maintaining equipment and crew at some distant location while they await suitable conditions. Because of these problems, plus the high cost of an aircraft and camera, many photogrammetric firms do not purchase their own photographic equipment, but instead subcontract their aerial photography through other organizations and then pass the costs along to the client. Subcontractors will figure their costs in the same manner as described above, but they will often simplify their price quotes to some systematic formula such as a general fee to mobilize the aircraft, a charge per mile to travel to and from the project site, a surcharge for extra-high flying altitudes, and a unit price per exposure for each negative.

16-13.2 Ground Control

Costs of establishing ground control can be among the most difficult of project elements to estimate due to the many variables that affect field surveys. Factors influencing the rates at which field surveys can proceed include topography and vegetation in the area, accessibility to the area, weather conditions, etc. The availability and quality of existing control in the area is also critical.

The usual procedures followed in estimating costs of ground control consist of first obtaining as much information about the area as possible. This involves research to locate existing maps and descriptions of ground control in the area. The existing control points are plotted on the maps, and then a field reconnaissance should be made in an attempt to recover as many of these control points as possible, as well as to assess topography, vegetation, etc.

Following field reconnaissance, the general configuration of ground control surveys can be planned and laid out on the maps, and field procedures selected. An important consideration in planning is the required number and location of new control points to be established, and this depends not only on project requirements but also upon whether or not aerotriangulation will be performed to densify the control. If aerotriangulation will be used, basic ground control surveys may have to precede the flight mission so that panels can be placed on the control points prior to photography.

When planned survey lines are laid out on the maps, the required number of miles of traverse and leveling can be measured. Costs can then be figured on the basis of the estimated number of miles of survey progress that can be made per day and a daily charge for survey crews. Experience shows that a three- or four-person crew, equipped with theodolites and EDM equipment, can run from 2 to 3 mi of traverse per day, while a two-person leveling crew can advance at approximately double this rate. These progress rates can vary considerably, depending upon project conditions, however, and should be carefully checked out with the most experienced surveyors in the organization.

In addition to the determination of hourly rates for the survey crews, ownership costs of surveying equipment (including vehicles) must be included, together with travel and subsistence costs for the crews. Office time for computing and adjusting the surveys must also be added, and if panels are to be set, costs for materials and time to place them must be included.

16-13.3 Aerotriangulation

If aerotriangulation will be performed on a project, costs of this operation can be estimated on the basis of a unit cost of time and materials for preparing the diapositives, plus the costs involved in locating and marking control on the diapositives, performing point transfer, making comparator readings, and doing the electronic data processing. In the preparation of diapositives, a factor that will affect material costs is whether glass or film will be used. Time estimates for marking control, point transfer, comparator reading, and data processing are best made on the basis of past experience with the personnel to be assigned these tasks. Many firms have been able to reduce the task of estimating aerotriangulation costs to that of simply applying a unit cost per diapositive processed.

16-13.4 Stereoplotting

Costs for stereoplotting consist basically of charges for operator time plus the cost of machine ownership. Operator time is usually figured in hours per model and should include time for plotting the grid and control points on the manuscript, plotter orientation time, and the actual time required for compilation of planimetry and contours. Like ground control, stereoplotting costs are very difficult to estimate accurately because of the many variables which influence this operation.

One of the most important factors affecting stereoplotting time is map compilation scale. In general, as map scale decreases, compilation time increases because a greater number of features must be shown per unit of area on the map. Contour interval is another important factor since the number and density of contours increases as the interval decreases. Other variables affecting compilation time include the enlargement factor from photo scale to map scale, the nature and density of planimetric details in the area, the ruggedness of the terrain, vegetation, operator experience, etc.

Actual times required for compiling planimetry and contours can vary from as low as 5 hr or less per model for sparse planimetry and contours compiled at large map scales with small enlargement factors, to as high as 30 or 40 hr per model for dense planimetry and contours at small compilation scales with big enlargement factors. As with ground control, the best basis for making estimates of stereoplotting time is past experience. The total number of models required to cover a given project can be obtained by preparing a scaled transparent template such as that shown in Fig. 16-10, overlaying it on a map of the project area, and counting the models.

As a part of stereoplotting, the cost of the map manuscript base material should be included, and this is usually figured as a cost per square foot.

16-13.5 Map Editing and Field Completion

Editing consists of careful and detailed checking of the compiled manuscripts. It includes examination for completeness, consistency of contours with planimetry, correctness of contour numbers and feature names, and overall adherence to proper cartographic representation. At this stage it is usually advantageous to allow the client an opportunity to inspect the maps for additions, deletions, or other changes.

Areas that may have been obscured or unclear for stereoplotting because of vegetation or other factors should be filled in by field completion surveys. At this time feature names that may be in question can be clarified, and other doubtful matters can be checked. As a part of field completion, the maps can be tested for their accuracy. Contour accuracy can be checked by running field profiles that are at least 5 in in length when plotted on the map and that cross at least 10 contours. Field elevations are then compared against elevations for the same points obtained by interpolation from the map. Planimetric accuracy can be checked by obtaining ground coordinates of several well-defined objects by field survey and then comparing these to the coordinates of the same objects as scaled from the map. Best results are obtained by actually taking the maps to the field for completion annd checking.

Costs for these operations must be based upon estimates of actual time involved, plus any expenses. Again they are relatively difficult values to estimate and can best be obtained through experience.

16-13.6 Orthophoto Production

Methods of estimating costs of producing orthophotos are similar to those for stereoplotting and normally are figured on the basis of charges for operator time plus machine ownership. In addition to the actual scanning time per model, operator charges must include time for preparation of the base manuscript and instrument orientation time. Although scanning time can vary somewhat, it is not as difficult to estimate as stereoplotting time because scanning rates are not significantly affected by scale, density of planimetric features, terrain ruggedness, etc. Usually a model can be scanned within about 2 hr or less.

If contours are to be superimposed onto the orthophotos, the cost of compiling the contours will normally have to be figured as a separate stereoplotting operation in accordance with procedures discussed in Sec. 16-13.4. Photographic laboratory costs of developing the orthonegatives, costs of splicing and annotating, and reproduction of the final orthophoto maps must also be determined on the basis of labor and materials and added into the total cost estimate.

16-13.7 Drafting

Drafting includes the processes of inking or scribing final maps from compiled manuscripts. Besides inking and scribing, however, time must also be included for layout of the manuscripts and edge-matching them to obtain the exact coverage of each individual map. This is necessary because individual model coverages will rarely coincide with the coverages of each final map. Drafting normally constitutes one of the major cost items in a photogrammetric mapping project.

Estimates for drafting time are often made on the basis of the number of hours required per square foot of manuscript. Like stereoplotting time, drafting time varies considerably with conditions and is difficult to estimate accurately. Factors that affect drafting time are map scale, contour interval, the number and type of planimetric features that exist per unit of area on the map, and the density and character of the contours. Depending upon conditions, drafting time can vary from as low as 2 hr per square foot for sparse planimetry and contours on large-scale maps, to as high as 10 hr or more per square foot for dense planimetry and contours on small-scale maps. In relation to stereoplotting time, drafting usually takes longer, and in some instances it can exceed stereoplotting time by factors of two or three. Because of the complexities involved in estimating drafting time, the most experienced draftspersons in the organization should be consulted to arrive at reliable figures.

Digitized tracing tables of the type described in Sec. 12-16 make it possible to automatically accomplish much of the final drafting during compilation. This will, of course, considerably reduce the expense and painstaking work involved in drafting.

In addition to drafting time, the material upon which the maps are drafted must be considered, and this is normally figured on a square foot basis. Also, if reproductions are to be made from the final maps, the laboratory time and materials for this must be added into the estimate.

16-13.8 Summary

From the foregoing, it should be apparent that accurate cost estimates can be rather difficult to prepare. Labor, which generally constitutes the major expense on photogrammetric projects, is also the most challenging to approximate.

In estimating, it is easy to omit small items, but enough of these over a period of time can accumulate to cause a significant loss of revenue. Therefore, care must be exercised to prevent these omissions, and the use of check lists is a good way to handle this problem.

As has been mentioned several times above, the most accurate estimates are likely to be made by those persons having the greatest amount of experience.

Prices for labor and materials have not been given above because they fluctuate with time and vary from one locale to another. Some representative prices are given, however, in the sample estimate which follows.

16-14 SAMPLE COST ESTIMATE

To illustrate the procedures involved, the following example cost estimate is presented. Assume that topographic maps are to be prepared photogrammetrically for the project area of Example 16-8. The aerial photography will be obtained by a subcontractor according to the specifications of the example problem. The project area is located 150 mi from the base of operations and from the photography subcontractor. Some ground control exists in the area, and aerotriangulation will be performed to densify the photo control. Delivery items consist of one set of double weight glossy contact prints and one set of inked maps on mylar. Map scale is 200 ft/in with 5 ft contour interval. Map sheet size is 30 × 36 in, with a 3 in border.

A. Preliminary calculations
   1. Project area

$$10 \text{ mi} \times 6.5 \text{ mi} = 65 \text{ mi}^2$$

$$65 \times 640 = 41{,}600 \text{ acres}$$

   2. Total photos = 114 (See Example 16-8)
   3. Total plates for aerotriangulation:

$$6 \text{ strips} \times \left[\frac{10 \times 5{,}280}{3{,}600} \right]^{\dagger} + 1 = 96$$

   4. Total models:

$$6 \text{ strips} \times \left[\frac{10 \times 5{,}280}{3{,}600} \right]\dagger = 90$$

   5. Map sheet areas:

$$\text{Sheet size} = \frac{30 \text{ in} \times 36 \text{ in}}{144} = 7.5 \text{ ft}^2$$

$$\text{Map area per sheet} = \frac{24 \text{ in} \times 30 \text{ in}}{144} = 5 \text{ ft}^2$$

$$\text{Ground area per sheet} = \frac{(24 \times 200) \times (30 \times 200)}{43{,}560} = 661 \text{ acres}$$

   6. Total Maps:

$$\text{No. rows} = \frac{5{,}280 \times 6.5\dagger}{24 \times 200} = 8$$

$$\text{No. columns} = \frac{5{,}280 \times 10\dagger}{30 \times 200} = 9$$

Total maps = 8 × 9 = 72

B. Cost estimates
   (*Note:* Prices used herein are approximate for 1982, and are given only for the sake of example. Current prices should be carefully checked prior to performing an actual cost estimate. All labor prices include overhead.)

---

†These values are rounded up to the nearest integer.

1. Aerial photography
   (Total photos: 114)

Mobilization of aircraft: lump sum	$ 200.00	
Travel:		
150 × 2 @ $0.50/mi	150.00	
Exposures:		
114 @ $7.00	798.00	
2 sets of contact prints:		
2 × 114 @ $2.00	456.00	
Subtotal:		$ 1,604.00

2. Ground control
   Research:
   2 hr @ $30.00/hr $ 60.00

   Prepare flight map:
   2 hr @ $30.00/hr   60.00

   Prepare survey plan:
   4 hr @ $30.00/hr   120.00

   Traverse:
   38 miles (from survey plan)
   3-person party @ $650.00/day

   $$\text{Party days} = \frac{38 \text{ mi}}{2 \text{ mi/day}} = 19 \text{ days}$$

   19 days @ $650.00/day $12,350.00

   Leveling:
   56 mi (from survey plan)
   2-person party @ $400.00/day

   $$\text{Party days} = \frac{56 \text{ mi}}{4 \text{ mi/day}} = 14 \text{ days}$$

   14 days @ $400.00/day $ 5,600.00

   Computing:
   5 days @ $250.00/day   1,250.00

   Travel and miscellaneous expenses:
   lump sum   1,000.00

   Lodging and subsistence:
   Person days = (19 × 3)
   + (14 × 2) = 85
   85 days @ $30.00/ day   2,550.00

   Panel material:
   28 points @ $5.00/point   140.00

   Panel placement:
   2-person crew @ $400.00/day
   2 days @ $400.00/day   800.00

   Subtotal: $ 23,930.00

3. Aerotriangulation
   (Total plates = 96)
   Prepare diapositives (film):
       96 @ $6.00 $ 576.00
   Basic charge:
       96 @ $60.00/plate   5,760.00
                       Subtotal: $ 6,336.00

4. Stereoplotting
   (Total models = 90)
   Prepare manuscripts:
       90×1 hr @ $25.00/hr $ 2,250.00
   Plotter orientation:
       $90 \times \frac{1}{2}$ hr @ $30.00/hr   1,350.00
   Compilation:
       90×20 hr/model @ $30.00/hr   54,000.00
   Material:
       90×12 ft^2/model @ $0.60/ft^2 540.00
                       Subtotal: $ 58,140.00

5. Drafting
   (Total maps = 72)
   Manuscript edge-matching:
       72×1 hr/map @ $25.00/hr $ 1,800.00
   Ink drafting:
       72×5 ft^2/map $\times 8$ hr/ft^2 @
       $25.00/hr   72,000.00
   Material:
       72×7.5 ft^2/sheet @ $0.60/ft^2 324.00
                       Subtotal: $ 74,124.00

6. Editing
       20% of drafting = .20 $\times$ 72,000   14,400.00
                            Subtotal: $14,400.00
                      Total charges: $178,534.00
                      Profit (10%): $ 17,853.40
                      Grand total: $196,387.40

16-15 SCHEDULING

Once the total number of labor hours has been estimated for each phase of a project, schedules for completion of the various operations can be planned on the basis of the number of instruments and personnel available to do the work. In addition to these factors, however, another important consideration is the amount of other work in progress and its status in relation to required completion dates.

To arrive at realistic schedules, additional time in excess of that actually needed to perform the work must be added to account for uncontrollable circumstances. As an example, schedules for aerial photography and ground control surveys must account for possible delays due to inclement weather.

Every reasonable attempt should be made to accommodate clients with stringent scheduling needs. In some cases, to meet critical new scheduling requirements and still adhere to delivery dates already agreed upon, it may be necessary to consider hiring additional staff and run more than one work shift. Of course the possibility of purchasing additional equipment also exists, but this should be done with caution and only when anticipated quantities of continued future work can justify the expenditures.

REFERENCES

American Society of Photogrammetry: "Manual of Photogrammetry," 3d ed., Falls Church, Va., 1966, chaps. 5 and 7.

————: "Manual of Photogrammetry," 4th ed., Falls Church, Va., 1980, chap. 7.

Aguilar, A. M.: Cost Analysis of Aerial Surveying, *Photogrammetric Engineering,* vol. 33, no. 1, p. 81, 1967.

————: Management Planning for Aerial Surveying, *Photogrammetric Engineering,* vol. 35, no. 10, p. 1047, 1969.

Graham, L. C.: Flight Planning for Stereo Radar Mapping, *Photogrammetric Engineering and Remote Sensing,* vol. 41, no. 9, p. 1131, 1975.

Hobbie, D.: Orthophoto Project Planning, *Photogrammetric Engineering,* vol. 40, no. 8, p. 967, 1974.

Lafferty, M. E.: Accuracy/Costs with Analytics, *Photogrammetric Engineering,* vol. 39, no. 5, p. 507, 1973.

Lund, H. G.: Factors in Computing Photo Coverage, *Photogrammetric Engineering,* vol. 35, no. 1, p. 61, 1969.

Moffitt, F. H.: Photogrammetric Mapping Standards, *Photogrammetric Engineering and Remote Sensing,* vol. 45, no. 12, p. 1637, 1979.

Paterson, G. L.: Photogrammetric Costing, *Photogrammetric Engineering,* vol. 37, no. 12, p. 1267, 1971.

Pryor, W. T.: Specifications for Aerial Photography and Mapping by Photogrammetric Methods for Highway Engineering Purposes, *Photogrammetric Engineering,* vol. 16, no. 3, p. 439, 1950.

Scott, L., et al.: Specification for Vertical Air Photography, Photogrammetric Record, vol. IX, no. 54, p. 739, 1979.

Ulliman, J. J.: Cost of Aerial Photography, *Photogrammetric Engineering and Remote Sensing,* vol. 41, no. 4, p. 491, 1975.

U.S. Dept. of Transportation, Federal Highway Administration: *Reference Guide Outline, Specifications for Aerial Surveys and Mapping by Photogrammetric Methods for Highways,* Washington, D.C., 1968.

Walker, P. M., and D. T. Trexler: Low Sun-Angle Photography, *Photogrammetric Engineering and Remote Sensing,* vol. 43, no. 4, p. 493, 1977.

Wood, G.: Photo and Flight Requirements for Orthophotography, *Photogrammetric Engineering,* vol. 38, no. 12, p. 1190, 1972.

Woodward, L. A.: Survey Project Planning, *Photogrammetric Engineering,* vol. 36, no. 6, p. 587, 1970.

Wright, M. S.: What Does Photogrammetric Mapping Really Cost?, *Photogrammetric Engineering,* vol. 26, no. 3, p. 452, 1960.

PROBLEMS

16-1 The air base of a stereopair of vertical photos is 3,890 ft and flying height above average ground is 6,450 ft. If the camera has a 6-in focal length and a 9-in-square format, what is the percent end lap?

16-2 Repeat Prob. 16-1, except that the air base is 235 m and flying height above ground is 395 m.

16-3 For Prob. 16-1, if adjacent flight lines are spaced at 6,810 ft, what is percent side lap?

16-4 For Prob. 16-2, if adjacent flight lines are spaced at 415 m, what is percent side lap?

16-5 An average photo scale of 1:15,000 is required of vertical photos. What air base is required to achieve 60 percent end lap if the camera has a 9-in-square format?

16-6 Repeat Prob. 16-5, except that required photo scale is 1:7,200 and average end lap must be 55 percent.

16-7 Vertical photographs are exposed from 7,500 ft above average ground. If a B/H' ratio of 0.65 is required, what should be the length of the air base? What will percent end lap be for these photos if the camera focal length is 6 in and the format is 9 in square?

16-8 Repeat Prob. 16-7, except that the photos were exposed from 1,500 m above ground, and the required B/H' ratio is 0.55.

16-9 What is the B/H' ratio for vertical photography exposed with 55 percent end lap using a camera having a 6-in focal length and a 9-in-square format?

16-10 Repeat Prob. 16-9, except that end lap is 60 percent and camera focal length is 210 mm.

16-11 A map to a scale of 1:9,000 is to be compiled from vertical aerial photographs using a Balplex (525) stereoplotting instrument. Optimum enlargement from photo to map is 3.5 for this instrument. If a 6-in-focal-length camera is used, what should be flying height above average ground for the photography?

16-12 Repeat Prob. 16-11, except that a Kelsh stereoplotting instrument with optimum enlargement of five is to be used to compile a map to a scale of 1 in/500 ft.

16-13 A stereoscopic plotting instrument having a C factor of 1,500 will be used to compile a map with a contour interval of 10 ft. What flying height is recommended, and what is corresponding photo scale if the camera has a 6-in focal length?

16-14 Repeat Prob. 16-13, except that the stereoplotter has a C factor of 1,200 and contour interval is 5 ft.

16-15 An engineering design map is to be compiled from aerial photography. The map is to have a scale of 1 in/200 ft and a 5-ft contour interval. The enlargement factor of the stereoscopic plotting instrument is five from photo to map, and the plotter's C factor is 1,000. If the camera focal length is 152.4 mm, what is the required flying height above average ground based upon required map scale? Based upon contour interval? Which conditions control flying height?

16-16 Repeat Prob. 16-15, except that map scale is 500 ft/in, contour interval is 10 ft, and the stereoplotter has a C factor rating of 1,500 with maximum enlargement ratio of 7.5.

16-17 Vertical aerial photographs are taken from a flying height of 9,000 ft above average ground using a camera with a 210-mm-focal-length lens and a 9-in-square format. End lap is 60 percent at average terrain elevation. How many acres of ground are covered in a single photograph? In the neat model? (Assume 15 percent side lap.)

16-18 For Prob. 16-17, if low, average, and high terrain are 1,500, 1,900, and 2,600 ft above datum, what is percent end lap at low terrain? At high terrain? What is percent side lap at low terrain? At high terrain?

16-19 An aerial camera is equipped with an intervalometer that can be set to the nearest second only. Vertical aerial photos at a scale of 1:7,200 and with 60 percent end lap are required. If the camera format is 9 in suare and aircraft velocity is maintained at 150 mi/hr during photography, what is the required intervalometer setting?

16-20 Repeat Prob. 16-19, except that photo scale is 1:10,000 and aircraft speed is 160 mi/hr.

16-21 A rectangular area 9 mi in the north-south direction by $5\frac{1}{2}$ mi in the east-west direction is to be covered with aerial photography having a scale of 1:6,000. End lap and side lap are to be 60 percent and

25 percent, respectively. A camera having a 9-in-square format is to be used. Compute the total number of photographs in the project, assuming that the flight strips are parallel with the east and west project boundaries and that the coverage of the first and last flight lines is 75 percent within the project boundary. Also add two photos at the ends of each strip to ensure complete coverage.

16-22 If a flight map is to be prepared for Prob. 16-21 on a base map having a scale of 1:24,000, what should be the spacing in inches of flight lines on the map? What is the map distance in inches between successive exposures along a flight line?

16-23 A transparent template of neat models, similar to that shown in Fig. 16-10, is to be prepared to overlay on a map having a scale of 1:12,000. What should be the dimensions of neat models on the template if the camera format is 9 in square, photo scale is 1:4,800, end lap is 60 percent and side lap is 30 percent?

16-24 Repeat Prob. 16-23, except that map scale is 1:100,000 and photo scale is 1:24,000.

16-25 For the photography of Prob. 16-21, inked topographic maps are to be prepared having a scale of 100 ft/in and 2-ft contour interval. The map sheet size must be 36 in × 42 in with 3-in borders. In addition to delivering one set of maps, a set of contact prints must also be delivered. Aerotriangulation will be performed on the project to densify photo control. Compute a total cost estimate for the project using time estimates and prices as given in Sec. 16-14. Assume that ground control surveys will comprise 25 percent of the total project estimate, and that the travel distance to the project for aerial photography is 250 miles.

16-26 A rectangular project area 8,800 ft in the north-south direction and 10,000 ft in the east-west direction is to be photographed at a scale of 1:2,400. End lap and side lap are to be 60 percent and 35 percent, respectively, and the camera format is 9 in square. From this photography, inked topographic maps are to be prepared at a scale of 1:480 with 1-ft contour interval. Map sheet size is to be 30 in × 36 in with 3-in borders. Aerotriangulation will be performed to densify photo control. Calculate the total cost estimate to deliver one set of contact prints and one set of maps. Use prices and time estimates as given in Sec. 16-14, except assume stereoplotter compilation time to be 10 hr per model and drafting time to be 4 hours per square foot. The distance to the project area from the aerial photo subcontractor is 300 miles. Estimate the cost of ground control surveys to be 30% of the total project cost.

SEVENTEEN
OBLIQUE AND PANORAMIC PHOTOGRAPHS

PART I   OBLIQUE PHOTOGRAPHS

17-1 INTRODUCTION

Oblique photographs are aerial photos taken with the camera axis intentionally inclined at an angle with the vertical. If the inclination angle is so great that the horizon shows in the pictures, the photos are termed *high obliques*. If the horizon does not show, the photos are called *low obliques*. Figure 17-1 is a high oblique photograph, and Fig. 17-2 is a low oblique.

Some of the advantages that oblique photos hold over vertical photos are demonstrated in Figs. 17-1 and 17-2. It is readily noticed, for example, that obliques (especially high obliques) provide far greater ground coverage than verticals exposed from the same altitude. Also, side views of objects afforded by obliques make them more valuable than vertical photos for some interpretation purposes. Tree species, for example, are often readily identified from obliques showing tree profiles, whereas identification may be difficult or even impossible on verticals. To the untrained person, imagery of vertical photos may be rather difficult to interpret because the vertical view is unfamiliar. Oblique photos, on the other hand, show objects in more easily recognized forms.

Oblique photos have some disadvantages in comparison to vertical photos also. Obliques are more difficult to analyze numerically, and they are not as readily adapted to map compilation. Also, objects in the foreground of obliques frequently obscure other objects from view.

17-2 LOW OBLIQUE PHOTOS

Low oblique photos are most frequently taken for nonmapping purposes. They are especially valuable for use in reconnaissance and planning. They are also used for various special purposes in photo interpretation. The pictorial qualities of low obliques

Figure 17-1 High oblique photograph exposed over Oakland, Calif. (*Courtesy Pacific Resources, Inc.*)

make them excellent for showing city skylines, layouts of large industries, proposed construction sites, construction progress and newly completed construction projects, transportation routes, etc.

Since low oblique photos are simply aerial photos that contain large amounts of tilt, the same equations developed in Chap. 11 for tilted photos may be used to analyze them. If the magnitude and direction of tilt are known, together with focal length and flying height above datum, it is possible to determine scale, relief displacement, tilt displacement, and ground point coordinates from measurements of image positions on low oblique photos.

Convergent photography, as illustrated in Fig. 17-3, consists of stereopairs of low oblique photos taken with the camera axes converging toward one another. Cameras such as the one shown in Fig. 4-7 have been designed specifically for taking convergent photography. A single aerial camera can also be used for taking convergent photos, in which case the camera must be quickly tilted *fore* and *aft* to make the exposures.

Advantages of convergent photography in mapping are that up to 100 percent end

Figure 17-2 Low oblique photograph showing the National Capitol in Washington, D.C. (*Courtesy Maps, Inc.*)

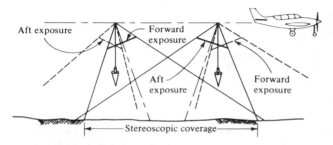

Figure 17-3 Convergent photography.

lap can be obtained, and this results in larger coverage per stereomodel. Also, greater base-height ratios can be achieved, which increases parallactic angles of intersection of corresponding rays and makes possible greater accuracy in locating point positions photogrammetrically. Although not all stereoscopic plotting instruments are capable of handling convergent photography, many of them are. Orientation procedures for stereopairs of convergent photos are similar to those described in Chap. 12 for vertical photos. With optical projection plotters, however, in order to achieve optimum focus throughout the entire stereomodel, the *Scheimpflug condition,* described in Sec. 2-8, must be satisfied. Analytical computations with low oblique photos are performed in the same manner as with tilted photos. When the utmost in accuracy is desired in analytical computations, convergent photos are used because of their improved parallactic angles.

17-3 HIGH OBLIQUE PHOTOS

High oblique photos are now most frequently taken for nonmapping purposes. During the 1940s and 1950s, however, *trimetrogon* photography which utilized high oblique photos was used extensively for small-scale charting of vast areas of the world. Trimetrogon photography was so called because three cameras with *metrogon* lenses were used. Figure 17-4 illustrates the orientation of the three cameras of a trimetrogon system. The center camera exposed a vertical photo while the two side cameras exposed high oblique photos. The three cameras fired simultaneously and obtained ground coverage from horizon to horizon transverse to the direction of flight. As the aircraft progressed along a flight strip, successive exposures overlapped preceding ones. The coverage obtained with trimetrogon photography enabled rapid, small-scale planimetric map compilation of an extensive area.

In recent years, modern aircraft and space vehicles have made possible the acquisition of extremely high-altitude vertical photographs (see Fig. 20-17). These have almost completely supplanted high oblique photos for small-scale mapping. Obliques still retain their value, however, for interpretation, reconnaissance, and intelligence purposes.

High oblique photography has been used to supplement the ground control needed

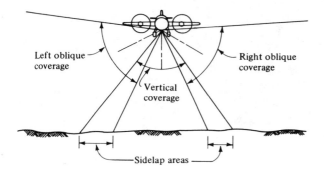

Figure 17-4 Orientation of three cameras used in trimetrogon photography.

for small-scale mapping with vertical photography, especially in areas with difficult access for ground surveying. By exposing two high oblique photos (called *horizon photos*) simultaneously with each vertical exposure, magnitudes and directions of tilt of the vertical photos can be determined. This is possible because the two high oblique cameras are mounted with their optical axes tilted in planes perpendicular to one another; their orientations with respect to the vertical camera axis are known; and the tilt angles of the oblique camera axes are readily determined (see Eqs. 17-1 through 17-4).

Since high oblique photos are merely severely tilted photos, they can be analyzed using the tilted photo equations of Chap. 11. A somewhat different approach is also convenient for analyzing them, however. Figure 17-5a illustrates the principal plane of a high oblique photograph, and Fig. 17-5b shows the resulting photo. The photograph was taken at exposure station L from a flying height H' above datum. The camera focal length was f. The ground nadir point N is the point on the ground vertically beneath the exposure station. The photographic nadir point n is the point where a vertical line from the exposure station pierces the photo plane. Because of the large tilts of high obliques, the photo nadir may not actually exist on the photo but rather fall on the extension of the photographic plane, as shown in Fig. 17-5a.

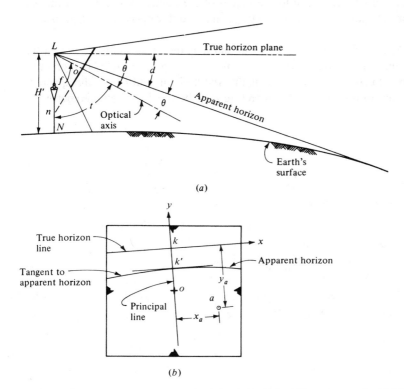

(a)

(b)

Figure 17-5 (a) Side view of the principal plane of a high oblique photograph. (b) Plan view of a high oblique photograph.

The following terms pertinent to a discussion of high obliques are described below and are illustrated in Figs. 17-5a and 17-5b:

Apparent horizon The actual line on the photograph indicating where the land and sky meet. Due to earth curvature, for small-scale high obliques covering large areas, the apparent horizon will appear as a slightly curved line.

True horizon The horizontal plane containing the exposure station. On the photograph, the imaginary line of intersection of the true horizon plane and the photo plane is the *true horizon line*.

Apparent depression angle The angle θ′ measured in the principal plane between the camera optical axis and the apparent horizon. If the apparent horizon is visible on the photograph, the distance *ok′* from the principal point to the apparent horizon can be measured and θ′ calculated as follows:

$$\theta' = \tan^{-1}\left(\frac{ok'}{f}\right) \tag{17-1}$$

Dip angle Angle *d* measured in the principal plane between the true and apparent horizons. The dip angle is due to the height of the camera above the earth. An equation for calculating its value in seconds may be developed with reference to Fig. 17-6.

In Fig. 17-6, *O* is the center of the earth and *R* is the mean radius of the earth. A high oblique photograph is exposed at *L* from a flying height *H′* above ground. *N* is the ground nadir point. Line *LK* is the true horizon, and *LK′* is the apparent horizon tangent to the earth at *K′*; therefore *d* is the dip angle. By geometry the

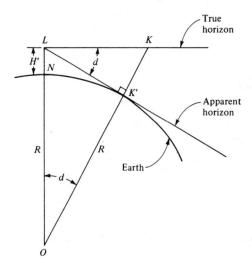

Figure 17-6 Dip angle of high oblique photograph.

angle at O subtended by arc NK' is also d. From the figure,

$$\tan d = \frac{LK'}{R} \qquad (a)$$

Also,

$$LK' = \sqrt{(R + H')^2 - R^2} \qquad (b)$$

Substituting (b) into (a) and reducing,

$$\tan d = \frac{\sqrt{2RH' + H'^2}}{R} \qquad (c)$$

Since H' is small compared to R, Eq. (c) reduces to

$$\tan d = \sqrt{\frac{2H'}{R}} \qquad (d)$$

Because the dip angle is always small, $\tan d$ is approximately d in radians. Also, due to atmospheric refraction, line LK' is bent such that d is reduced in size. Multiplying Eq. (d) by the constant 0.9216 to account for atmospheric refraction in a standard atmosphere, substituting into the equation the mean radius of the earth 20.9×10^6 ft, and converting radians to seconds, the following equation for dip angle in seconds results:

$$d'' = 58.8 \sqrt{H'} \qquad (17\text{-}2)$$

In Eq. (17-2), H' is the height of the camera above ground in feet.

True depression angle Angle θ measured in the principal plane between the camera optical axis and the true horizon. The true depression angle on Fig. 17-5a is seen to be the sum of the apparent depression angle and the dip angle, or

$$\theta = \theta' + d \qquad (17\text{-}3)$$

Tilt angle Angle t measured in the principal plane between the camera axis and the vertical line through the exposure station. It is also the angle between the horizontal plane and the plane of the photo. The true depression angle and the tilt angle are related by the following expression:

$$t = 90° - \theta \qquad (17\text{-}4)$$

17-4 PHOTOCOORDINATE SYSTEM FOR HIGH OBLIQUES

Locations of image points on oblique photos are commonly given with respect to the xy coordinate system shown in Fig. 17-5b. In this system, the x and y axes coincide with the true horizon line and principal line, respectively, where the principal line is in the line of intersection of the principal plane and photo plane. The photographic position of any image point a may be given by its rectangular coordinates x_a and y_a, as shown in Fig. 17-5b. Points above the x axis have positive y coordinates and those below have negative y coordinates. Also, points to the right of the y axis have positive x coordinates and those to the left have negative x coordinates. With this sign convention, except for tall objects in the foreground, all points will have negative y coordinates on high oblique photos.

The x and y axes may be readily located on a high oblique photo if the apparent horizon shows as a nearly straight line such as that of Fig. 17-1. First, point k' of Fig. 17-5b is located by construction at the point where a tangent to the apparent horizon is perpendicular to a line passing through the principal point. The principal line (y axis) is drawn through principal point o and point k'. Length ok' (from principal point to apparent horizon) is measured, and from Eq. (17-1) the apparent depression angle θ' is calculated. From Eq. (17-2) the dip angle is computed, and the true depression angle θ is then determined from Eq. (17-3). Length ok from principal point to true horizon is then calculated using the following equation:

$$ok = f \tan \theta \tag{17-5}$$

Distance ok is laid off from the principal point to locate point k, and finally the true horizon line (x axis) is drawn through k perpendicular to the principal line. If the apparent horizon is not visible on the photo, the *vanishing point,* or *nadir point,* methods described in Sec. 17-11 might possibly by used to locate the xy axes and calculate angle θ.

17-5 SCALE OF HIGH OBLIQUE PHOTOS

With high oblique photos, object distances vary from minimum distances in the foreground to extremely long distances near the apparent horizon. These extreme variations in object distances cause corresponding variations in photo scale. The scale at any point in a high oblique photo in the x direction differs from its scale in the y direction. On flat terrain scale is constant along any line in the x direction (along any line parallel to the true horizon), while it continually varies along any line in the y direction. The closer a point is to the true horizon—i.e., the smaller its y coordinate—the smaller will be its scale.

On Fig. 17-7 scale in the x direction at point a is the ratio of photo distance $a'a$ to corresponding ground distance $A'A$, or

$$S_{x_a} = \frac{a'a}{A'A} \tag{e}$$

From similar triangles in Fig. 17-7,

$$\frac{a'a}{A'A} = \frac{La'}{LA'} = \frac{La''}{LA''} = S_{x_a} \tag{f}$$

Substituting $y_a \cos \theta$ for La'' and $(H - h)$ for LA'' into (f) and dropping subscripts:

$$S_x = \frac{|y| \cos \theta}{(H - h)} \tag{17-6}$$

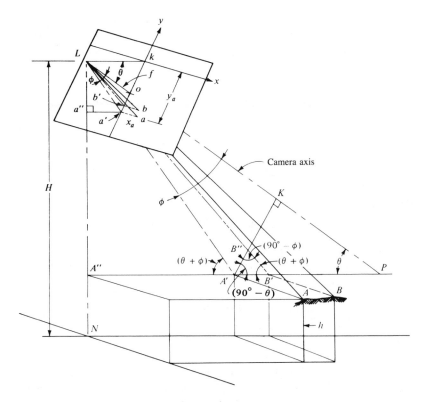

Figure 17-7 Scale of high oblique photograph.

In Eq. (17-6), S_x is the scale at any point on a high oblique photo in the x direction, $|y|$ is the absolute value of the y photocoordinate of the point measured from the true horizon of the photo, θ is the true depression angle, H is flying height above datum, and h is the elevation of the point above datum.

Again referring to Fig. 17-7, an expression for the scale in the y direction for any point on a high oblique photo may be developed. On the figure ground distance AB is considered infinitesimally small, and the scale in the y direction at point a may be expressed as the ratio of photo distance ab to corresponding ground distance AB. However, since $a'b' = ab$, and $A'B' = AB$,

$$S_{y_a} = \frac{a'b'}{A'B'} \tag{g}$$

In Fig. 17-7 let ϕ be angle oLa', the vertical angle between the camera axis and ray LA'. Also, line $A'K$ is constructed perpendicular to the camera axis so that $A'K$ is parallel to $a'k$. In triangle $A'PK$ the angle at P is equal to θ, and the angle at A' is therefore $(90° - \theta)$. Since $A'B'$ is infinitesimally short, lines LA' and LB' are assumed parallel, and hence in triangle $A'B'B''$ the angle at B' is $(\theta + \phi)$. Therefore the angle at B'' is $(90° - \phi)$. By the law of sines in triangle $A'B'B''$,

$$A'B' = \frac{A'B'' \cos \phi}{\sin(\theta + \phi)} \qquad (h)$$

Substituting (h) into (g),

$$S_{y_a} = \frac{a'b' \sin(\theta + \phi)}{A'B'' \cos \phi} \qquad (i)$$

From similar triangles of Fig. 17-7,

$$\frac{a'b'}{A'B''} = \frac{La'}{LA'} \qquad (j)$$

Substituting (j) into (i),

$$S_y = \frac{La' \sin(\theta + \phi)}{LA' \cos \phi} \qquad (k)$$

But from (f), $La'/LA' = S_{x_a}$. Also from Fig. 17-7, $La' = f/\cos \phi$ and $LA' = (H - h)/\text{Cos}(90° - \theta - \phi)$. Therefore,

$$S_{x_a} = \frac{f \sin(\theta + \phi)}{(H - h)\cos \phi} \qquad (m)$$

Rearranging (m),

$$\frac{S_{x_a}(H - h)}{f} = \frac{\sin(\theta + \phi)}{\cos \phi} \qquad (n)$$

Substituting $La'/LA' = S_{x_a}$, and Eq. (n) into (k),

$$S_{y_a} = \frac{(S_{x_a})S_{x_a}(H - h)}{f} \qquad (p)$$

Finally, substituting Eq. (17-6) into (p), dropping subscripts, and reducing,

$$S_y = \frac{y^2 \cos^2 \theta}{f(H - h)} \qquad (17\text{-}7)$$

In Eq. (17-7), S_y is the scale in the y direction at any point on a high oblique photo, and the remaining terms are as previously described.

Example 17-1 A high oblique photo was exposed from a flying height of 10,000 ft above ground with a camera having a focal length of 6 in. The distance ok' was measured as 2.92 in. Find the scale in the x and y directions at the principal point.

SOLUTION From Eqs. (17-1), (17-2), and (17-3),

$$\theta' = \tan^{-1}\left(\frac{2.92}{6.00}\right) = 25°57'$$

$$d = 58.8 \sqrt{10,000} = 5,880'' = 1°38'$$

$$\theta = 25°57' + 1°38' = 27°35'$$

From Eq. (17-5),

$$ok = 6.00 \tan 27°35' = 3.13 \text{ in}$$

(ok is also the y coordinate of the principal point.)
From Eqs. (17-6) and (17-7),

$$S_x = \frac{(3.13) \cos 27°35'}{10,000} = 1 \text{ in}/3,605 \text{ ft}$$

$$S_y = \frac{(3.13)^2 \cos^2 27°35'}{(6.00)10,000} = 1 \text{ in}/7,800 \text{ ft}$$

17-6 HORIZONTAL AND VERTICAL ANGLES FROM HIGH OBLIQUE PHOTOS

Horizontal and vertical angles to points whose images appear in oblique photos may be readily determined provided the reference xy axes can be located and θ calculated. As depicted in Fig. 17-8, an oblique photo exposed from L has its camera axis inclined downward at a true depression angle θ. Image point a in the photo has coordinates x_a and y_a as illustrated. Line aa' is constructed vertical, a' being in the horizontal plane containing the exposure station. Plane Lko is the principal plane (a vertical plane), and plane Laa' is also vertical because it contains vertical line aa'. Line $a'a''$ is parallel to the x axis and is horizontal.

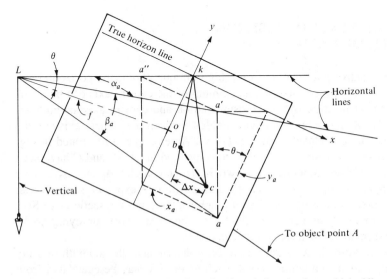

Figure 17-8 Lengths of lines, and horizontal and vertical angles from a high oblique photograph.

Angle α_a between vertical planes Lko and Laa' (horizontal angle between camera axis and ray to object point A) may be calculated as

$$\alpha_a = \tan^{-1}\left(\frac{a'a''}{La''}\right) = \tan^{-1}\left(\frac{x_a}{Lk - a''k}\right)$$

Substituting for $a'a''$ and La'',

$$\alpha_a = \tan^{-1}\left(\frac{x_a}{f \sec \theta + y_a \sin \theta}\right) \qquad (17\text{-}8)$$

In Eq. 17-8 note that the term $y_a \sin \theta$ will be negative by virtue of the negative sign of y_a.

After α_a has been calculated, vertical angle β_a between horizontal line La' and ray La may be calculated as follows:

$$\beta_a = \tan^{-1}\left(\frac{aa'}{La'}\right) = \tan^{-1}\left(\frac{y_a \cos \theta}{La'' \sec \alpha_a}\right)$$

Substituting for La'',

$$\beta_a = \tan^{-1}\left(\frac{y_a \cos \theta}{(f \sec \theta + y_a \sin \theta) \sec \alpha_a}\right) \qquad (17\text{-}9)$$

In both Eqs. (17-8) and (17-9), due regard must be given the algebraic signs of x_a and y_a. Horizontal angles therefore have negative signs if x_a is negative, and vertical angles have negative algebraic signs if y_a is negative. The algebraic sign of the true depression angle θ is considered positive in all equations of this chapter.

17-7 EXPOSURE STATION POSITION AND AZIMUTH OF CAMERA AXIS

If images of three control points appear on a high oblique photo, the horizontal and vertical position of the exposure station and azimuth of the camera axis can be determined. The horizontal position of the exposure station and camera axis azimuth are determined first. In this procedure horizontal angles between the camera axis and the three control points are calculated using Eq. (17-8). An example is illustrated in Fig. 17-9, where horizontal angles α_a, α_b, and α_c to control points A, B, and C have been calculated. From the three α angles, the two horizontal angles δ_1 and δ_2 can be determined. Using angles δ_1 and δ_2, the horizontal position of exposure station L can be determined by three-point resection. Either graphical three-point resection (see Sec. 9-8) may be performed, or numerical methods may be used. Most surveying books describe numerical methods of three-point resection.

When ground coordinates X_L and Y_L have been determined, the azimuth of a ray from the exposure station to a control point, such as ray LA may be calculated from

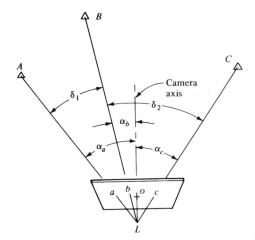

Figure 17-9 A high oblique photograph in space, illustrating three-point resection method of locating exposure station.

coordinate differences. By adding α_a to the azimuth of LA, the azimuth of the camera axis may be determined. A check can be obtained by repeating this procedure with another control point.

If the vertical angle is computed for one vertical control point whose horizontal position is also known, the height of the camera above the control point may be determined. This presumes exposure station coordinates X_L and Y_L are known. In calculating camera height, it may be necessary to account for atmospheric refraction and curvature of the earth. Figure 17-10 illustrates a high oblique photograph taken at a flying height H above datum. The image a of vertical control point A appears in the photo. The elevation of A above datum is h_A. Vertical angle β_a of the ray from the exposure station to the control point may be calculated using Eq. (17-9). With the horizontal positions of the exposure station and control point known, horizontal distance D from exposure station L to control point A can be calculated from coordinate differences. Then vertical distance ($V = D \tan \beta_a$) may be calculated.

Due to atmospheric refraction, calculated vertical angle β_a is too small by an amount $\Delta\beta_a$, so that calculated V is not the true difference in elevation between L and A. Instead, due to atmospheric refraction, V is too small by an amount R. If R is added to V to obtain V', this is still not the true difference in elevation ($H - h_A$), but is now too large by an amount C, which is due to earth curvature. The value of C is the amount of departure between a tangent to the earth's surface and the curved earth. It is proportional to the square of the distance D and is equal to $0.067 M^2$, where M is the distance D in miles. The refraction correction R is approximately one-seventh the value of C but is in the opposite direction. The following expression yields the combined correction for the effects of refraction and curvature:

$$(R - C) = -0.574 M^2 \qquad (17\text{-}10)$$

For small values of D, refraction and curvature is negligible; e.g., its value is only 0.57 ft for D equal to 1 mi. For large values of D, however, it becomes significant; e.g., it is 57 ft in 10 mi. The refraction and curvature correction, when added

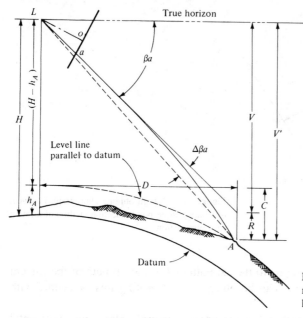

Figure 17-10 Earth curvature and atmospheric refraction corrections with high oblique photography.

to computed V, yields the following equation for calculating flying height for a high oblique photo:

$$H = D \tan \beta_a + (R - C) + h_A \qquad (17\text{-}11)$$

Example 17-2 Suppose that on a high oblique photo having a true depression angle of $30°00'$ and a focal length of 6.00 in, image a of vertical control point A has x and y coordinates of 3.83 in and -2.94 in, respectively. Calculate the flying height of the photo above datum if h_A is 1,460 ft and if the horizontal distance D from exposure station to point A is 6,550 ft.

SOLUTION From Eq. (17-8), horizontal angle α_a is first calculated:

$$\alpha_a = \tan^{-1}\left(\frac{3.83}{(6.00 \times \sec 30°00') - (2.94 \times \sin 30°00')}\right) = 35°03'$$

From Eq. (17-9), vertical angle β_a to control point A is

$$\beta_a =$$

$$\tan^{-1}\left(\frac{-2.94 \times \cos 30°00'}{[(6.00 \times \sec 30°00') - (2.94 \times \sin 30°00')] \sec 35°03'}\right)$$

$$= -20°54'$$

From Eq. (17-10), the refraction and curvature correction is:

$$(R - C) = -0.574\left(\frac{6,550}{5,280}\right)^2 = -0.9 \text{ ft}$$

Finally, from Eq. (17-11), flying height for the photo is

$$H = 6,550 \tan 20°54' - 0.9 + 1,460 = 3,960 \text{ ft above datum}$$

17-8 LOCATING POINTS FROM OVERLAPPING HIGH OBLIQUE PHOTOS

If images of an object point appear on two or more oblique photos whose positions and camera axis orientations are known, the horizontal and vertical position of the object point can be determined by the method of intersection. In Fig. 17-11, for example, images a_1 and a_2 of object point A appear on oblique photos exposed at L_1 and L_2. If the positions and camera axis directions of both exposures are known, then the length and direction of the airbase L_1L_2 may be calculated from differences in ground coordinates. Horizontal angles α_1 and α_2 may be calculated from Eq. (17-8), from which horizontal angles ϕ_1 and ϕ_2 may be determined. The oblique triangle L_1AL_2, with its angles ϕ_1 and ϕ_2 and length L_1L_2 known, may be solved by the law

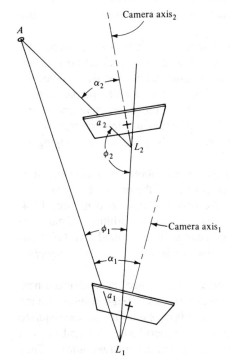

Figure 17-11 An overlapping pair of high oblique photographs in space, illustrating method of intersection for locating new points.

of sines to obtain the horizontal ground position of A. Instead of using numerical techniques described above, graphical procedures can also be applied to obtain the position of point A.

If the flying height for the oblique photo is known, the elevation h_A of point A of Fig. 17-11 may be determined by first calculating vertical angle β_a using Eq. (17-9) and then solving for h_A in Eq. (17-11), or

$$h_A = H - D \tan \beta_a - (R - C) \qquad (17\text{-}12)$$

To solve Eq. (17-12), it is first necessary to determine the horizontal position of point A so that horizontal distance D from the exposure station to point A is available.

17-9 PERSPECTIVE GRIDS

To introduce the concept of a perspective grid, suppose a high oblique photo was exposed over flat terrain upon which existed a square grid pattern such as section lines of the U.S. Public Land System. If the camera were pointing in a northerly direction during exposure, all north-south section lines would appear to converge (vanish) at a point on the true horizon line of the photo. Such a point of convergence of parallel lines on an oblique photo is called the *vanishing point*. The square sections of land would appear on the photo as a pattern of quadrilateral grids. This grid pattern is properly called a *perspective grid*, because it represents a perspective view of a ground grid. In this case each perspective grid quadrilateral would represent a ground grid square of 1-mi dimensions.

Given flying height, focal length, and depression angle, it is possible to prepare a perspective grid for an oblique photo even though no ground grid actually exists. Ground dimensions of grid squares represented by quadrilaterals of the perspective grid can be selected at any convenient value. The grid is usually prepared on a transparent medium, and when superimposed on the photo, the perspective grid shows photo locations of imaginary ground squares. This provides a reference rectangular coordinate system for determining ground distances or for planimetric mapping by direct tracing.

In constructing a perspective grid for a high oblique photo, it is convenient to make one of the grid lines coincide with the principal line of the photo. The procedure involves first locating the true horizon line of the photo as described in Sec. 17-4. Then, as shown in Fig. 17-12, principal line ko is extended an arbitrary distance to establish point e. Because it corresponds to the principal line, ground line EE is the so-called *ground principal meridian,* and grid line ke is called the *grid principal meridian.*

A *base parallel*, perpendicular to the grid principal meridian, is established next by constructing a line parallel to the true horizon which passes through point e. Having selected M, the desired dimension of ground squares, it is necessary to find the required spacing m along the base parallel between adjacent grid meridians. (All grid lines of Fig. 17-12 that converge at k, such as ck, dk, fk, etc., are called *grid meridians*. They

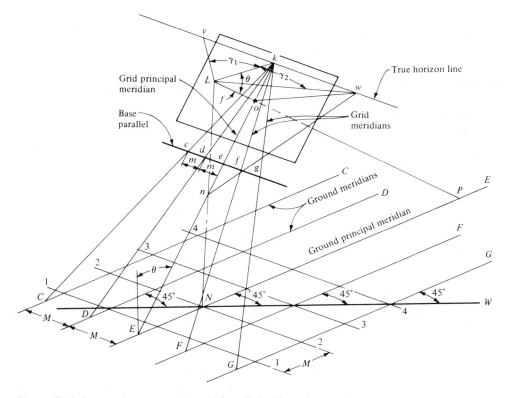

Figure 17-12 Constructing a perspective grid for a high oblique photograph.

have corresponding ground lines *CC*, *DD*, *FF*, etc., called *ground meridians*, which are parallel to the ground principal meridian. Point *k* is the vanishing point for all ground meridians.) From similar triangles *kde* and *kDE* of Fig. 17-12,

$$\frac{m}{ke} = \frac{M}{kE}$$

Substituting (H' sec θ) for *kE* into the above equation and solving for *m*,

$$m = \frac{M(ke)}{H' \sec \theta} \tag{17-13}$$

In Eq. (17-13), *m* is the required spacing between adjacent grid meridians measured along the base parallel, *M* is a selected value which is the dimension of ground squares represented by grid quadrilaterals, *ke* is the length of the grid principal meridian from *k* to *e*, and *H'* is flying height above ground. The units of *M* and *H'* must be the same, and the units of *m* will be the same as the units of *ke*. Having calculated *m*, segments of length *m* are marked along the base parallel to locate points *c*, *d*, *f*, etc., and grid meridians *ck*, *dk*, *fk*, etc., are drawn. The spacing in the *x* direction on the oblique photo between adjacent meridians, no matter where measured, represents a ground distance of M.

It is next required to draw on the perspective grid a system of lines parallel to the base parallel. These lines are the so-called *grid parallels*. Consider a diagonal ground line NW of Fig. 17-12 which crosses ground principal meridian $ENPE$ at a 45° angle. Line NW also crosses all ground meridians at 45°. Perpendiculars 1-1, 2-2, 3-3, etc., erected to ground meridians CC, DD, EE, etc., at the points where line NW intersects them forms a square ground grid of dimensions M.

The key to drawing parallels on the perspective grid is to find the perspective grid location of a diagonal ground line, such as NW, that makes a 45° angle with the principal plane. Perpendiculars (grid parallels) are then erected to the grid meridians where this diagonal crosses them. This produces the grid parallels. In Fig. 17-12, if distance kw is made equal to Lk where Lk is equal to $f \sec \theta$, then angle γ_2 is a horizontal angle of 45° and line Lw intersects the true horizon line at point w. The vertical plane containing line Lw also contains lines nw and NW. Line nw in the photo plane is therefore the perspective grid location of ground line NW, and point w is the vanishing point for line NW. According to perspective geometry, any system of lines that are parallel in the flat ground plane all vanish at the same vanishing point, provided they are not parallel to the horizon line. Point w is therefore the vanishing point for all lines in the ground plane that make an angle of 45° with the ground principal meridian. Hence grid line wo also represents a diagonal ground line that crosses ground meridians at 45°.

By similar analysis, the vanishing point for another set of parallel ground lines that make a 45° degree angle γ_1 with the ground principal meridian occurs at v, where distance kv is equal to $f \sec \theta$.

In applying the above principles to the construction of grid parallels, points v and w are first located on the true horizon line, as illustrated in Fig. 17-13, where

$$vk = wk = f \sec \theta \qquad (17\text{-}14)$$

Diagonal lines vv' and ww' are then drawn, and where these diagonals intersect grid meridians, grid parallels are constructed, as shown in Fig. 17-13. Meridians and parallels are drawn until the area of interest on the photo is covered. Once completed, the grid may be used to interpolate distances, or it may be used for planimetric mapping. It should be noted that accuracy is highest in the picture foreground, and it diminishes rapidly toward the apparent horizon.

Diagonals vv' and ww' could be drawn at any random location as long as they converge at v or w. It is convenient, however, to draw them through the principal point, as this locates the ground principal point explicitly at a grid intersection on the map. The ground nadir point may then be located on the map by measuring back along the principal meridian on the map a distance PN, where

$$PN = \frac{H'}{\tan \theta} \qquad (17\text{-}15)$$

As noted earlier, perspective grids are valuable for determining the relative locations of points and for planimetric mapping. By counting n_x and n_y, the number of

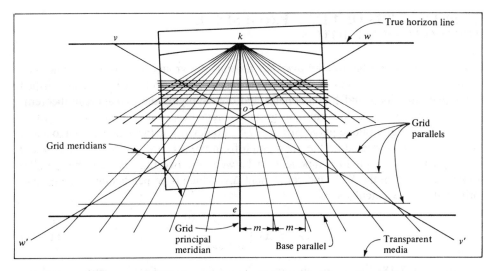

Figure 17-13 Constructing parallels of a perspective grid for a high oblique photograph.

grid quadrilaterals by which two points are separated in the x and y directions, respectively, the Δx and Δy ground distance components between the two points can be calculated as

$$\Delta x = n_x M \qquad (17\text{-}16)$$

$$\Delta y = n_y M$$

In arriving at the numbers n_x and n_y for use in Eq. (17-16), partial grid spacings between points can be estimated. The components Δx and Δy will be ground components perpendicular and parallel, respectively, to the ground principal meridian and their units will be the same as the units selected for M. With the two ground components known, the ground length of the line D, between the two points, can be calculated by the pythagorean theorem as

$$D = \sqrt{(\Delta x)^2 + (\Delta y)^2} \qquad (17\text{-}17)$$

Planimetric maps can also be drawn using perspective grids. In this technique a manuscript map is first prepared by drawing squares to scale having the dimensions of M. Details that appear within each grid quadrilateral are then transferred manually to their proper locations within their corresponding manuscript squares. This technique was used extensively in the past for small-scale planimetric mapping of much of Canada's flat wilderness area and was called the *Canadian Grid* method. The procedure involved preparing perspective grids for oblique photos such that each quadrilateral on the perspective grid represented a ground square of $\frac{1}{8}$-mi dimensions. Map compilation was done on a base sheet having a square grid of $\frac{1}{8}$-in dimensions, so that the resulting map scale was 1 in/1 mi.

17-10 LENGTHS OF LINES FROM SINGLE HIGH OBLIQUE PHOTOS

For a high oblique photo *taken over flat terrain,* the ground length of any line may be determined if the images of its end points appear on the photo. This is accomplished by determining its X and Y components and then applying the pythagorean theorem. Consider, for example, determining ground length BC from images b and c which appear in the high oblique photo of Fig. 17-8. To determine the X and Y components, lines are first constructed on the photograph from point k through image points b and c. At point c, the x photo component Δx between the two lines is measured parallel to the x axis. Then the X ground component ΔX, which is perpendicular to the ground principal line, is calculated as

$$\Delta X = \Delta x \left(\frac{H'}{|y_c| \cos \theta} \right) \tag{17-18}$$

In Eq. (17-18), $|y_c|$ is the absolute value of the y coordinate of point c, and the expression in parentheses is the inverse of scale in the x direction at point c given by Eq. (17-6). Similarly, ΔY, the Y ground component, which is parallel to the ground principal line, is calculated as

$$\Delta Y = \left(\frac{1}{y_c} - \frac{1}{y_b} \right) \frac{fH'}{\cos^2 \theta} \tag{17-19}$$

Finally, the ground length of line BC is determined from its X and Y components by applying Eq. (17-17).

Example 17-3 The focal length, flying height above average ground, and true depression angle for a high oblique photograph are 152 mm, 6,000 ft, and 30°00′, respectively. Find ground length AB of a line whose images a and b appear on the photo. The Δx photo component measured at b is 6.21 mm, and the y coordinates of a and b are − 100.38 mm and − 104.81 mm, respectively.

SOLUTION From Eq. (17-18),

$$\Delta X = 6.21 \left(\frac{6,000}{104.81 \cos 30°00′} \right) = 410 \text{ ft}$$

From Eq. (17-19),

$$\Delta Y = \left(\frac{1}{-104.81} - \frac{1}{-100.38} \right) \frac{152 \times 6,000}{\cos^2 30°00′} = 512 \text{ ft}$$

From Eq. (17-17),

$$AB = \sqrt{(410)^2 + (512)^2} = 656 \text{ ft}$$

17-11 TRUE HORIZON AND TRUE DEPRESSION ANGLE BY VANISHING POINT AND NADIR POINT METHODS

If a square or rectangular ground object such as a 1-mi square *section* of land appears on an oblique photo taken over flat terrain, the true horizon and true depression angle may be found (even though the apparent horizon is not visible on the photo) using the *vanishing point* method.

In Fig. 17-14, a square field, *abcd*, appears on an oblique photo as shown. Since parallel lines vanish at a point on the true horizon, lines *ab* and *cd* are extended until they intersect to locate vanishing point *v*. Likewise lines *ad* and *bc* are extended until they intersect to locate vanishing point *w*. The true horizon line *vw* may then be constructed. Principal point *o* is located from fiducial marks, and a perpendicular to the true horizon line is constructed which passes through the principal point. Point *k* where the perpendicular intersects the true horizon line is the origin of the oblique photocoordinate system. The *y* axis coincides with line *ok*, and the *x* axis coincides with line *vw*. Length *ok* is measured and the true depression angle calculated using Eq. (17-5).

If images of vertical objects such as tall buildings, smokestacks, or telephone poles appear on an oblique photo, the true horizon and true depression angle may be determined by the *nadir point* method. In this technique, as illustrated in Fig. 17-14, image lines of vertical objects are extended to their intersection. This locates the

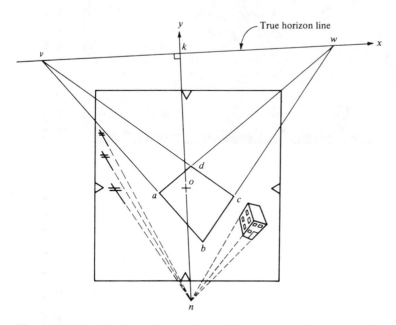

Figure 17-14 Locating the true horizon of an oblique photograph by vanishing points method and by nadir point method.

Figure 17-15 Location of oblique photocoordinate axes and determination of depression angle by the vanishing point method. (*Photo courtesy State of Wisconsin, Department of Transportation.*)

Figure 17-16 Location of oblique photocoordinate axes and determination of depression angle by the nadir point method. (*Photo courtesy State of Wisconsin, Department of Transportation.*)

photographic nadir point. Line *no* is then constructed (this is the *y* axis). Distance *no* is measured and the true depression angle is calculated as

$$\theta = 90° - \tan^{-1}\left(\frac{no}{f}\right) \tag{17-20}$$

With θ known, distance *ok* is calculated using Eq. (17-5), it is laid off along the *y* axis to locate point *k*, and the true horizon is constructed by erecting a perpendicular to the *y* axis at point *k*.

Figures 17-15 and 17-16 show examples of the construction of the photocoordinate systems of high oblique photos by the vanishing point and nadir point methods, respectively.

PART II   PANORAMIC PHOTOGRAPHS

17-12 INTRODUCTION

A panoramic photograph is a picture of a strip of terrain taken transverse to the direction of flight. The exposure is made by a specially designed camera which scans laterally from one side of the flight path to the other. The lateral scan angle may be as great as 180°, in which case the photograph contains a panorama of the terrain from horizon to horizon. The longitudinal field of view of the scan (angular coverage in the direction of flight) is narrow, but the camera is capable of rapid cycling so that successive photos overlap the previous ones, thereby providing continuous stereoscopic coverage of the terrain.

Panoramic photos may be exposed either as *verticals* or *forward-looking obliques*. Figure 17-17*a* and Fig. 17-17*b* are front and side views illustrating lateral and longitudinal angular coverages of a vertical panoramic photograph, while Fig. 17-17*c* shows actual ground coverage of overlapping vertical panoramic photos. Figure 17-18 illustrates camera orientation for forward-looking oblique photos. Figure 17-19 is an example of a vertical panoramic photo. Note that scale decreases toward the horizons from its maximum scale directly beneath the flight path. This is, of course, expected, since scale decreases with an increase in object distance. This helps explain the "butterfly" appearance of ground coverage of a vertical panoramic photo shown in Fig. 17-17*c*. Figure 17-20 is a forward-looking oblique panoramic photo. The extreme curvature of the horizon in this photo is primarily due to the geometric nature of forward-looking oblique panoramic photos and only very slightly due to earth curvature.

In recent years increasing interest has been shown in panoramic photography. One reason for this interest is that compared to frame camera photos, panoramic photos cover a much greater area. Also because they utilize only the center portion of the camera lens, image resolution provided by panoramic photos may be as much as five

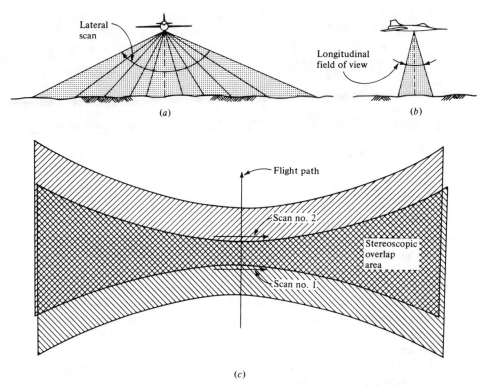

Figure 17-17 (*a*) Lateral scan of a panoramic photograph. (*b*) Longitudinal field of view of vertical panoramic photograph. (*c*) Ground coverage of overlapping vertical panoramic photos.

times greater than that of frame camera photography. These advantages, together with more uniform illumination over the entire format, make panoramic photos exceptionally well suited for photo interpretation and for intelligence and reconnaissance purposes. The U.S. Forest Service has found panoramic photography useful to monitor timber stands and check diseased trees and forest fires. They recently photographed the entire State of Pennsylvania in 4 hr; an important consideration in terms of cost, and in cases where weather conditions may not permit extended periods of time needed for normal vertical photo coverage. Panoramic photos have the disadvantage that they lack the geometric fidelity of frame camera photos, a characteristic that is important for mapping.

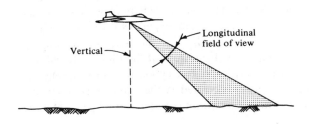

Figure 17-18 Longitudinal field of view of forward-looking oblique panoramic photograph.

Figure 17-19 Vertical panoramic photograph taken over Marietta, Ohio, with a Perkin-Elmer KS-68A panoramic camera. (*Courtesy Perkin-Elmer.*)

17-13 PANORAMIC CAMERAS

Panoramic cameras are generally one of two types: those which scan by rotating the camera lens, and those which scan by rotating a prism in front of the camera lens. Two rotating-lens types of scanning systems are illustrated in simplified diagrams of Figs. 17-21 and 17-22, and a rotating prism system is illustrated in Fig. 17-24. With the rotating-lens type shown in Fig. 17-21, the film is curved in the shape of a cylinder, and the camera lens is on the axis of this cylinder. The radius of the cylinder is equal to the focal length of the camera lens. The exposure is made on the film through a thin slit in a scanning arm with moves with the lens as it rotates in scanning from side to side. During exposure, the film is held fixed in position, except that it may be moved slowly backward opposite the flight direction to compensate for image motion caused by forward movement of the aircraft. After the scan is completed, the film is advanced into position for the next exposure.

A second rotating-lens type of panoramic camera is diagrammed in Fig. 17-22. In this system, designed for the recent lunar Apollo missions, the entire optical assembly rotates about a horizontal axis. Light rays entering the objective lens are folded (deflected) by two mirrors, and focussed through a narrow slit onto the film which travels around a roller. This manner of folding the ray path enables a very long focal length lens to be used (24 in) and yet retain a relatively compact instrument. A camera similar to the one used for the Apollo program has been designed for civilian use. Figure 17-23 shows the Apollo camera. It provides up to 140° of lateral coverage, and has a photo frame width of $4\frac{1}{2}$ in with film capacity of 6,500 ft (2,000 m).

With the rotating-prism type panoramic camera illustrated in Fig. 17-24, a double dove prism mounted in front of the camera lens rotates about its longitudinal axis, thereby scanning the terrain from side to side. The camera lens remains fixed during exposure. As the prism rotates, the film, which is stretched between two rollers, advances at a speed which is synchronized with the scanning rate of the rotating prism. The exposure is made through a narrow slit in front of the film. Although the film in rotating-prism cameras is not actually curved in cylindrical form during exposure, this cylindrical geometrical equivalent is created by the scanning action of the prism. As with the rotating-lens type, image motion due to aircraft advance may be

Figure 17-20 Forward-looking oblique panoramic photograph showing New York's World Fair grounds taken with a Fairchild F-415 panoramic camera. (*Courtesy Fairchild Space and Defense Systems.*)

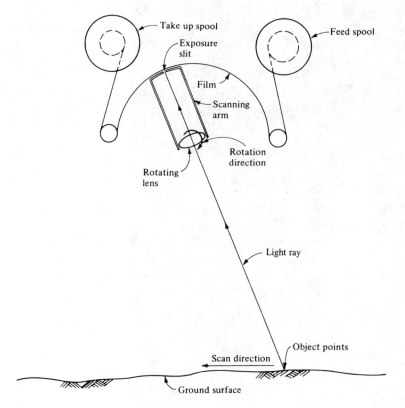

Figure 17-21 Schematic diagram of operation of rotating lens type of panoramic camera.

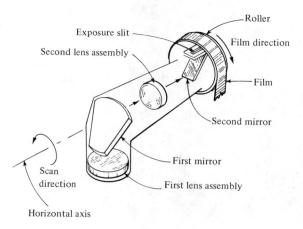

Figure 17-22 Schematic diagram of Itek Optical Bar panoramic camera. (*Courtesy Itek Optical Systems, Inc.*)

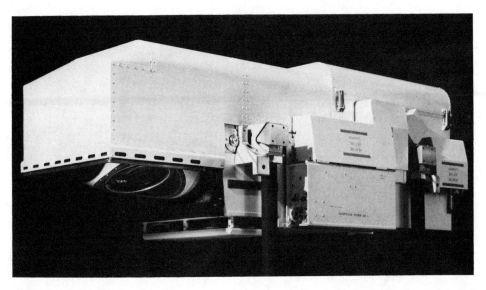

Figure 17-23 Itek Optical Bar panoramic camera used on NASA Apollo missions. (*Courtesy Itek Optical Systems, Inc.*)

compensated for by shifting the film slowly backward along the flight line as the exposure is made. Figure 17-25 shows a rotating-prism panoramic camera having a 6-in focal length, 180° lateral coverage, and 40° longitudinal field of view. This camera makes exposures having a $4\frac{1}{2}$-in width.

With both the rotating-lens and rotating-prism systems, the narrow scanning slit allows only light rays which pass through the center of the lens to expose the film. This accounts for the high resolutions and uniform light intensities which are characteristic of panoramic photos.

17-14 GEOMETRY OF VERTICAL PANORAMIC PHOTOGRAPHS

Figure 17-26a is an isometric view illustrating the geometry of a vertical panoramic photo taken from exposure station L. The camera focal length is f and flying height above datum is H. The cylindrical surface of the photo positive is shown in Fig. 17-26a, while Fig. 17-26b shows the photo positive laid out flat. As shown in Fig. 17-26a, the scan angle ξ_a for any object point A may be defined as the angle between the vertical line Lo and line L'_a which is the projection of ray La onto the vertical plane containing the y axis.

For purposes of specifying image point positions on a vertical panoramic photo, an xy photocoordinate axis system is adopted. In this system the x axis is taken in the direction of flight passing through the position of zero scan angle. The y axis is taken through the center of the picture format and perpendicular to x. If the camera has no fiducial marks with which to locate the origin o of the xy axis system, it may be

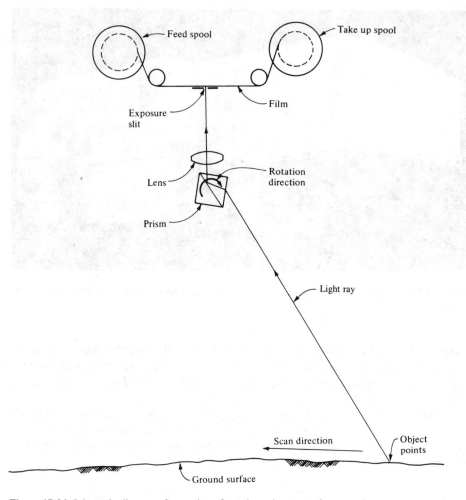

Labels in figure: Feed spool, Take up spool, Exposure slit, Film, Rotation direction, Lens, Prism, Light ray, Scan direction, Object points, Ground surface

Figure 17-24 Schematic diagram of operation of rotating prism type of panoramic camera.

Figure 17-25 Fairchild KA-77 rotating prism type panoramic camera. (*Courtesy Fairchild Space and Defense Systems.*)

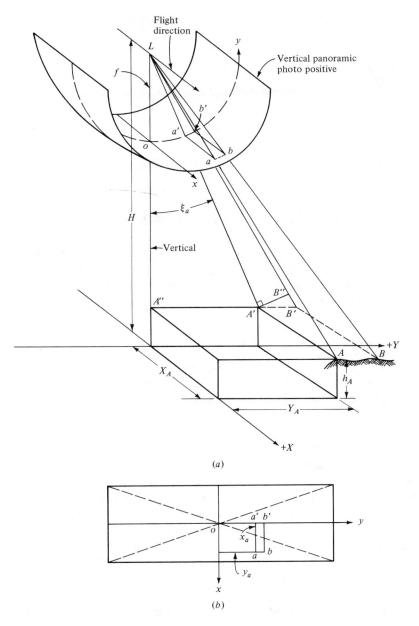

(a)

(b)

Figure 17-26 (a) Geometry of a vertical panoramic photograph. (b) Photocoordinate system of a vertical panoramic photograph.

roughly located at the intersection of diagonals from the four corners of the format, as shown by the dashed lines of Fig. 17-26b. The y axis is then constructed through o parallel to the long side of the format, and x is perpendicular to y passing through o. This reference xy axis system is shown in Fig. 17-26b.

The scan angle ξ_a corresponding to any image point a may be expressed in terms of the camera focal length and measured y photocoordinate of the point as

$$\xi_a = \frac{y_a}{f} \frac{180°}{\pi} \tag{17-21}$$

In Eq. (17-21), y_a and f must be in the same units and ξ_a is in degree units.

17-15 SCALE OF A VERTICAL PANORAMIC PHOTOGRAPH

In Figs. 17-26a and b, a' is the orthogonal projection of image point a onto the y axis. In Fig. 17-26a, A' is in the vertical plane containing the y photo axis. From similar triangles La'a and LA'A, the photo scale in the x direction at any point a on a vertical panoramic photo may be expressed as

$$S_{x_a} = \frac{aa'}{AA'} = \frac{f}{LA'}$$

But LA' is equal to $(H - h_A)$ sec ξ_a. Substituting this into the above equation and dropping subscripts, the following general equation is obtained for scale in the x direction at any point of a vertical panoramic photo:

$$S_x = \frac{f \cos \xi}{(H - h)} \tag{17-22}$$

In Eq. (17-22), S_x is the scale in the x direction at any point on a vertical panoramic photo, f is the camera focal length, ξ is the scan angle computed from Eq. (17-21), H is flying height above datum, and h is the elevation of the object point above datum.

Also from Fig. 17-26a, an expression for the scale in the y direction for any point on a panoramic photo may be obtained. Consider ground distance AB as being infinitesimally small. Then the scale in the y direction at image point a may be given as

$$S_{y_a} = \frac{ab}{AB} = \frac{a'b'}{A'B'}$$

Line A'B'' is constructed perpendicular to LA', and therefore, from similar triangles La'b' and LA'B'',

$$\frac{La'}{LA'} = \frac{a'b'}{A'B''}$$

In the above expression, La'/LA' is S_{x_a}, which from Eq. (17-22) is equal to f cos ξ/(H − h). Also, A'B'' is equal to (A'B') cos ξ_a. Substituting these values into the above expression, dropping subscripts, and reducing,

$$S_y = \frac{f \cos^2 \xi}{(H - h)} = S_x \cos \xi \qquad (17\text{-}23)$$

In Eq. (17-23), S_y is the scale at any point in the y direction on a panoramic photo. The other terms in the equation are as described for Eq. (17-22).

From an analysis of Eqs. (17-22) and (17-23), it is seen that as ξ increases, scale decreases. Near the horizons, where ξ approaches $90°$, $\cos \xi$ approaches zero and photo scale becomes infinitely small. Most reliable quantitative information on a panoramic photo is therefore obtained near the center of the photo where scale is largest.

17-16 GROUND COORDINATES FROM MEASUREMENTS ON A VERTICAL PANORAMIC PHOTO

Positions of points whose images appear in a vertical panoramic photo may be calculated in a rectangular ground coordinate system from photocoordinate measurements. The ground coordinate system has its origin in the datum plane vertically beneath the exposure station. The X and Y ground axes are in the vertical planes containing the x and y photo axes, respectively, and they are positive in the same directions as the photo axes. From similar triangles $La'a$ and $LA'A$ of Fig. 17-26a,

$$X_A = \frac{x_a(H - h_A)}{f \cos \xi_a} \qquad (17\text{-}24)$$

Also, from triangle $LA''A'$,

$$Y_A = (H - h_A) \tan \xi_a \qquad (17\text{-}25)$$

In Eqs. (17-24) and (17-25), X_A and Y_A are ground coordinates in the aforementioned arbitrary coordinate system and the other terms are as previously described. It should be noted that no provision is made in these equations for earth curvature or atmospheric refraction. Furthermore, unless corrections are made for the several types of distortions present in panoramic-imagery, the accuracy with which points can be located from panoramic photos is limited. Near the horizons, scale becomes extremely small, so that accuracy is severely curtailed in those areas. Consequently the above scale and ground coordinate equations should not generally be considered as acceptable for mapping, but they can be used a guides to assist in interpretation from panoramic photography.

Example 17-4 A vertical panoramic photo was exposed from a flying height of 10,000 ft above datum with a panoramic camera having a 6-in-focal-length lens. Two points, A and B, whose elevations are both 1,200 ft above datum, have the following photocoordinates measured on the positive with respect to a photo axis system pictured in Fig. 17-26b:

$$x_a = 27.08 \text{ mm} \qquad y_a = 98.43 \text{ mm}$$

$$x_b = -19.14 \text{ mm} \qquad y_b = -63.79 \text{ mm}$$

Calculate the scale of the photo at points a and b and calculate ground distance AB.

SOLUTION From Eq. (17-21) the scan angles are

$$\xi_a = \frac{(98.43)(180°)}{6(25.4)3.1416} = 37°00'$$

$$\xi_b = \frac{(-63.79)(180°)}{6(25.4)3.1416} = -24°00'$$

Photo scales from Eqs. 17-22 and (17-23) are

$$S_{x_a} = \frac{6.00 \cos 37°00'}{10,000 - 1,200} = 1 \text{ in}/1,840 \text{ ft}$$

$$S_{y_a} = \frac{1 \text{ in}}{1,840 \text{ ft}} \cos 37°00' = 1 \text{ in}/2,305 \text{ ft}$$

$$S_{x_b} = \frac{6.00 \cos 24°00'}{10,000 - 1,200} = 1 \text{ in}/1,610 \text{ ft}$$

$$S_{y_b} = \frac{1 \text{ in}}{1610 \text{ ft}} \cos 24°00' = 1 \text{ in}/1765 \text{ ft}$$

Ground coordinates from Eqs. (17-24) and (17-25) are

$$X_A = \frac{27.08(10,000 - 1,200)}{6(25.4) \cos 37°00'} = 1,960 \text{ ft}$$

$$Y_A = (10,000 - 1,200) \tan 37°00' = 6,630 \text{ ft}$$

$$X_B = \frac{-19.14(10,000 - 1,200)}{6(25.4) \cos 24°00'} = -1,210 \text{ ft}$$

$$Y_B = (10,000 - 1,200) \tan -24°00' = -3,920 \text{ ft}$$

(Note that negative signs of X_B and Y_B correspond to negative x and y photo-coordinates.)

By the pythagorean theorem,

$$AB = \sqrt{(-1,210 - 1,960)^2 + (-3,920 - 6,630)^2} = 11,020 \text{ ft}$$

17-17 IMAGE DISTORTIONS IN PANORAMIC PHOTOS

Panoramic photos contain several types of distortions not present in frame photography. First of all, there is a displacement of images called *panoramic distortion* which occurs because of the cylindrical shape of the film surface and because of the nature of scanning. If a square grid on flat ground is photographed with a panoramic camera in vertical orientation, panoramic distortions produce the characteristic trapezoidal grid pattern shown in Fig. 17-27.

Scan positional distortion (also sometimes referred to as *sweep positional distortion*) is a displacement of images caused by the forward motion of the aircraft during the time of exposure. It causes a relative displacement in *x* coordinates for points which are separated in the *y* direction on the photo. Relative scan positional distortion for any two points is the product of photo scale and ground distance traveled in the elapsed time interval between exposure of the two points.

If the film or lens is translated during exposure to compensate for image motion due to aircraft advance, a relative displacement of images known as *image motion compensation distortion* results. Image motion compensation distortion tends to cancel the effects of scan positional distortion, although it does not compensate for it completely. The dashed line of Fig. 17-27 illustrates the nature of the resultant effects of scan positional distortion and image motion compensation distortion for images along the *y* photo axis. Relative positions of images throughout the entire photo undergo similar distortions.

A distortion called *tipped panoramic distortion* results if the camera axis is tipped either forward or backward during exposure. Forward tipping is the case with the forward-looking oblique panoramic photos such as the one shown in Fig. 17-20. The extreme curvature of the apparent horizon in this photo is due primarily to tipped panoramic distortion.

The distortions described above can be removed and image positions rectified if enough information is known about camera and flight conditions. Necessary data include camera focal length, flying height, angular velocity of scanning, ground speed of the aircraft, and tip angle (angle camera axis makes with vertical). Rectification for selected image points may be accomplished analytically or, if performed by electronic rescanning or optical projection, all photo imagery may be rectified. If reliable quantitative photogrammetric data is to be taken from panoramic photos, image po-

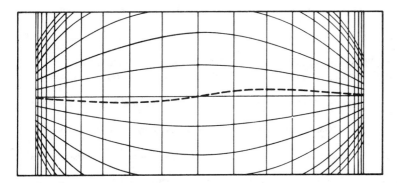

Figure 17-27 Panoramic distortion.

sitions must be rectified. It is impossible to remove all distortions, however, because of the many unpredictable variables in panoramic photography. Film slippage, irregularities in angular scanning, film expansion or contraction, etc., create distortions which are difficult if not impossible to detect. New instrumentation is currently being developed which should, however, help to alleviate some of these uncertanties.

REFERENCES

Abraham, V.: Relative Geometric Strength of Frame, Strip, and Panoramic Cameras, *Photogrammetric Engineering*, vol. 27, no. 5, p. 755, 1961.

American Society of Photogrammetry: "Manual of Photogrammetry," 3d ed., Falls Church, Va., 1966, chaps. 4 and 18.

———: "Manual of Photogrammetry," 4th ed., Falls Church, Va., 1980, chap. 4.

Arena, A.: The KA-92 Panoramic Camera System, *Photogrammetric Engineering*, vol. 40, no. 10, p. 1225, 1974.

——— and M. Umlas: A New Panoramic Camera Development, *Photogrammetric Engineering*, vol. 34, no. 2, p. 169, 1968.

Donnelly, C. B. C.: Trimetrogon Photogrammetry, Some Usages in the Preparation of the Canadian Aeronautical Chart, *Photogrammetric Engineering*, vol. 15, no. 1, p. 22, 1949.

Gay, S. P., Jr.: Measurement of Vertical Heights from Single Oblique Aerial Photographs, *Photogrammetric Engineering*, vol. 23, no. 5, p. 900, 1957.

Griffin, E. P.: 20 Degree Convergent Versus Vertical Photography, *Photogrammetric Engineering*, vol 26, no. 1, p. 59, 1960.

Hovey, S. T.: Panoramic Possibilities and Problems, *Photogrammetric Engineering*, vol. 31, no. 4, p. 727, 1965.

Itek Laboratories: Panoramic Progress, Part I, *Photogrammetric Engineering*, vol. 27, no. 5, p. 747, 1961.

———: Panoramic Progress, Part II, *Photogrammetric Engineering*, vol. 28, no. 1, p. 99, 1962.

Kawachi, D. A.: Image Motion and its Compensation for the Oblique Frame Camera, *Photogrammetric Engineering*, vol. 31, no. 1, p. 154, 1965.

———: Image Geometry of Vertical and Oblique Panoramic Photography, *Photogrammetric Engineering*, vol. 32, no. 2, p. 298, 1966.

Le Resche, J.: Analysis of the Panoramic Aerial Photograph, *Photogrammetric Engineering*, vol. 24, no. 5, p. 772, 1958.

McCash, D. K.: Apollo 15 Panoramic Photographs, *Photogrammetric Engineering*, vol. 39, no. 1, p. 65, 1973

McNeil, G. T.: Oblique Plotting Scale, *Photogrammetric Engineering*, vol. 15, p. 455, 1949.

———: A Wide-Field Underwater Panoramic Camera, *Photogrammetric Engineering*, vol. 32, no. 1, p. 37, 1966.

Raisz, E.: Direct Use of Oblique Air Photos for Small Scale Maps, *Surveying and Mapping*, vol. 13, no. 4, p. 496, 1953.

Schmutter, B., and U. Etrog: Analysis of Panoramic Photos, *Photogrammetric Engineering*, vol. 40, no. 4, p. 489, 1974.

Skiff, E. W.: "Analytical Treatment of Strip and Pan Photos," *Photogrammetric Engineering*, vol. 33, no. 11, p. 1290, 1967.

Stewart, R. A.: The Application of the Balplex Plotter to Trimetrogon Obliques, *Photogrammetric Engineering*, vol. 23, no. 4, p. 697, 1957.

PROBLEMS

17-1 A high oblique photograph was taken from a flying height of 6,450 ft above datum with a camera having a 210-mm focal length. Average ground elevation was 1,325 ft above datum. Distance ok' measured along the principal line from principal point to apparent horizon is 69.7 mm. What is the true depression angle for the photo?

17-2 Repeat Prob. 17-1, except that flying height is 3,480 m above datum, average ground is 675 m above datum, camera focal length is 152 mm, and measured distance ok' is 2.74 in.

17-3 For the photograph of Prob. 17-1, what is photo scale at the principal point in the x and y directions? (Assume average ground elevation at the principal point.)

17-4 For the photograph of Prob. 17-1, what are the scales in the x and y directions at the isocenter? (Assume average ground elevation at the isocenter.) What conclusion do you draw about scale at the isocenter?

17-5 For the photograph of Prob. 17-2, what are the scales in the x and y directions at the principal point? At the isocenter?

17-6 A high oblique photograph taken from a flying height of 7,650 ft above datum with a 152-mm-focal-length camera has a true depression angle of $27°20'$. Images a and b have photocoordinates with respect to the xy high oblique photocoordinate system of $x_a = 77.5$ mm, $y_a = -43.4$ mm, $x_b = -54.9$ mm, and $y_b = -115.8$ mm. Calculate photo scale in the x and y directions for points a and b if ground points A and B have elevations of 895 ft and 785 ft above datum, respectively.

17-7 Calculate the horizontal angle ALB at the exposure station between points A and B of Prob. 17-6.

17-8 Calculate the vertical angles from the exposure station to points A and B of Prob. 17-6.

17-9 A high oblique photograph is exposed with a 152.4-mm-focal-length camera. An altimeter reading at the exposure station provided an approximate flying height above datum of 4,950 ft. Average ground elevation was 1,065 ft. Images a, b, and c of horizontal control points A, B, and C appear on the photo. Their photocoordinates with respect to the xy high oblique photocoordinate axis system and their ground coordinates are given below. Photo distance ok' measured along the principal line from principal point to the apparent horizon is 74.0 mm. On a map, plot the control points A, B, and C. Compute horizontal angles ALB and BLC and determine the X and Y coordinates of the exposure station by graphical three-point resection (see Sec. 9-8).

	Photocoordinates		Ground coordinates	
Point	x, mm	y, mm	X, ft	Y, ft
A	101.4	-88.2	1,829,562	77,718
B	17.2	-64.8	1,824,579	78,184
C	-71.7	-69.2	1,819,618	77,488

17-10 Determine the camera axis direction for the photograph of Prob. 17-9.

17-11 Determine flying height above datum for the photograph of Prob. 17-9 if the elevation of control point A is 1,075 ft above datum.

17-12 Construct a perspective grid for the high oblique photograph of Prob. 17-1. Make the principal line coincide with one of the grid lines, and make perspective grid quadrilaterals represent ground squares of 500-ft dimensions. Make the photographic principal point occur at one of the perspective grid intersections and complete the grid from the picture foreground to about 1 in above the principal point. Assume the photo has a 9-in-square format.

17-13 Repeat Prob. 17-12, except that the perspective grid for the photograph of Prob. 17-2 should be constructed making grid quadrilaterals represent ground squares of 200-m dimensions.

17-14 The images of a rectangular field are imaged on a high oblique aerial photograph in a manner similar to that shown in Fig. 17-14. The camera focal length was 152.44 mm, and the photocoordinates of points a, b, c, and d were measured with respect to the fiducial axis system and are given below. Plot these points, graphically locate the true horizon and principal line by the vanishing point method, and calculate the true depression angle for the photo.

Point	x, mm	y, mm
a	−19.30	−4.83
b	12.45	−42.55
c	52.83	9.91
d	20.32	−25.02

17-15 The images of two vertical ratio towers appear on a high oblique photograph, and their photocoordinates with respect to the fiducial axis system are given below. The camera had a focal length of 152.35 mm. Plot these points, graphically locate the true horizon and principal line by the nadir point method, and calculate the true depression angle of the photo.

Point	x, mm	y, mm
Tower A, top	−78.74	55.88
Tower A, bottom	−57.15	−25.40
Tower B, top	84.07	27.94
Tower B, bottom	70.87	−43.18

17-16 How do panoramic photographs differ from frame camera aerial photos?

17-17 List the advantages of panoramic photographs as compared to frame camera aerial photos. List the disadvantages.

17-18 Describe the different types of panoramic cameras.

17-19 A vertical panoramic photograph was taken with a 310-mm-focal-length camera from a flying height of 10,500 ft above ground. Calculate photographic scale in the x and y directions at points a and b, whose x and y photocoordinates are $x_a = 0.00$ mm, $y_a = 0.00$ mm, $x_b = 35.27$ mm, and $y_b = 116.85$ mm.

17-20 Calculate the ground length of line AB from image positions a and b of the panoramic photograph of Prob. 17-19. Assume level ground throughout the coverage of the panoramic photo.

17-21 Images c and d of ground points C and D appear on a vertical panoramic photograph taken with a 152-mm-focal-length camera from a flying height of 6,450 ft above datum. Points C and D have image coordinates of $x_c = −39.25$ mm, $y_c = 17.65$ mm, $x_d = 29.48$ mm, and $y_d = −132.73$ mm. Calculate the ground length of line CD if the elevations of C and D are 1,960 ft and 1,740 ft above datum, respectively.

17-22 Discuss the various image distortions present in panoramic photographs.

EIGHTEEN

TERRESTRIAL AND
CLOSE-RANGE PHOTOGRAMMETRY

18-1 INTRODUCTION

Terrestrial photogrammetry is an important branch of the science of photogrammetry. It deals with photographs taken with cameras located on the surface of the earth. The cameras may be hand-held, mounted on tripods, or suspended from towers or other specially designed mounts. The term ''close-range photogrammetry'' is generally used for terrestrial photographs having object distances of up to about 300 m. Unlike aerial photography, with terrestrial photography the cameras are usually accessible, so that direct measurements can be made to obtain exposure station positions. Camera angular orientation can also usually be measured or set to fixed values, so that all elements of exterior orientation of a terrestrial photo are commonly known and need not be calculated. These known exterior orientation parameters are a source of control for terrestrial photos, replacing in whole or in part the necessity for locating control points in the object space.

Terrestrial photography may be *static* (photos of stationary objects) or *dynamic* (photos of moving objects). For static photography, slow, fine-grained, high-resolution films may be used and the pictures taken with long exposure times. Stereopairs can be obtained by using a single camera and making exposures at both ends of a base line. In taking dynamic terrestrial photos, fast films and rapid shutter speeds are necessary. If stereopairs of dynamic occurrences are required, two cameras located at the ends of a base line must make simultaneous exposures.

18-2 APPLICATIONS OF TERRESTRIAL
AND CLOSE-RANGE PHOTOGRAMMETRY

Historically the science of photogrammetry had its beginning with terrestrial photography, and topographic mapping was among its early applications. Terrestrial photos were found especially useful for mapping rugged terrain which was difficult to map by conventional field-surveying methods. Although it was known that topographic mapping could be done more conveniently using aerial photos, no practical method was available for taking aerial photographs until the airplane was invented. Following the invention of the airplane, emphasis in topographic mapping shifted from terrestrial to aerial methods. Terrestrial photogrammetry is still used in topographic mapping, but its application is usually limited to small areas and special situations such as deep gorges or rugged mountains that are difficult to map from aerial photography. Other topographic applications of terrestrial photogrammetry are in mapping construction sites, areas of excavation, borrow pits, material stockpiles, etc.

Through the years terrestrial photogrammetry has continued to gain prominence in numerous diversified nontopographic applications. Examples of nontopographic applications occur in such areas as agriculture, conservation, ecology, forestry, archaeology, anthropology, architecture, geology, geography, engineering, mining, industry, criminology, oceanography, medicine, dentistry, and many more. In the field of medicine, x-ray photogrammetry has been utilized advantageously for measuring sizes and shapes of body parts, recording tumor growth, studying the development of fetuses, locating foreign objects within the body, etc.

Terrestrial photogrammetry has been used to great advantage as a reliable means of investigating traffic accidents. Photos which provide all information necessary to reconstruct the accident may be rapidly obtained. Time-consuming sketches and ground measurements, which all too often are erroneous, are not needed, and normal traffic flow can more quickly be restored. Terrestrial photogrammetry has been widely practiced in accident investigation for many years in several European countries.

Time-lapse terrestrial photos have been used to record speeds of automobiles, directions and velocities of water currents, rate and manner of plant growth, etc. A unique and highly specialized application of terrestrial photogrammetry was in connection with the establishment of a worldwide satellite triangulation network. In this program, positions of orbiting satellites were recorded as they passed through the fields of view of highly precise ballistic cameras, such as the one shown in Fig. 1-3. The camera stations, which actually serve as a framework for a worldwide geodetic datum, were widely distributed and their positions accurately determined on the basis of measurements taken from the photos. In the space program terrestrial photos are also used to calibrate large parabolic antennas used in tracking spacecraft.

Terrestrial photogrammetry has become a very useful tool in many areas of scientific and engineering research for several reasons. One reason is that it makes possible measurements of objects which are inaccessible for direct measurement. Also, measurements can be obtained without actually touching delicate objects. In some experiments, such as measurements of water waves and currents, physical contact during measurement would upset the experiment and render it inaccurate. Cameras which freeze the action at a particular instant of time make possible measurements of dynamic occurrences such as deflections of beams under impact loads. Because of the many advantages and conveniences offered by terrestrial photogrammetry, its importance in the future seems assured.

18-3 TERRESTRIAL CAMERAS

A variety of cameras are used in terrestrial photography. All of them fall into one of two general classifications: *metric* or *nonmetric*. The term "metric camera," as used herein, includes those cameras manufactured expressly for photogrammetric applications. They have fiducial marks built into their focal planes, which enable accurate recovery of their principal points. Metric cameras are stably constructed and completely calibrated before use. Their calibration values for focal length, principal point coordinates, and lens distortions can be applied with confidence over long periods of time.

Nonmetric cameras are manufactured for amateur or professional photography where pictorial quality is important but geometric accuracy requirements are generally not considered paramount. These cameras do not contain fiducial marks, but they can be modified to include them. Nonmetric cameras can be calibrated and used with satisfactory results for many terrestrial photogrammetric applications. Examples of nonmetric cameras are shown in Figs. 3-3 and 4-5.

Aerial cameras can be used for terrestrial applications, but this requires special tripod mounts. Aerial cameras used for terrestrial photography would, of course, be considered as metric cameras because of their stability and calibration characteristics. Most metric terrestrial cameras are fixed for focus at infinity and cannot produce sharp geometrically correct images at very short ranges. These cameras can be modified, however, to permit sharp focus at short ranges. Exposures with metric cameras are often made directly on glass plates; therefore, film-flattening devices are not necessary and the utmost in dimensional stability of the image is achieved. Many metric cameras use polyester-base film having high dimensional stability.

Phototheodolites and *stereometric* cameras are two special types of terrestrial camera systems in the metric classification which are described in the following two sections. Of the many available terrestrial cameras, only a few are described and pictured herein as examples. Omission of other comparable cameras is not intended to imply any inferiority of these instruments.

18-4 PHOTOTHEODOLITES

The term "phototheodolite" generally applies to a combination camera and theodolite. In a broader sense, however, a phototheodolite is any camera, usually metric, equipped for orienting its optical axis in a known direction with respect to a base line. Phototheodolites are generally mounted on tripods and centered over a desired camera station by means of a plumb bob.

The Zeiss TMK of Fig. 18-1 is a terrestrial camera of the phototheodolite classification. It has a telescope with cross hairs for sighting along a base line, and by means of a right-angle prism, the camera axis may be oriented at 90° to a base line. If the camera optical axis is directed at 90° from both ends of a base line, as shown in Fig. 18-2, the photos of the resulting stereopair have their optical axes parallel.

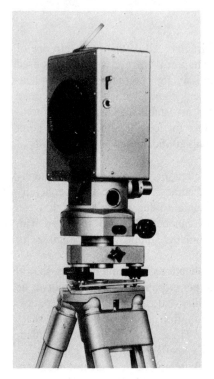

Figure 18-1 Zeiss TMK phctotheodolite. (*Courtesy Carl Zeiss, Oberkochen.*)

The TMK has a nominal focal length of 60 mm, and its pictures are exposed on glass plates of dimensions 90 by 120 mm.

The Wild P-30 which is shown in Fig. 1-1, fits the classical description of a phototheodolite. It combines a Wild T-2 theodolite and a metric camera. The theodolite has an upper and lower motion. With its upper motion locked and lower motion open, the camera and theodolite turn in azimuth as a unit. With the upper motion

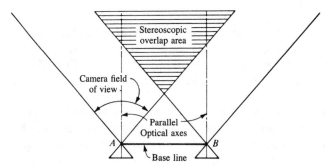

Figure 18-2 Stereo coverage obtained by exposing terrestrial photos at 90° to a base line from the ends of the base line.

open and lower motion locked, the theodolite can be turned in azimuth while the camera retains its direction. The horizontal angle between the theodolite telescope axis and camera axis may be read from the horizontal circle. When the horizontal circle of the theodolite reads 0°00′00″, the theodolite telescope axis and the camera axis point in the same direction. The camera axis can be set in any desired azimuth with respect to a base line. Stereopairs with parallel optical axes, as shown in Fig. 18-2, can be obtained if desired by setting 90° on the horizontal circle with the upper motion and then sighting on the opposite end of the base line with the lower motion. The theodolite of the P-30 also has a vertical circle which facilitates determining camera station differences in elevation.

By leveling the theodolite, the camera of the P-30 is simultaneously leveled. The optical axis of the camera can be set horizontal, it can be tilted upward at an elevation angle of 7^g (6°18′), or it can be tilted downward at depression angles of 7^g, 14^g, 21^g, or 28^g. The camera's nominal focal length is 165 mm and exposures are made on glass plates of dimensions 100 by 150 mm. A terrestrial photo taken with the P-30 is shown in Fig. 1-2.

The Wild P-32 terrestrial camera shown in Fig. 18-3 has been designed to attach directly onto a standard surveying theodolite. The camera axis can be set at any desired vertical angle by making the appropriate setting on the theodolite's vertical circle, and it can be aimed at any horizontal angle with respect to the camera base line by means of the theodolite's horizontal circle. The P-32 camera has a 64-mm focal length and

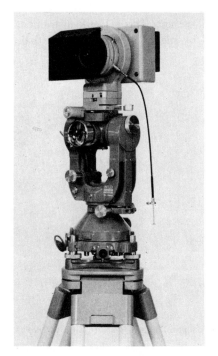

Figure 18-3 Wild P-32 terrestrial camera mounted on a T-2 theodolite. (*Courtesy Wild Heerbrugg Instruments, Inc.*)

can make exposures either on glass plates of dimensions 65 by 90 mm or on roll film, in which case a film flattening device is used.

The Jena Photheo 19/1318 of Fig. 18-4 is a phototheodolite of slightly different characteristics. With this camera the lens may be shfted vertically upward or downward to obtain the desired vertical coverage. An orientation attachment on top of the instrument makes it possible to align the camera axis in any desired azimuth and to measure vertical angles for determining camera station elevations. The Photheo 19/1318 camera has a nominal focal length of 190 mm and the dimensions of its photographic plates are 130 by 180 mm.

18-5 STEREOMETRIC CAMERAS

A stereometric camera consists of two identical metric cameras mounted rigidly at the ends of a fixed base so that their optical axes are parallel to one another. This means that relative orientation of the cameras is known after calibration, and it remains constant for all stereopairs taken. The shutters of both cameras are actuated simultaneously—a condition enabling dynamic occurrences to be photographed stereoscopically. A variety of stereometric systems are available, affording a wide choice in field of view, format dimensions, focusing distances, and fixed base lengths. Lengths of the fixed bases vary with different cameras, but 200-, 120-, and 40-cm lengths are common. Shorter base systems apply for photographing at closer range. The Zeiss SMK stereocamera, consisting of two TMK cameras mounted at a fixed base of 120 cm, is shown in Fig. 18-5.

Figure 18-6 shows the Kelsh K-490 stereometric camera. This versatile instrument has a variable range in base-length settings, and by means of a micrometer dial,

Figure 18-4 Jenoptik Photheo ¹⁹/₁₃₁₈. (*Courtesy Jenoptik Jena.*)

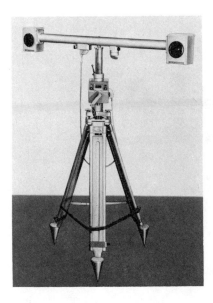

Figure 18-5 Zeiss SMK stereometric camera with 120-centimeter fixed base. (*Courtesy Carl Zeiss, Oberkochen.*)

principal distances (image distances) can be varied to provide sharp focus for object distances from about 2 ft up to infinity. The camera's focal length is 90 mm, and it makes exposures either on glass plates or roll film of 105-mm by-178-mm size.

In terrestrial photogrammetry, *depth* is the distance from the camera base to photographed objects. *Base-depth ratios* (similar to base-height ratios of aerial stereopairs) provide a means of assessing relative geometric strengths of terrestrial stereopairs. Geometric strength is greater for large base-depth ratios because intersection angles between corresponding rays are greater. Because their bases are relatively short, stereometric cameras can attain strong base-depth ratios only when objects being photographed are at a rather close range. For longer ranges, stereopairs should be obtained by exposing single photos at the ends of a measured base line using cameras of the type described in Sec. 18-4.

Stereometric cameras are leveled for normal horizontal photography by means of level vials. Besides horizontal orientation of the cameras, however, special attachments available with most stereometric cameras permit the two cameras to be aimed vertically downward, vertically upward, or at various oblique angles in between. In each case the two camera axes remain parallel. Figure 18-7 shows a horizontal stereopair of terrestrial photos taken with a Wild C-120 stereometric camera. The stereopair was used to assist in reconstructing a traffic accident.

18-6 HORIZONTAL AND VERTICAL ANGLES FROM A HORIZONTAL PHOTO

A horizontal terrestrial photo is obtained if the camera axis is horizontal when the exposure is made. The plane of a horizontal photo is then vertical, and if the camera

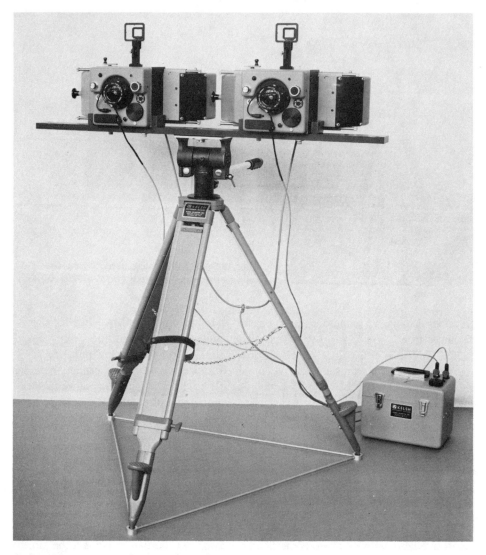

Figure 18-6 Kelsh K-490 Stereometric Camera. (*Courtesy Kelsh Instrument Division, Danko Arlington, Inc.*)

is properly leveled before exposure, the x and y photo axes (defined by the fiducial lines) are horizontal and vertical lines, respectively. As with aerial photos, lines joining opposite fiducial marks should intersect at or very near the photographic principal point.

Figure 18-8 illustrates the positive of a horizontal terrestrial photo exposed at camera station L. The focal length of the camera is f, and o is the photographic principal point. Points A and B in the object space are imaged at a and b on the

Figure 18-7 Horizontal stereopair of terrestrial photos taken for an auto accident investigation with a Wild C-120 stereometric camera. (*Courtesy Dr. T.M. Lillesand.*)

positive, and their photocoordinates are x_a, y_a, x_b, and y_b, respectively. These photocoordinates may be measured by any of the techniques described in Chap. 5. The horizontal angle α_a between the optical axis Lo and the ray to object point A is

$$\alpha_a = \tan^{-1}\left(\frac{x_a}{f}\right) \tag{18-1}$$

Also, the vertical angle of inclination β_a between the horizontal plane and the ray to object point A is

$$\beta_a = \tan^{-1}\left(\frac{y_a}{(x_a^2 + f^2)^{1/2}}\right) \tag{18-2}$$

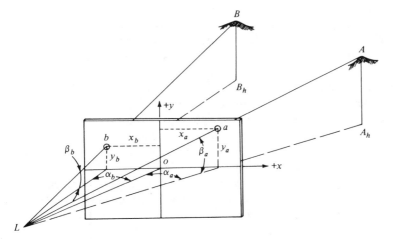

Figure 18-8 Horizontal and vertical angles from measurements on a horizontal terrestrial photo.

In similar manner, the horizontal and vertical angles to point B are

$$\alpha_b = \tan^{-1}\left(\frac{x_b}{f}\right) \tag{18-3}$$

and

$$\beta_b = \tan^{-1}\left(\frac{y_b}{(x_b^2 + f^2)^{1/2}}\right) \tag{18-4}$$

Correct algebraic signs for these angles are automatically obtained by affixing proper algebraic signs to measured photocoordinates according to conventional procedure for rectangular coordinate systems. A horizontal angle to the left of the optical axis is therefore considered negative and one to the right is considered positive. Also, vertical angles above horizontal are positive and those below horizontal are negative. Horizontal angle A_hLB_h subtended at the exposure station by the two object points A and B is

$$A_hLB_h = \alpha_a - \alpha_b \tag{18-5}$$

Horizontal and vertical angles determined by the above equations are the same angles that would be measured using a surveyor's theodolite at the camera station. An advantage of the photographic approach is that one click of the shutter instantaneously produces an infinite number of points to which angles may later be determined. The photo measurements may be made in the comfort of an office, an important consideration if the alternative requires field surveying during times of adverse weather.

Horizontal and vertical angles may also be determined from horizontal terrestrial photos by graphical procedures. The results are likely to be less accurate than angles determined by analytical procedures, but in many cases they are satisfactory. Figure 18-9 illustrates the graphical construction of angles α_a, α_b, β_a, and β_b from the terrestrial photo in Fig. 18-8. To construct horizontal angle α_a, point a_h is located on the x axis by transferring point a parallel to the y axis. Distance oa_h therefore equals photocoordinate x_a. From point o distance Lo (equal to the camera focal length) is laid off perpendicular to the x axis. Line La_h is then drawn, and horizontal angle α_a is thus constructed.

Once α_a has been constructed, vertical angle β_a may be readily determined graphically. A perpendicular to line a_hL is constructed from point a_h and extended a distance equal to coordinate y_a to locate point a_v. Line La_v is then drawn, which yields angle a_vLa_h, the required vertical angle β_a. Angles α_b and β_b are constructed in similar manner, as shown in Fig. 18-9.

The photo may be used directly in graphical construction by overlaying it with a piece of transparent mylar. Another approach is to perform the construction on a separate piece of drafting paper after transferring measured x and y coordinates from the photo to the drafting paper, using dividers or an engineer's scale. Either method avoids defacing the photo.

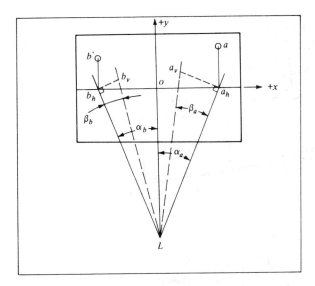

Figure 18-9 Horizontal and vertical angles from a horizontal terrestrial photo by graphical contruction.

18-7 LOCATING POINTS BY INTERSECTION FROM TWO OR MORE HORIZONTAL PHOTOS

If images of an object point appear in two or more horizontal photos, the position and elevation of the point can readily be determined, provided the camera positions and directions of the optical axes are known. Figure 18-10a illustrates two horizontal photos taken from exposure stations L and L'. Figure 18-10b is a plan view of the situation. Images of object point A appear at a and a' on the two photos. Assume that angles δ and δ' of Fig. 18-10b have been measured with respect to the base line, so that relative directions of the optical axes of the two exposure stations are known. Assume also that the horizontal length of the base line has been measured and that the camera station elevations are known. An arbitrary XY object-space coordinate system is adopted with origin at exposure station L and the X axis in the plane of the base line. It is required to determine the X and Y coordinates and elevation of point A.

This problem may be solved analytically or graphically. In the analytical solution, angles α_a, α_a', β_a, and β_a' are calculated from Eqs. (18-1) and (18-2). Then angles ϕ, ϕ', and ϕ'' of Fig. 18-10b are calculated as follows:

$$\phi = \delta - \alpha_a$$

$$\phi' = \delta' + \alpha_a' \qquad (18\text{-}6)$$

$$\phi'' = 180° - \phi - \phi'$$

Applying the law of sines, distance LA in Fig. 18-10b may be calculated as

$$LA = \frac{B \sin \phi'}{\sin \phi''} \qquad (18\text{-}7)$$

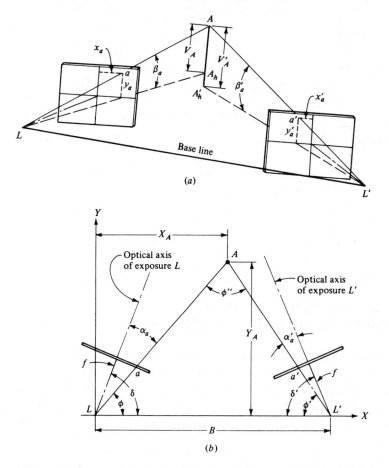

Figure 18-10 (a) Locating points by intersection from two horizontal terrestrial photos. (b) Plan view of intersection from two horizontal terrestrial photos.

Also, by the law of sines,

$$L'A = \frac{B \sin \phi}{\sin \phi''} \tag{18-8}$$

Coordinates X_A and Y_A may then be calculated as

$$X_A = (LA) \cos \phi$$
$$Y_A = (LA) \sin \phi \tag{18-9}$$

A check on these coordinates may be obtained as follows:

$$X_A = B - (L'A) \cos \phi'$$
$$Y_A = (L'A) \sin \phi' \tag{18-10}$$

With reference to Fig. 18-10a, the elevation of point A is determined as

$$\text{Elev } A = \text{elev } L + V_A \qquad (18\text{-}11)$$

where $V_A = LA_h \tan \beta_a$, and LA_h is equal to horizontal length LA found in Eq. (18-7).

A check may also be obtained on the elevation of A as follows:

$$\text{Elev } A = \text{elev } L' + V_A' \qquad (18\text{-}12)$$

where $V_A' = L'A_h' \tan \beta_\alpha'$, and $L'A_h'$ is equal to horizontal length $L'A$ found in Eq. (18-8).

If images of an object point appear on more than two photos, additional check values on position and elevation may be obtained and the averages of several solutions can be adopted.

Example 18-1 As illustrated in Fig. 18-10a and Fig. 18-10b, two horizontal terrestrial photos were taken with a phototheodolite having a focal length of 164.95 mm. The horizontal base line length was 250.0 ft and the exposure station elevations were 862.7 and 855.4 ft above mean sea level for the left and right exposures, respectively. Angles δ and δ' were measured as $69°30'$ and $66°10'$, respectively. Images of object point A were measured on both photos with the following results: $x_a = 46.23$ mm, $y_a = 41.07$ mm, $x_a' = -17.83$ mm, and $y_a' = 48.20$ mm. Calculate the X and Y coordinates of point A in a rectangular coordinate system with origin at L and with the X axis in the plane of the base line, as shown in Fig. 18-10b.

SOLUTION Horizontal angles α_a and α_a' are calculated by Eq. (18-1),

$$\alpha_a = \tan^{-1} \left(\frac{46.23}{164.95} \right) = 15°39'$$

$$\alpha_a' = \tan^{-1} \left(\frac{-17.83}{164.95} \right) = -6°10'$$

Vertical angles β_a and β_a' are calculated by Eq. (18-2),

$$\beta_a = \tan^{-1} \left(\frac{41.07}{[(46.23)^2 + (164.95)^2]^{1/2}} \right) = 13°29'$$

$$\beta_a' = \tan^{-1} \left(\frac{48.20}{[(-17.83)^2 + (164.95)^2]^{1/2}} \right) = 16°12'$$

By Eqs. (18-6), angles ϕ, ϕ', and ϕ'' are

$$\phi = 69°30' - 15°39' = 53°51'$$

$$\phi' = 66°10' - 6°10' = 60°00'$$

$$\phi'' = 180°00' - 53°51' - 60°00' = 66°09'$$

Horizontal distances LA and $L'A$ by Eqs. (18-7) and (18-8) are

$$LA = \frac{250.0 \sin 60°00'}{\sin 66°09'} = 236.7 \text{ ft}$$

$$L'A = \frac{250.0 \sin 53°51'}{\sin 66°09'} = 220.7 \text{ ft}$$

The X_A and Y_A coordinates of point A by Eq. (18-9) are

$$X_A = 236.7 \cos 53°51' = 139.6 \text{ ft}$$

$$Y_A = 236.7 \sin 53°51' = 191.1 \text{ ft}$$

The elevation of point A by Eq. (18-11) is

Elev $A = 862.7 + 236.7 \tan 13°29' = 919.5$ ft above mean sea level

Applying Eqs. (18-10) and (18-12), checks may be obtained for the X_A and Y_A coordinates and for the elevation of A as follows:

$$X_a = 250.0 - 220.7 \cos 60°00' = 139.6 \text{ ft (Check!)}$$

$$Y_A = 220.7 \sin 60°00' = 191.1 \text{ ft (Check!)}$$

$$\text{Elev } A = 855.4 + 220.7 \tan 16°12' = 919.5 \text{ (Check!)}$$

The map position and elevation of point A may also be determined graphically. In this approach camera stations L and L' are plotted on a manuscript map with their optical axes also drawn in their correct directions, as illustrated in Fig. 18-11. With the apex of construction taken at the two exposure stations, horizontal angles α_a and α_a' are constructed as described in Sec. 18-6. Corresponding rays La_h and L_{ah}' are extended to their intersection, which locates the map position of point A.

Vertical angles β_a and β_a' may also be constructed, as shown in Fig. 18-11, and rays La_v and L_{av}' extended, as illustrated. A perpendicular to line LA is established at A and extended until it intersects with extended line La_v. The distance from A to this intersection is V_A, the elevation of point A above camera station L. The elevation of point A is obtained by adding the scaled value of V_A to the elevation of station L. A check on the elevation of A is obtained by graphically determining V_A' and adding its scaled value to the elevation of camera station L'.

18-8 PARALLAX EQUATIONS

If a stereopair of *horizontal* terrestrial photos is exposed with the two camera axes perpendicular to the base line, parallax equations—similar to Eqs. (8-5) through (8-7) for vertical aerial photos—may be developed for calculating positions and elevations of points in the overlap area. Two cases must be considered in the terrestrial situation: (1) where the exposing cameras have equal elevations, and (2) where they have unequal elevations.

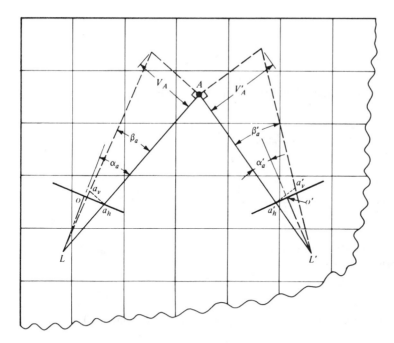

Figure 18-11 Map location of points by graphical intersection from two horizontal terrestrial photos.

18-8.1 Equal Elevations of Camera Stations

Figure 18-12 illustrates a stereopair of horizontal photos exposed from camera stations L and L' *having the same elevations.* An arbitrary XYZ object-space coordinate system is adopted with origin at exposure station L. The Y axis is horizontal and coincident with the optical axis of the left photo. X is horizontal and coincides with the base line, and Z points vertically upward.

Object point A appears in the overlap area of the stereopair. Its photocoordinates measured with respect to the fiducial axis systems are x_a and y_a on the left photo and x_a' and y_a' on the right photo. Note that for the adopted image and object axes, object X and image x are parallel, object Y and image z are parallel, and object Z and image y are parallel.

Parallax equations for calculating object space coordinates X_A, Y_A, and Z_A may be developed by equating similar triangles of Fig. 18-12. From similar triangles Lom and LOM,

$$\frac{X_A}{Y_A} = \frac{x_a}{f} \quad \text{from which } X_A = \frac{x_a}{f} Y_A \qquad (a)$$

Also, from similar triangles $L'o'm'$ and $L'O'M'$,

$$\frac{B - X_A}{Y_A} = \frac{-x_a'}{f} \quad \text{from which } X_A = B + \frac{x_a'}{f} Y_A \qquad (b)$$

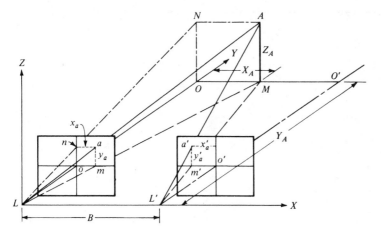

Figure 18-12 Stereopair of horizontal terrestrial photos exposed with camera axes parallel from exposure stations of equal elevation.

Equating (a) and (b), substituting $p_a = x_a - x_a'$, and reducing,

$$Y_A = \frac{Bf}{p_a}$$ (18-13)

Now substituting Eq. (18-13) into (a),

$$X_A = \frac{Bx_a}{p_a}$$ (18-14)

From similar triangles *Lon* and *LON*,

$$\frac{Z_A}{Y_A} = \frac{y_a}{f}$$ (c)

Substituting Eq. (18-13) into (c),

$$Z_A = \frac{By_a}{p_a}$$ (18-15)

Equations (18-13), (18-14), and (18-15) yield X, Y, and Z object-space coordinates of any points whose parallaxes have been measured from a stereopair of terrestrial photos. If the assumed conditions of (1) horizontal photos, (2) equal camera station elevations, and (3) camera axes perpendicular to the base line are met, these equations are exact and very convenient. In practice, a properly leveled stereometric camera exposes stereopairs which automatically meet the assumed conditions. Also, if the terrain is nearly flat, these conditions could be established using a phototheodolite. The parallax equation approach has the advantages of simplified calculations and ease of measuring parallax, especially if a parallax bar is used. When a parallax bar is used, the photos are oriented with their x axes along a common line, and parallax measurements are taken as described in Sec. 8-5.

Example 18-2 A horizontal stereopair was exposed with a stereometric camera having a fixed base of 120 cm and a focal length of 64.00 mm. Images of object point A have photocoordinates $x_a = 32.41$ mm and $y_a = 23.74$ mm on the left photo and $x_{a'} = 28.06$ mm and $y_{a'} = 23.73$ mm on the right photo. Calculate the X, Y, and Z coordinates of point A. Find the elevation of point A if the cameras were at elevation 941.8 when the exposure was made.

SOLUTION The parallax of point a is

$$p_a = x_a - x_{a'} = 32.41 - 28.06 = 4.35 \text{ mm}$$

From Eqs. (18-13) through (18-15),

$$Y_A = \frac{(120.0)(64.00)}{4.35} = 1,766 \text{ cm} = 17.66 \text{ m}$$

$$X_A = \frac{(120.0)(32.41)}{4.35} = 894 \text{ cm} = 8.94 \text{ m}$$

$$Z_A = \frac{(120.0)(23.74)}{4.35} = 655 \text{ cm} = 6.55 \text{ m}$$

Elev $A = 941.8 + 6.55 \times 3.28 \text{ ft/m} = 963.3$ ft above datum

18-18.2 Unequal Elevations of Camera Stations

In many cases, usually because of terrain variations, equal camera heights are impossible. A stereopair taken with unequal camera heights is illustrated in Fig. 18-13. Even though the cameras are not at equal elevations, if the photos are horizontal and exposed with their optical axes parallel, parallax Eqs. (18-13) through (18-15) may still be used. Like the case of equal camera height, the object-space coordinate system has its origin at the left exposure station. As shown in Fig. 18-13, the Y axis is coincident with the optical axis of the left exposure, X is horizontal and in the vertical plane containing the base line, and Z points vertically upward.

Equations (18-13) through (18-15) may be used without modification for horizontal stereopairs taken with unequal camera elevations. Parallaxes and photocoordinates, however, must be calculated with respect to the fiducial axis system. These are not true parallaxes but rather projections of true parallaxes onto the X axis. If a parallax bar is used for parallax measurement, the photos must be oriented as illustrated in Fig. 18-14. The line joining principal points and conjugate principal points makes an angle ξ with the x photo axes. Angle ξ is the inclination of the base line

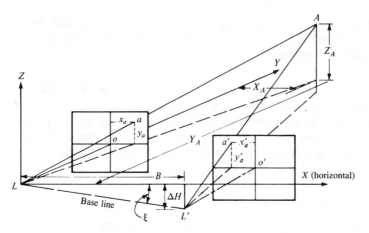

Figure 18-13 Stereopair of horizontal terrestrial photos exposed with camera axes parallel but from exposure stations of unequal elevation.

LL' with the horizontal, and it may either be scaled from the photos or calculated as follows:

$$\xi = \tan^{-1}\left(\frac{\Delta H}{B}\right) \tag{18-16}$$

In Eq. (18-16), ΔH is the difference in elevation of the two exposure stations and B is the horizontal distance between them. Parallaxes measured with a parallax bar must be multiplied by the cosine of ξ to obtain projections of true parallaxes onto the X axis. Photocoordinates x_a and y_a for Eqs. (18-14) and (18-15) are measured with respect to the fiducial x axis, and the value of B used in Eqs. (18-13) through (18-15) is the horizontal distance between the two exposure stations.

If the camera axes for a stereopair are perpendicular to the base line but the camera x axes are not horizontal, parallax equations can still be used. In this case, however, principal points and conjugate principal points must be located to define the "photo base line" axes in the same manner that "photo flight line" axes are located for the aerial photographic case. Parallaxes and photocoordinates are then measured with respect to the photo base line axes. If the camera elevations are unequal, these parallaxes and photocoordinates must be projected onto an x axis parallel to the horizontal ground axis by multiplying them by the cosine of ξ, which has been determined by Eq. (18-16).

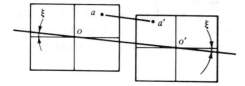

Figure 18-14 Orientation of photos of Fig. 18-13 for parallax measurement with a parallax bar.

18-9 OBLIQUE TERRESTRIAL PHOTOS

To center objects in the field of view, it is frequently necessary to incline the camera axis either up or down from the horizontal. The resulting photos are oblique terrestrial photos. With many terrestrial cameras the angle of inclination can be set or measured so that it is a known quantity. Figure 18-15 illustrates an oblique terrestrial photo taken with the camera axis inclined downward at a *depression* angle θ. To conform with usual sign convention, depression angles are considered negative in algebraic sign. If the camera axis is inclined upward, θ would be an *elevation* angle and considered positive.

In Fig. 18-15, L is the exposure station, and Lo is f, the camera focal length. Lk is a horizontal line intersecting the photo at k. If the camera is leveled before taking a photo, the x axis will be horizontal. A line in the plane of the photo through k and parallel to the x axis is also horizontal and termed the *horizon line*. The horizon line is the x' axis, and the line to which y' photocoordinates are referred. The y'_a coordinate of image point a is given by

$$y'_a = y_a + f \tan \theta \qquad (18\text{-}17)$$

In Eq. (18-17) the correct algebraic sign must be applied to θ. Also if an image point lies above the x' axis, its y' coordinate is considered positive; if it is below the x axis, its y' coordinate is given a negative algebraic sign. In Fig. 18-15, aa' is a vertical line and a' lies in the horizontal plane of the exposure station and horizon line. Horizontal angle α_a between the vertical plane containing image point a and the vertical plane containing the camera axis is

$$\alpha_a = \tan^{-1} \left(\frac{ha'}{Lk - hk} \right) = \tan^{-1} \left(\frac{x_a}{f \sec \theta - y'_a \sin \theta} \right) \qquad (18\text{-}18)$$

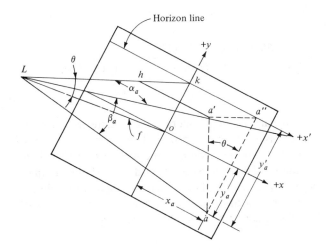

Figure 18-15 Horizontal and vertical angles from measurements on an oblique terrestrial photo.

In Eq. (18-18) it should also be noted that correct algebraic signs must be applied to x_a, y'_a, and θ. Algebraic signs of α angles are positive if they are clockwise from the optical axis and negative if they are counterclockwise.

After the horizontal angle α_a, has been determined vertical angle β_a to image point a can be calculated from the following equation:

$$\beta_a = \tan^{-1}\left(\frac{aa'}{La'}\right) = \tan^{-1}\left(\frac{aa'}{(Lk - hk)\sec\alpha_a}\right)$$

$$= \tan^{-1}\left(\frac{y'_a \cos\theta}{(f\sec\theta - y'_a \sin\theta)\sec\alpha_a}\right) \qquad (18\text{-}19)$$

The algebraic signs of β angles are automatically obtained from the signs of the y' coordinates used in Eq. (18-19).

Example 18-3 An oblique terrestrial photo was exposed with the camera axis depressed at an angle of 30^g $(-27°00')$. The camera focal length was 60.00 mm. Compute the horizontal and vertical angles to an object point A whose image has photocoordinates $x_a = 27.41$ mm and $y_a = -34.90$ mm measured with respect to the fiducial axes.

SOLUTION From Eq. (18-17) the y' coordinate of point a is

$$y'_a = -34.90 + 60.00 \tan(-27°00') = -65.47 \text{ mm}$$

From Eq. (18-18),

$$\alpha_a = \tan^{-1}\left(\frac{27.41}{60.00 \sec(-27°00') - (-65.47)\sin(-27°00')}\right) = 36°05'$$

From Eq. (18-19),

$$\beta_a = \tan^{-1}$$

$$\left(\frac{-65.47 \cos(-27°00')}{[60.00 \sec(-27°00') - (-65.47)\sin(-27°00')]\sec 36°05'}\right)$$

$$= -51°25'$$

If images of an object point appear on two or more oblique terrestrial photos whose camera positions and orientations are known, horizontal and vertical positions of the object point may be determined. The procedure is the same as described in Sec. 17-8 for high oblique aerial photos. Also the perspective grid method of mapping, as described in Sec. 17-9 for high oblique aerial photos, may be applied to oblique terrestrial photos. The procedure assumes a flat object-space plane and requires that the true horizon can be located. If the apparent horizon is not visible on the photo to enable locating the true horizon, it may be possible to locate the true horizon using the "vanishing point" or "nadir point" methods described in Sec. 17-11. The perspective grid method of mapping has proved very useful in certain areas such as mapping highway collisions. In this case a flat roadway may be taken as the object *XY* plane, and parallel road edges and cross joints may often be used for obtaining the true horizon by the vanishing point method.

18-10 EXPOSURE STATION LOCATION AND CAMERA AXIS DIRECTION

Sometimes the location of the exposure station and camera axis direction are unknown for a terrestrial photo and must be determined. A simple and convenient method for locating the horizontal position of the exposure station and the direction of the optical axis is *three-point resection*. It may be done either graphically or numerically, but in order to obtain the solution, the θ angle must be known and images of at least three horizontal control points must appear in the photo. Using Eq. (18-1), or (18-18) if the photo is oblique, horizontal angles between the camera axis and rays to the three control points are calculated. In the graphical three-point resection procedure described in Sec. 9-8, the three control points are plotted to scale on a base map according to their ground coordinates. A transparent template containing the three rays and the camera axis is prepared based on the calculated horizontal angles. The template is placed on the base map and adjusted in position until the three rays simultaneously pass through their respective plotted control points. The exposure station position and direction of the optical axis are then marked on the map by pinpricking through the template. Numerical techniques of three-point resection are discussed in most textbooks on surveying and shall not be discussed herein.

The elevation of the exposure station is the height of the camera lens above datum. If the elevation of the occupied station is known, the elevation of the camera lens is usually determined by measuring the vertical distance from the ground point up to the camera lens and adding it to the elevation of the ground point. If the exposure station elevation is unknown, it may be determined from a vertical control point, provided that the horizontal position of the camera and direction of its optical axis are known.

Assume that the position and elevation of point *A* in Fig. 18-16 are known. Vertical angle β_a to control point *A* is calculated from Eq. (18-2), or Eq. (18-19) if the photo is oblique. With the horizontal distance LA_h known, the camera station elevation may be calculated from the following equation:

$$\text{Elev } L = \text{elev } A - LA_h \tan \beta_a \qquad (18\text{-}20)$$

If more than one control point is available, the average of exposure station elevations determined from all control points is adopted. If possible, it is best to measure the position and orientation of the camera at the time of photography so that the calculations described in this section are unnecessary.

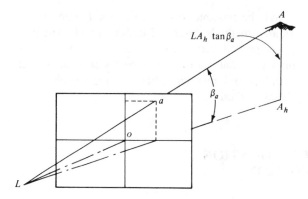

Figure 18-16 Determining elevation of camera station of terrestrial photo using one vertical control point.

18-11 STEREOSCOPIC PLOTTING INSTRUMENTS

In addition to analytical and graphical techniques, stereoscopic plotting instruments may be used to determine object point positions from stereopairs of terrestrial photos. Although terrestrial photos can be used in all plotters described in Chap. 12, there are certain critical limitations with many of them. One of the basic problems is that Y coordinates (depth) of horizontal and near-horizontal terrestrial stereomodels correspond to Z coordinates (elevation) of aerial stereomodels. Normally the depth range required in terrestrial stereomodels far exceeds the elevation range encountered with aerial stereomodels. Many stereoplotters (including all direct optical projection instruments) designed specifically for aerial photography lack the depth range required for terrestrial photos because of both optical and mechanical limitations. If the depth range of a terrestrial stereopair is small, instruments designed for aerial photos may be satisfactory.

Another problem in accommodating terrestrial photos in direct optical projection plotters is that terrestrial camera focal lengths are not generally within the range of projector principal distance accommodation. Diapositives can be prepared with proper principal distances regardless of camera focal length (see Sec. 12-7.1). Printers are not commonly available, however, which handle the large variations in enlargement or reduction ratios that can result from the large disparities between terrestrial camera focal lengths and projector principal distances.

Most mechanical projection instruments are capable of accommodating a wide range in principal distances, so that diapositives are generally more easily prepared for them. In some cases contact-printed diapositives may be used. A number of stereoscopic plotters are designed to handle both aerial and terrestrial photos. With these instruments the Y and Z drives can be interchanged.

If the stereoplotter has dials for setting the projector rotations and translations and if the relative orientation and base line length are known for the stereopair, relative and absolute orientation can be accomplished by setting the dials to the known values. Difficulties will be encountered in relative and absolute orientation if the instrument lacks dials and if a large portion of the stereomodel lies above the horizon and is void of imagery. In these cases model points for relative orientation, and control points for absolute orientation, will not exist in the conventionally desirable locations of the stereomodel. With laboratory photography, this problem can often be overcome by placing artificial targets so that their images appear near the principal points and corners of the stereomodels.

Some stereoplotters have been designed specifically to handle terrestrial photography. The Wild *A-40* of Fig. 18-17 and the Jenoptik *Technocart* of Fig. 18-18 are examples. These mechanical projection instruments will handle only stereopairs taken with stereometric cameras or those obtained from single-camera setups at the ends of a base line with their optical axes aligned parallel. This is because their carriers provide for no rotational motions. The A-40 will accommodate principal distances in the range of from 54 to 100 mm and will accept diapositives up to 92 by 125 mm in size. The Technocart has a principal distance range of from 50 to 215 mm and can handle diapositives of a size up to 9-in square. The Zeiss *Terragraph* is another instrument designed specifically for plotting from terrestrial photos.

18-12 CONTROL FOR TERRESTRIAL PHOTOGRAMMETRY

In terrestrial photogrammetry the object space is often relatively close to the camera; in fact, in many cases the photographs are taken in laboratories. Object distances vary

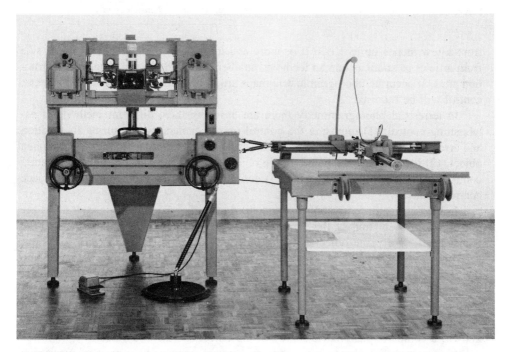

Figure 18-17 Wild A-40 Autograph for plotting from terrestrial stereopairs. (*Courtesy Wild Heerbrugg Instruments, Inc.*)

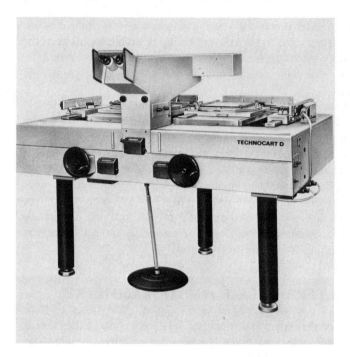

Figure 18-18 Jenoptik Technocart-D Terrestrial Stereoplotter. (*Courtesy Jenoptik Jena.*)

from a few inches up to 1,000 ft or more and the objects photographed vary in size from articles as small as human teeth and smaller, to very large buildings or construction sites. If accurate photogrammetric maps are to be made of photographed objects, control will be required.

In terrestrial photogrammetry there are basically three different methods of establishing control: (1) imposing the control on the camera by measuring its position and orientation with respect to a coordinate system or with respect to the photographed object, (2) locating control points in the object space in a manner similar to locating control for aerial photography, and (3) combining camera control and object space control points.

In the first method, no control points need appear in the object space. Rather, the position and orientation of the camera or cameras are measured with respect to the object itself. If a plane object is being photographed from a single camera station, control requirements may be satisfied by measuring the distance from the camera to the plane surface and orienting the camera optical axis perpendicular to the surface. Perpendicular orientation can be accomplished by mounting a plane-surfaced mirror parallel to the object plane and then moving the camera about until the reflection of the camera lens occupies the center of the field of view. If the camera focal length is known, a complete planimetric survey of the object can then be made.

If stereopairs of photos are taken, the control survey can consist of measuring the horizontal distance and difference in elevation between the two camera stations and also determining the orientations of the camera optical axis for each photo. Phototheodolites enable a complete determination of camera orientation and direction of optical axis. Stereometric cameras automatically provide control by virtue of their known base-line length and relative orientation. In exposing stereopairs with less elaborate cameras, horizontal orientation can be enforced by using level vials and parallel orientation of the camera axes can be accomplished by reflection from parallel mirrors.

In the second method of controlling terrestrial photos, points should be selected in the object space which provide sharp and distinct images in favorable locations in the photographs. Their positions in the object space should then be carefully measured. If no satisfactory natural points can be found in the object space, artificial targets may be required. Targets should be designed so that their images appear sharp and distinct in the photos. White crosses on black cards may prove satisfactory. If the object space is small and the control points are close together, measurements for locating the targets may be made directly by means of graduated scales. If the object space is quite large or if the control points are inaccessible for direct measurement, triangulation with precise theodolites set up at the ends of a carefully measured base line may be necessary. In some cases a premeasured grid pattern may be placed in the object space and photographed along with the object, thereby affording control.

If the object being photographed is stationary, control points may be located on the object. Corners of window frames, for example, may be used if a building is being photographed. If a dynamic occurrence is being photographed at increments of time, e.g., photographing beam deflections under various loads, then targets may have to be mounted on some stationary framework apart from the object. By means of engineer's levels, targets may be set at equal elevations, thereby providing a horizontal line in the object space. Vertical lines may be easily established by hanging plumb bobs in the object space and attaching targets to the string.

The third method of controlling terrestrial photography is a combination of the first two methods. This third approach is generally regarded as prudent because it provides redundancy in the control which prevents mistakes from going undetected and also enables increased accuracy to be obtained.

18-13 A GENERAL SOLUTION

A general solution to any problem in terrestrial or close-range photogrammetry can be obtained by applying collinearity condition equations. Collinearity, as explained and developed in Appendix C, expresses the condition that an exposure station, any object point, and its corresponding image all lie on a straight line. The terrestrial collinearity condition is illustrated in Fig. 18-19. The equations for terrestrial photos are similar to Eqs. (C-5) and (C-6) of Appendix C for aerial photos; in fact, they can be written by inspection from these equations after comparing axis systems of the aerial and terrestrial cases.

If the terrestrial image and object space are defined as shown in Fig. 18-19, and if rotation angles omega, phi, and kappa are taken as counterclockwise rotations about

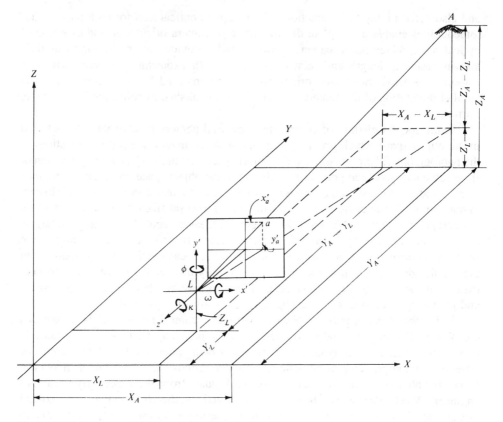

Figure 18-19 Geometry of the terrestrial collinearity condition.

image axes x, y, and z, respectively, then there is only a slight difference between the terrestrial collinearity equations and the aerial collinearity equations. As shown in Fig. C-2, in developing collinearity for an aerial photograph, transformed image space axes x', y', and z' are parallel to the X, Y, and Z object-space axes, respectively. In the terrestrial case, however, image axis x' is parallel to object axis X, but image axes y' and z' are parallel to object axes Z and Y, respectively, as shown in Fig. 18-19. Following a development similar to that given for the aerial photo in Sec. C-3 of Appendix C, the terrestrial collinearity equations are obtained as follows:

$$x_a = -f \frac{m_{11}(X_A - X_L) + m_{12}(Z_A - Z_L) + m_{13}(Y_L - Y_A)}{m_{31}(X_A - X_L) + m_{32}(Z_A - Z_L) + m_{33}(Y_L - Y_A)} \quad (18\text{-}21)$$

$$y_a = -f \frac{m_{21}(X_A - X_L) + m_{22}(Z_A - Z_L) + m_{23}(Y_L - Y_A)}{m_{31}(X_A - X_L) + m_{32}(Z_A - Z_L) + m_{33}(Y_L - Y_A)} \quad (18\text{-}22)$$

In Eqs. (18-21) and (18-22), the m's are elements of the rotation matrix and their values are given by Eq. (B-21) of Appendix B.

Terrestrial collinearity Eqs. (18-21) and (18-22) are nonlinear and must be linearized using Taylor's theorem. Linearization follows the same steps as outlined in Sec. C-4 for linearizing the aerial collinearity equations and is left to the student.

The terrestrial collinearity equations are general and contain the six elements of exterior orientation (omega, phi, kappa, X_L, Y_L, and Z_L) as well as object-space coordinates X_A, Y_A, and Z_A of points whose images appear in the photograph. These equations are applicable to almost any problem in terrestrial or close-range photogrammetry, including *space resection* to determine the position and orientation of a photograph or *relative orientation* and *absolute orientation* to determine coordinates of points whose images appear in the overlap area of a stereopair. The equations are applicable for overlapping photos regardless of camera orientation.

18-14 X-RAY PHOTOGRAMMETRY

X-rays are invisible energy rays having a wavelength about 10,000 times shorter than visible energy. Because of their short wavelength, x-rays are capable of penetrating substances which normally absorb or reflect visible light. The shorter the x-rays, the greater their penetrating ability. X-rays are produced electronically in an x-ray tube. As illustrated in Fig. 18-20, when x-rays are directed toward an object such as the human body, and if they pass through, they create a latent image on the emulsion of a film held in a cassette below the object. Upon developing the exposed film, an x-ray negative called a *radiograph* is obtained. Bones or other solid objects which do not easily allow passage of x-rays cause less energy to reach the film; hence these areas show up light on the radiograph negative. Monoscopic measurements may be taken from individual radiographs, or objects may be viewed stereoscopically and

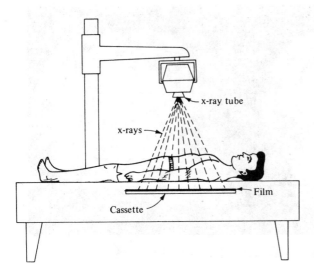

Figure 18-20 Exposing x-ray radiographs.

measured three-dimensionally from stereopairs of radiographs. A stereoradiograph is shown in Fig. 18-21.

X-rays are used extensively in the medical profession to locate broken bones; to find foreign objects in the body such as bullets, pins, or tacks; to detect the presence of tumors, ulcers, and gallstones; and to examine the heart, lungs, liver, stomach, and digestive tract for disease or defects. Through stereoradiography, the position and depth in the body of defects can be accurately determined. X-rays are commonly used in dentistry for locating and determining sizes of cavities, for examining the roots of teeth, and for determining jawbone structure. X-rays are also used by orthodontists to plan and monitor corrective treatments. In industry, x-rays have a host of applications, including examination of radioactive fuels, castings, and welds for defects; locating pipes and wires in buildings; and inspecting tires, radio tubes, etc.

Although radiographs may be analyzed using photogrammetric principles, there are several basic differences between radiographs and photographs. In radiography, for example, energy radiates through the object and onto the film, while in photography energy reflects from objects, is collected by a lens, and is focused onto the film. Also, images of radiographs are usually near the actual size of their corresponding objects, whereas in photography images are generally much smaller than the object they represent.

Figure 18-22*a* illustrates a side view of the geometry of the *normal case* of a stereopair of radiographs, and Fig. 18-22*b* and *c* show the left and right radiographs, respectively. In the normal case, x-ray tubes at T and T' are equidistant from the film plane and the object reference plane and film plane are parallel. These conditions can be established in x-ray photography. *Object distance D* (perpendicular distance from the object reference plane to the focal spots of the x-ray tubes) and *image distance d* (perpendicular distance from the film plane to the focal spots of the x-ray tubes) are normally known. The focal spots of the x-ray tubes from which the x-rays emanate

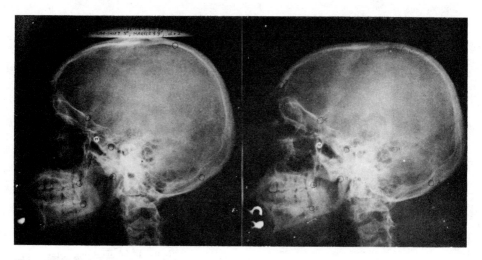

Figure 18-21 Stereopair of radiographs of human skull. (*Courtesy Dr. R.S. Singh.*)

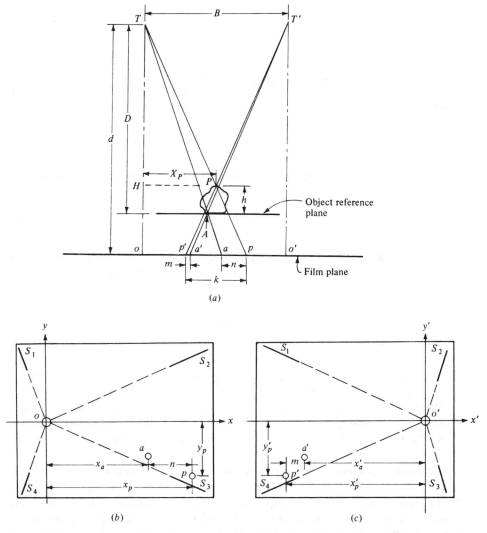

Figure 18-22 (a) Geometry of the normal case of stereoradiographs. (b) Left radiograph of stereopair; (c) right radiograph of stereopair.

are separated by a base B, which is also assumed known. In Fig. 18-22b and c, points o and o' are the principal points of the radiographs (points where a perpendicular to the film plane from the focal spots intersect the film plane). Point A is a reference point situated on the object reference plane, and point P is any point in the object whose three-dimensional position is desired. From similar triangles TAT' and $a'Aa$ of Fig. 18-22a,

$$\frac{B}{D} = \frac{k - (m + n)}{d - D} \qquad \text{from which} \quad k = \frac{B(d - D)}{D} + (m + n) \qquad (d)$$

Also, from similar triangles TPT' and $p'Pp$,

$$\frac{B}{D - h} = \frac{k}{h + d - D} \qquad \text{from which } k = \frac{B(h + d - D)}{D - h} \qquad (e)$$

Let Δp equal $m + n$, the difference in parallax between points P and A. Equating (d) and (e), substituting Δp for $m + n$, and reducing, the following equation is obtained for the height h of point P above the object reference plane:

$$h = \frac{D\Delta p}{Bd/D + \Delta p} \qquad (18\text{-}23)$$

In practice, since d is approximately equal to D and Δp is generally small compared to B, the following approximate relationship is obtained:

$$h = \frac{D\Delta p}{B} \text{ (approx)} \qquad (18\text{-}24)$$

The following expression for calculating the X object space coordinate of point P is obtained from similar triangles TPH and Tpo of Fig. 18-22a:

$$X_P = \frac{x_p(D - h)}{d} \qquad (18\text{-}25)$$

Similarly, an expression for the Y object space coordinate of point P is

$$Y_P = \frac{y_p(D - h)}{d} \qquad (18\text{-}26)$$

In Eqs. (18-25) and (18-26), the object space coordinate system is in the object reference plane, the origin occurs where line To of the left radiograph intersects the object reference plane, and the X axis is the intersection of plane $Too'T'$ with the object reference plane.

Radiographs contain no fiducial marks for locating principal points and coordinate axes, and therefore they must be located by some other means if Eqs. (18-25) and (18-26) are to be used. [Note that Eqs. (18-23) and (18-24) require measurement only of parallax differences from the radiographs and therefore it is not necessary to accurately locate principal points and coordinate axes if a parallax bar is used.] One simple method of locating principal points is by relief displacements. In this method, four thin wire studs are mounted perpendicular to the object reference plane so that they appear in the corners of the radiographs. As illustrated in Fig. 18-22b and c, these studs appear as lines on the radiographs at S_1, S_2, S_3, and S_4. Since the four lines are radial from the principal point, the principal points are located by either graphically or numerically extending the four lines to their common intersection. The intersection of any two of these lines will uniquely locate the principal point, but if all four lines are used in a numerical solution, least squares techniques can be used to obtain an improved solution. The x and x' axes are taken along the line joining o and o'.

In practice, stereoradiographs have been taken in the following four different ways: (1) radiograph of a stationary object using two stationary x-ray tubes; (2) exposing the left radiograph of a stationary object using one x-ray tube, then translating the tube to the right for the second exposure: (3) using one stationary x-ray tube and translating the object between exposures; and (4) using one stationary x-ray tube and rotating the object (which is on a tilting reference object platform) between exposures. Regardless of the method used, unless a simultaneous double exposure is made in method 1, there will be some elapsed time between exposures while the cassete is removed and replaced with a cassette containing unexposed film for the second radiograph, and while the x-ray tube or object is moved into position for the second radiograph. If the object moves with respect to its reference platform during the elapsed time between exposures, y parallax and other errors will be introduced into the system. Methods are available to compensate for these errors; however, they are rather complex and are beyond the scope of this text.

Radiograph image coordinates and parallaxes and parallax differences may be measured monoscopically to discrete points using any of the scales described in Chap. 5. Parallaxes and parallax differences may also be measured stereoscopically by parallax bar with the advantages that principal points and coordinate axes are not needed and nondiscrete points may be measured. Instruments have been designed specifically for making measurements of radiographs. The Zeiss STR 1-3 shown in Fig. 18-23 is

Figure 18-23 Zeiss STR 1-3 Stereoscopic System for measuring radiographs. (*Courtesy Carl Zeiss, Oberkochen.*)

such an instrument. Its stages will accommodate radiographs up to 40 cm square, and they can be viewed stereoscopically under magnification. The instrument is capable of making accurate measurements of x and y coordinates as well as parallax.

It is impossible to do the subject of x-ray photogrammetry justice in a short section such as this. Nevertheless this brief introduction explains some basic approaches in determining space positions of points and indicates some problem areas in x-ray photogrammetry. Students interested in further study in this area are referred to some of the references cited at the end of this chapter.

18-15 HOLOGRAMMETRY

Hologrammetry, a relatively new development within the discipline of photogrammetry, uses *holograms* rather than photographs. Unlike the photographic approach, which requires a stereopair of photos to achieve a three-dimensional effect, with hologrammetry a three-dimensional image can be obtained from a single hologram. These three-dimensional images, like the photogrammetric stereomodel, may be measured and mapped. Most applications of hologrammetry have been directed primarily to close-range (laboratory) applications, and for this reason the subject is presented in this chapter.

The science of hologrammetry is based upon the wave theory of light, which states that light is transmitted in regular sinusoidal oscillations, as described in Sec. 2-1. Because light transmits sinusoidally, it is possible to produce from two light sources brighter light by adding wave crest to wave crest, and it is also possible to produce darkness from the same two light sources by adding wave crest to wave trough. This principal, called *interference,* is illustrated in Fig. 18-24. For light to be capable of interference, it must be coherent and monochromatic; i.e., it must consist of energy of a uniform wavelength. Coherent light used in hologrammetry is commonly produced by *lasers*.

To produce holograms, two coherent light beams are produced from a single source by means of a beam splitter. In Fig. 18-25a, for example, coherent light is

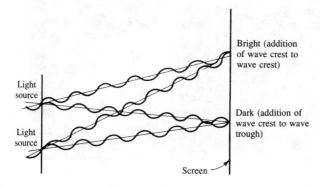

Figure 18-24 Principle of light interference.

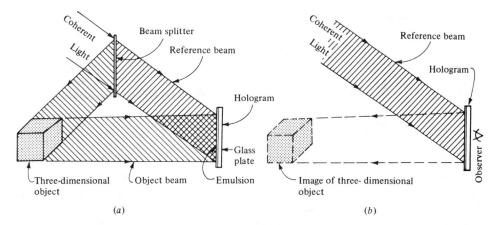

Figure 18-25 (*a*) Recording a hologram. (*b*) Reconstructing a three-dimensional virtual image from a hologram.

directed toward a beam splitter. Part of the light (reference beam) travels directly to the hologram, and part reflects from the object (object beam) and also impinges onto the hologram. The interference patterns created by interactions of reference and object beams are recorded on the hologram emulsion. The emulsion is developed in a darkroom to obtain a usable hologram.

Figure 18-25*b* illustrates reconstruction of the three-dimensional image of an object from the hologram. In reconstruction, a duplicate of the reference beam is directed toward the hologram at the same angle that it had in recording the hologram. An observer behind the hologram perceives a three-dimensional image of the object from the combination of the reference beam and interference patterns of the hologram. To an observer, the image appears exactly the same as if the original object were directly viewed from behind a window. The image may be measured or mapped through the use of a self-illuminated measuring mark and tracing system such as that of an optical projection stereoplotter.

The physical sizes of objects that can be directly recorded on holograms are limited by the size of the hologram plate. This is one of the reasons why holography has to date been used primarily in mapping small objects. If the same wavelength reference beam is used in recording and in reconstruction, the virtual image will be the same size as the original object. Magnification of the reconstructed image is possible if longer-wavelength coherent light is used in reconstruction than was used to record the hologram. Hologrammetry may be used for topographic mapping, but this procedure consists of first obtaining conventional stereophotography and then recording holograms of stereomodels created from these overlapping photos.

Hologrammetry possesses both advantages and disadvantages over conventional photography for mapping. Principal among its advantages is that, since there are no lenses involved in either recording or in reconstruction, there are practically no optical limitations such as lens distortions, aberrations, focusing problems, or depth of field limitations. The last two advantages make hologrammetry especially advantageous over conventional photogrammetry for very close-range work. Other advantages of hologrammetry are that the three-dimensional model can be recorded on a single plate without relative orientation except for positioning the hologram relative to the reconstruction beam. If the hologram plate should be accidentally broken, the entire hologrammetric model can be recovered from any of the pieces with only a slight loss of resolution. Also, since the hologram contains only interference patterns and no imaged points, scratches or blemishes on the plate cause no particular difficulties.

Disadvantages of holographic recording systems are that they must use coherent monochromatic light, they require extreme stability, and their exposure times are relatively long. These requirements confine holographic systems to laboratories. Photographic systems, on the other hand, are very mobile and can operate in a variety of environments. Emulsions are available which can record energy from the ultraviolet to near-infrared ranges of the spectrum, and any light source, including incoherent and monochromatic, can be used. Exposure times may be very short and cameras can be readily carried in aircraft for photographing terrain from high vantage points. Of course, one great advantage of photography is that the resulting images appear true and recognizable and can therefore be readily interpreted.

REFERENCES

Adamec, A.: Let's Not Forget Terrestrial Photogrammetry, *Australian Surveyor*, vol. 26, no. 3, p. 172, 1974.

Agnard, J.: Canadian Contribution to Hologrammetry, *Photogrammetric Engineering and Remote Sensing*, vol. 42, no. 3, p. 343, 1976.

American Society of Photogrammetry: "Manual of Photogrammetry," 3d ed., Falls Church, Va., 1966, chap. 19.

——: "Handbook of Non-Topographic Photogrammetry," Falls Church, Va., 1979.

——: "Manual of Photogrammetry," 4th ed., Falls Church, Va., 1980, chap. 16.

Bopp, H., and H. Krauss: An Orientation and Calibration Method for Non-Topographic Applications, *Photogrammetric Engineering and Remote Sensing*, vol. 44, no. 9, p. 1191, 1978.

Borchers, P. E.: The Photogrammetric Study of Structural Movements in Architecture, *Photogrammetric Engineering*, vol. 30, no. 5, p. 809, 1964.

Brown, D. C.: Close-Range Camera Calibration, *Photogrammetric Engineering*, vol. 37, no. 8, p. 855, 1971.

Burgess, G. H., and J. Zulqar-Nain: Dental Research Using a Close-Range System, *Photogrammetric Engineering*, vol. 34, no. 7, p. 677, 1968.

Eastman Kodak Co.: "Fundamentals of Radiography," Rochester, N.Y., 1960.

Fraser, C. S.: Atmospheric Refraction Compensation in Terrestrial Photogrammetry, *Photogrammetric Engineering and Remote Sensing*, vol. 45, no. 9, p. 1281, 1979.

Fryer, J. G., et al.: Underwater 35 mm Photogrammetry, *Australian Surveyor*, vol. 29, no. 7, p. 461, 1979.

Garbor, D.: Holography, 1948–1971, *Science*, vol. 177, no. 4046, p. 299, 1972.

Gates, J. W.: Position and Displacement Measurement by Holography and Related Techniques, *Photogrammetric Record*, vol. VIII, no. 46, p. 389, 1975.

Ghosh, S. K.: Photogrammetry for Police Use: Experience in Japan, *Photogrammetric Engineering and Remote Sensing*, vol. 46, no. 3, p. 329, 1980.

—— and H. Nagaraja: Scanning Electron Micrography and Photogrammetry, *Photogrammetric Engineering and Remote Sensing*, vol. 42, no. 5, p. 649, 1976.

Hallert, B.: Determination of Interior Orientation of Cameras for Non-Topographic Photogrammetry, Microscopes, X-ray Instruments and Television Images, *Photogrammetric Engineering*, vol. 26, no. 5, p. 748, 1960.

————: "X-ray Photogrammetry, Basic Geometry and Quality," Elsevier Publishing Co., Amsterdam, 1970.

Halsman, J.: Stereoscopic Medical Photography, *Photogrammetric Engineering*, vol. 22, no. 2, p. 374, 1956.

Karara, H. M.: Universal Stereometric Systems, *Photogrammetric Engineering*, vol. 33, no. 11, p. 1303, 1967.

————: Simple Cameras for Close Range Applications, *Photogrammetric Engineering*, vol. 38, no. 5, p. 447, 1972.

————: Accuracy Aspects of Non-Metric Imageries, *Photogrammetric Engineering*, vol. 40, no. 9, p. 1107, 1974.

————: Non-Topographic Photogrammetry, 1972–1976, *Photogrammetric Engineering and Remote Sensing*, vol. 42, no. 1, p. 37, 1976.

Kobelin, J.: Mapping Street Intersections Using Close Range Photogrammetry, *Photogrammetric Engineering and Remote Sensing*, vol. 42, no. 8, p. 1083, 1976.

Konecny, G.: Structural Engineering Application of the Stereometric Camera, *Photogrammetric Engineering*, vol. 31, no. 1, p. 96, 1965.

Kratky, V.: Analytical X-Ray Photogrammetry in Scoliosis, *Photogrammetria*, vol. 31, no. 6, p. 195, 1975.

Leydolph, W. K.: Stereophotogrammetry in Animal Husbandry, *Photogrammetric Engineering*, vol. 20, no. 5, p. 804, 1954.

Linkwitz, K.: A Precision Test Field for Close Range Photogrammetry, *Photogrammetric Record*, vol. VIII, no. 46, p. 501, 1975.

Malhotra, R. C.: Holography as Viewed by a Photogrammetrist, *Photogrammetric Engineering*, vol. 36, no. 2, p. 152, 1970.

McNeil, G. T.: X-ray Stereo Photogrammetry, *Photogrammetric Engineering*, vol. 32, no. 6, p. 993, 1966.

Moellman, D., and H. Karara: Close-Range Photogrammetric Data Reduction Scheme, *ASCE Journal of the Surveying and Mapping Division*, vol. 94, no. SU2, p. 211, 1968.

Mikhail, E. M., and G. Glaser: Mensuration Aspects of Holograms, *Photogrammetric Engineering*, vol. 37, no. 3, p. 267, 1971.

Newton, I.: Dimensional Quality Control of Large Ship Structures by Photogrammetry, *Photogrammetric Record*, vol. VIII, no. 44, p. 139, 1974.

Oshima, T.: Recent Development of Industrial Photogrammetry in Japan, *Photogrammetric Engineering and Remote Sensing*, vol. 42, no. 3, p. 339, 1976.

Salley, J. R.: Close Range Photogrammetry: A Useful Tool in Traffic Accident Investigation, *Photogrammetric Engineering*, vol. 30, no. 4, p. 568, 1964.

Schernhorst, J. N.: Close Range Instrumentation, *Photogrammetric Engineering*, vol. 33, no. 4, p. 377, 1967.

Scott, P. J.: Structural Deformation Measurement of a Model Box Girder Bridge, *Photogrammetric Record*, vol. IX, no. 51, p. 361, 1978.

Singh, R. S.: Radiographic Measurements, *Photogrammetric Engineering*, vol. 36, no. 11, p. 1137, 1970.

Veress, S. A., et al.: An Analytical Approach to X-Ray Photogrammetry, *Photogrammetric Engineering and Remote Sensing*, vol. 43, no. 12, p. 1503, 1977.

PROBLEMS

18-1 Discuss some of the uses of terrestrial or close-range photogrammetry.

18-2 Explain the differences between a phototheodolite and a stereometric camera.

18-3 A horizontal terrestrial photo was exposed with a phototheodolite having a focal length of 60.00 mm. Find the horizontal angle ALB at the exposure station subtended by points A and B if corresponding images a and b have photocoordinates of $x_a = 32.45$ mm, $y_a = -17.69$ mm, $x_b = -22.24$ mm, and $y_b = 29.73$ mm.

18-4 For the data of Prob. 18-3, calculate the vertical angles from the exposure station to points A and B.

18-5 Solve Probs. 18-3 and 18-4 graphically.

18-6 Repeat Prob. 18-3, except that the camera focal length is 190.04 mm and the measured photocoordinates are $x_a = -79.28$ mm, $y_a = 39.84$ mm, $x_b = 45.00$ mm, and $y_b = 21.92$ mm.

18-7 Calculate the vertical angles for points A and B of Prob. 18-6.

18-8 Solve Probs. 18-6 and 18-7 graphically.

18-9 A stereopair of oblique terrestrial photos was exposed as illustrated in Fig. 18-10. The camera had a 164.96-mm focal length and the camera axis was inclined upward from horizontal at an angle of 6°18' for both photos. Horizontal angles δ and δ' measured from the base line were 82°25' and 76°42', respectively. The horizontal length of the base line was 85.74 ft and the elevations of camera stations L and L' were 104.93 ft and 103.57 ft, respectively. Calculate X and Y ground coordinates of point A if photocoordinates of image a on the left and right photo were $x_a = -1.61$ mm, $y_a = 26.17$ mm, $x_a' = -63.57$ mm, and $y_a' = 23.24$ mm. Assume the origin of ground coordinates to be at camera station L and that the X axis coincides with the base line, as illustrated in Fig. 18-10.

18-10 Calculate the elevation of point A of Prob. 18-9.

18-11 On the overlapping pair of terrestrial photos of Prob. 18-9, a second point B has image coordinates on the left and right photos of $x_b = 63.42$ mm, $y_b = 22.51$ mm, $x_b' = 2.78$ mm, and $y_b' = 26.37$ mm. Calculate the horizontal length of line AB.

18-12 A horizontal stereopair of terrestrial photos was exposed using a stereometric camera having a base of 2.000 m. The focal lengths of both cameras were 90.00 mm. Image coordinates of a church spire on the left and right photos were $x = 6.15$ mm, $y = 53.36$ mm, $x' = 0.79$ mm, and $y' = 53.35$ mm. Using the parallax equations, calculate X, Y, and Z coordinates of the church spire in a ground coordinate system with origin at the left camera station, as indicated in Fig. 18-12.

18-13 Repeat Prob. 18-12, except that the cameras had focal lengths of 64.00 mm, the camera base was 80 cm, and images of point A had measured photocoordinates on the left and right photos of $x_a = -8.12$ mm, $y_a = 4.83$ mm, $x_a' = -13.97$ mm, and $y_a' = 4.83$ mm.

18-14 A horizontal stereopair of terrestrial photos was exposed with unequal camera station elevations, as illustrated in Fig. 18-13. The camera axes were oriented parallel, and the phototheodolite had a focal length of 164.96 mm. The horizontal length of the camera base line was 53.28 ft. Camera elevation was 105.88 ft at station L and 100.28 ft at L'. Using a parallax bar, parallaxes of 58.36 mm and 41.29 mm were obtained for image points a and b, with the photos oriented as illustrated in Fig. 18-14. Photocoordinates of images a and b on the left photo measured with respect to the fiducial axes were $x_a = 9.64$ mm, $y_a = 19.26$ mm, $x_b = 27.21$ mm, and $y_b = 38.84$ mm. Calculate X, Y, and Z coordinates of points A and B in the ground coordinate system of Fig. 18-13 using the parallax equations. What is the length of line AB?

18-15 An oblique terrestrial photo was exposed with the camera axis depressed at an angle 12°36'. The camera focal length was 194.95 mm. Calculate the horizontal and vertical angles between the rays from the camera station to object points A and B if their images have photocoordinates measured with respect to the fiducial axes of $x_a = -85.72$ mm, $y_a = 19.40$ mm, $x_b = 51.88$ mm, and $y_b = -46.27$ mm.

18-16 Repeat Prob. 18-15, except that the depression angle was 10°30', the camera focal length was 90.01 mm, and the image coordinates measured with the respect to the fiducial axes system were $x_a = 1.70$ mm, $y_a = -39.95$ mm, $x_b = 57.22$ mm, and $y_b = 40.86$ mm.

18-17 Discuss three basic approaches in establishing control for terrestrial or close-range photogrammetry.

18-18 Consult the references listed at the end of this chapter and write a brief report on one of the successful applications of x-ray photogrammetry.

18-19 Describe the different methods used in obtaining stereopairs of radiographs.

18-20 Measured image coordinates of the tops and bottoms of studs S_1 through S_4 on a radiograph such as that of Fig. 18-22*b* are given below. Plot these coordinates and graphically determine the coordinates of the principal point of the radiograph.

Stud	Top		Bottom	
	x, mm	y, mm	x, mm	y, mm
S_1	12.49	237.53	18.02	212.48
S_2	237.58	237.52	192.96	212.51
S_3	237.54	12.53	193.03	37.50
S_4	12.56	12.50	17.97	37.46

18-21 Solve Prob. 18-20 numerically using S_1 and S_3. Check by solving the problem using S_2 and S_4.

18-22 A stereopair of radiographs was exposed by shifting the x-ray tube 3.6 in between exposures. Distances d and D from the x-ray tube to film plane and object reference plane are 36.5 in and 36.0 in, respectively. Δp for image point p is measured as 0.22 in. What is the height of object point P above the object reference plane? (Use both the exact and approximate formulas and compare the results.)

18-23 For Prob. 18-22, if measured x and y coordinates of the image of p on the left radiograph are $x_p = 85.75$ mm and $y_p = 37.22$ mm, calculate the X and Y coordinates of the point.

18-24 Discuss the advantages and disadvantages of hologrammetry as compared to conventional photographic photogrammetry for mapping.

NINETEEN

PHOTOGRAPHIC INTERPRETATION†

19-1 INTRODUCTION

Photographic interpretation is the act of examining photographic images for the purpose of identifying objects and judging their significance. The process, however, is not restricted to making decisions concerning what objects appear in photographs; it also usually includes a determination of their relative locations and extents. This requires the application of at least some of the elementary photogrammetric measurement and mapping techniques discussed in earlier chapters. Although this chapter concentrates on the use of conventional aerial photographs for performing image interpretation, many other sources of information are also available. These include images obtained from satellite sensor systems, thermal scanners, multispectral scanners, side-looking airborne radar systems, and passive microwave instruments. Images obtained from these sources are treated under the general heading of *remote sensing,* and are described in Chap. 20.

Aerial photographs contain a detailed record of the ground at the time of exposure. A photo interpreter systematically examines the photos, but in addition may also study other materials such as maps and reports of field trips. Based on this study an interpretation is made as to the objects appearing on the photos. Success in photo interpretation will vary with the training and experience of the interpreter, the nature of objects being interpreted, and the quality of the photographs being used. Generally the most capable photo interpreters have keen powers of observation plus imagination and a great deal of patience.

Photo interpretation has been applied successfully in many fields, including agriculture, archaeology, conservation, engineering, ecology, forestry, geography, geology, meteorology, military intelligence, natural resource management, oceanography, soil science, and urban and regional planning. By consulting references cited at the end of this chapter, the student can gain an appreciation of the many ways in which photo interpretation has been used to solve a wide variety of practical problems.

†By Dr. Ralph W. Kiefer, Professor of Civil and Environmental Engineering, University of Wisconsin, Madison, Wis.

19-2 BASIC CHARACTERISTICS OF PHOTOGRAPHIC IMAGES

A systematic study of aerial photographs usually involves a consideration of the basic characteristics of photographic images. Seven of these characteristics are shape, size, pattern, shadow, tone, texture, and site.

Shape relates to the general form, configuration or outline of an individual object. Shape is probably the most important single factor in recognizing objects from their photographic images. A railroad is usually readily distinguished from a highway, for example, because its shape consists of long straight tangents and gentle curves as opposed to the curvy shape of a highway.

Size of objects on photographs will vary with photographic scale. Objects can be misinterpreted if their sizes are not evaluated properly. A dog house, for example, might be misinterpreted as a barn if size were not considered.

Pattern relates to the spatial arrangement of objects. The repetition of certain general forms or relationships is characteristic of many objects, both natural and man-made, and gives objects a pattern which aids the photo interpreter in recognizing them. An outdoor drive-in theater, for example, has a particular layout and pattern of parking spaces which aid in its identification.

Shadows are of importance to photo interpreters in two opposing respects: (1) the shape or outline of a shadow affords a profile view of objects, which aids interpretation; and (2) objects within shadows reflect little light and are difficult to discern on photographs, which hinders interpretation.

Tone refers to color or relative shades of gray of images on the photographs. It is related to reflectance of light from objects. Water, which absorbs nearly all incident light, photographs black, whereas a portland cement concrete highway reflects a high percentage of light and consequently produces very light tones. Without tonal differences, shapes, patterns, and texture of objects could not be discerned.

Texture is the frequency of tone change in the photographic image. Texture is produced by an aggregate of unit features which may be too small to be clearly discerned individually on the photograph. It is a product of their individual shape, size, pattern, shadow, and tone. As the scale of the photograph is reduced, the texture of a given object becomes progressively finer and eventually disappears. On large-scale photographs, for example, a stand of large-leaf tree species such as basswood or oak could likely be distinguished from small-leaf species such as poplar on the basis of their coarser texture.

Site or location of objects in relation to other features may be very helpful in identification. As an example, a ferris wheel might be difficult to identify if standing in a field near a barn, whereas it would be easy to identify if it were in an area recognized as an amusement park.

Figure 19-1 is a stereogram of aerial photos showing part of the Madison campus of the University of Wisconsin. The numbers on the figure refer to objects which can be identified by the amalgamation of their particular characteristics of shape, size, pattern, shadow, tone, texture, and site. Some of the objects which can be identified are (1) lake, (2) deciduous trees, (3) parked cars, (4) multistory buildings, (5) smokestack, (6) sidewalks, (7) building shadows, and (8) tree shadows.

While a consideration of basic image characteristics is usually sufficient to identify rather obvious objects such as those specifically noted in Fig. 19-1, successful photo interpretation of complex situations requires specialized techniques. Determining tree

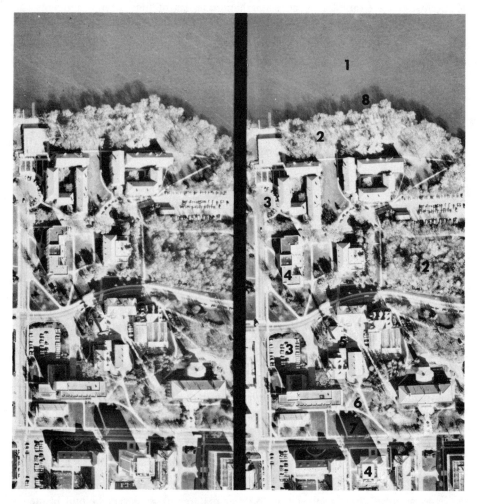

Figure 19-1 Stereogram of the University of Wisconsin, Madison Campus. Objects identified are (1) lake, (2) deciduous trees, (3) parked cars, (4) multistory buildings, (5) smokestack, (6) sidewalks, (7) building shadows, and (8) tree shadows. Photo scale, 1:4,800. (*Courtesy State of Wisconsin, Department of Transportation.*)

species or estimating terrain conditions such as bedrock type, soil texture, and drainage characteristics are examples of complex tasks that cannot be solved by merely considering basic image characteristics.

19-3 PHOTOGRAPHIC INTERPRETATION FOR FORESTRY USES

Photographic interpretation can aid in mapping tree species; determining age, density, and sizes of trees; and it can be used in solving many other forestry related problems such as appraising fire, insect, and disease damage. The use of photographic interpretation for tree species identification will be illustrated here.†

Basic image characteristics of shape, size, pattern, shadow, tone, texture and site described in Sec. 19-2 vary for different kinds of trees and can therefore be used by interpreters to aid in identification of tree species. Individual tree species have their own characteristic *size* and crown *shape*. Some species such as oak have rounded crowns, some such as balsam have cone-shaped crowns, and others such as white pine have star-shaped crowns. The arrangement of tree crowns produces a stand *pattern* that is quite distinct for many species. Tree *shadows* often provide a profile image of trees and can assist in species identification. *Tone* in aerial photographs depends on many factors, and it is not possible to correlate "absolute" tone values with individual tree species. "Relative" tones on a single photograph or strip of photographs, however, may be of great value in tree species identification. Variations in crown *texture* are important in species identification; e.g., some species have a tufted appearance, others appear smooth, and still others look billowy. *Site* can also play a key role in tree identification, as certain species occur only in highlands or on slopes while others are found only in lowlands or in swamps.

The success of photographic interpretation for tree species identification may depend on various photographic parameters involved, such as time of year, time of day of photography, photographic scale, and film-filter combination used. The illustrations included here are vertical stereograms using panchromatic film.

A pure stand of black spruce (outlined area) surrounded by aspen is shown in Fig. 19-2. Black spruce are coniferous trees with very slender crowns and pointed tops. In pure stands, the canopy is regular in pattern and the tree height is even or changes gradually with the quality of the site. The crown texture of dense black spruce is carpetlike in appearance. In contrast to black spruce, aspen are deciduous trees with rounded crowns and there may be open space between trees. The striking difference in photo texture between black spruce and aspen is very clear in Fig. 19-2.

Stands of balsam fir and black spruce are shown in Fig. 19-3. Balsam fir are very symmetrical coniferous trees with sharply pointed tops. Since the crown widens rapidly toward the base with dense branching, balsam fir usually seems a thicker tree than the slender black spruce. Area 2 is a pure stand of black spruce. Area 1 is a mixed stand containing 60 percent balsam fir and 40 percent black spruce. Balsam fir

---

†Sec. 19-3 is based on material contained in the report by Victor G. Zsilinszky, "Photographic Interpretation of Tree Species in Ontario," 2d ed., revised, The Ontario Department of Lands & Forests, Canada, 1966. Mr. Zsilinszky supplied Figs. 19-2 and 19-3 and permission to use materials from his report.

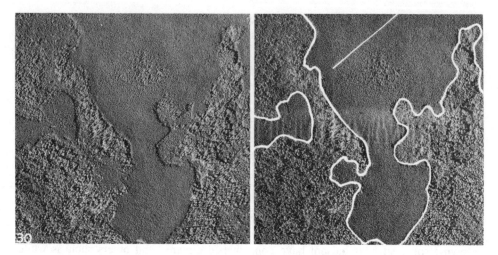

Figure 19-2 Black spruce (outlined area) and aspen, Ontario, Canada. Photo scale, 1:16,000. (*Courtesy Ontario Dept. Of Lands and Forests.*)

stands often have erratic changes in size, forming an uneven stand profile and an irregular stand pattern. Note the contrast in Fig. 19-3 between the smooth, fine-textured (almost carpetlike) pattern of the black spruce as compared with the coarser-textured, more erratic pattern of the balsam fir.

The process of tree species identification using aerial photographic interpretation is not generally as simple as might be implied by the straightforward examples of Figs. 19-2 and 19-3. Results of photo interpretation work in Ontario have shown that

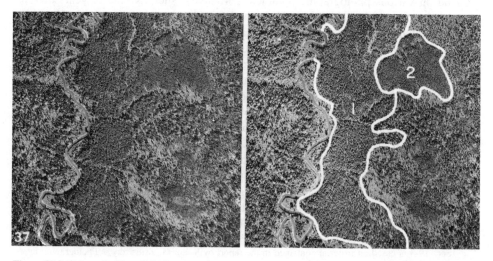

Figure 19-3 Balsam fir (1) and black spruce (2). Ontario, Canada. Photo scale, 1:16,000. (*Courtesy Ontario Dept. of Lands and Forests.*)

it is difficult to identify tree species until the trees are over 20 years of age, although conifers can be distinguished from hardwoods at a younger age. In addition, individual stands are variable in appearance depending on age, site conditions, geographic location, geomorphic setting, and other factors. Other variables which complicate tree species identification are time of day of photography (direction and intensity of illumination), time of year (season), soil types, moisture content in the soil and topography. In spite of these variations and complications, however, identification of tree species has generally been very successful when practiced by skilled and experienced interpreters.

19-4 BASIC ELEMENTS IN PHOTOGRAPHIC INTERPRETATION FOR TERRAIN ANALYSIS

Various terrain characteristics are important to soil scientists, geologists, geographers, civil engineers, urban and regional planners, landscape architects, and others seeking to understand the nature of soils and rocks. The principal terrain characteristics that can be estimated by means of air photo interpretation are geologic landform,† bedrock type, soil texture and plasticity, site drainage conditions, susceptibility to flooding, and depth of soil cover over bedrock.

To estimate the above terrain conditions, photo interpreters study individual characteristics of landscape elements separately and in relation to each other and attempt to ferret out their meaning and interrelationship. A great deal of deductive and inductive reasoning is required to achieve success. To estimate terrain conditions, the following five basic elements are studied stereoscopically from aerial photos: (1) topographic form, (2) drainage, (3) erosion, (4) photo tone, and (5) vegetation and land use. These are briefly explained as follows:

1. *Topographic form* The size and shape of a landform are probably its most important identifying characteristics. There is often a distinct topographic change at the boundary between two landforms, as can be seen in several of the illustrative example air photos that follow.
2. *Drainage* The drainage pattern and texture seen on aerial photographs are indicators of landform and bedrock type and also suggest soil characteristics and site drainage conditions. Figure 19-4 shows coarse- and fine-textured "dendritic" (treelike) drainage patterns. The dendritic drainage pattern is the most common drainage pattern found in nature. It develops under many terrain conditions, including homogeneous unconsolidated materials and rocks with a uniform resistance to erosion, such as horizontally bedded sedimentary rocks and granitic rocks. In a given climatic area, coarse-textured patterns would tend to develop where

---

†The term "landform" as used herein implies a specific topographic form, geologic origin, and soil or rock material. As an example, sand dunes have distinct topographic forms which can be recognized on aerial photographs; their geologic origin is *eolian* (wind-deposited); and the soil particles in sand dunes are almost entirely sand-sized owing to the efficient wind sorting of particles during the formation of the sand dune. Other examples of geologic landforms are river terraces, drumlins, eskers, glacial moraines, etc.

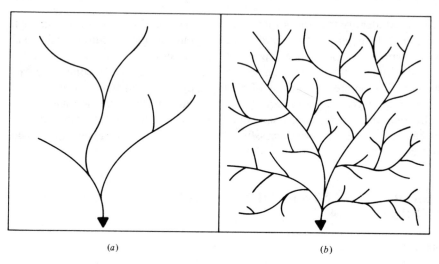

(a)   (b)

Figure 19-4 Illustrative drainage patterns. (*a*) Coarse-textured dendritic pattern; (*b*) fine-textured dendritic pattern.

the soils and rocks have good internal drainage with little surface runoff, whereas a fine-textured pattern would tend to develop where the soils and rocks have poor internal drainage and surface runoff is high. Also, soft, easily eroded rocks, such as shale, would tend to develop fine-textured patterns, whereas hard, massive rocks, such as granite, would tend to develop coarse-textured patterns.

3. *Erosion* This term refers to the size and shape of the gullies (smallest drainage features that can be seen on aerial photographs). Gullies may be as small as a few feet deep and a few hundred feet long. Through an inspection of the *cross section, plan,* and *profile* characteristics of gullies, it is possible to estimate whether the soil is predominantly sand and gravel, silt, or clay. Figure 19-5 shows three typical gully cross-section shapes. Short gullies with V-shaped cross sections (Fig. 19-5*a*) develop in sand and gravel, gullies with U-shaped cross sections (Fig. 19-5*b*) develop in silty soils, and long gullies with gently rounded cross sections (Fig. 19-5*c*) develop in silty clay and clay soils.

4. *Photo tone* The absolute value of the photo tone depends not only on certain terrain characteristics such as topography and soil moisture content but also on photographic factors such as film-filter combination, exposure, and photographic processing. Tone also depends on meteorological and climatological factors such as atmospheric haze, sun angle, and cloud shadows. Thus, as in the case of photo interpretation for forestry uses, photo interpretation for estimating terrain conditions relies on an evaluation of ''relative'' tone values. These relative tone values are very important because they form photographic patterns which may be of great significance. The striking mottled tonal patterns characteristic of fine-grained glacial till soils illustrated in Fig. 19-7 are good examples of the significance of photo tone patterns in the air photo interpretation process.

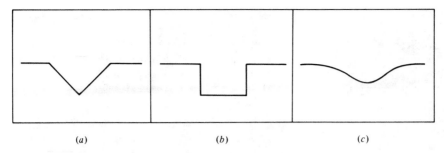

(a)	(b)	(c)

Figure 19-5 Illustrative gully cross sections. (a) Gully in sand and gravel; (b) gully in silt; (c) gully in silty-clay or clay.

5. *Vegetation and land use* Differences in natural or cultivated vegetation often indicate differences in terrain conditions. As an example, orchards and vineyards are generally located on well-drained soils, whereas truck farming activities often take place on highly organic soils such as muck and peat.

19-5 INTERPRETATION OF LANDFORMS

As previously stated, one of the principal terrain characteristics that can be estimated by means of air photo interpretation is geologic landform. This section includes example air photos illustrating three different geologic landforms. Each of these is briefly described, and the basic elements described in Sec. 19-4 that are used in interpreting them are discussed.

19-5.1 Esker

Figure 19-6 shows an air photo stereogram of an *esker* (A) winding across an area of *glacial till* (B). An esker is a deposit of stratified sand and gravel formed by streams flowing on, within, or under glacial ice. After the glacial ice melted, sand and gravel remained in the form of a sinuous ridge. In many geographic areas, eskers are important sources of sand and gravel for construction materials. Glacial till is an unsorted mixture of varying amounts of gravel, sand, silt, and clay which were carried on and within glacial ice and deposited over the underlying materials as the glacial ice melted.

The *topographic form* of an esker is its single most important characteristic. The esker of Fig. 19-6 is a sinuous ridge 150 to 300 ft wide which stands 20 to 40 ft above the surrounding glacial till. Because internal *drainage* is excellent through the sand and gravel, no drainage pattern has developed and *erosion* is virtually nonexistent. The *photo tone* on the esker is uniformly light as compared with the more mottled pattern on the glacial till. The *vegetation* and *land use* on the esker are different than those on the glacial till. The esker is not cultivated because of its steep slopes and excessively well-drained soils. However, the surrounding glacial till area is cultivated.

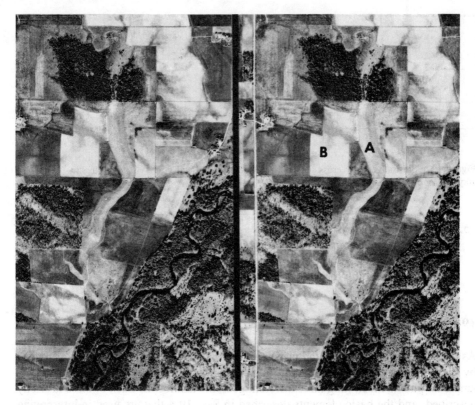

Figure 19-6 Esker in northern Wisconsin. Photo scale, 1:20,000. (*Courtesy U.S. Dept of Agriculture, ASCS.*)

19-5.2 Glacial Till

Figure 19-7 shows poorly drained, fine-grained (silty clay) glacial till soils in Madison County, Indiana. The *topography* has very little relief. There is less than 20 ft difference in elevation between the lowest and highest points in the 1-square-mi area shown in the figure. Because of the flat topography and the young age (geologically) of this landform, there is no well-developed surface *drainage* pattern and very little *erosion*. Artificial surface drainage (ditches) and subsurface drainage (buried tile drains) that appear in many places on this photograph are clues to the poor internal drainage of the soil. The *photo tone* differences create a mottled pattern which is characteristic of glacial till areas with fine-grained soils. The tonal differences, especially well developed in fields *C* and *D*, are caused by differences in sunlight reflection due to the varying moisture content of the soil. The soils on the small rises (2 to 3 ft high) are dryer and lighter in tone. The surrounding areas are wetter and darker in tone. Figure 19-7 was photographed when the crop cover was not heavy (mid-June). The *vegetation* and *land use* are characteristic of a rural area with nearly all of the area under cultivation. Fine-grained glacial till soils, especially those in the darker-toned

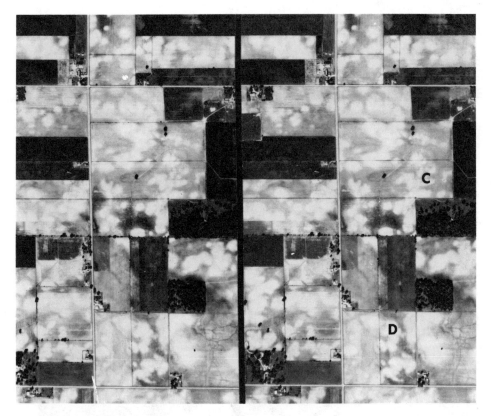

Figure 19-7 Fine-grained glacial till in northern Indiana. Photo scale 1:20,000. (*Courtesy U.S. Dept. of Agriculture, ASCS.*)

depressions, have high moisture content and low strength. In these areas careful planning and design will be required to avoid highway pavement failures or unstable homesite foundation conditions.

19-5.3 Flood Plain

Figure 19-8 shows a river flood plain. Formation of flood plains is a normal part of the fluvial geomorphic cycle. Young streams have steep gradients and are actively engaged in the process of deepening and widening their valleys through erosion. Their courses are relatively straight, and they do not construct flood plains. Mature streams have flatter gradients and actively deposit flood plain materials during periods of flooding. Their courses are meandering, with the width of the meander belt about equal to the valley width. Old streams have gentle gradients and the width of their flood plains is several times that of the meander belt width.

The construction of flood plains is a depositional process. Streams transport materials primarily through the traction (bed load) and suspension (suspended load) of

Figure 19-8 Flood plain in Indiana. The present channel of the stream can be seen at *PC,* an abandoned Channel at *AC,* point bar deposits at *PB,* an oxbow lake at *OX,* and slack water at *SW.* Photo scale, 1:27,000. (*Courtesy U.S. Dept. of Agriculture, ASCS.*)

materials. When a stream cannot transport the load imposed upon it (either because the total load exceeds the capacity of the stream or the particle sizes exceed the competence of the stream), deposition will occur. Flood plains are constructed through the normal processes of meander development and by over-the-bank flooding. Meander development results in the formation of a variety of features, including point bar deposits, channel scars, abandoned channels, and oxbow lakes. These features have a wide variety of associated soil types, from the sands and gravels of point bar deposits to the silts, clays, and organic soils of oxbows. Over-the-bank flooding deposits silts and clays in slack water (stillwater) and sands and coarse silts on and near the natural levees along the streams. Soils with a high organic content may develop in the slack water deposits.

Flood plain terrain has overall level relief with minor irregularities and a gentle downstream gradient. Soil deposits are complex and variable. Depth to bedrock is also variable, but wider flood plains generally have deeper soil deposits. Internal soil drainage is commonly poor due to the high water table.

The engineering significance of flood plains is that they are characterized by great variation in soils in both the horizontal and vertical direction and are subject to flooding at periodic intervals. Development and new construction on flood plains should proceed only when absolutely necessary, and then only after a careful study has been made of the soils present, the internal soil drainage and groundwater conditions, and the expected frequency and severity of flooding.

The flood plain shown on the airphoto of Fig. 19-8 is of the White River in Knox and Daviess Counties, Indiana. The topography is very flat on this flood plain (illustration is not a stereopair). Most of the flood plain features previously mentioned can be seen in this figure. The present channel of the stream can be seen at *PC*. An abandoned channel is seen at *AC*. Point bar deposits are shown at *PB*. The point bar deposits are small ridges of sand and gravel which photograph quite light in tone due to their sandy and well-drained nature. Shown at *OX* is an oxbow lake which contains shallow standing water that is gradually filling with organic growth. Fine-grained, poorly drained slack water deposits are shown at *SW*. There is a well-developed artificial drainage system, consisting of open ditches and "dead furrows," in the area of the slack water deposits. Without this drainage system, the soils would be too wet for cultivation.

In flood plain analysis, as with other areas of photo interpretation, many significant features that can be seen on the air photos may be completely overlooked by an observer on the ground. A limitation of air photos, however, is that they reveal primarily the surface soil conditions, and in the case of flood plain soils, there are often significant differences in soil characteristics at different depths. Buried granular deposits (e.g., point bar deposits) or buried organic soils (e.g., abandoned channels or oxbows) may occur on flood plains. It is usually necessary, and always desirable, to supplement the air photo interpretation process with selective fieldwork, especially when it is important to learn the characteristics of soils at depths.

19-6 INTERPRETATION OF BEDROCK TYPES

Through careful study of aerial photographs, skilled interpreters are able to identify different types of rocks. Figures 19-9 and 19-10 illustrate the striking differences in appearance of two different rock types. These differences in appearance are the result

Figure 19-9 Sandstone in southern Utah. Photo scale, 1:20,000. (*Courtesy U.S. Geological Survey.*)

of variations in bedding, jointing, internal drainage, and resistance to erosion of these rocks. An interpreter must consider topography, drainage, and erosion characteristics to distinguish between different kinds of rocks.

Figure 19-9 shows sandstone, a sedimentary rock composed of cemented sand-sized particles arranged in horizontal layers. It is relatively pervious and resistant to erosion, and it is heavily "jointed" (cracked vertically) in two approximately perpendicular directions. Because of these characteristics, few surface streams have developed and the drainage is coarse-textured (see Fig. 19-4a). Those streams which have formed have cut deeply into the sandstone rock in directions which tend to align with the two principal directions of jointing.

Figure 19-10 shows shale, a sedimentary rock composed of cemented silt and clay-sized particles arranged in horizontal layers which are thinner than the sandstone layers. Shale is impervious and easily eroded as compared with sandstone. Because of these characteristics, the streams have formed a fine-textured dendritic drainage pattern (see Fig. 19-4b) and a minutely dissected landscape has developed.

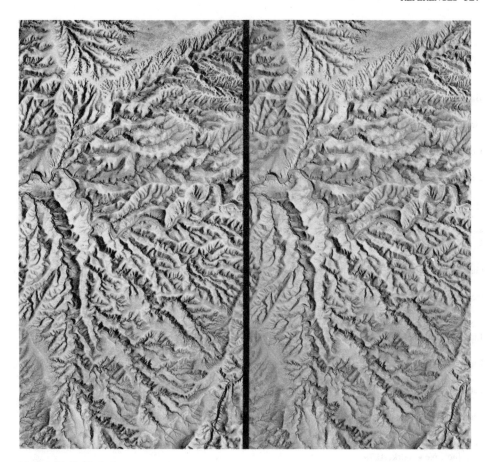

Figure 19-10 Shale in southern Utah. Photo scale, 1:20,000. (*Courtesy U.S. Geological Survey.*)

Many other bedrock types such as limestone, granite, and basalt can also be identified using air photo interpretation approaches similar to those discussed above. However the student should not be misled into concluding that all rock identifications are as easy as those in the above examples. Indeed, many of them are extremely complex and require a great deal of study and background experience. For further study, some of the references cited at the end of this chapter may be consulted.

REFERENCES

American Society of Photogrammetry: "Manual of Photographic Interpretation," Falls Church, Va., 1960.

Anson, A.: Color Photo Comparison, *Photogrammetric Engineering*, vol. 32, no. 2, p. 286, 1966.

Avery, T. E.: Evaluating the Potential of Photo Interpreters, *Photogrammetric Engineering*, vol. 31, no. 6, p. 1051, 1965.

————: "Forester's Guide to Aerial Photo Interpretation," Agriculture Handbook No. 308, U.S. Department of Agriculture, Forest Service, December 1969.

————: *Interpretation of Aerial Photographs,* 3d ed., Burgess Publishing Company, Minneapolis, 1977.

———— and D. M. Richter: An Airphoto Index to Physical and Cultural Features in Eastern U.S., *Photogrammetric Engineering,* vol. 31, no. 5, p. 896, 1965.

———— and J. Canning: Airphoto Measurements of New Zealand Pines, *Photogrammetric Engineering,* vol. 40, no. 8, p. 957, 1974.

Baker, R. D., et al.: Land-Use/Land-Cover Mapping from Aerial Photographs, *Photogrammetric Engineering and Remote Sensing,* vol. 45, no. 5, p. 661, 1979.

Branch, M. C.: "City Planning and Aerial Information," Harvard University Press, Cambridge, Mass., 1971.

Colwell, R. N.: Aids for the Selection and Training of Photo Interpreters, *Photogrammetric Engineering,* vol. 31, no. 12, p. 327, 1965.

Eastman Kodak Co.: "Photointerpretation and Its Uses," Kodak Publication No. M-42, Rochester, New York, 1968.

————: "Photointerpretation for Land Managers," Kodak Publication No. M-76, Rochester, New York, 1970.

————: "Photointerpretation for Planners," Kodak Publication No. M-81, Rochester, New York, 1972.

Fezer, F.: Photo Interpretation Applied to Geomorphology—A Review, *Photogrammetria,* vol. 27, no. 1, p. 7, 1971.

Gautam, N. C.: Aerial Photo-Interpretation Techniques for Classifying Urban Land Use, *Photogrammetric Engineering and Remote Sensing,* vol. 42, no. 6, p. 815, 1976.

Henderson, F. M.: Effects of Interpretation Techniques on Land-Use Mapping Accuracy, *Photogrammetric Engineering and Remote Sensing,* vol. 46, no. 3, p. 359, 1980.

Howard, John A.: *Aerial Photo-Ecology,* American Elsevier Publishing Co., New York, 1970.

Jones, A. D.: Computers and the Teaching of Airphoto Interpretation, *Photogrammetric Engineering and Remote Sensing,* vol. 44, no. 10, p. 1267, 1978.

Kiefer, R. W.: Landform Features in the United States, *Photogrammetric Engineering,* vol. 33, no. 2, p. 174, 1967.

Lattman, L. H., and R. G. Ray: "Aerial Photographs in Field Geology," Holt, Rinehart and Winston, Inc., New York, 1965.

Leachtenauer, J. C.: Photo Interpretation Test Development, *Photogrammetric Engineering,* vol. 39, no. 11, p. 1187, 1973.

Lillesand, T. M., and R. W Kiefer: "Remote Sensing and Image Interpretation," John Wiley & Sons, Inc., New York, 1979.

Lueder, D. R.: *Aerial Photographic Interpretation,* McGraw-Hill Book Company, New York, 1959.

Miller, V. C.: *Photogeology,* McGraw-Hill Book Company, New York, 1961.

Morgan, K. M., et al.: Airphoto Analysis of Erosion Control Practices, *Photogrammetric Engineering and Remote Sensing,* vol. 46, no. 5, p. 637, 1980.

Olson, C. E., Jr.: Photographic Interpretation in the Earth Sciences, *Photogrammetric Engineering,* vol. 29, no. 6, p. 968, 1963.

Parry, J. T.: The Development of Air Photo Interpretation in Canada, *Canadian Surveyor,* vol. 27, no. 4, p. 320, 1973.

Rib, H. T., and R. D. Miles: Automatic Interpretation of Terrain Features, *Photogrammetric Engineering,* vol. 35, no. 2, p. 153, 1969.

Richter, D. M.: An Airphoto Index to Physical and Cultural Features in Western U.S., *Photogrammetric Engineering,* vol. 33, no. 12, p. 1402, 1967.

Sadacca, R.: Human Factors in Image Interpretation, *Photogrammetric Engineering,* vol. 29, no. 6, p. 978, 1963.

Scovell, et. al.: *Atlas of Landforms,* John Wiley & Sons, New York, 1966.

Siegal, B. S., and A. R. Gillespie, (eds.): "Remote Sensing in Geology," John Wiley & Sons, Inc., New York, 1980.

Strandberg, C. H.: *Aerial Discovery Manual,* John Wiley & Sons, New York, 1967.

Van Lopik, J., T. M. Merifield et al.: Photo Interpretation in the Space Sciences, *Photogrammetric Engineering,* vol. 31, no. 6, p. 1060, 1965.

Way, D.: *Terrain Analysis, A Guide to Site Selection Using Aerial Photographic Interpretation,* Dowden, Hutchinson & Ross, Inc., Stroudsburg, Pa., 1973.

Whitcher, G. H.: Canada's Air Photo Library, *Photogrammetric Engineering,* vol. 31, no. 5, p. 807, 1965.

Zsilinszky, V. G.: "Photographic Interpretation of Tree Species in Ontario," 2d ed., The Ontario Department of Lands and Forests, Ottawa, Ontario, Canada, 1966.

————: The Practice of Photo Interpretation for a Forest Inventory, *Photogrammetria,* vol. 19, no. 5, 1964.

PROBLEMS

19-1 List the seven basic characteristics of photographic images that are considered in photographic interpretation and give an example of how each may be used to identify a particular object.

19-2 Briefly describe the techniques involved in tree species identification using aerial photographs.

19-3 Describe the different gully cross sections that occur in (*a*) sand and gravel, (*b*) silt, and (*c*) clay.

19-4 List and briefly discuss the five principal elements studied stereoscopically in terrain analysis.

19-5 Discuss the basic differences between sandstone and shale that assist in interpreting these types of bedrock.

19-6 Consult the references listed at the end of this chapter. Write a brief report on one successful application of airphoto interpretation.

TWENTY

REMOTE SENSING†

20-1 INTRODUCTION

Broadly defined, *remote sensing* is any methodology employed to study the characteristics of objects from a distance. Human sight, smell, and hearing are examples of rudimentary forms of remote sensing. A bat's guidance system is another type of remote sensor. Photographic interpretation, (see Chap. 19), is considered a form of remote sensing since it is used for identifying objects and judging their significance without physically touching them. Photo interpretation, however, is limited to a study of images recorded on photographic emulsions. These materials are sensitive only to energy in or near the visible portion of the electromagnetic spectrum. This chapter treats sensor systems which record energy in more quantifiable formats over a much broader range of the electromagnetic spectrum. Many of the sensors discussed record image data electronically, making these data inherently amenable to computer processing. The ability of these systems to ''see'' or sense energy outside of the visible portion of the spectrum and to provide image data in a digital form has greatly increased the earth resource information provided by remote sensing.

The technology of remote sensing has developed most rapidly during a time when people have become increasingly conscious of the need to strike the appropriate balance between resource development and environmental preservation. Today, remote sensing affords a practical means for frequent and accurate monitoring of the earth's resources on literally a global basis. It is aiding in assessing the impact of human activities on our air, water, and land. Data obtained from remote sensors have provided information necessary for making sound decisions and formulating policy in a host of resource-development and land-use applications. Remote sensing techniques have also been used in numerous special applications. Expediting geologic investigations, locating forest fires, detecting diseased crops and trees, monitoring population growth and distribution, determining the locations and extents of oil spills and other water pollutants, and positioning icebergs are but a few of the many varied applications of remote sensors that have benefited humankind.

†By Dr. Thomas M. Lillesand, University of Wisconsin, Madison, Wis.

20-2 ELECTROMAGNETIC RADIATION

As discussed in Sec. 3-7, the sun and various artificial sources radiate electromagnetic energy over a range of wavelengths. Light is a particular type of electromagnetic radiation that can be seen or sensed by the human eye. All electromagnetic radiation, whether visible or invisible, propagates in the form of sinusoidal waves traveling at the speed of light.

Electromagnetic energy does not interact with itself; rather, it can be perceived or sensed only through its interaction with matter. As an example, when one sees rays of light in a dark room, the perceived phenomenon is actually a manifestation of the interaction of light radiation with dust and other particulate matter in the air. When a unit of electromagnetic energy strikes an object on the earth's surface, it can interact with the object in any or all of three different ways. The incident energy can be *reflected, transmitted,* or *absorbed.* (The absorbed component is subsequently re-emitted from the object.) The particular mix of these three possible interactions is dependent upon the physical nature of the object. For example, a green shirt can be distinguished from a red shirt because incident light is reflected highest in the green portion of the spectrum from the green shirt and highest from the red portion of the spectrum from the red shirt. If electromagnetic energy were to interact with all objects in the same manner, electromagnetic discrimination, including human sight, would not exist. As it is, different energy interactions are generally associated with different types of objects. Remote sensors record these variations of energy interaction to discriminate between earth surface features and to assist in quantifying their condition.

20-3 IDEALIZED REMOTE SENSING SYSTEMS

Practical remote sensing systems can perhaps be best understood by first considering an idealized remote sensing system or sequence. Such a sequence is illustrated in Fig. 20-1, wherein

1. Electromagnetic energy of all wavelengths and of known uniform radiance (intensity) is produced by an ideal source.
2. The energy propagates from the source, without atmospheric modification or loss, to a homogeneous object.
3. The energy of various wavelengths selectively interacts with the object, resulting in a unique return signal of reflected and emitted energy.
4. The returned signal propagates, again without atmospheric modification, to a sensor that responds to energy of all wavelengths of any radiance level.
5. In real time, a radiance versus wavelength response is recorded, processed into an interpretable format, and recognized as being unique to the particular object type observed in its particular physical-chemical-biological state. That is, the sensor *data* are accurately processed into resource *information.*
6. The information obtained about the particular earth surface feature is made readily available, in a useful form, to users.

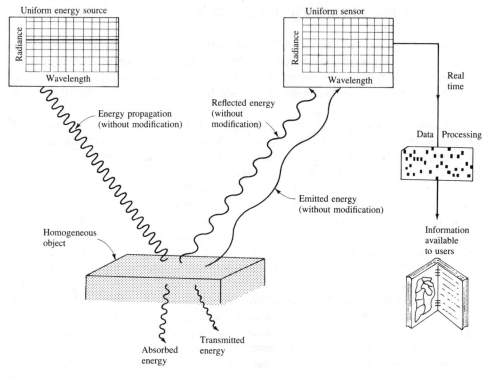

Figure 20-1 Idealized remote sensing system.

Unfortunately, an ideal remote sensing system as described above does not exist. The constraints of nature and limitations of technology complicate each element of the ideal sequence. These complications become apparent upon considering the characteristics of energy sources, the numerous possible forms of energy interaction with earth surface features, effects due to the atmosphere, the responsiveness of various sensor hardware, and the varied formats of sensor responses.

20-4 ENERGY SOURCES

All materials at temperatures above absolute zero ($-273°C$) continually emit electromagnetic radiation. The intensity and spectral character of this radiation is a function of the surface temperature of the emitting object. An object which would absorb all energy incident upon it is called a *blackbody*. Since by definition a blackbody absorbs all incident energy, no energy is reflected and the body does indeed appear perfectly black. A blackbody is also a perfect radiator; i.e., it reemits all of its absorbed energy. Inasmuch as a blackbody absorbs all incident energy and in turn emits all absorbed energy, it is in a state of equilibrium, growing neither hotter nor colder. A blackbody radiates energy according to the *Stefan-Boltzmann law*, which is as follows:

$$W_B = \sigma T^4 \tag{20-1}$$

In Eq. (20-1), W_B is spectral radiant emittance per unit area for a blackbody source, T is temperature in degrees Kelvin, and σ is a proportionality constant. A radiating blackbody emits electromagnetic energy over a wide range of wavelengths. The dominant wavelength, or wavelength at which maximum energy is radiated, is related to temperature by *Wien's displacement law,* which is as follows:

$$\lambda_m = \frac{A}{T} \tag{20-2}$$

In Eq. (20-2), λ_m is the wavelength of the emitted energy having maximum radiance, T is temperature in degrees Kelvin, and A is a constant. The relationship between temperature of a blackbody source and intensity and spectral character of its radiant emittance is depicted in Fig. 20-2. In the figure, radiant emittance curves are shown for blackbody sources at various temperatures. Note that with increasing temperature there is a marked increase in radiant emittance. Also, there is a progressive shift of the dominant radiated wavelength toward the lower end of the spectrum as the temperature of the source increases. Note for example that the peak of the 6000° curve occurs much to the left of the peak of the 200° curve. This phenomenon can be observed when a metal body such as a piece of iron is heated. As the iron becomes progressively hotter, it begins to glow and its color changes successively from dull red to orange to yellow and finally to white. As stated in Wien's displacement law, these changing colors are the result of dominance of shorter-wavelength emission as the iron is heated.

Sunlight has the character of blackbody radiation at about 6000°K. Objects at temperatures higher than 6000°K, such as certain stars, appear blue (their dominant wavelengths are in the blue region of the electromagnetic spectrum). Most incandescent lamps emit radiation typified by the 3000°K curve. (Note that the 3000° curve of Fig. 20-2 displays a lower level of blue energy than the sun, and this characteristic is readily detected by observing the lamps' orange color. Because of this condition, blue filters or blue flashbulbs must be used for indoor photography with daylight or outdoor film.)

The ambient temperature of most earth surface features (soil, water, vegetation, rock) is about 300°K. They radiate energy that peaks in the 8- to 14-μm range, which is far beyond the visible or photographic part of the electromagnetic spectrum. To sense this energy, *thermal radiometers* and *scanners* (described in Secs. 20-10 and 20-11) must be employed.

It is important to bear in mind that no perfect blackbody radiators exist in nature. Rather, most radiators are so-called "gray" bodies that emit varying proportions of the energy emitted by a theoretical blackbody. Radiant energy emittance for a gray-body source is given by the following equation:

$$W_G = \xi \sigma T^4 \tag{20-3}$$

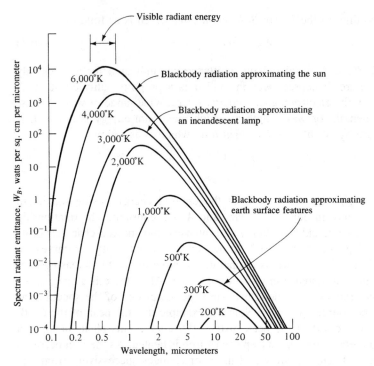

Figure 20-2 Spectral radiant emittance of blackbodies at various temperatures.

In Eq. (20-3), W_G is the radiant energy emitted from a graybody, T is temperature in degrees Kelvin, σ is a proportionality constant, and ξ is the *emissivity factor* of the source material. Note that Eq. (20-3) for graybody emittance differs from Eq. (20-1) for blackbody emittance only by the addition of the emissivity factor. Emissivity factors are simply ratios of the emittance of a particular material to the emittance of a blackbody. If, for example, the radiant emittance for a blackbody at temperature T is W_B, and radiant emittance for a graybody at the same temperature is W_G, then the emissivity factor ξ for that graybody is equal to W_G/W_B. Emissivity factors depend upon the composition and condition of the material involved. Typical emissivity factors for common materials are water, 0.98; wet soil, 0.95; dry soil, 0.92; sand, 0.92; snow, 0.85; and polished gold, 0.02. Polished gold is a good reflector but absorbs very little energy; consequently it is a poor radiator.

20-5 ENERGY INTERACTIONS WITH EARTH SURFACE FEATURES

As previously stated, electromagnetic energy can interact with an object in three ways. The nature of these possible interactions was schematically indicated in Fig. 20-1. The three interactions include

1. *Reflection,* wherein the incident energy is returned to the medium of propagation fundamentally unchanged.
2. *Transmission,* where the energy propagates through the object.
3. *Absorption,* where incident radiation is converted to some other form of energy such as heat.

As indicated above, all objects are also continually emitting energy and the particular quantitative mix of reflection and emission displayed by any given object is dependent upon wavelength. The amount of energy at various wavelengths that is returned to the sensor from a given object defines a theoreticallly unique *spectral response pattern*. Figure 20-3 shows typical spectral response envelopes (ranges of values) for deciduous trees and coniferous trees based upon their reflection of sunlight over the visible and near-infrared portions of the spectrum. Such graphs are termed *spectral reflectance curves*. From the curves it can be seen that through all visible wavelengths, particularly in the green region, the spectral response envelopes overlap, and therefore both types of trees can display nearly identical reflectances; i.e., they have essentially the same color. In the near-infrared region, however, the energy reflected by the deciduous trees is much greater than that of the conifers. This property makes infrared photography extremely useful for discriminating between these tree types. The aerial photos of Fig. 20-4*a* and *b* were exposed on panchromatic film (sensitive to only the visible portion of the electromagnetic spectrum) and black-and-white infrared film, respectively. On the infrared photo, note how easy it is to discern coniferous trees (dark areas) from deciduous trees (light areas). Except to a highly trained interpreter, these differences between tree types are not easily detected on the panchromatic photo. This example illustrates how variation in spectral response patterns between feature types can be used to discriminate between the types. This is the fundamental principle upon which all sensors operate. Some record energy variations

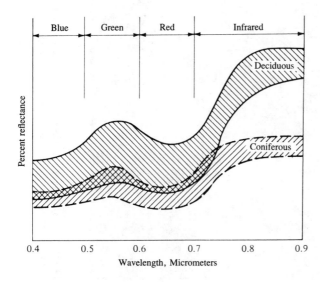

Figure 20-3 Spectral response envelopes for deciduous and coniferous trees.

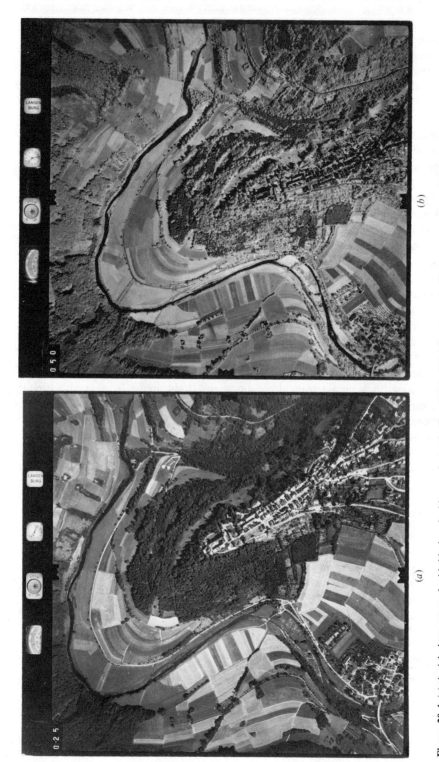

Figure 20-4 (*a*) Aerial photo exposed with black-and-white panchromatic film. (*b*) Aerial photo exposed with black-and-white infrared film. On this photo note how easily coniferous trees (dark areas) can be discerned from deciduous trees (light areas). (*Courtesy Carl Zeiss, Oberkochen.*)

in only the reflected portion of the spectrum (e.g., photographic systems); some record emitted energy (e.g., thermal scanners); and some record variation in both forms of energy (e.g., multispectral scanners).

20-6 ATMOSPHERIC EFFECTS

Discussion has thus far avoided the issue of how the atmosphere influences the spectral response patterns recorded by a sensor. Since the atmosphere contains a wide variety of suspended particles, it offers energy interaction capabilities just as "ground" objects do. In fact, in many meteorological applications of remote sensing, the atmosphere is the object of primary interest. For most engineering, planning, and resource management work, however, it is desirable to look *through* and not *at* the atmosphere. The extent to which the atmosphere transmits electromagnetic energy is dependent upon wavelength, as can be seen in the central portion of Fig. 20-5. This is a plot of percent atmospheric transmission of electromagnetic energy versus wavelength. Wavelengths in cross-hatched areas are essentially not transmitted by the atmosphere. *Atmospheric windows* are said to exist in wavelength areas of high percentage transmission (non-cross-hatched areas). Such a window exists in the 0.4- to 0.7- μm region of the electromagnetic spectrum—a situation enabling passage of the energy to which our eyes are sensitive. Note the low percentage of transmission in the range of 0.3 μm (ultraviolet) and shorter. This shows that the atmosphere effectively blocks passage of this short-wavelength energy.

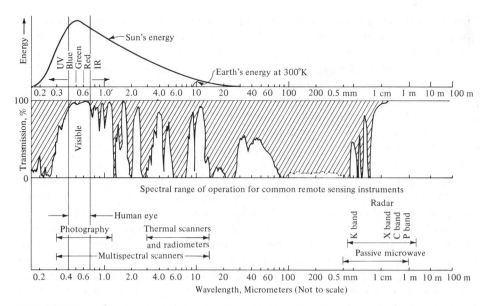

Figure 20-5 Composite chart showing source energy, atmospheric transmission, and range of operation of various sensors.

The blocking effect of the atmosphere is the result of many complex interactions. These interactions consist of scatter (diffusion) and absorption of electromagnetic energy by particles that range in size and type from gas molecules, to aerosols and vapor, and up to water droplets. The effect of these interactions depends upon the application, the type of sensor, and the amount of atmosphere through which the energy must pass. In photographic remote sensing, for example, scatter from gas molecules predominates at low wavelengths in accordance with the *Rayleigh effect*, which states that scatter is inversely proportional to the fourth power of wavelength. Accordingly, scatter increases rapidly as wavelength decreases. A blue sky is a manifestation of the Rayleigh effect, as we actually see the scattered blue wavelengths of sunlight. Without these scattering effects, the sky would appear black. A red sun in the early morning or late evening results from the increased atmospheric path length of solar radiation. At these times the blue and green components of sunlight are nearly completely scattered, so that red is the only remaining portion of the visible spectrum penetrating the atmosphere.

All energy that finally reaches a sensing system must travel through the atmosphere. The effect of the atmosphere upon this energy is an important element that must be taken into account in the design of a practical remote sensing system. On the one hand, atmospheric windows dictate where sensors can "look" at earth surface features within the spectrum. On the other hand, the atmosphere can return extraneous energy to a sensor and thereby modify the observed energy signal from ground objects.

20-7 PRACTICAL REMOTE SENSING SYSTEMS

Considering the varied nature of electromagnetic radiation, the character of its sources, its interaction with objects, and its interaction with the atmosphere, it can be appreciated that the design and utility of practical remote sensing systems also vary. Subsequent discussion in this chapter presents the salient characteristics of currently available sensor hardware. Multiband photographic systems, radiometers, scanners, side-looking airborne radar systems, and passive microwave units are included. Figure 20-5 can be treated as a graphic frame of reference for discussion of these various systems. It shows the spectral regions in which these various sensors operate, gives relative transparency of the atmosphere within these regions, and illustrates the relationship that these regions bear to energy from the sun at 6000°K and the earth at 300°K.

Before discussing individual sensor systems, a few characteristics illustrated by Fig. 20-5 are worthy of particular note. Photographic systems efficiently "look" through the atmosphere at energy supplied by the sun and *reflected* from objects. The atmospheric window within which photography operates, as shown on the lower portion of Fig. 20-5, extends from wavelengths slightly shorter than visible, out to about 1 μm in length. "Thermal" instrumentation senses energy *emitted* from the earth through the 3- to 5-μm and 8- to 14-μm atmospheric windows. Multispectral scanners can respond to both *reflected and emitted* radiation. Cameras and scanners are said to be *passive* systems in that they do not generate their own energy source, but rather respond to an externally supplied source of energy (from the sun, or the radiating features themselves).

Radar systems, which operate in the 1-mm to 1-m range of the electromagnetic spectrum, are *active* systems because they incorporate their own energy source. As will be discussed later, radar systems beam energy toward the ground and record the reflected return signal. Passive microwave units operate essentially in the radar range, but respond to extremely low levels of naturally emitted energy.

The device used to transport a sensor is termed a *platform*. Platforms can be anything from a stepladder to the space shuttle. To date, the most common platforms used in remote sensing have been low-altitude aircraft, high-altitude aircraft, and earth-orbiting satellites. Regardless of the platform used to carry any given sensor, the operating principles involved are fundamentally the same. For convenience each sensor is discussed in the context of airborne operation first. Near the end of this chapter a summary is given of the operation of these sensors from space platforms—with emphasis on the Landsat series of satellites. Discussion herein is centered on the operating principles of the most common sensing systems, rather than on the various applications of remote sensing. These applications are so numerous and varied that the literature abounds with books, articles, and reports on the subject. The bibliography at the end of this chapter represents a small cross section of this literature.

20-8 REFERENCE DATA

Before describing the various sensor systems available, it should be emphasized that rarely—if ever—is remote sensing employed without the use of some form of *reference data*. This normally refers to conventional information about the objects being studied in a remote sensing analysis. Reference data can range from laboratory reports about soil or water samples, to field observation of such things as tree identification, crop vigor, temperature, etc. Historically, reference data have been referred to by the term *ground truth*.

Reference data is used to aid in the interpretation of remotely sensed data, and it often forms the basis for evaluating the accuracy of a remote sensing analysis. In some cases, reference data are actually used to calibrate a sensor—such as in a study of earth surface temperature. Ground control in a photogrammetric analysis is a form of reference data. In many remote sensing applications both geometric and nongeometric ground observations are needed. As with the establishment of ground control, the collection of reference data is very labor intensive, costly, and essential. In years to come, as hardware and the state of the art improve, the dependency of remote sensing analyses upon reference data will likely diminish, but it will never be eliminated. Accordingly, remote sensing should be perceived as a supplement to, not a replacement for, conventional resource inventory and monitoring techniques.

20-9 MULTIBAND PHOTOGRAPHIC SYSTEMS

Multiband photographs are photographs taken simultaneously from the same geometric vantage point, but with different film-filter combinations. Multiband images can be taken with a multilens frame camera or a multicamera system (see Sec. 4-2.2

and Fig. 4-6). Figure 20-6 illustrates a bank of 70-mm format cameras which can be used to acquire multiband photography in four spectral bands. Typically, multiband photographs depict the identical scene, imaged on black-and-white infrared film, but filtered for discrete wavelength bands in the blue, green, red, and reflected infrared portions of the spectrum. The "best" image or combination of images for discriminating a given scene object varies with the spectral response pattern for that object. The separation, or "taking apart" of object reflectances through multiband photography normally yields enhanced contrast between different terrain feature types and between different conditions of the same feature type. To optimize this contrast, film-filter combinations are chosen for the specific features of interest in spectral regions where the maximum spectral reflectance differences are known, or are anticipated to exist.

One basic problem in using multiband photography is the fact that simultaneous analysis of multiple images of a single ground scene is inherently difficult. *Color additive viewers* are designed to assist in the interpretation of multiband photography. These devices normally incorporate four projectors which are aimed at a single viewing screen. Each projector has a variable brightness and color filter control. In the operation of the viewer, the image analyst uses up to four black-and-white multiband images which are in a positive transparency format. The transparency for a particular spectral band is placed in a projector and projected through the color filter (blue, green, or red) assigned to that band by the analyst. Optical superposition of multiple bands in this fashion results in the production of *color composite images* on the viewer screen in accordance with color additive principles. Normally three projectors are used simultaneously. Optical combination of spectral positives from the blue, green, and red portions of the spectrum results in a "true"-color display. Projection of positives taken in the green, red, and photographic infrared results in a "false"-color display similar to color infrared photography. Through arbitrary assignment of positives and color filters, "exotic"-color displays can be created which often enhance discrimi-

Figure 20-6 Multiband photographic system consisting of a bank of four 70-millimeter Hasselblad cameras. (*Courtesy Paillard, Inc.*)

nation of features of interest. For example, the system might be adjusted to display a particular crop type in a unique, readily discriminated color.

Though they are relatively simple and economical tools, multiband photographic systems naturally have the inherent limitation of sensing only in the photographic portion of the spectrum. Likewise, the multiband photographic image recording format produces data which are not inherently amenable to automated data processing. Multispectral scanners, to be described later in this chapter, circumvent these shortcomings—but at the expense of the spatial resolution and geometric fidelity of photographic systems.

20-10 RADIOMETERS

A *radiometer* is a nonimaging sensor that responds to radiant energy electronically. The spectral sensitivity of radiometers ranges from 0.3 to 14 μm. Their physical characteristics vary with application. In general the following three components are common to all radiometers:

1. Energy collection optics
2. Wavelength selecting unit
3. Signal converting unit

The function of the energy collection optics is to establish the field of view and effective aperture of the radiometer and to transmit incoming energy to the wavelength selecting unit. Collecting optics of radiometers range from fiber optic devices for laboratory or close-range field work, to telescopic collectors for sensing at great distances.

The wavelength selecting unit isolates the incoming electromagnetic energy into discrete spectral bands. Incident energy in the 0.3- to 1.2-μm range is typically separated by refractive techniques using prisms (see Fig. 3-7) or diffraction gratings. Discrete energy bands in the 1.2- to 14-μm range are isolated using interference techniques because refractive optics are ineffective at these wavelengths. Both techniques separate or "fan out" incoming energy, and by placing an energy detector in the proper geometric position in the fan, only the energy band of interest is sensed.

Energy from the wavelength selecting unit is radiated to a detector in the signal conversion unit. The detector may be a photomultiplier tube for energy in the 0.3- to 1.2-μm-wavelength range, or a heat sensitive transducer for energy in the middle or far-infrared ranges. In any case, the energy is converted to an electrical signal, amplified, and recorded. For *static* radiometry, the signal is displayed as a meter reading whose variation with wavelength (when referred to a standard) describes the spectral response pattern for the target. In *dynamic airborne* radiometry, the detector signal is generally recorded as a continuous radiance profile on a strip chart recorder or other recording device.

Of particular importance in airborne applications are "thermal radiometers," which provide a profile of thermal infrared emittance generally in the 8- to 14-μm region of the electromagnetic spectrum. Figure 20-7 illustrates the basic operation of such a system. In Fig. 20-7*a*, an aircraft carrying a radiometer flies over an area. A plan view of the narrow path being sampled beneath the aircraft is shown in Fig. 20-7*b*. Figure 20-7*c* illustrates the recorded output from the system's detector. This output can be converted to temperatures in accordance with Eq. (20-3) if the emissivities of the surface features along the flight path are known and atmospheric effects are taken into account. It is important to note that radiometric output does not indicate true temperatures directly. Rather, levels of emitted radiation are recorded, and these, when properly reduced and related to ground reference temperatures, yield temperatures accurate to within 1°C. For this reason thermal radiometers are often termed *airborne profiling thermometers*. The continuous record provided from thermal radiometers is an ideal calibration reference for other radiometric systems such as *thermal scanners*.

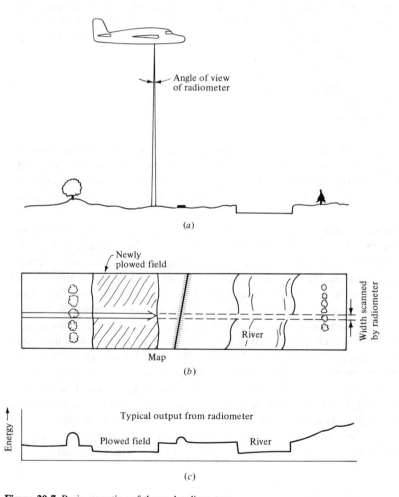

Figure 20-7 Basic operation of thermal radiometers.

20-11 SCANNERS

A scanner is conceptually an airborne radiometer that senses a continuous strip of terrain beneath the path of the aircraft in which it is carried. The strip is composed of contiguous *scan lines* repeatedly taken transverse to the direction of flight, as illustrated in Fig. 20-8. A rotating mirror is used to scan the system's field of view from one side of the aircraft to the other. The line-scanning process is analogous in many respects to the acquisition of panoramic photography (see Chap. 17). As the instrument scans the object space, responses are amplified and recorded at a rate proportional to the speed of the aircraft. The result is a series of electrical response profiles for the contiguous scan lines.

Because detector signals are in electrical formats, they are conveniently utilized in a number of ways. In-flight monitoring is accomplished by displaying a time versus amplitude trace of detector responses on a cathode ray tube (CRT). The CRT display can be photographed using a strip camera, thereby producing "imagery" with tonal variations representing differential levels of reflected or emitted energy within the scanned scene. Simultaneous recording of the detector signal on magnetic tape permits further manipulation of the sensor data. Such manipulation may include, for example, partial geometric rectification of the scanner display and image enhancement using electronic computers.

The collection and recording sequence employed in a "single channel" scanner is schematically represented by Fig. 20-9. The wavelength band sensed by a single-channel scanner is a function of the spectral sensitivity of its single detector. A *multispectral scanner* (MSS), on the other hand, employs a wavelength selecting subsystem and multiple detectors to sense energy in numerous wavelength bands simultaneously. Imagery prepared from the response of each spectral channel shows the same scene viewed at the same time under the same conditions.

Figure 20-10 illustrates the output from six channels of an MSS operated over the range from 0.38 to 0.86 μm. Note how the appearance of earth surface features changes as a function of the wavelength band of sensing. For example, the stream in

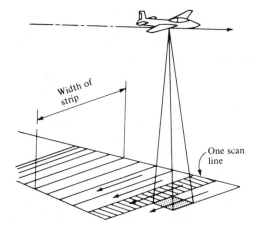

Figure 20-8 Scanner coverage.

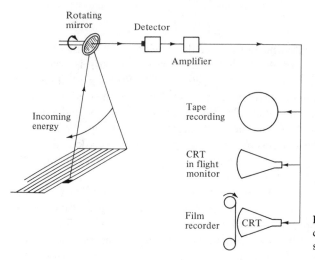

Figure 20-9 Schematic diagram of collection and recording sequence of single-channel scanners.

the upper right portion of the scene is extremely dark in channel 6 relative to the surrounding vegetated areas. In other bands, the reflectance of these two feature types is nearly equal. Note also how many buildings appear much brighter than the grass and trees in channel 1 and the reverse is true on channel 6. Such differences in spectral response can be used to discriminate between the different feature types appearing in the imagery.

The interpretation of multichannel imagery is normally accomplished using the computer-assisted procedure of *spectral pattern recognition*. While the details of this process are outside the scope of this discussion, suffice it to say that the procedure employs the image data in digital form rather than in pictorial form. The signal for each detection is sampled along each scan line and expressed digitally on computer compatible tape (CCT). The "image" is then expressed to the computer as a matrix of contiguous picture elements, or *pixels*. The numerical response of each detector within each pixel is then available for use in various algorithms designed for automatic interpretation of the image data. Such techniques have proven to be extremely powerful in applications such as crop identification. (It should be noted that these computer-assisted procedures are not completely "automatic" in that they require some form of reference data to "train" the computer to interpret the data.)

A particular type of scanner commonly applied in remote sensing work is a single-channel *thermal scanner*, which operates generally in the 8- to 14-μm range of the electromagnetic spectrum. Thermal scanners depict variations in the thermal energy *emitted* from the scene below the aircraft. They have been particularly valuable in studying such things as thermal effluents from industrial processes. Figure 20-11a is a portion of a strip of thermal imagery showing the thermal plume of a power plant cooling effluent being discharged into a lake. In the figure, the lighter the tone, the hotter the water. This thermal imagery was analyzed in conjunction with thermal radiometer data to study the thermal structure of the plume and its related effect on the receiving body of water. Since emitted energy is a surface phenomenon, the

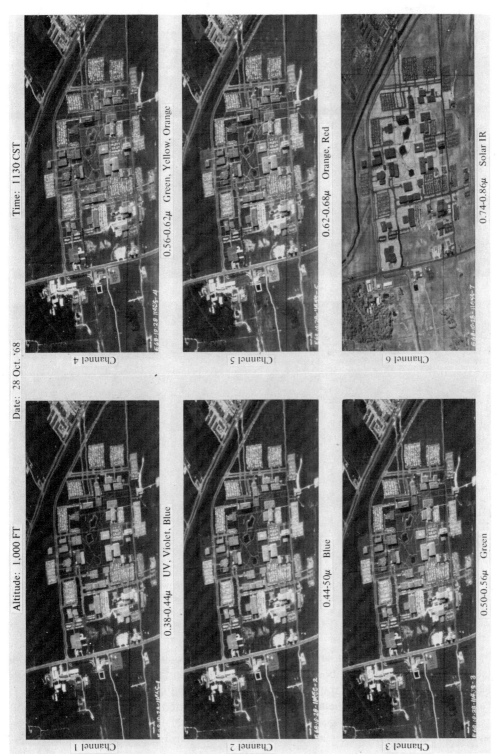

Altitude: 1,000 FT   Date: 28 Oct. '68   Time: 1130 CST

Channel 1   0.38-0.44μ   UV, Violet, Blue
Channel 2   0.44-50μ   Blue
Channel 3   0.50-0.56μ   Green
Channel 4   0.56-0.62μ   Green, Yellow, Orange
Channel 5   0.62-0.68μ   Orange, Red
Channel 6   0.74-0.86μ   Solar IR

Figure 20-10 Six simultaneously derived images of the same scene in different spectral ranges. (*Courtesy Bendix Aerospace Systems Division, Ann Arbor, Michigan.*)

Figure 20-11 (*a*) Thermal scanner image of the discharge plume of a power plant cooling effluent into a lake. Lighter tones indicate hotter water. (*b*) Black-and-white panchromatic aerial photo taken over the area of the plume. Note that the plume is not visible because panchromatic film does not respond to thermal energy. (*Courtesy Remote Sensing Program, University of Wisconsin, Madison.*)

indicated response is that of surface water temperatures only. Note the dark lines in the plume caused by power boats circulating cooler subsurface water to the surface. Note also on the figure that rough geometric outlines of buildings, cars, streets, etc., can be identified from their temperature differences detected by the thermal scanner.

Figure 20-11*b* is a black-and-white panchromatic aerial photo of the same area shown in Fig. 20-11*a*, taken at the same time. Since panchromatic film is sensitive only to the visible portion of the electromagnetic spectrum, it was unable to record the presence of the thermal plume.

Because they sense energy emitted from objects, thermal infrared systems can operate day or night; in fact, diurnal coverage of an area often yields differences in images that facilitate interpretation and identification. Objects which undergo relatively rapid cooling during the night, for example, will generally appear drastically different on day imagery than on night imagery, and identifications can often be made based on knowledge of their cooling and heating rates as compared to other objects. An example of thermal scanner imagery obtained at night is shown in Fig. 20-12. Note that details are as readily interpreted from this image as they are from the daytime image of Fig. 20-11*a*.

At this point it is worth reiterating the fundamental difference between sensing *reflected* infrared energy and *emitted* infrared energy. While the former can be detected using infrared-sensitive film (Fig. 20-4), the latter involves the use of a thermal radiometer or scanner. Because both forms of energy are invisible and are termed "infrared," it is a common misconception that one can photograph normal levels of heat with infrared film. Such heat sensing, however, can only be done using a thermal radiometer or scanner.

As with aerial photography, imagery obtained from scanners is subject to scale variations and geometric distortions, but more severely. A complete discussion of these scale variations and image distortions is beyond the scope of this text, but Fig. 20-13*a* illustrates in a qualitative manner the nature of scale variations typical of scanner images. The figure shows the decrease in scale that occurs with an increase in distance

Figure 20-12 Thermal scanner image obtained at night. Cars can be identified at *A*, commercial buildings at *B*, a highway interchange at *C*, sandy soil at *D*, bare soil at *E*, trees at *F*, grass at *G*, and residential area at *H*. (*Courtesy Remote Sensing Program, University of Wisconsin, Madison.*)

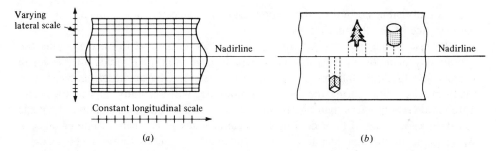

Figure 20-13 (*a*) Qualitative nature of scale variation of scanner imagery. (*b*) Relief displacement of scanner imagery.

from the *nadir line* (line on imagery representing terrain vertically beneath the aircraft's path). Figure 20-13*b* indicates the nature of relief displacement of scanner imagery. Note that in all cases vertical lines in the object space are recorded as lines which diverge *perpendicularly* to the nadir line. Thus, relief displacement on a scanner image is one-dimensional.

From the examples presented herein, it should be apparent that scanner images depart radically from their photographic counterparts in terms of geometry, resolution, and fidelity. As remote sensing technology evolves, and as military systems become declassified, resolution and geometric fidelity of scanners will continue to improve.

20-12 SIDE-LOOKING AIRBORNE RADAR

Radar is an acronym for *ra*dio *d*etecting *a*nd *r*anging. Some familiar forms of radar are the *P*lan *P*osition *I*ndicator (PPI) system typical of local weather station installations, and doppler radar used for vehicle speed control. In many respects radar is similar to photographing with a flashbulb. The flashbulb illuminates the object and the camera captures the rays of light that are reflected from the object. With radar, instead of using light, a narrow beam of microwave radio energy is sent out by a directional antenna and a radio receiver captures the reflected radio waves. Again, because it generates and transmits its own energy, radar is an *active* system—as opposed to a *passive* system which generates no energy of its own but instead senses only natural radiation.

Of particular interest and utility in aerial remote sensing are *S*ide-*L*ooking *A*irborne *R*adar (SLAR) systems. As with all radar systems, SLAR units transmit radio signals in the 1-mm- to 1-m-wavelength range and monitor the travel time for them to return as reflections from objects within the field of transmission. Since the propagation rate of the source energy is a known constant, travel times can be used directly to determine distances to reflecting targets. As shown in Fig. 20-14, the transmitting and receiving antenna *A* for a SLAR system directs and receives signals in thin slices approximately normal to the direction of flight. As the antenna advances in the direction of flight, transmitted pulses initiate the sweep of a CRT trace. Return signals corresponding to

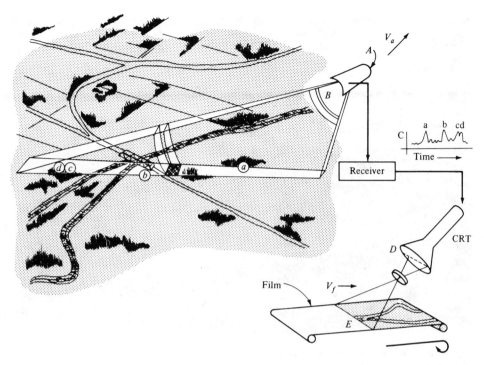

Figure 20-14 Recording sequence of side-looking airborne radar. (*Courtesy Westinghouse Electric Corporation.*)

various terrain features appear on the trace and have their geometric positions on the CRT discretely defined on the basis of their respective travel times. Luminances of points recorded on the CRT are determined by the physical character and configuration of the objects in the field of view. Individual CRT traces can be recorded on film by means of a strip camera.

In Fig. 20-14 the antenna A is being repositioned longitudinally at aircraft velocity V_a. Each transmitted pulse B returns "echoes" from objects within the antenna beam. The time-versus-amplitude video signal C is imaged on a trace line D of the CRT. Ground objects such as the river at E are imaged on the film as they advance past the tube at a velocity V_f, which is proportional to the velocity of the aircraft. The resulting images have photographic clarity and reasonably good spatial resolution, as illustrated by the example in Fig. 20-15. Lateral ground coverages of SLAR strips are extensive, on the order of from 10 to 12 mi and greater. SLAR images contain numerous geometric distortions which are outside the realm of this discussion.

Because they are active systems, SLAR units can operate day and night and in a variety of transmitting and receiving modes. Source wavelength and polarization can be chosen to maximize object discrimination. Recorded returns are in an electrical format and therefore can be conveniently stored and manipulated. Depending on wavelength, SLAR possesses the ability to penetrate clouds, precipitation, and light vege-

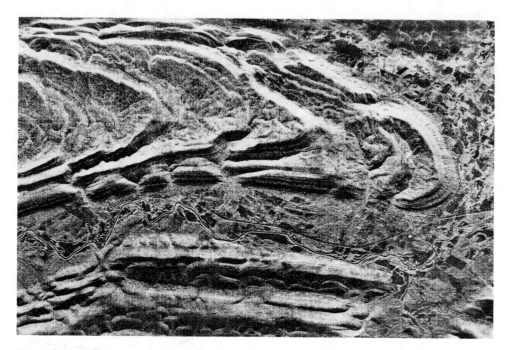

Figure 20-15 Side-looking airborne radar image showing the Ouchita Moutains, Oklahoma. (*courtesy Westinghouse Electric Corporation.*)

tation. For this reason it was used to map the perpetually overcast Panama Canal region of Central America. By flying the same region from opposite directions, or from two flying heights, stereoscopic SLAR coverage can be achieved. Geologic structural features such as faults, lineations, and landforms that are often missed in ground observations or masked on other imagery may be detected from SLAR images.

20-13 PASSIVE MICROWAVE SYSTEMS

As previously mentioned, the earth is a source of emitted radiation that is closely approximated by a blackbody radiator at a temperature of 300°K. The radiation curve of a blackbody heated to 300°K peaks at a wavelength of about 10 μm, but energy of lesser intensity is emitted in wavelengths out to 1 m and longer. As previously discussed, thermal radiometers and thermal scanners are used to sense energy in the areas of relatively intense earth radiation near the peak of the curve. Passive microwave systems are employed to sense the extremely low-intensity energy of wavelengths from 1 mm to 1 m. This is the same spectral region in which radar systems operate: however, passive microwave systems, as implied by their name, are passive systems, whereas radar systems are active.

As with radar, passive microwave units employ antennas to detect radiation. The output from these devices is essentially the same as that from thermal radiometers. The major difference is that the longer-wavelength radiation sensed by passive microwave systems often contains information on subsurface conditions of objects. For this reason, passive microwave systems have been used to obtain soil temperature data and related moisture conditions.

The fact that relatively little energy is radiated in the spectral region in which passive microwave operates would be of little consequence if the sensor could be held stationary so it could gather energy over a long period of time. Because of their forward motion, however, airborne passive microwave units cannot concentrate or "dwell" on a given area long enough to gather sufficient energy to achieve a satisfactory reading. To overcome this problem, the size of the *resolution cell* (area sensed at any instant of time) is increased to gather enough energy to get an adequate signal from each area sensed. As a consequence of the large resolution cells, the overall resolution of the passive microwave record is poor.

Passive microwave units may be of the *nonscanning* or *scanning* types. Nonscanning units sense only a strip of terrain directly below the aircraft, and the output is simply a profile indicating energy emitted by ground objects. This is essentially the same technique as that employed with thermal radiometers. Each system produces a temperature-related profile, the only difference being the wavelength of energy sensed.

Scanning passive microwave units differ from nonscanning units in that the antenna scans from side to side as the plane passes over the ground. The resulting series of profiles is usually recorded on magnetic tape and processed later in a computer. The computer can assign colors to different energy levels, thereby producing a false-color picture in which particular colors correspond to given ranges of temperatures on the scanned terrain.

20-14 REMOTE SENSING FROM SPACE

The use of satellites as sensor platforms has made possible the acquisition of repetitive, high-resolution multispectral data of the earth's surface on a global basis. Satellite images not only cover very large areas, but many are also virtually free of image distortions due to relief because of the typically high flying heights. Figure 20-16 is an image obtained from the multispectral scanner system (MSS) on board the first Landsat satellite (Landsat-1). It shows the San Francisco Bay area and a large portion of the Sacramento Valley as recorded from a flying height of approximately 570 mi. Note the significant amount of detail that appears in the image. The city of San Francisco in the lower left portion of the picture is partially obscured by a fog bank. It is of interest to compare the coverage of this single image (with coverage of approximately 115 mi square) to the coverage of the 2,000-photo mosaic of the same area that is shown in Fig. 10-1.

Landsat sensor systems, to date, have included a combination of return beam vidicon (RBV) cameras and multispectral scanning (MSS) systems. The Landsat-1 satellite system, for example, included three multispectral RBV cameras and a four-channel multispectral scanner. Neither Landsat-1 nor Landsat-2 is in operation at the

Figure 20-16 Imagery obtained from Landsat-1 satellite. This image, made from an approximately 570-mile flying height, shows the entire San Francisco Bay area and a major portion of the Sacramento Valley. (*Courtesy NASA.*)

current time. Landsat-3 is currently active and it carries two single-band RBV cameras and a five-channel MSS system (of which only four channels are operative due to technical problems). Film records of the detector signals from each of the MSS bands can be used separately or in any combination. Figure 20-16 is actually a black-and-white reproduction of a "color composite" picture made from three bands of the multispectral scanner of Landsat-1. In accordance with the color additive process, each of the three primary colors blue, green, and red were assigned to three of the satellite's recording bands and later recombined in a single composite image. Blue on the composite image corresponds to channel 4, a green-sensitive channel (0.5- to 0.6-μm wavelengths); green corresponds to channel 5, a red-sensitive channel (0.5- to 0.6-μm wavelengths); and red corresponds to channel 6, a near-infrared sensitive channel (0.7- to 0.8-μm wavelengths). The combination of these channels produces a color composite image very similar to an ordinary color-infrared photograph.

Figures 20-17a and 20-17b are portions of black-and-white reproductions made from the responses of Landsat-1 MSS channels 5 (red channel), and 6 (near-IR channel), respectively. Some rather striking differences in the two responses can be seen. Note, for example, that the lakes at 1 and 2 in a can hardly be discerned. This is primarily due to suspended material in the water which reflects red energy nearly the same as the surrounding land area. These lakes absorb almost all of the near-IR energy, however, and therefore appear the same as the lakes at 3 in b. There is also a marked difference in the responses obtained from urban areas. The city of Madison, Wisconsin, located at 3, for example, is visible as a lighter toned area on a, while there is practically no indication of urbanization in that area on b.

Landsat MSS data have been used extensively in computer compatible tape format. In this form the data are amenable not only to the automated techniques of spectral pattern recognition and numerical enhancement, but also to geometric and radiometric correction. These data can also be merged digitally with other data sources (such as digital terrain data) in geographic information systems. To date, Landsat data have been used in tasks ranging from crop forecasting and mineral exploration, to applications as diverse as pollution detection, rangeland monitoring, and commercial fishing. Landsat data are a primary source of basic resource information in many developing nations throughout the world. At the current time, non-United States Landsat receiving stations are in operation in Argentina, Australia, Brazil, Canada, India, Italy, Japan, South Africa, and Sweden. Stations are under development in China and Thailand.

Numerous future space remote sensing missions are in various stages of planning and development. Landsat-D is scheduled for launch in 1982 and it is to contain an advanced seven-channel multispectral scanner, the *thematic mapper*. Relative to previous Landsats, this system should offer better spatial resolution (30 m vs. 80 m), maximized vegetation discrimination for agricultural applications, and a thermal sensing capability. The Space Shuttle also affords a new launch vehicle for placing earth resource satellites in orbit. The Shuttle itself can also be used as a platform for sensors. A host of engineering studies involving a range of remote sensors is planned for the Shuttle. Among other things, the Shuttle will enable a detailed assessment of the utility of high-resolution radar data collected from space. The Shuttle is also scheduled to carry the Large Format Camera (see Sec. 4-2.1 and Fig. 4-4). This camera has a 305-mm-focal-length lens and a 230 × 460 mm film format. From an altitude of 300 km, each frame will cover 225 × 450 km at a scale of 1:1,000,000 and a resolution of 15 m.

Also planned for the future are a number of space remote sensing activities by many non-United States agencies. For example, the first French *Satellite Probatoire pour l'Observation de la Terre* (SPOT-1) will carry two high-resolution visible (HRV)

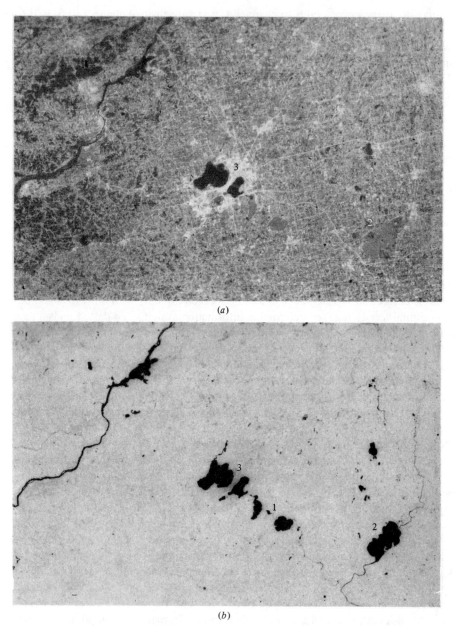

(a)

(b)

Figure 20-17 Imagery obtained from the multispectral scanner (MSS) or the Landsat-1 satellite. (a) is the response obtained in the red-sensitive channel 5, and (b) is the response from the near infrared-sensitive channel 6. Note that due to suspended material in the water, the lakes at 1 and 2 can hardly be discerned in (a,) but these same lakes are readily discernible in (b). Note also that the city of Madison, Wisconsin, which is located at 3, is visible as a lighter toned area in (a), while there is practically no indication of urbanization in that area in (b). These examples illustrate how responses from different parts of the electromagnetic spectrum may be useful in accentuating different objects, thereby aiding in their interpretation and identification. (*Courtesy NASA.*)

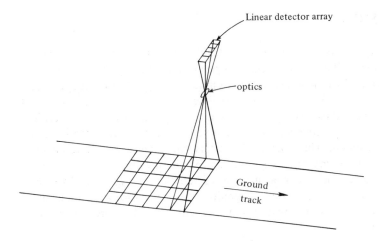

Figure 20-18 Geometry of pushbroom scanning procedure using a linear detector array.

scanners of advanced design. The spatial resolution of the system will be selectable at either 20 m or 10 m, the field of view will be pointable within an 800-km range, and orbit to orbit stereoscopic coverage can be produced. This system, and others under development, will acquire data using a process called *pushbroom scanning*. Instead of a single detector for each spectral band, these future sensors will contain thousands of individual detectors arranged in *multispectral linear arrays* (MLA). Figure 20-18 illustrates how the pushbroom scanning process operates. For each spectral band, a linear array of detectors is located in the focal plane of the sensor's optical system and oriented perpendicular to the ground track of the satellite. A line of data is obtained by recording the response of the detector elements along the array. Successive lines of ground coverage are thereby obtained as the satellite moves over the earth.

Linear array technology has certain advantages over the conventional line scanning procedure. Pushbroom scanners have no moving parts, are lighter in weight, use less power, present a simpler data handling problem, and have a longer life expectancy. The current disadvantage to such systems is that available detectors are not suitable for use with wavelengths longer than about 1.05 μm. However, detectors are currently under development which will extend this sensitivity out to the thermal portion of the spectrum.

A principal advantage of linear array technology is the elimination of the need for a moving scan mirror, which virtually eliminates the distortions and displacements that hinder the geometric integrity of current MSS data. Because of this inherent advantage, MLAs are being proposed for various systems which could obtain stereoscopic coverage of sufficient geometric integrity to permit accurate topographic mapping at scales on the order of 1:50,000.

20-15 CONCLUSION

Space has necessarily limited this discussion. Basic concepts that dictate the design and utility of common civilian remote sensing systems have been presented. Most of the sensors discussed look at the terrestrial environment through "different eyes" and with ever increasing rates of data-generation capabilities. Processing of remotely sensed data into formats useful to those who monitor and manage the environment is a critical step in the remote sensing sequence and undoubtedly one that will receive much attention in future research. The state of this art is changing at a very rapid rate. Continued evolution of automated storage and interpretation techniques, advances in sensor hardware, and an increasing awareness of remote sensing capabilities and applications, combined with a growing recognition of our need to better understand the impact of our activities on the natural environment, should ensure a significant future for remote sensing.

REFERENCES

Alfoldi, T. T., and J. C. Munday, Jr.: Water Quality Analysis by Digital Chromaticity Mapping of Landsat Data, *Canadian Journal of Remote Sensing,* vol. 4, no. 2, p. 108, 1978.
American Society of Photogrammetry: "Manual of Photographic Interpretation," Falls Church, Va., 1960.
———: "Manual of Color Photography," Falls Church, Va., 1968.
———: "Manual of Remote Sensing," Falls Church, Va., 1975.
Barr, D. J., and R. D. Miles: SLAR Imagery and Site Selection, *Photogrammetric Engineering,* vol. 36, no. 11, p. 1155, 1970.
Barrett, E. C., and L. F. Curtis: "Introduction to Environmental Remote Sensing," Halsted Press, John Wiley & Sons, Inc., New York, 1976.
Berrill, A. R., and E. Clerici: Statistical Tests of Digital Rectification of Landsat Image Data, *Australian Surveyor,* vol. 28, no. 8, p. 497, 1977.
Bryant, E., et al.: Landsat for Practical Forest Type Mapping, a Test Case, *Photogrammetric Engineering and Remote Sensing,* vol. 46, no. 12, p. 1575, 1980.
Chevrel, M., et al.: SPOT Satellite Remote Sensing Mission, *Photogrammetric Engineering and Remote Sensing,* vol. 47, no. 8, p. 1163, 1981.
Colwell, R. N., et al.: Basic Matter and Energy Relationships Involved in Remote Reconnaissance, *Photogrammetric Engineering,* vol. 34, no. 5, p. 761, 1963.
———: Some Significant Elements in the New Remote Sensing Panorama, *Surveying and Mapping,* vol. 34, no. 2, p. 133, 1974.
———: Remote Sensing as an Aid to the Inventory and Management of Natural Resources, *Canadian Surveyor,* vol. 32, no. 2, p. 183, 1978.
Derenyi, E. E.: Planimetric Accuracy of Infrared Line Scan Imagery, *Canadian Surveyor,* vol. 28, no. 3, p. 247, 1974.
——— and S. C. Macritchie: Photogrammetric Application of Skylab Photography, *Canadian Surveyor,* vol. 34, no. 2, p. 123, 1980.
Doyle, F. J.: The Next Decade of Satellite Remote Sensing, *Photogrammetric Engineering and Remote Sensing,* vol. 44, no. 2, p. 155, 1978.
Eastman Kodak Co.: "Kodak Wratten Filters for Scientific and Technical Use," 1st ed., Kodak Scientific and Technical Data Book No. B-3, Rochester, New York, 1970.
———: "Applied Infrared Photography," Kodak Publication M-28, Rochester, New York, 1977.
Estes, J. E., et al.: Measuring Soil Moisture with an Airborne Imaging Passive Microwave Radiometer, *Photogrammetric Engineering and Remote Sensing,* vol. 43, no. 10, p. 1273, 1977.

—— and L. W. Senger (eds.): "Remote Sensing: Techniques for Environmental Analysis," Hamilton Publishing Co., Santa Barbara, Calif., 1974.

Fritz, N. L.: Optimum Methods for Using Infrared-Sensitive Color Film, *Photogrammetric Engineering*, vol. 33, no. 10, p. 1128, 1967.

Gammon, P. T., and V. Carter: Vegetation Mapping with Seasonal Color Infrared Photographs, *Photogrammetric Engineering and Remote Sensing*, vol. 45, no. 1, p. 87, 1979.

Gerbermann, A. H., et al.: Color and Color I. R. Films for Soil Identification, *Photogrammetric Engineering*, vol. 32, no. 4, p. 359, 1971.

Gregory, A. F.: Remote Sensing: A New Look at the Canadian Environment, *Canadian Surveyor*, vol. 25, no. 2, p. 131, 1971.

Henderson, F. M.: Land-Use Analysis of Radar Images, *Photogrammetric Engineering and Remote Sensing*, vol. 45, no. 3, p. 295, 1979.

Holz, R. K. (ed.): "The Surveillant Science: Remote Sensing of the Environment," Houghton Mifflin Company, Boston, 1973.

Hudson, R. D., Jr.: "Infrared System Engineering," John Wiley & Sons, Inc., New York, 1969.

Jackson, M. J., et al.: Urban Land Mapping from Remotely Sensed Data, *Photogrammetric Engineering and Remote Sensing*, vol. 46, no. 8, p. 1041, 1980.

Jensen, N.: "Optical and Photographic Reconnaissance Systems," John Wiley & Sons, Inc., New York, 1968.

Klemos, V., and W. D. Philpot: Drift and Dispersion Studies of Ocean-Dumped Waste Using Landsat Imagery and Current Drogues, *Photogrammetric Engineering and Remote Sensing*, vol. 47, no. 4, p. 333, 1981.

Klooster, S. A., and J. P. Scherz: Water Quality by Photographic Analysis, *Photogrammetric Engineering*, vol. 40, no. 8, p. 927, 1974.

Kloostermann, B., et al.: Computer-Assisted Color Separation for the Production of Thematic Maps, *Canadian Surveyor*, vol. 28, no. 1, p. 31, 1974.

Leberl, F.: Imaging Radar Applications to Mapping and Charting, *Photogrammetria*, vol. 32, no. 3, p. 75, 1976.

——: Accuracy Analysis of Stereo Side-Looking Radar, *Photogrammetric Engineering and Remote Sensing*, vol. 45, no. 8, p. 1083, 1979.

Lillesand, T. M., and R. W. Kiefer: "Remote Sensing and Image Interpretation," John Wiley & Sons, Inc., New York, 1979.

Lintz, J., and D. S. Simonnett (eds.): "Remote Sensing of Environment," Addison-Wesley Publishing Company, Reading, Mass., 1976.

MacDonald, R. B., and F. G. Hall: Global Crop Forecasting, *Science*, vol. 208, p. 670, 1980.

Madden, J. D.: Coastline Delineation by Aerial Photography, *Australian Surveyor*, vol. 29, no. 2, p. 76, 1978.

Mead, R. A.: Occupational Preparation in Remote Sensing, *Photogrammetric Engineering and Remote Sensing*, vol. 45, no. 11, p. 1513, 1979.

Merideth, R. W.: Doctoral Dissertations Pertaining to Remote Sensing and Photogrammetry—A Selected Bibliography, *Photogrammetric Engineering and Remote Sensing*, vol. 47, no. 5, p. 617, 1981.

Mott, P. G., and H. J. Chismon: The Use of Satellite Imagery for Very Small Scale Mapping, *Photogrammetric Record*, vol. VIII, no. 46, p. 458, 1975.

Payne, D. P.: "Aerial Photography and Image Interpretation for Resource Management," John Wiley & Sons, Inc., New York, 1981.

Rudd, R. D.: "Remote Sensing: A Better View," Duxbury Press, North Scituate, Mass., 1974.

Sabins, F. F., Jr.: "Remote Sensing: Principles and Interpretation," Freeman Press, San Francisco, 1978.

Sayn-Wittgenstein, L., et al.: The ERTS Experiments of the Canadian Forestry Service, *Canadian Surveyor*, vol. 28, no. 2, p. 110, 1974.

Scarpace, F. L., et al.: Scanning Thermal Plumes, *Photogrammetric Engineering and Remote Sensing*, vol. 41, no. 10, p. 1223, 1975.

—— and B. K. Quirk: Land Cover Classification Using Digital Processing of Aerial Imagery, *Photogrammetric Engineering and Remote Sensing*, vol. 46, no. 8, p. 1059, 1980.

Scherz, J. P., et al.: Photographic Characteristics of Water Pollution, *Photogrammetric Engineering,* vol. 35, no. 1, p. 38, 1969.

Schmugge, T. J.: Microwave Approaches in Hydrology, *Photogrammetric Engineering and Remote Sensing,* vol. 46, no. 4, p. 495, 1980.

Siegel, B. S., and A. R. Gillespie (eds.): "Remote Sensing in Geology," John Wiley & Sons, Inc., New York, 1980.

Slater, P. N.: A Re-Examination of the Landsat MSS, *Photogrammetric Engineering and Remote Sensing,* vol. 45, no. 11, p. 1479, 1979.

———: "Remote Sensing Optics and Optical Systems," Addison-Wesley Publishing Company, Reading, Mass., 1980.

Smith, W. L. (ed.): "Remote Sensing Applications for Mineral Exploration," Dowden, Hutchinson & Ross Publishers, Stroudsburg, Pa., 1977.

Strandberg, C. H.: "Aerial Discovery Manual," John Wiley & Sons, Inc., New York, 1967.

Swain, P. H., and S. M. Davis (eds.): "Remote Sensing: The Quantitative Approach," McGraw-Hill Book Company, New York, 1978.

Thompson, L. L.: Remote Sensing Using Solid-State Array Technology, *Photogrammetric Engineering and Remote Sensing,* vol. 45, no. 1, p. 47, 1979.

Warne, D. K., et al.: Landsat Imagery as a Tool in Regional Planning, *Australian Surveyor,* vol. 28, no. 3, p. 128, 1976.

Wolfe, E. W.: Thermal IR for Geology, *Photogrammetric Engineering,* vol. 37, no. 1, p. 43, 1971.

Yost, E., and S. Wenderoth: Multispectral Color for Agriculture and Forestry, *Photogrammetric Engineering,* vol. 32, no. 6, p. 590, 1971.

PROBLEMS

20-1 Define the term "remote sensing" and briefly discuss some of its applications.

20-2 In what ways can incident electromagnetic energy interact with earth surface features?

20-3 Explain why blue flashbulbs are used for indoor photography with daylight film.

20-4 Describe the term "blackbody" as it relates to remote sensing.

20-5 Why is the aerial photographer concerned about scattered energy? What steps can be taken to circumvent the problems caused by scattered energy?

20-6 It is frequently desirable to use photographic imagery in conjunction with thermal scanners. Why?

20-7 Compare radiometers and scanners.

20-8 What are the advantages of analyzing multispectral scanner data in computer compatible tape format rather than in pictorial form?

20-9 Explain the difference between active and passive remote sensing systems. Classify each of the sensors discussed in this chapter into either the active or passive category.

20-10 Discuss the advantages and disadvantages of side-looking airborne radar (SLAR).

20-11 What is reference data? Discuss how its character might differ for a photographic mapping project as opposed to a thermal scanner pollution study.

20-12 List the advantages and disadvantages of linear array technology for multispectral scanning.

20-13 Consult the references listed at the end of the chapter. Write a brief report on one successful application of remote sensing.

RANDOM ERRORS
AND LEAST SQUARES ADJUSTMENT

A-1 CLASSIFICATION OF ERRORS

In the processes of measuring any quantity, factors such as human limitations, instrumental imperfections, and instabilities in nature render the measured values inexact. Due to these factors, no matter how carefully a measurement is performed, it will always contain some error. Photogrammetry is a science that frequently requires measurements and therefore an understanding of errors—including how they occur and how they are treated in computation—is important. Before proceeding with an analysis of the treatment of errors, it will be helpful to classify errors as follows:

Systematic error. An error in a measurement which follows some mathematical or physical law. If the conditions causing the error are measured, a correction can be calculated and the systematic error eliminated. Systematic errors will remain constant in magnitude and algebraic sign if the conditions producing them remain the same. Because their algebraic sign tends to remain the same, systematic errors accumulate, and consequently they are often referred to as *cumulative* errors. Examples of systematic errors in photogrammetry are shrinkage or expansion of photographs, camera lens distortions, and atmospheric refraction distortions.

Random error. After systematic errors have been eliminated, the errors that remain are called random or *accidental* errors. Random errors are generally small, but they can never be avoided entirely in measurements. They do not follow physical laws like systematic errors, and therefore they must be dealt with according to the mathematical laws of probability. Random errors are as likely to be positive as negative; hence they tend to compensate each other and consequently are often called *compensating* errors. Examples where random errors occur in photogrammetry are in estimating between least graduations of a measuring scale or in indexing a scale.

Note that *mistakes* or blunders resulting from carelessness or confusion are not classified as errors. These must be eliminated by careful procedures and alertness. Henceforth in this discussion it will be assumed that all mistakes and systematic errors have been removed from measured values and only random errors remain. The treatment of these random errors is the subject of this appendix.

A-2 DEFINITIONS

The following definitions of terms necessarily must precede a discussion of random errors:

Observations. Directly observed (or measured) quantities which contain random errors.

True value. The theoretically correct or exact value of a quantity. In measurements, however, the true value can never be determined, because no matter how much care is exercised in measurement, small random errors will always be present.

Error. The difference between any measured quantity and the true value for that quantity. Since the true value of a measured quantity can never be determined, errors are likewise indeterminate, and hence they are strictly theoretical quantities.

Most probable value. That value for a measured quantity which, based upon the observations, has the highest probability. The most probable value is determined through *least squares adjustment*, which is based upon the mathematical laws of probability. The most probable value for a quantity which has been *directly* and independently measured several times with observations of equal weight is simply the mean, or

$$MPV = \frac{\Sigma x}{m} \tag{A-1}$$

In Eq. (A-1), Σx is the sum of the individual measurements and m is the number of observations. Methods for calculating most probable values of quantities determined through *indirect* observations which may or may not be equally weighted are described in later sections of this Appendix.

Residual. The difference between any measured quantity and the most probable value for that quantity. It is the value which is dealt with in adjustment computations, since errors are indeterminate. The term "error" is frequently used when "residual" is in fact meant, and although they are very similar, there is this theoretical distinction.

Degrees of freedom. The number of redundant observations (those in excess of the number actually needed to calculate the unknowns). Redundant observations reveal discrepancies in observed values and make possible the practice of least squares adjustment for obtaining most probable values.

Weight. The relative worth of an observation compared to any other observation. Measurements may be weighted in adjustment computations according to their precisions. A very precisely measured value should logically be weighted heavily in an adjustment so that the correction it receives is smaller than that received by a less precise quantity. If the same equipment and procedures are used on a group of measurements, each observation would be given an equal weight. Weights are discussed further in Sec. A-6.

Standard deviation. A quantity used to express the precision of a group of measurements. Standard deviation is sometimes called the *root mean square error.* It may also be called the *68 percent error,* since according to the theory of probability, 68 percent of the observations in a group should have residuals smaller than the standard deviation. An expression for the standard deviation of a quantity for which a number of direct equally weighted observations have been made is

$$S = \sqrt{\frac{\Sigma v^2}{r}} \tag{A-2}$$

In Eq. (A-2), S is standard deviation, Σv^2 is the sum of the squares of the residuals, and r is the number of degrees of freedom. In the case of m repeated measurements of the same quantity (a common occurrence in photogrammetry), the first measurement establishes a value for the unknown and all additional measurements, $(m - 1)$ in number, are redundant. The units of standard deviation are the same as those of the original measurements.

Example A-1 The 10 values listed in column (*a*) below were obtained in measuring a photographic distance with a glass scale. Each value was measured using the same instrument and procedures; thus equal weights shall be assumed. What is the most probable value and standard deviation of the group of measurements?

(a) Measured values, mm	(b) Residuals,* mm	(c) Squared residuals, mm^2
105.27	-0.005	25×10^{-6}
105.26	-0.015	225
105.29	0.015	225
105.29	0.015	225
105.30	0.025	625
105.27	-0.005	25
105.26	-0.015	225
105.28	0.005	25
105.28	0.005	25
105.25	-0.025	625
$\Sigma = 1,052.75$		$\Sigma v^2 = 2,250 \times 10^{-6}$

1. By Eq. (A-1) the most probable value for a quantity which has been directly and independently measured several times is

$$MPV = \frac{1,052.75}{10} = 105.275 \text{ mm}$$

*2. Residuals listed in column (*b*) above are obtained by subtracting the MPV from each measurement. The squared residuals are listed in column (*c*).

3. By Eq. (A-2), standard deviation is

$$S = \sqrt{\frac{0.002250}{10 - 1}} = \pm 0.016 \text{ mm}$$

A-3 HISTOGRAMS

A *histogram* is a graphical representation of the distribution of a group of measurements or of the residuals for a group of measurements. It illustrates in an "easily digestible" form the nature of occurrence of random errors. A histogram is simply a bar graph of the size of measured values, or the size of residuals, as abscissas versus their frequency of occurrence as ordinates. An example histogram of residuals for 50 measurements of a photo distance is shown in Fig. A-1.

Histograms graphically display the following information about a particular group of measurements:

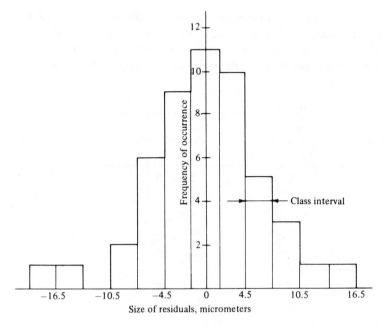

Figure A-1 Histogram for fifty measurements of a photographic distance.

1. Whether the measurements or residuals are symmetrically distributed about some central value.
2. The total spread or dispersion in the measured values or in the residuals.
3. The precision of the measured values. (A tall, narrow histogram represents good precision, while a short, wide one represents poor precision.)

Histograms of varying shape will be obtained for varying personnel and for variations in equipment used. The histogram of many comparator measurements of a photographic distance, for example, would likely produce a very narrow histogram with a high ordinate at the center. A histogram of the same distance measured the same number of times with an engineers scale, and plotted at the same scale, would be wider, with a much lower ordinate value at the center.

When a histogram of residuals is plotted, the residuals are calculated and classified into groups or *classes* according to their sizes. The range of residual size in each class is called the *class interval*. The width of bars on a histogram is equal to the class interval, and for the example of Fig. A-1 it is 3 μm. It is important to select a class interval that portrays the distribution of residuals adequately, and usually a class interval that produces about 15 bars (classes) on the graph is ideal. The number of residuals within each class (frequency of residuals) is then counted and plotted on the ordinate scale versus the residual size for the class on the abscissa scale.

A-4 NORMAL DISTRIBUTION OF RANDOM ERRORS

In the remaining discussion it shall be assumed that all error distributions are *normal*. This is a good assumption in photogrammetry since most distributions are in fact normal or very nearly normal.

A normal distribution, or *Gaussian distribution* as it is often called, is one which graphs as the typical bell-shaped curve shown in Fig. A-2. It is symmetrical about the ordinate for a residual of zero. For a very large group of measurements, this curve may be obtained in much the same way as a histogram except that the sizes of residuals are plotted on the abscissa versus their *relative frequencies* of occurrence on the ordinate. The relative frequency of residuals occurring within an interval Δv is simply the ratio of the number of residuals within that interval to the total number of residuals. If the number of measurements in the group is infinitely large and if the size of the interval is taken infinitesimally small, the resulting curve becomes smooth and continuous, as shown in Fig. A-2.

The equation of the normal distribution curve is

$$y = \frac{he^{-h^2v^2}}{\sqrt{\pi}} \tag{A-3}$$

where y = the ordinate of the normal distribution curve and is equal to the relative frequency of occurrence of residuals between the size of v and $(v + \Delta v)$

e = the base of natural logarithms

h = a constant which depends on the precision of the measurements.

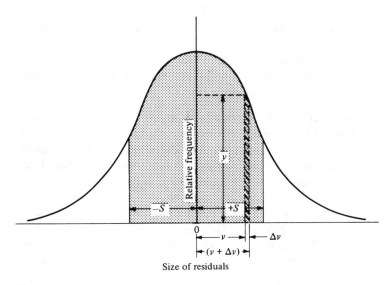

Figure A-2 Normal distribution curve.

The derivation of Eq. (A-3) is beyond the scope of this text, but it can be found in references listed at the end of this appendix. In Fig. A-2 the probability of a residual occurring between the limits of v and $(v + \Delta v)$ is equal to the cross-hatched area under the curve between those limits. It is the product of the ordinate y times the interval Δv. For any single measurement of a group of measurements, the probability that its residual occurs between any two abscissas on the curve (such as between $-S$ and $+S$ of Fig. A-2) is equal to the area under the normal distribution curve between those abscissas. Since for a group of measurements all residuals must fall somewhere on the abscissa scale of the normal distribution curve, the total area under the curve represents total probability and is therefore equal to 1.0. The area under the curve between any two abscissas may be found by integrating Eq. (A-3) between those two abscissa limits. The integration is beyond the scope of this text, but it is pertinent to point out that the area between $-S$ and $+S$ (shown shaded on Fig. A-2) is 68 percent of the total area under the curve. Hence S is called the 68 percent error, as previously mentioned.

A-5 INTRODUCTION TO LEAST SQUARES

Least squares is a procedure for adjusting observations containing random errors. It is by no means a new method. Karl Gauss, a German mathematician, used the method as early as the latter part of the eighteenth century. Until the invention of computers, however, it was used rather sparingly because of the lengthy calculations involved.

For a group of equally weighted observations, the fundamental condition that is enforced in least squares adjustment is that *the sum of the squares of the residuals is minimized*. This condition, which has been developed from the equation for the normal distribution curve, provides most probable values for the adjusted quantities. Suppose a group of m equally weighted measurements were taken having residuals v_1, v_2, v_3, ..., v_m. Then in equation form, the fundamental condition of least squares is expressed as

$$\sum_{i=1}^{m} (v_i)^2 = (v_1)^2 + (v_2)^2 + (v_3)^2 + \cdots + (v_m)^2 = \text{minimum} \qquad \text{(A-4)}$$

Some basic assumptions which underlie least squares theory are that the number of observations being adjusted is large and the frequency distribution of the errors is *normal*. Although these basic assumptions are not always met, least squares adjustment still provides the most rigorous error treatment available, and hence it has become very popular and important in many areas of modern photogrammetry. Besides yielding most probable values for the unknowns, least squares adjustment also enables precisions of adjusted quantities to be determined, and it reveals the presence of large errors and mistakes so that steps can be taken to eliminate them.

A-6 WEIGHTED OBSERVATIONS

Weights of individual observed values may be assigned according to a priori estimates, or they may be obtained from the standard deviations of the observations if they are available. An equation expressing the relation between standard deviation and weights is

$$p_i = \frac{1}{(S_i)^2} \qquad \text{(A-5)}$$

In Eq. (A-5), p_i is the weight of the ith observed quantity and $(S_i)^2$ is the square of the standard deviation or *variance* of that observation. Equation (A-5) states that *weights are inversely proportional to variances*. If measured values are to be weighted in least squares adjustment, then the fundamental condition to be enforced is that the *sum of the weights times their corresponding squared residuals is minimized*, or, in equation form,

$$\sum_{i=1}^{m} p_i(v_i)^2 = p_1(v_1)^2 + p_2(v_2)^2 + p_3(v_3)^2 + \cdots + p_m(v_m)^2 = \text{minimum} \qquad \text{(A-6)}$$

A-7 APPLYING LEAST SQUARES

In the "observation equation" method of least squares adjustment, *observation equations* are written which relate measured values to their residual errors and the unknown parameters. One observation equation is written for each measurement. For a unique solution the number of equations must equal the number of unknowns. If redundant observations are made, then more observation equations can be written than are needed for a unique solution, and most probable values of the unknowns can be determined by the method of least squares. For a group of equally weighted observations, an equation for each residual error is obtained from each observation equation. The residuals are squared and added to obtain the function $\sum_{i=1}^{m} (v_i)^2$. To minimize the function, partial derivatives are taken with respect to each unknown variable and set equal to zero. This yields a set of equations called *normal equations* which are equal in number to the number of unknowns. The normal equations are solved to obtain the most probable values for the unknowns.

Example A-2 As an elementary example illustrating the method of least squares adjustment by the observation equation method, consider the following three equally weighted measurements taken between points A, B, and C of Fig. A-3:

$$x + y = 3.0$$

$$x = 1.5$$

$$y = 1.4$$

These three equations relate the two unknowns x and y to the observations. Values for x and y could be obtained from any two of these equations so that the remaining equation is redundant. Notice however, that the values obtained for x and y will differ, depending upon which two equations are solved. It is therefore apparent that the measurements contain errors. The equations may be rewritten as observation equations by including residual errors as follows:

$$x + y = 3.0 + v_1$$

$$x = 1.5 + v_2$$

$$y = 1.4 + v_3$$

To arrive at the least squares solution, the observation equations are rearranged to obtain expressions for the residuals; these are squared and added to form the function $\sum_{i=1}^{m} (v_i)^2$ as follows:

$$\sum_{i=1}^{m} (v_i)^2 = (x + y - 3.0)^2 + (x - 1.5)^2 + (y - 1.4)^2$$

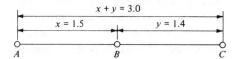

Figure A-3 Measurements for least squares adjustment example, Prob. A-2.

The above function is minimized, enforcing the condition of least squares, by taking partial derivatives with respect to each unknown and setting them equal to zero. This yields the following two equations:

$$\frac{\partial \Sigma v^2}{\partial x} = 0 = 2(x + y - 3.0) + 2(x - 1.5)$$

$$\frac{\partial \Sigma v^2}{\partial y} = 0 = 2(x + y - 3.0) + 2(y - 1.4)$$

The above equations are called *normal equations*. The reduced normal equations are

$$2x + y = 4.5$$

$$x + 2y = 4.4$$

Solving the reduced normal equations simultaneously yields $x = 1.533$ and $y = 1.433$. According to the theory of probability, these values have the highest probability. Having the most probable values for the unknowns, the residuals can be calculated by substitution back into the original observation equations, or

$$v_1 = 1.533 + 1.433 - 3.000 = -0.033$$

$$v_2 = 1.533 - 1.500 = 0.033$$

$$v_3 = 1.433 - 1.400 = 0.033$$

The above example is indeed simple, but it serves to illustrate the method of least squares without complicating the mathematics. Least squares adjustment of large systems of observation equations is performed in the same manner.

A-8 SYSTEMATIC FORMULATION OF NORMAL EQUATIONS

In large systems of observation equations it is helpful to utilize systematic procedures to formulate normal equations. Consider the following system of m linear observation equations of equal weight containing n unknowns:

$$\left. \begin{aligned} a_1A + b_1B + c_1C + \cdots + n_1N - L_1 &= v_1 \\ a_2A + b_2B + c_2C + \cdots + n_2N - L_2 &= v_2 \\ \cdots\cdots\cdots\cdots\cdots\cdots\cdots\cdots\cdots\cdots\cdots\cdots \\ a_mA + b_mB + c_mC + \cdots + n_mN - L_m &= v_m \end{aligned} \right\} \quad \text{(A-7)}$$

In Eqs. (A-7), the a's, b's, c's, etc., are coefficients of unknowns A, B, C, etc.; the L's are constants; and the v's are residuals. By squaring the residuals and summing them, the function Σv^2 is formed. Taking partial derivatives of Σv^2 with respect to each unknown A, B, C, etc., yields n normal equations. After reducing and factoring the normal equations, the following generalized system for expressing normal equations results:

$$\left.\begin{array}{l} [aa]A + [ab]B + [ac]C + \cdots + [an]N = [aL] \\ [ba]A + [bb]B + [bc]C + \cdots + [bn]N = [bL] \\ [ca]A + [cb]B + [cc]C + \cdots + [cn]N = [cL] \\ \cdots\cdots\cdots\cdots\cdots\cdots\cdots\cdots\cdots\cdots\cdots \\ [na]A + [nb]B + [nc]C + \cdots + [nn]N = [nL] \end{array}\right\} \quad \text{(A-8)}$$

In Eqs. (A-8), the symbol [ ] signifies the sum of the products; for example $[aa] = a_1a_1 + a_2a_2 + a_3a_3 + \ldots + a_ma_m$; $[ab] = a_1b_1 + a_2b_2 + a_3b_3 + \ldots + a_mb_m$; etc.

It may be similarly shown that normal equations may be systematically formed from weighted observation equations in the following manner:

$$\left.\begin{array}{l} [paa]A + [pab]B + [pac]C + \cdots + [pan]N = [paL] \\ [pba]A + [pbb]B + [pbc]C + \cdots + [pbn]N = [pbL] \\ [pca]A + [pcb]B + [pcc]C + \cdots + [pcn]N = [pcL] \\ \cdots\cdots\cdots\cdots\cdots\cdots\cdots\cdots\cdots\cdots\cdots \\ [pna]A + [pnb]B + [pnc]C + \cdots + [pnn]N = [pnL] \end{array}\right\} \quad \text{(A-9)}$$

In Eqs. (A-9) the terms are as described previously, except that the p's are the relative weights of the individual observations. Examples of the bracket terms are $[paa] = p_1a_1a_1 + p_2a_2a_2 + \ldots + p_ma_ma_m$; $[pbL] = p_1b_1L_1 + p_2b_2L_2 + \ldots + p_mb_mL_m$; etc.

The formulation of normal equations from observation equations may be further systematized by handling the systems of Eqs. (A-8) or (A-9) in a tabular manner.

Example A-3 Using the tabular method, form the normal equations for Example A-2.

SOLUTION

Equation no.	a	b	L	aa	ab	bb	aL	bL
1	1	1	3.0	1	1	1	3.0	3.0
2	1	0	1.5	1	0	0	1.5	0
3	0	1	1.4	0	0	1	0	1.4
				2	1	2	4.5	4.4
				$[aa]$	$[ab]$	$[bb]$	$[aL]$	$[bL]$

For this example, normal equations are formed by satisfying Eqs. (A-8) as follows:

$$[aa]x + [ab]y = [aL]$$

$$[ab]x + [bb]y = [bL]$$

Substituting the appropriate values from the above table yields the required normal equations as follows:

$$2x + y = 4.5$$

$$x + 2y = 4.4$$

A-9 MATRIX METHODS IN LEAST SQUARES ADJUSTMENT

It has been previously mentioned that least squares computations are quite lengthy and are therefore most economically performed on a computer. The algebraic approach—Eqs. (A-8) and (A-9)—for forming normal equations, and for obtaining their simultaneous solution, can be programmed for computer solution. The procedure is much more easily adapted to matrix methods, however.

In developing matrix equations for least squares computations, analogy will be made to the algebraic approach presented in Sec. A-8. First of all, observation Eqs. (A-7) may be represented in matrix form as:

$$_mA_n \, _nX_1 = \, _mL_1 + \, _mV_1 \tag{A-10}$$

where

$$A = \, _m\begin{bmatrix} a_1 & b_1 & c_1 & \cdots & n_1 \\ a_2 & b_2 & c_2 & \cdots & n_2 \\ \cdot & \cdot & \cdot & \cdots & \cdot \\ \cdot & \cdot & \cdot & \cdots & \cdot \\ a_m & b_m & c_m & \cdots & n_m \end{bmatrix}^n \qquad X = \, _n\begin{bmatrix} A \\ B \\ C \\ \cdot \\ \cdot \\ N \end{bmatrix}^1$$

$$L = \, _m\begin{bmatrix} L_1 \\ L_2 \\ L_3 \\ \cdot \\ \cdot \\ L_m \end{bmatrix}^1 \qquad V = \, _m\begin{bmatrix} v_1 \\ v_2 \\ v_3 \\ \cdot \\ \cdot \\ v_m \end{bmatrix}^1$$

Upon studying the following matrix representation, it will be noticed that normal Eqs. (A-8) are obtained as follows:

$$A^T A X = A^T L \qquad (A\text{-}11)$$

In the above equation, $A^T A$ is the matrix of normal equation coefficients of the unknowns. Premultiplying both sides of Eq. (A-11) by $(A^T A)^{-1}$ and reducing, there results:

$$(A^T A)^{-1}(A^T A)X = (A^T A)^{-1}A^T L$$

$$IX = (A^T A)^{-1}A^T L$$

$$X = (A^T A)^{-1}A^T L \qquad (A\text{-}12)$$

In the above reduction, I is the identity matrix. Equation (A-12) is the basic least squares matrix equation for *equally weighted* observations. The matrix X consists of most probable values for unknowns, A, B, C, . . ., N. For a system of *weighted* observations, the following matrix equation provides the X matrix of most probable values for the unknowns:

$$X = (A^T P A)^{-1}A^T P L \qquad (A\text{-}13)$$

In Eq. (A-13) the matrices are identical to those of the equally weighted equations, except that the P matrix is a diagonal matrix of weights and is defined as follows:

$$P = {}_{m}\begin{bmatrix} p_1 & & & & & \\ & p_2 & & & \text{zeros} & \\ & & p_3 & & & \\ & & & \cdot & & \\ \text{zeros} & & & & \cdot & \\ & & & & & \cdot \\ & & & & & p_m \end{bmatrix}^{m}$$

In the above P matrix, all off-diagonal elements are shown as zeros. This is proper when the individual observations are independent and uncorrelated; e.g., they are not dependent upon each other. This is almost always the case in photogrammetry.

Example A-4 Solve Example A-2 using matrix methods.
(*a*) The observation equations of Example A-2 may be expressed in matrix form as follows:

$$_3A_2\ _2X_1 = {}_3L_1 + {}_3V_1$$

where

$$A = {}_3\begin{bmatrix} 1 & 1 \\ 1 & 0 \\ 0 & 1 \end{bmatrix}^2 \qquad X = {}_2\begin{bmatrix} x \\ y \end{bmatrix}^1 \qquad L = {}_3\begin{bmatrix} 3.0 \\ 1.5 \\ 1.4 \end{bmatrix}^1 \qquad V = {}_3\begin{bmatrix} v_1 \\ v_2 \\ v_3 \end{bmatrix}^1$$

(*b*) Solving matrix Eq. (A-12),

$$(A^TA) = \begin{bmatrix} 1 & 1 & 0 \\ 1 & 0 & 1 \end{bmatrix} \begin{bmatrix} 1 & 1 \\ 1 & 0 \\ 0 & 1 \end{bmatrix} = \begin{bmatrix} 2 & 1 \\ 1 & 2 \end{bmatrix}$$

$$(A^TA)^{-1} = \frac{1}{3} \begin{bmatrix} 2 & -1 \\ -1 & 2 \end{bmatrix} \qquad A^TL = \begin{bmatrix} 4.5 \\ 4.4 \end{bmatrix}$$

$$X = (A^TA)^{-1}A^TL = \frac{1}{3} \begin{bmatrix} 2 & -1 \\ -1 & 2 \end{bmatrix} \begin{bmatrix} 4.5 \\ 4.4 \end{bmatrix} = \begin{bmatrix} 1.533 \\ 1.433 \end{bmatrix}$$

Note that this solution yields exactly the same values for x and y as were obtained through the algebraic approach of Example A-2.

A-10 MATRIX EQUATIONS FOR PRECISIONS OF ADJUSTED QUANTITIES

The matrix equation for calculating residuals after adjustment, whether the adjustment is weighted or not, is

$$V = AX - L \qquad \text{(A-14)}$$

The standard deviation of unit weight for an unweighted adjustment is

$$S_o = \sqrt{\frac{(V^TV)}{r}} \qquad \text{(A-15)}$$

The standard deviation of unit weight for a weighted adjustment is

$$S_o = \sqrt{\frac{(V^TPV)}{r}} \qquad \text{(A-16)}$$

In Eqs. (A-15) and (A-16), r is the number of degrees of freedom and equals the number of observation equations minus the number of unknowns, or $r = (m - n)$.

Standard deviations of the adjusted quantities are

$$S_{X_i} = S_o \sqrt{(Q_{X_iX_i})} \qquad \text{(A-17)}$$

In Eq. (A-17), S_{X_i} is the standard deviation of the ith adjusted quantity, e.g., the quantity in the ith row of the X matrix; S_o is the standard deviation of unit weight as calculated by Eqs. (A-15) or (A-16); and $Q_{X_iX_i}$ is the element in the ith row and the ith column of the matrix $(A^TA)^{-1}$ in the unweighted case, or the matrix $(A^TPA)^{-1}$ in the weighted case. The $(A^TA)^{-1}$ and $(A^TPA)^{-1}$ matrices are the so-called *covariance* matrices.

Example A-5 Calculate the standard deviation of unit weight and the standard deviations of the adjusted quantities x and y for the unweighted problem of Example A-4.

(a) By Eq. (A-14) the residuals are

$$V = \begin{bmatrix} 1 & 1 \\ 1 & 0 \\ 0 & 1 \end{bmatrix} \begin{bmatrix} 1.533 \\ 1.433 \end{bmatrix} - \begin{bmatrix} 3.0 \\ 1.5 \\ 1.4 \end{bmatrix} = \begin{bmatrix} -0.033 \\ 0.033 \\ 0.033 \end{bmatrix}$$

(b) By Eq. (A-15) the standard deviation of unit weight is

$$V^T V = \begin{bmatrix} -0.033 & 0.033 & 0.033 \end{bmatrix} \begin{bmatrix} -0.033 \\ 0.033 \\ 0.033 \end{bmatrix} = 0.0033$$

$$S_o = \sqrt{\frac{0.0033}{3 - 2}} = \pm 0.057$$

(c) Using Eq. (A-17), the standard deviations of the adjusted values for x and y are

$$S_x = \pm 0.057 \sqrt{\tfrac{2}{3}} = \pm 0.046$$

$$S_y = \pm 0.057 \sqrt{\tfrac{2}{3}} = \pm 0.046$$

In part (c) above, the numbers $\tfrac{2}{3}$ under the radicals are the (1,1) and (2,2) elements of the $(A^T A)^{-1}$ matrix of Example A-4. The interpretation of the standard deviations computed under part (c) is that there is a 68 percent probability that the adjusted values for x and y are within ± 0.046 of their true value. Note that for this simple example problem the three residuals calculated in part (a) were equal, and also the standard deviations of x and y were equal in part (c). This is due to the symmetric nature of this particular problem (illustrated in Fig. A-3), but it is not generally the case with more complex problems.

A-11 PRACTICAL EXAMPLE

The following example is presented to illustrate a practical application of least squares in photogrammetry. The example also shows the method of calculating coefficients of a polynomial which approximates the radial-lens-distortion curve for an aerial camera (see Sec. 5-12).

Example A-6 From the radial-lens-distortion calibration data of an aerial camera given in the table below, calculate the coefficients of a polynomial which approximates the radial-lens distortion curve.

Radial distance r, mm	Radial-lens distortion Δr, mm
0.000	0.000
20.072	0.004
40.855	0.018
63.155	0.047
88.034	0.062
116.995	0.035
152.472	−0.064

SOLUTION As presented in Sec. 5-12, a polynominal of the following form satisfactorily defines radial-lens distortions:

$$\Delta r = k_1 r + k_2 r^3 + k_3 r^5 + k_4 r^7 \qquad \text{(A-18)}$$

In Eq. (A-18), Δr is the radial-lens distortion at a radial distance r from the principal point. The k's are coefficients which define the shape of the radial-lens distortion curve. One equation of the form of Eq. (A-18) can be written for each radial distance at which the distortion is known from calibration. Since there are four k's, four equations are required to obtain a unique solution for them. From calibration, radial-lens distortions are known for seven radial distances; hence seven equations can be written and the k's may be computed by least squares.

Based on the calibration data, the following seven observation equations may be written (note that residual errors are added for consistency):

$$0.000 + v_1 = (\ 0.000)k_1 + (\ 0.000)^3 k_2 + (\ 0.000)^5 k_3 + (\ 0.000)^7 k_4$$
$$0.004 + v_2 = (\ 20.072)k_1 + (\ 20.072)^3 k_2 + (\ 20.072)^5 k_3 + (\ 20.072)^7 k_4$$
$$0.018 + v_3 = (\ 40.855)k_1 + (\ 40.855)^3 k_2 + (\ 40.855)^5 k_3 + (\ 40.855)^7 k_4$$
$$0.047 + v_4 = (\ 63.155)k_1 + (\ 63.155)^3 k_2 + (\ 63.155)^5 k_3 + (\ 63.155)^7 k_4$$
$$0.062 + v_5 = (\ 88.034)k_1 + (\ 88.034)^3 k_2 + (\ 88.034)^5 k_3 + (\ 88.034)^7 k_4$$
$$0.035 + v_6 = (116.995)k_1 + (116.995)^3 k_2 + (116.995)^5 k_3 + (116.995)^7 k_4$$
$$-0.064 + v_7 = (152.472)k_1 + (152.472)^3 k_2 + (152.472)^5 k_3 + (152.472)^7 k_4$$

Using least squares Eq. (A-12) and a computer, these seven equations were solved and the following values were obtained for the four coefficients:

$$k_1 = 1.99697 \times 10^{-4} \qquad k_3 = -1.97388 \times 10^{-11}$$
$$k_2 = 1.94801 \times 10^{-7} \qquad k_4 = 0.43934 \times 10^{-15}$$

Using these k's in Eq. (A-18), radial-lens distortions for any value of r may be readily calculated.

REFERENCES

American Society of Photogrammetry: "Manual of Photogrammetry," 3d ed., Falls Church, Va., 1966, chap. 2.

————: "Manual of Photogrammetry," 4th ed., Falls Church Va., 1980, chap. 2

Benjamin, J. R., and C. A. Cornell: "Probability, Statistics and Decision for Civil Engineers," McGraw-Hill Book Company, New York, 1970.

Crandall, K. C., and R. W. Seabloom: "Engineering Fundamentals in Measurements, Probability and Dimensions," McGraw-Hill Book Company, New York, 1970.

Gale, L. A.: Theory of Adjustments by Least Squares, *Canadian Surveyor*, vol. 19, no. 1, p. 42, 1965.

Hallert, B.: "Photogrammetry," McGraw-Hill Book Company, New York, 1960.

Hamilton, W. C.: "Statistics in Physical Science," The Ronald Press Company, New York, 1964.

Hardy, R. L.: Least Squares Prediction; *Photogrammetric Engineering and Remote Sensing,* vol. 43, no. 4, p. 475, 1977.

Hirvonen, R. A.: "Adjustment by Least Squares in Photogrammetry and Geodesy," Frederick Ungar Publishing Co., New York, 1971.

Konecny, G.: Classical Concepts of Least Squares Adjustment, *Canadian Surveryor,* vol. 19, no. 1, p. 16, 1965.

Linnik, Y. V.: "Method of Least Squares and Principles of the Theory of Observation," Pergamon Press, New York, 1961.

Mikhail, E. M.: Parameter Constraints in Least Squares, *Photogrammetric Engineering,* vol. 36, no. 12, p. 1277, 1970.

————: "Observations and Least Squares," Harper & Row Publishers, Incorporated, New York, 1976.

Rainsford, H. F.: "Survey Adjustments and Least Squares," Frederick Ungar Publishing Company, New York, 1958.

Rampal, K. K.: Least Squares Collocation in Photogrammetry, *Photogrammetric Engineering and Remote Sensing,* vol. 42, no. 5, p. 659, 1976.

Richardus, P.: "Project Surveying," North-Holland Publishing Company, Amsterdam, 1966.

Wolf, P. R.: "Adjustment Computations: Practical Least Squares for Surveyors," P.B.L. Publishers, Madison, Wis., 1980.

Wong, K. W.: Propagation of Variance and Covariance, *Photogrammetric Engineering and Remote Sensing,* vol. 41, no. 1, p. 75, 1975.

Zimmerman, D. S.: Least Squares by Diagonal Partitioning, *Canadian Surveyor,* vol. 28, no. 5, p. 677, 1974.

PROBLEMS

A-1 A photographic distance was measured 10 times using the same equipment and procedures with the following results: 95.76, 95.68, 95.70, 95.72, 95.69, 95.75, 95.72, 95.77, 95.70, and 95.71 mm. Calculate the most probable value for the photo distance and the standard deviation of the group of measurements.

A-2 Repeat Prob. A-1, except that the following 15 measurements were obtained: 64.29, 64.26, 64.31, 64.29, 64.34, 64.28, 64.30, 64.30, 64.33, 64.29, 64.35, 64.31, 64.28, 64.32, and 64.33 mm.

A-3 Compute the most probable values of unknowns x_1, x_2, and x_3 for the following observation equations using the tabular least squares method, and calculate the standard deviations of the adjusted quantities.

$$2x_1 + 3x_2 + x_3 = 16$$

$$x_1 - 2x_2 + 3x_3 = -9$$

$$7x_1 + x_2 - 2x_3 = 21$$

$$-x_1 - x_2 - x_3 = -5$$

A-4 Suppose the constant terms 16, -9, 21, and -5 of the four equations of Prob. A-3 represent measurements having relative weights of 2, 3, 1, and 4, respectively. Using weighted least squares, calculate most probable values for x_1, x_2, and x_3 and determine standard deviations of these values.

A-5 Repeat Prob. A-3, except that the four equations are as follows:

$$5x_1 + 3x_2 - 2x_3 = -15$$
$$2x_1 - x_2 + 6x_3 = 25$$
$$x_1 + 3x_2 + x_3 = 13$$
$$-4x_1 + 3x_2 - 3x_3 = 1$$

A-6 If the constant terms -15, 25, 13, and 1 of Prob. A-5 represent measurements having relative weights of 4, 2, 1, and 2, respectively, calculate the least squares solution for the unknowns and determine their standard deviations.

B

COORDINATE TRANSFORMATIONS

B-1 INTRODUCTION

A problem frequently encountered in photogrammetric work is conversion from one rectangular coordinate system to another. This is because photogrammetrists commonly determine coordinates of unknown points in convenient *arbitrary* rectangular coordinate systems. These arbitrary coordinates may be read from comparators or stereoscopic plotters, or they may result from analytical computation. The arbitrary coordinates must then be converted into a *final* system such as the camera photo coordinate system in the case of comparator measurements, or a ground coordinate system such as the state plane coordinate system in the case of stereoplotter or analytically derived arbitrary model coordinates. The procedure for converting from one coordinate system to another is known as *coordinate transformation*. The procedure requires that some points have their coordinates known in both the arbitrary and the final coordinate systems. Such points are called *control points*.

B-2 TWO-DIMENSIONAL CONFORMAL COORDINATE TRANSFORMATION

The term "two-dimensional" means that the coordinate systems lie on plane surfaces. A *conformal* transformation is one in which true shape is preserved after transformation. To perform a two-dimensional conformal coordinate transformation, it is necessary that coordinates of at least two points be known in both the arbitrary and final coordinate systems. Accuracy in the transformation is improved by choosing the two points as far apart as possible. If more than two control points are available, an improved solution may be obtained by applying the method of least squares.

A two-dimensional conformal coordinate transformation consists of three basic steps: (1) scale change, (2) rotation, and (3) translation. The example illustrated in Fig. B-1 will be used to demonstrate the procedure. This example uses the minimum of two control points. Section B-4 describes the procedure when more than two control

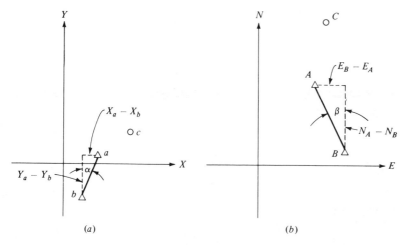

Figure B-1 (*a*) Arbitrary *XY* two-dimensional coordinate system. (*b*) Ground *EN* two-dimensional coordinate system.

points are available. Figure B-1*a* shows the positions of points *a* through *c*, whose coordinates are known in an arbitrary *XY* system. Figure B-1*b* illustrates the positions of the same points, labeled *A* through *C* in a (ground) *EN* system. The coordinates of *A* and *B* are known in the ground system and it is required to determine the coordinates of *C* in the ground system.

Step 1: Scale Change

Comparing Fig. B-1*a* and *b*, it is evident that the lengths of lines *ab* and *AB* are unequal, hence the scales of the two coordinate systems are unequal. The scale of the *XY* system is made equal to the scale of the *EN* system by multiplying each *X* and *Y* coordinate by a scale factor *s*. The scaled coordinates are labeled *X'* and *Y'*. By use of the two control points, the scale factor is calculated in relation to the two lengths *AB* and *ab* as

$$s = \frac{AB}{ab} = \frac{[(E_B - E_A)^2 + (N_B - N_A)^2]^{1/2}}{[(X_b - X_a)^2 + (Y_b - Y_a)^2]^{1/2}} \tag{B-1}$$

Step 2: Rotation

If the scaled *X'Y'* coordinate system is superimposed over the *EN* system of Fig. B-1*b* so that line *AB* in both systems coincides, the result is as shown in Fig. B-2. An auxiliary axis system *E'N'* is constructed through the origin of the *X'Y'* axis system parallel to the *EN* axes. It is necessary to rotate from the *X'Y'* system to the *E'N'* system, or in other words, to calculate *E'N'* coordinates for the unknown points from their *X'Y'* coordinates. The *E'N'* coordinates of point *C* may be calculated in terms

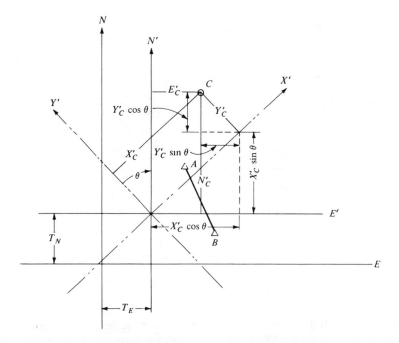

Figure B-2 Scaled $X'Y'$ coordinate system superimposed onto the EN ground coordinate system.

of the clockwise rotation angle θ by using the following equations:

$$E'_C = X'_c \cos \theta - Y'_c \sin \theta$$

$$N'_C = X'_c \sin \theta + Y'_c \cos \theta$$

(B-2)

Rotation angle θ, shown in Fig. B-2, is the sum of angles α and β which are indicated on Fig. B-1a and b. From the coordinates of the two control points, these angles are calculated as

$$\alpha = \tan^{-1}\left(\frac{X_a - X_b}{Y_a - Y_b}\right)$$

$$\beta = \tan^{-1}\left(\frac{E_B - E_A}{N_A - N_B}\right)$$

(B-3)

Step 3: Translation

The final step in the coordinate transformation is a translation of the origin of the $E'N'$ system to the origin of the EN system. The translation factors required are T_E and T_N, which are illustrated in Fig. B-2. Final E and N ground coordinates for point C then are

$$E_C = E'_C + T_E$$ (B-4)

$$N_C = N'_C + T_N$$

Translation factors T_E and T_N are calculated as

$$T_E = E_A - E'_A = E_B - E'_B$$ (B-5)

$$T_N = N_A - N'_A = N_B - N'_B$$

Note from Eqs. (B-5) that these translation factors may be calculated in two different ways by using either control point A or B. It is advisable to calculate them using both points to obtain a computation check.

Working sketches are recommended in computing coordinate transformations to aid in reducing the likelihood of mistakes. Caution should be exercised to ensure that correct algebraic signs are applied to the coordinates used in the transformation equations.

Example B-1 Assume that in Fig. B-1a and b the arbitrary and ground coordinates of points A through C are as follows:

Point	Arbitrary coordinates		Ground coordinates	
	X	Y	E	N
A	632.17	121.45	1,100.64	1,431.09
B	355.20	-642.07	1,678.39	254.15
C	1,304.81	596.37		

It is required to compute the coordinates of point C in the ground EN system.

SOLUTION (a) The scale factor is calculated from Eq. (B-1) as

$$s = \frac{[(1,678.39 - 1,100.64)^2 + (254.15 - 1,431.09)^2]^{1/2}}{[(355.20 - 632.17)^2 + (-642.07 - 121.45)^2]^{1/2}}$$

$$= \frac{1,311.10}{812.20} = 1.61426$$

The arbitrary coordinates are then expanded to the $X'Y'$ system, which is equal in scale to the ground coordinate system, by multiplying each of the arbitrary coordinates by the scale factor. After multiplication, the $X'Y'$ coordinates are

Point	Scaled coordinates	
	X'	Y'
A	1,020.49	196.05
B	573.39	$-1,036.47$
C	2,106.30	962.70

(b) The rotation angle, calculated by Eqs. (B-3), is

$$\tan \alpha = \frac{632.17 - 355.20}{121.45 + 642.07} = 0.362753 \qquad \alpha = 19°56'18''$$

$$\tan \beta = \frac{1,678.39 - 1,110.64}{1,431.09 - 254.15} = 0.490892 \qquad \beta = 26°08'46''$$

$$\theta = 19°56'18'' + 26°08'46'' = 46°05'04''$$

Rotation Eqs. (B-2) are then solved to obtain E' and N' coordinates. The solution in tabular form is as follows (with $\sin \theta = 0.720363$, $\cos \theta = 0.693597$):

Point	$X' \cos \theta$	$Y' \sin \theta$	$X' \sin \theta$	$Y' \cos \theta$	E'	N'
A	707.81	141.23	735.12	135.98	566.58	871.10
B	397.70	−746.63	413.05	−718.89	1,144.33	−305.84
C	1,460.92	693.49	1,517.30	667.73	767.43	2,185.03

(c) The translation factors T_E and T_N are calculated next, using Eqs. (B-5) as follows:

$$T_E = E_A - E'_A = 1,100.64 - 566.58 = 534.06$$

also

$$T_E = E_B - E'_B = 1,678.39 - 1,144.33 = 534.06 \ Check!$$

$$T_N = N_A - N'_A = 1,431.09 - 871.10 = 559.99$$

also

$$T_N = N_B - N'_B = 254.15 + 305.84 = 559.99 \ Check!$$

By adding the translation factors to E' and N', the following final transformed E and N coordinates are obtained for point C:

$$E_C = 767.43 + 534.06 = 1301.49$$

$$N_C = 2185.03 + 559.99 = 2745.02$$

In the above example, although only one unknown point (point C) was transformed, any number of points could have been transformed using just the two control points.

B-3 ALTERNATE METHOD OF TWO-DIMENSIONAL CONFORMAL TRANSFORMATION

If a computer is available, it is advantageous to compute coordinate transformations by an alternate method. In this method, equations involving four transformation coefficients are formulated in terms of the coordinates of the two or more points whose positions are known in both coordinate systems. The formation of the equations follows the same three steps discussed in Sec. B-2. The procedure consists first of multiplying each of the original coordinates of points a and b by a scale factor. The following four equations result:

$$X'_a = sX_a$$
$$Y'_a = sY_a \qquad \text{(B-6)}$$
$$X'_b = sX_b$$
$$Y'_b = sY_b$$

Equations (B-6) are now substituted into Eqs. (B-2), except that subscripts of Eqs. (B-2) are changed to be applicable for points A and B. This substitution yields

$$E'_A = sX_a \cos \theta - sY_a \sin \theta$$
$$N'_A = sX_a \sin \theta + sY_a \cos \theta \qquad \text{(B-7)}$$
$$E'_B = sX_b \cos \theta - sY_b \sin \theta$$
$$N'_B = sX_b \sin \theta + sY_b \cos \theta$$

Finally, translation factors T_E and T_N, as described previously, are added to Eqs. (B-7) to yield the following equations:

$$E_A = sX_a \cos \theta - sY_a \sin \theta + T_E$$
$$N_A = sX_a \sin \theta + sY_a \cos \theta + T_N \qquad \text{(B-8)}$$
$$E_B = sX_b \cos \theta - sY_b \sin \theta + T_E$$
$$N_B = sX_b \sin \theta + sY_b \cos \theta + T_N$$

Let $a = s \cos \theta$ and $b = s \sin \theta$. Making the substitution of a and b into Eqs. (B-8) and rearranging, there results

$$E_A = aX_a - bY_a + T_E$$
$$N_A = aY_a + bX_a + T_N \qquad \text{(B-9)}$$
$$E_B = aX_b - bY_b + T_E$$
$$N_B = aY_b + bX_b + T_N$$

Because both the XY and EN coordinates for points A and B are known, Eqs. (B-9) contain only four unknowns, the transformation factors a, b, T_E, and T_N. The four equations may be solved simultaneously to obtain the unknowns. When the four transformation factors have been computed, an E and an N equation of the form of Eqs. (B-9) may be solved to obtain the final coordinates of each point whose coordinates were known only in the XY coordinate system.

Example B-2 Solve Example B-1 using the alternate method.
(a) Formulate Eqs. (B-9) for the points whose coordinates are known in both systems:

$$1,100.64 = 632.17a - 121.45b + T_E$$

$$1,431.09 = 121.45a + 632.17b + T_N$$

$$1,678.39 = 355.20a + 642.07b + T_E$$

$$254.14 = -642.07a + 355.20b + T_N$$

(b) The simultaneous solution of the above four equations yields the following:

$$a = 1.11965$$

$$b = 1.16285$$

$$T_E = 534.06$$

$$T_N = 559.99$$

(c) Using the four transformation factors, the final EN ground coordinates of point C are calculated as follows:

$$E_C = (1.11965)(1,304.81) - (1.16285)(596.37) + 534.06 = 1,301.49$$

$$N_C = (1.11965)(596.37) + (1.16285)(1,304.81) + 559.99 = 2,745.02$$

B-4 COORDINATE TRANSFORMATIONS WITH REDUNDANCY

In some instances more than two control points are available with coordinates known in both the arbitrary and final systems. In that case redundancy exists and the transformation can be computed using a least squares solution. In this method, as discussed in Appendix A, the sum of the squares of the residuals in the measurements are minimized, which, according to the theory of probability, produces the most probable solution. The least squares method has the additional advantages that any mistakes in the coordinates may be detected and the precisions of the transformed coordinates may be obtained. For these reasons it is advisable to use redundancy in coordinate transformations if possible.

In the least squares procedure, it is convenient to use the alternate method discussed in Sec. B-3. Two observation equations similar to those of Eqs. (B-9) are formed for each point whose coordinates are known in both systems. Residuals v are included in the equations to make them consistent, as follows:

$$aX - bY + T_E = E + v_E \tag{B-10}$$

$$bX + aY + T_N = N + v_N$$

If n points are available whose coordinates are known in both systems, $2n$ equations may be formed containing the four unknown transformation factors. The equations are solved by the method of least squares to obtain the most probable transformation factors. Transformed coordinates of all required points may then be found by using the transformation factors as illustrated in (c) of Example B-2.

It is theoretically correct in least squares to associate residuals with actual observations. In Eqs. (B-10), however, the X and Y coordinates are observed, yet the residuals have been associated with the E and N control coordinates. This is an easier approach and one that has been commonly used and found to yield entirely satisfactory results.

B-5 MATRIX METHODS IN COORDINATE TRANSFORMATIONS

Coordinate transformations involve rather lengthy calculations and are therefore best handled on a computer. Matrix algebra is ideal for computer calculations and is therefore convenient for performing transformations.

To illustrate the application of matrix algebra in coordinate transformations, assume that coordinates of three control points, A through C, are known in both the XY system and the EN system. Let their coordinates be of equal reliability so that their weights are equal. Assume also that transformation into the EN system is required for points D through N, whose coordinates are known only in the XY system.

First of all, six observation equations of the form of Eqs. (B-10) are developed, two for each control point A, B, and C, as follows:

$$X_A a - Y_A b + T_E = E_A + v_{E_A}$$

$$Y_A a + X_A b + T_N = N_A + v_{N_A}$$

$$X_B a - Y_B b + T_E = E_B + v_{E_B}$$

$$Y_B a + X_B b + T_N = N_B + v_{N_B} \tag{B-11}$$

$$X_C a - Y_C b + T_E = E_C + v_{E_C}$$

$$Y_C a + X_C b + T_N = N_C + v_{N_C}$$

In matrix representation, the above six equations are

$$_6A_4 \, _4X_1 = \, _6L_1 + \, _6V_1 \tag{B-12}$$

In matrix Eq. (B-12), A is the matrix of coefficients of the unknown transformation factors, X is the matrix of unknown transformation factors, L is the matrix of constant terms which is made up of control point coordinates, and V is the matrix of residual discrepancies in those coordinates brought about by measurement errors. More specifically these matrices are

$$A = \begin{bmatrix} X_A & -Y_A & 1 & 0 \\ Y_A & X_A & 0 & 1 \\ X_B & -Y_B & 1 & 0 \\ Y_B & X_B & 0 & 1 \\ X_C & -Y_C & 1 & 0 \\ Y_C & X_C & 0 & 1 \end{bmatrix}_6^4 \qquad X = \begin{bmatrix} a \\ b \\ T_E \\ T_N \end{bmatrix}_4^1$$

$$L = \begin{bmatrix} E_A \\ N_A \\ E_B \\ N_B \\ E_C \\ N_C \end{bmatrix}_6^1 \qquad V = \begin{bmatrix} v_{E_A} \\ v_{N_A} \\ v_{E_B} \\ v_{N_B} \\ v_{E_C} \\ v_{N_C} \end{bmatrix}_6^1$$

As discussed in Appendix A, matrix Eq. (A-12) is used to solve this equally weighted system for the transformation factors. The final transformation of all points D through N into the EN system is performed as discussed in (c) of Example B-2. This phase of the computation is also readily adapted to matrix methods.

B-6 TWO-DIMENSIONAL AFFINE COORDINATE TRANSFORMATION

The two-dimensional *affine* coordinate transformation is only a slight modification of the two-dimensional conformal transformation to include different scale factors in the x and y directions. A photogrammetry problem commonly solved using the two-dimensional affine transformation is conversion of measured photocoordinates from an arbitrary comparator axis system to the conventional xy fiducial axis system. Assume for example that comparator measurements have been taken with a diapositive oriented on the comparator, as illustrated in Fig. B-3. Besides correcting for shrinkage by means of scale factors, the affine coordinate transformation also applies translations of X_o and Y_o to shift the origin from the XY comparator axis system to the origin o of the xy photo system, and it applies a rotation through angle θ (plus a small angular correction for nonorthogonality) to orient the axes in the xy photo system. It is entirely logical to use an affine transformation for this particular problem because it accounts for different magnitudes of film shrinkage or expansion that generally exist in the x and y directions. Following are the affine equations which transform from the XY

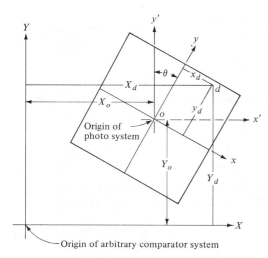

Figure B-3 Affine coordinate transformation from arbitrary comparator XY axis system to fiducial xy axis system.

comparator axes of Fig. B-3 to the xy photo system:

$$x = a_1 + a_2 X + a_3 Y$$

$$y = b_1 + b_2 X + b_3 Y$$

(B-13)

As with the two-dimensional conformal transformation, the application of the affine transformation is a two-step procedure of (1) determining the a and b coefficients using points whose coordinates are known in both the XY and xy systems and (2) applying these coefficients to calculate transformed xy coordinates for all other points from their XY coordinates. In correcting photocoordinates, the fiducial marks are used to perform step 1, since their calibrated xy coordinates are known from camera calibration, and their XY coordinates are available from comparator measurements. For a given photographic plate, a pair of equations of the form of Eq. (B-13) can be written for each fiducial mark. If there are four fiducials, four x and four y equations are obtained. Any three of the x equations will yield a solution for the three unknown a coefficients, and any three of the y equations will yield a solution for the three unknown b's. An improved solution may be obtained, however, if each set of four equations is solved simultaneously using least squares.

Example B-3 Calibrated coordinates and comparator measured coordinates of the four fiducial marks for a certain photographic plate are given in the following table. The comparator-measured coordinates of other points 1, 2, and 3 are also given. It is required to compute the corrected coordinates of points 1, 2, and 3 using the affine transformation.

Point	Comparator coordinates		Calibrated coordinates	
	X, mm	Y, mm	x, mm	y, mm
Fiducial A	55.149	159.893	−113.000	0.000
Fiducial B	167.716	273.302	0.000	113.000
Fiducial C	281.150	160.706	113.000	0.000
Fiducial D	168.580	47.299	0.000	−113.000
1	228.498	105.029		
2	270.307	199.949		
3	259.080	231.064		

SOLUTION Equations of the form Eq. (B-13), with residuals added for consistency, are formulated first for the x coordinates of the four fiducial marks as

$$-113.000 + v_1 = a_1 + (\ 55.149)a_2 + (159.893)a_3$$

$$0.000 + v_2 = a_1 + (167.716)a_2 + (273.302)a_3$$

$$113.000 + v_3 = a_1 + (281.150)a_2 + (160.706)a_3$$

$$0.000 + v_4 = a_1 + (168.580)a_2 + (\ 47.299)a_3$$

Any three of the above four equations could be solved to obtain the three unknown a's. In this example, however, the v's were included and least squares Eq. (A-12) of Appendix A was used to obtain the three coefficients with the following results:

$$a_1 = -168.759 \qquad a_2 = 0.999982 \qquad a_3 = 0.00382289$$

Now the corrected coordinates of unknown points 1 through 3 are calculated using these a's in Eq. (B-13) as follows:

$$x_1 = -168.759 + (0.999982)(228.498)$$

$$+ (0.00382289)(105.029) = 60.137$$

$$x_2 = -168.759 + (0.999982)(270.307)$$

$$+ (0.00382289)(199.949) = 102.307$$

$$x_3 = -168.759 + (0.999982)(259.080)$$

$$+ (0.00382289)(231.064) = 91.199$$

In a similar manner, equations are written for the y coordinates of each of the four fiducial marks. These equations are solved to obtain the b's, whereupon the y-corrected coordinates of points 1 through 3 are calculated. These b's and corrected y coordinates are

$$b_1 = -159.691 \qquad b_2 = -0.00359723 \qquad b_3 = 0.999973$$

$$y_1 = -55.487 \qquad y_2 = 39.281 \qquad y_3 = 70.435$$

If the fiducial marks are in the four corners, as is the case with Wild cameras, similar procedures may be followed. Of course, it is required that calibrated coordinates and comparator coordinates be known for each mark. If eight fiducial marks are available, all of them may be used and an improved solution obtained.

B-7 THREE-DIMENSIONAL CONFORMAL COORDINATE TRANSFORMATION

As implied by its name, a three-dimensional conformal coordinate transformation involves converting from one three-dimensional system to another. In the transformation, true shape is retained. This type of coordinate transformation is essential in analytical or computational photogrammetry for two basic problems: (1) to convert coordinates of points from a tilted photographic coordinate system to an equivalent vertical photographic system which is parallel to a ground or arbitrary object space system, and (2) to form continuous three-dimensional "strip models" from independent stereomodels. Three-dimensional conformal coordinate transformation equations are developed here in general, while their application to specific photogrammetry problems is described elsewhere in the text where appropriate.

In Fig. B-4 it is required to transform coordinates of points from an xyz system to an XYZ system. As illustrated in the figure, the two coordinate systems are not parallel. The necessary transformation equations can be expressed in terms of seven independent transformation factors: three rotation angles, omega (ω), phi (ϕ), and kappa (κ); a scale factor s; and three translation factors T_x, T_y, and T_z. Before proceeding with the development of the transformation equations, it is important to define sign conventions. All coordinate systems shall be defined as right-handed, i.e., systems in which positive X, Y, and Z are defined as shown in Fig. B-4. Rotation angles ω, ϕ, and κ are defined as positive if they are counterclockwise when viewed from the positive end of their respective axes. Positive ω rotation about the x' axis, for example, is shown in Fig. B-4.

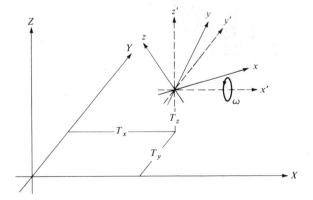

Figure B-4 XYZ and xyz right-handed three-dimensional coordinate systems.

The transformation equations shall be developed in the following two basic steps: (1) rotation and (2) scaling and translation.

Step 1: Rotation

In Fig. B-4 an $x'y'z'$ coordinate system parallel to the XYZ object system is constructed with its origin at the origin of the xyz system. In the development of rotation formulas, it is customary to consider the three rotations as taking place so as to convert from the $x'y'z'$ system to the xyz system. The rotation equations are developed in a sequence of three independent two-dimensional rotations. These rotations, illustrated in Fig. B-5, are first, ω rotation about the x' axis which converts coordinates from the $x'y'z'$ system into an $x_1y_1z_1$ system; second, ϕ rotation about the once rotated y_1 axis which converts coordinates from the $x_1y_1z_1$ system into an $x_2y_2z_2$ system; and third, κ rotation about the twice rotated z_2 axis which converts coordinates from the $x_2y_2z_2$ system into the xyz system of Fig. B-4. The exact amount and direction of the rotations for any three-dimensional coordinate transformation will depend upon the orientation relationship between the xyz and XYZ coordinate systems.

The development of the rotation formulas is as follows:

First, rotation through the angle ω about the x' axis, as illustrated in Fig. B-6. The coordinates of any point A in the once rotated $x_1y_1z_1$ system, as shown graphically in Fig. B-6, are

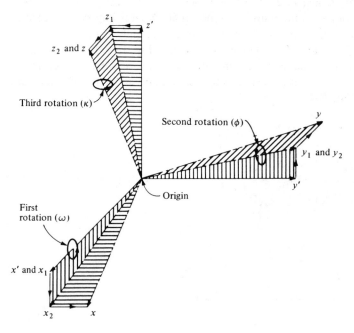

Figure B-5 The three sequential angular rotations.

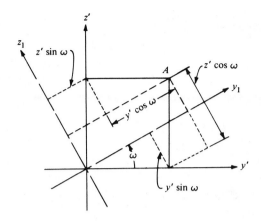

Figure B-6 Omega rotation about the x' axis.

$$x_1 = x'$$

$$y_1 = y' \cos \omega + z' \sin \omega \qquad \text{(B-14)}$$

$$z_1 = -y' \sin \omega + z' \cos \omega$$

Since this rotation was about x', the x' and x_1 axes are coincident and therefore the x coordinate of A is unchanged.

Second, rotation through ϕ about the y_1 axis, as illustrated in Fig. B-7. The coordinates of A in the twice rotated $x_2 y_2 z_2$ coordinate system, as shown graphically in Fig. B-7, are

$$x_2 = -z_1 \sin \phi + x_1 \cos \phi$$

$$y_2 = y_1 \qquad \text{(B-15)}$$

$$z_2 = z_1 \cos \phi + x_1 \sin \phi$$

In this rotation about y_1, the y_1 and y_2 axes are coincident and therefore the y coordinate of A is unchanged. Substituting Eqs. (B-14) into (B-15):

$$x_2 = -(-y' \sin \omega + z' \cos \omega) \sin \phi + x' \cos \phi$$

$$y_2 = y' \cos \omega + z' \sin \omega \qquad \text{(B-16)}$$

$$z_2 = (-y' \sin \omega + z' \cos \omega) \cos \phi + x' \sin \phi$$

Third, rotation through κ about the z_2 axis, as illustrated in Fig. B-8. The coordinates of A in the three times rotated coordinate system, which has now become the xyz system as shown graphically in Fig. B-8, are

$$x = x_2 \cos \kappa + y_2 \sin \kappa$$

$$y = -x_2 \sin \kappa + y_2 \cos \kappa \qquad \text{(B-17)}$$

$$z = z_2$$

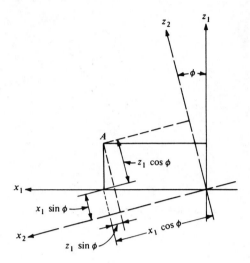

Figure B-7 Phi rotation about the y_1 axis.

In this rotation about z_2, the z_2 and z axes are coincident and therefore the z coordinate of A is unchanged. Substituting Eqs. (B-16) into (B-17):

$$x = [(y' \sin \omega - z' \cos \omega) \sin \phi + x' \cos \phi] \cos \kappa$$
$$+ (y' \cos \omega + z' \sin \omega) \sin \kappa$$
$$y = [(-y' \sin \omega + z' \cos \omega) \sin \phi - x' \cos \phi] \sin \kappa \qquad \text{(B-18)}$$
$$+ (y' \cos \omega + z' \sin \omega) \cos \kappa$$
$$z = (-y' \sin \omega + z' \cos \omega) \cos \phi + x' \sin \phi$$

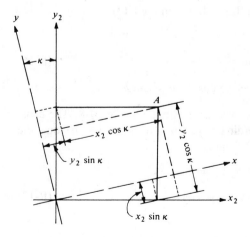

Figure B-8 Kappa rotation about the z_2 axis.

Factoring Eqs. (B-18),

$$x = x'(\cos \phi \cos \kappa) + y'(\sin \omega \sin \phi \cos \kappa + \cos \omega \sin \kappa)$$
$$+ z'(-\cos \omega \sin \phi \cos \kappa + \sin \omega \sin \kappa)$$
$$y = x'(-\cos \phi \sin \kappa) + y'(-\sin \omega \sin \phi \sin \kappa + \cos \omega \cos \kappa) \quad \text{(B-19)}$$
$$+ z'(\cos \omega \sin \phi \sin \kappa + \sin \omega \cos \kappa)$$
$$z = x'(\sin \phi) + y'(-\sin \omega \cos \phi) + z'(\cos \omega \cos \phi)$$

Substituting m's for the coefficients of x', y', and z' in Eqs. (B-19), these equations are

$$x = m_{11}x' + m_{12}y' + m_{13}z'$$
$$y = m_{21}x' + m_{22}y' + m_{23}z' \quad \text{(B-20)}$$
$$z = m_{31}x' + m_{32}y' + m_{33}z'$$

where

$$m_{11} = \cos \phi \cos \kappa$$
$$m_{12} = \sin \omega \sin \phi \cos \kappa + \cos \omega \sin \kappa$$
$$m_{13} = -\cos \omega \sin \phi \cos \kappa + \sin \omega \sin \kappa$$
$$m_{21} = -\cos \phi \sin \kappa$$
$$m_{22} = -\sin \omega \sin \phi \sin \kappa + \cos \omega \cos \kappa \quad \text{(B-21)}$$
$$m_{23} = \cos \omega \sin \phi \sin \kappa + \sin \omega \cos \kappa$$
$$m_{31} = \sin \phi$$
$$m_{32} = -\sin \omega \cos \phi$$
$$m_{33} = \cos \omega \cos \phi$$

Equations (B-20) may be expressed in matrix form as

$$X = MX' \quad \text{(B-22)}$$

where

$$X = \begin{bmatrix} x \\ y \\ z \end{bmatrix} \quad M = \begin{bmatrix} m_{11} & m_{12} & m_{13} \\ m_{21} & m_{22} & m_{23} \\ m_{31} & m_{32} & m_{33} \end{bmatrix} \quad \text{and} \quad X' = \begin{bmatrix} x' \\ y' \\ z' \end{bmatrix}$$

The matrix M is commonly called the *rotation matrix*. The individual elements of the rotation matrix are *direction cosines* which relate the two axis systems. These matrix elements, expressed in terms of direction cosines are

$$M = \begin{bmatrix} \cos xx' & \cos xy' & \cos xz' \\ \cos yx' & \cos yy' & \cos yz' \\ \cos zx' & \cos zy' & \cos zz' \end{bmatrix} \qquad \text{(B-23)}$$

In the above matrix, $\cos xx'$ is the direction cosine relating the x and x' axes, $\cos xy'$ relates the x and y' axes, etc. Direction cosines are simply the cosines of the angles in space between the respective axes, the angles being taken between $0°$ and $180°$. It is an important property that the sum of the squares of the three direction cosines of any straight line is unity. This property may be used to check the computed elements of the rotation matrix for correctness. The check is obtained if the sum of the squares of the elements of any row or of any column of the M matrix is equal to 1.

The rotation matrix is an *orthogonal* matrix, which has the property that its inverse is equal to its transpose, or

$$M^{-1} = M^T \qquad \text{(B-24)}$$

Using this property, Eq. (B-22) may be rewritten expressing $x'y'z'$ coordinates in terms of xyz coordinates as follows:

$$X' = M^T X \qquad \text{(B-25)}$$

In expanded form, this equation is:

$$x' = m_{11}x + m_{21}y + m_{31}z$$

$$y' = m_{12}x + m_{22}y + m_{32}z \qquad \text{(B-26)}$$

$$z' = m_{13}x + m_{23}y + m_{33}z$$

Step 2: Scaling and Translation

To arrive at the final three-dimensional coordinate transformation equations, i.e., equations which yield coordinates in the XYZ system of Fig. B-4, it is necessary to multiply each of Eqs. (B-26) by a scale factor s and add the translation factors T_x, T_y, and T_z. [Recall that the $x'y'z'$ coordinates given by Eqs. (B-26) are in a system that is parallel to the XYZ system.] This makes the lengths of any lines equal in both coordinate systems and translates from the origin of $x'y'z'$ to the origin of the XYZ system. Performing this step,

$$X = sx' + T_x = s(m_{11}x + m_{21}y + m_{31}z) + T_x$$

$$Y = sy' + T_y = s(m_{12}x + m_{22}y + m_{32}z) + T_y \qquad \text{(B-27)}$$

$$Z = sz' + T_z = s(m_{13}x + m_{23}y + m_{33}z) + T_z$$

In matrix form, Eqs. (B-27) are

$$\overline{X} = sM^T X + T \qquad \text{(B-28)}$$

In Eq. (B-28) matrixes M and X are as previously defined, s is the scale factor, and

$$\bar{X} = \begin{bmatrix} X \\ Y \\ Z \end{bmatrix} \quad \text{and} \quad T = \begin{bmatrix} T_x \\ T_y \\ T_z \end{bmatrix}$$

In Eqs. (B-27), the nine m's are not independent of each other, but rather, as seen in Eqs. (B-21), they are functions of the three rotation angles ω, ϕ, and κ. In addition to these three unknown angles, there are also three unknown translations and one scale factor in Eqs. (B-27), which makes a total of seven unknowns. A unique solution is obtained for the unknowns if the x and y coordinates of two horizontal points and the Z coordinates of three vertical points are known in both coordinate systems. If more than the minimum of seven coordinates are known in both systems, redundant equations may be written which make possible an improved solution through least squares techniques.

There are different approaches to the solution for the unknowns. One method which eliminates the three translation factors by subtraction, and thus reduces the number of equations that must be solved to four, was described in the first edition of this book. This method is advantageous when working with small computers. An improved solution, described herein, simultaneously solves all seven unknowns, although it requires a greater computational effort. Suppose there are three points, p, q, and r, whose coordinates are known in both systems. Then a total of nine equations of the type of Eqs. (B-27) may be written as follows:

$$
\begin{aligned}
(1) \qquad & X_P = s(m_{11}x_p + m_{21}y_p + m_{31}z_p) + T_x \\[4pt]
(2) \qquad & Y_P = s(m_{12}x_p + m_{22}y_p + m_{32}z_p) + T_y \\[4pt]
(3) \qquad & Z_P = s(m_{13}x_p + m_{23}y_p + m_{33}z_p) + T_z \\[4pt]
(4) \qquad & X_Q = s(m_{11}x_q + m_{21}y_q + m_{31}z_q) + T_x \\[4pt]
(5) \qquad & Y_Q = s(m_{12}x_q + m_{22}y_q + m_{32}z_q) + T_y \qquad \text{(B-29)} \\[4pt]
(6) \qquad & Z_Q = s(m_{13}x_q + m_{23}y_q + m_{33}z_q) + T_z \\[4pt]
(7) \qquad & X_R = s(m_{11}x_r + m_{21}y_r + m_{31}z_r) + T_x \\[4pt]
(8) \qquad & Y_R = s(m_{12}x_r + m_{22}y_r + m_{32}z_r) + T_y \\[4pt]
(9) \qquad & Z_R = s(m_{13}x_r + m_{23}y_r + m_{33}z_r) + T_z
\end{aligned}
$$

Equations (B-29) are nonlinear equations involving the seven unknowns, s, ω, ϕ, κ, T_x, T_y, and T_z. To solve these equations, they are linearized using Taylor's theorem. In accordance with Taylor's theorem, the linearized form of the first three of Eqs. (B-29) which relate to point P is

$$X_P = (X_P)_0 + \left(\frac{\partial X_P}{\partial s}\right)_0 ds + \left(\frac{\partial X_P}{\partial \omega}\right)_0 d\omega + \left(\frac{\partial X_P}{\partial \phi}\right)_0 d\phi$$

$$+ \left(\frac{\partial X_P}{\partial \kappa}\right)_0 d\kappa + \left(\frac{\partial X_P}{\partial T_x}\right)_0 dT_x + \left(\frac{\partial X_P}{\partial T_y}\right)_0 dT_y + \left(\frac{\partial X_P}{\partial T_z}\right)_0 dT_z$$

$$Y_P = (Y_P)_0 + \left(\frac{\partial Y_P}{\partial s}\right)_0 ds + \left(\frac{\partial Y_P}{\partial \omega}\right)_0 d\omega + \left(\frac{\partial Y_P}{\partial \phi}\right)_0 d\phi$$

$$\text{(B-30)}$$

$$+ \left(\frac{\partial Y_P}{\partial \kappa}\right)_0 d\kappa + \left(\frac{\partial Y_P}{\partial T_x}\right)_0 dT_x + \left(\frac{\partial Y_P}{\partial T_y}\right)_0 dT_y + \left(\frac{\partial Y_P}{\partial T_z}\right)_0 dT_z$$

$$Z_P = (Z_P)_0 + \left(\frac{\partial Z_P}{\partial s}\right)_0 ds + \left(\frac{\partial Z_P}{\partial \omega}\right)_0 d\omega + \left(\frac{\partial Z_P}{\partial \phi}\right)_0 d\phi$$

$$+ \left(\frac{\partial Z_P}{\partial \kappa}\right)_0 d\kappa + \left(\frac{\partial Z_P}{\partial T_x}\right)_0 dT_x + \left(\frac{\partial Z_P}{\partial T_y}\right)_0 dT_y + \left(\frac{\partial Z_P}{\partial T_z}\right)_0 dT_z$$

By simply changing subscripts, expressions similar to Eqs. (B-30) are written for points q and r, giving a total of nine linearized equations. In Eqs. (B-30), $(X_P)_0$, $(Y_P)_0$, and $(Z_P)_0$ are the right-hand sides of the first three Eqs. (B-29) evaluated at the initial approximations; $(\partial X_P/\partial s)_0$, $(\partial X_P/\partial \omega)_0$, etc., are partial derivatives with respect to the indicated unknowns evaluated at the initial approximations; and ds, $d\omega$, $d\phi$, $d\kappa$, dT_x, dT_y, and dT_z are corrections to the initial approximations. The units of $d\omega$, $d\phi$, and $d\kappa$ are radians.

Substituting letters for partial derivative coefficients, and adding residuals to make this redundant system of equations consistent, all nine linear equations of the form of Eqs. (B-30) are given below:

$$a_{11}ds + a_{12}d\omega + a_{13}d\phi + a_{14}d\kappa + a_{15}dT_x$$
$$+ a_{16}dT_y + a_{17}dT_z = [X_P - (X_P)_0] + v_{X_P}$$

$$a_{21}ds + a_{22}d\omega + a_{23}d\phi + a_{24}d\kappa + a_{25}dT_x$$
$$+ a_{26}dT_y + a_{27}dT_z = [Y_P - (Y_P)_0] + v_{Y_P}$$

$$a_{31}ds + a_{32}d\omega + a_{33}d\phi + a_{34}d\kappa + a_{35}dT_x$$
$$+ a_{36}dT_y + a_{37}dT_z = [Z_P - (Z_P)_0] + v_{Z_P}$$

$$a_{41}ds + a_{42}d\omega + a_{43}d\phi + a_{44}d\kappa + a_{45}dT_x$$
$$+ a_{46}dT_y + a_{47}dT_z = [X_Q - (X_Q)_0] + v_{X_Q}$$

$$a_{51}ds + a_{52}d\omega + a_{53}d\phi + a_{54}d\kappa + a_{55}dT_x \quad \text{(B-31)}$$

$$+ a_{56}dT_y + a_{57}dT_z = [Y_Q - (Y_Q)_0] + v_{Y_Q}$$

$$a_{61}ds + a_{62}d\omega + a_{63}d\phi + a_{64}d\kappa + a_{65}dT_x$$

$$+ a_{66}dT_y + a_{67}dT_z = [Z_Q - (Z_Q)_0] + v_{Z_Q}$$

$$a_{71}ds + a_{72}d\omega + a_{73}d\phi + a_{74}d\kappa + a_{75}dT_x$$

$$+ a_{76}dT_y + a_{77}dT_z = [X_R - (X_R)_0] + v_{X_R}$$

$$a_{81}ds + a_{82}d\omega + a_{83}d\phi + a_{84}d\kappa + a_{85}dT_x$$

$$+ a_{86}dT_y + a_{87}dT_z = [Y_R - (Y_R)_0] + v_{Y_R}$$

$$a_{91}ds + a_{92}d\omega + a_{93}d\phi + a_{94}d\kappa + a_{95}dT_x$$

$$+ a_{96}dT_y + a_{97}dT_z = [Z_R - (Z_R)_0] + v_{Z_R}$$

Equations (B-31) may be expressed in matrix form as

$$A \quad X = L + V \quad \text{(B-32)}$$

where

$$A = \begin{bmatrix} a_{11} & a_{12} & a_{13} & a_{14} & a_{15} & a_{16} & a_{17} \\ a_{21} & a_{22} & a_{23} & a_{24} & a_{25} & a_{26} & a_{27} \\ a_{31} & a_{32} & a_{33} & a_{34} & a_{35} & a_{36} & a_{37} \\ a_{41} & a_{42} & a_{43} & a_{44} & a_{45} & a_{46} & a_{47} \\ a_{51} & a_{52} & a_{53} & a_{54} & a_{55} & a_{56} & a_{57} \\ a_{61} & a_{62} & a_{63} & a_{64} & a_{65} & a_{66} & a_{67} \\ a_{71} & a_{72} & a_{73} & a_{74} & a_{75} & a_{76} & a_{77} \\ a_{81} & a_{82} & a_{83} & a_{84} & a_{85} & a_{86} & a_{87} \\ a_{91} & a_{92} & a_{93} & a_{94} & a_{95} & a_{96} & a_{97} \end{bmatrix}_{9}^{7}$$

$$X = \begin{bmatrix} ds \\ d\omega \\ d\phi \\ d\kappa \\ dT_x \\ dT_y \\ dT_z \end{bmatrix}_{7}^{1} \quad L = \begin{bmatrix} X_P - (X_P)_0 \\ Y_P - (Y_P)_0 \\ Z_P - (Z_P)_0 \\ X_Q - (X_Q)_0 \\ Y_Q - (Y_Q)_0 \\ Z_Q - (Z_Q)_0 \\ X_R - (X_R)_0 \\ Y_R - (Y_R)_0 \\ Z_R - (Z_R)_0 \end{bmatrix}_{9}^{1} \quad V = \begin{bmatrix} v_{X_P} \\ v_{Y_P} \\ v_{Z_P} \\ v_{X_Q} \\ v_{Y_Q} \\ v_{Z_Q} \\ v_{X_R} \\ v_{Y_R} \\ v_{Z_R} \end{bmatrix}_{9}^{1}$$

Equation (B-32) may be solved using least squares Eq. (A-12). The solution is iterated, due to the use of the Taylor series, until negligibly small values are obtained for the corrections to the initial approximations of the unknown transformation parameters. Once the transformation factors have been found, the transformed coordinates of all points whose coordinates are known only in the original system may be found by applying equations of the type of Eqs. (B-29).

To clarify the coefficients of Eqs. (B-31), the partial derivatives for the first three equations are

$$a_{11} = m_{11}(x_p) + m_{21}(y_p) + m_{31}(z_p)$$

$$a_{12} = 0$$

$$a_{13} = [(-\sin \phi \cos \kappa)(x_p) + \sin \phi \sin \kappa (y_p)$$
$$+ \cos \phi(z_p)]s$$

$$a_{14} = [m_{21}(x_p) - (m_{11})(y_p)]s$$

$$a_{15} = a_{26} = a_{37} = 1$$

$$a_{16} = a_{17} = a_{25} = a_{27} = a_{35} = a_{36} = 0$$

$$a_{21} = m_{12}(x_p) + m_{22}(y_p) + m_{32}(z_p)$$

$$a_{22} = [-m_{13}(x_p) - m_{23}(y_p) - m_{33}(z_p)]s$$

$$a_{23} = [(\sin \omega \cos \phi \cos \kappa)(x_p) + (-\sin \omega \cos \phi \sin \kappa)(y_p)$$
$$+ (\sin \omega \sin \phi)(z_p)]s$$

$$a_{24} = [m_{22}(x_p) - m_{12}(y_p)]s$$

$$a_{31} = m_{13}(x_p) + m_{23}(y_p) + m_{33}(z_p)$$

$$a_{32} = [m_{12}(x_p) + m_{22}(y_p) + m_{32}(z_p)]s$$

$$a_{33} = [(-\cos \omega \cos \phi \cos \kappa)(x_p) + (\cos \omega \cos \phi \sin \kappa)(y_p)$$
$$+ (-\cos \omega \sin \phi)(z_p)]s$$

$$a_{34} = [m_{23}(x_p) - m_{13}(y_p)]s$$

The coefficients for the other six equations from Eq. (B-31) are exactly the same as those for the first three, with the exception that the subscripts p are replaced with subscripts q and r. In general, if there are n points whose X, Y, and Z coordinates are known in both systems, then $3(n)$ equations of the type of Eqs. (B-31) may be formed. Three-dimensional conformal coordinate transformations involve lengthy calculations and therefore they are practical only when performed on computer.

B-8 TWO-DIMENSIONAL PROJECTIVE COORDINATE TRANSFORMATION

Two-dimensional projective transformation equations enable the analytical computation of the XY coordinates of points after they have been projected into a plane from
another nonparallel plane. The most common use of these equations is in analytical
rectification, e.g., calculating coordinates of points in a rectified-ratioed photo plane
based upon their coordinates in a tilted photo. This situation is illustrated in Fig.
B-9.

In Fig. B-9, a tilted photo with its xyz fiducial coordinate axis system (shown
dashed) is illustrated. The projection center is at L, and the projection of points a, b,
c, and d from the tilted photo onto the plane of a rectified-ratioed photo occur at A,
B, C, and D, respectively. Positions of projected points in the rectified-ratioed photo
plane are expressed in the XYZ coordinate system shown in the figure.

In the most simplified development of the equations for the two-dimensional

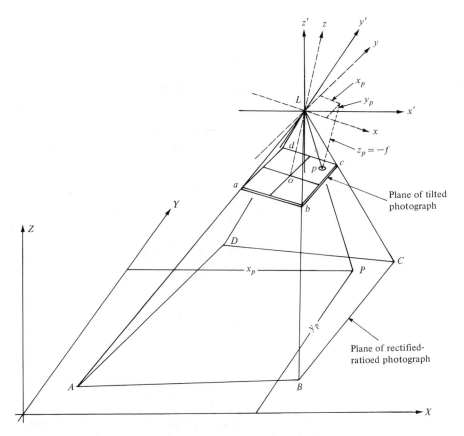

Figure B-9 Geometry of two-dimensional projective transformation.

projective transformation, an $x'y'z'$ coordinate system is adopted which is parallel to the XYZ system, and has its origin at L. Using Eqs. (B-20), which were developed in the preceeding section, the x, y, and z coordinates of any point, such as p of Fig. B-9, can be expressed in terms of $x'y'z'$ coordinates as follows:

$$x_p = m_{11}x'_p + m_{12}y'_p + m_{13}z'_p$$

$$y_p = m_{21}x'_p + m_{22}y'_p + m_{23}z'_p \tag{B-33}$$

$$z_p = m_{31}x'_p + m_{32}y'_p + m_{33}z'_p$$

In Eqs. (B-33) the m's are functions of rotation angles omega, phi, and kappa which define the tilt relationships between the two planes. These are described in the preceeding section. The other terms in Eqs. (B-33) are coordinates as previously described. Consider now Fig. B-10, which shows the parallel relationships between the $x'y'z'$ and XYZ planes after rotation. From similar triangles of Fig. B-10,

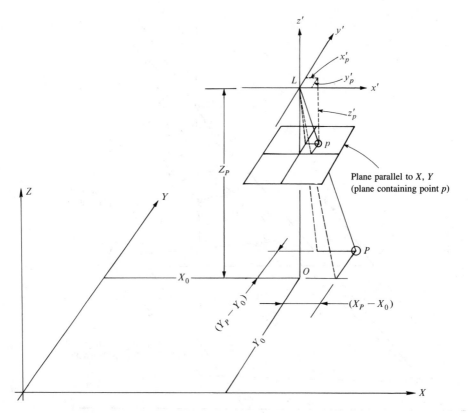

Figure B-10 Parallel relationships that exist after rotation between $x'y'z'$ and XYZ planes in two-dimensional projective transformation.

$$\frac{x'_p}{X_P - X_O} = \frac{z'_p}{Z_P}$$

from which

$$x'_p = \frac{z'_p (X_P - X_O)}{Z_P} \qquad (a)$$

In Eq. (a), Z_P is LO, or the orthogonal distance from L to the plane of the rectified-ratioed photo. The other terms are as previously defined. Again from similar triangles of Fig. B-10,

$$\frac{y'_p}{Y_P - Y_O} = \frac{z'_p}{Z_P}$$

from which

$$y'_p = \frac{z'_p (Y_P - Y_O)}{Z_P} \qquad (b)$$

Also, intuitively the following equation can be written:

$$z'_p = \frac{z'_p (Z_P)}{Z_P} \qquad (c)$$

Substituting Eqs. (a), (b), and (c) into Eq. (B-33),

$$x_p = m_{11} \frac{X_P - X_O}{Z_P} z'_p + m_{12} \frac{Y_P - Y_O}{Z_P} z'_p + m_{13} \frac{Z_P}{Z_P} z'_p$$

$$y_p = m_{21} \frac{X_P - X_O}{Z_P} z'_p + m_{22} \frac{Y_P - Y_O}{Z_P} z'_p + m_{23} \frac{Z_P}{Z_P} z'_p \qquad \text{(B-34)}$$

$$z_p = m_{31} \frac{X_P - X_O}{Z_P} z'_p + m_{32} \frac{Y_P - Y_O}{Z_P} z'_p + m_{33} \frac{Z_P}{Z_P} z'_p$$

Factoring Eqs. (B-34),

$$x_p = \frac{z'_p}{Z_P} [m_{11}(X_P - X_O) + m_{12}(Y_P - Y_O) + m_{13}Z_P] \qquad (d)$$

$$y_p = \frac{z'_p}{Z_P} [m_{21}(X_P - X_O) + m_{22}(Y_P - Y_O) + m_{23}Z_P] \qquad (e)$$

$$z_p = \frac{z'_p}{Z_P} [m_{31}(X_P - X_O) + m_{32}(Y_P - Y_O) + m_{33}Z_P] \qquad (f)$$

Dividing Eqs. (*d*) and (*e*) by Eq. (*f*),

$$x_p = \frac{z_p m_{11}(X_P - X_O) + z_p m_{12}(Y_P - Y_O) + z_p m_{13} Z_P}{m_{31}(X_P - X_O) + m_{32}(Y_P - Y_O) + m_{33} Z_P} \tag{g}$$

$$y_p = \frac{z_p m_{21}(X_P - X_O) + z_p m_{22}(Y_P - Y_O) + z_p m_{23} Z_P}{m_{31}(X_P - X_O) + m_{32}(Y_P - Y_O) + m_{33} Z_P} \tag{h}$$

If both the numerators and denominators of the right-hand sides of Eqs. (*g*) and (*h*) are divided by $m_{33}(Z_P)$, the following equations result:

$$x_p = \frac{\dfrac{z_p m_{11}}{m_{33} Z_P}(X_P - X_O) + \dfrac{z_p m_{12}}{m_{33} Z_P}(Y_P - Y_O) + \dfrac{z_p m_{13}}{m_{33} Z_P} Z_P}{\dfrac{m_{31}}{m_{33} Z_P}(X_P - X_O) + \dfrac{m_{32}}{m_{33} Z_P}(Y_P - Y_O) + \dfrac{m_{33} Z_P}{m_{33} Z_p}}$$

$$y_p = \frac{\dfrac{z_p m_{21}}{m_{33} Z_P}(X_P - X_O) + \dfrac{z_p m_{22}}{m_{33} Z_P}(Y_P - Y_O) + \dfrac{z_p m_{23}}{m_{33} Z_P} Z_P}{\dfrac{m_{31}}{m_{33} Z_P}(X_P - X_O) + \dfrac{m_{32}}{m_{33} Z_P}(Y_P - Y_O) + \dfrac{m_{33} Z_P}{m_{33} Z_p}} \tag{B-35}$$

In projecting points from one plane to another, the m's, z_p (which is equal to $-f$), and Z_P, X_O, and Y_O are all constants. Therefore Eqs. (B-35) can be simplified into the following form:

$$x_p = \frac{a_1 X_P + b_1 Y_P + c_1}{a_3 X_P + b_3 Y_P + 1}$$

$$y_p = \frac{a_2 X_P + b_2 Y_P + c_2}{a_3 X_P + b_3 Y_P + 1} \tag{i}$$

Equations (*i*) are those applied in performing a two-dimensional projective transformation. As developed, however, they yield x and y tilted photo coordinates from X and Y rectified-ratioed coordinates. It is customary to perform rectification in the opposite sense, i.e., to compute X and Y rectified-ratioed coordinates from x and y coordinates measured on a tilted photo. Since Eqs. (*i*) are general and simply express the projectivity between any two nonparallel planes, after dropping subscripts, they can be written in the following form to enable computing X and Y rectified-ratioed coordinates in terms of x and y tilted photo coordinates:

$$X = \frac{a_1 x + b_1 y + c_1}{a_3 x + b_3 y + 1}$$

$$Y = \frac{a_2 x + b_2 y + c_2}{a_3 x + b_3 y + 1} \tag{B-36}$$

In using Eqs. (B-36) for rectification, X and Y are ground coordinates of control points, and x and y are coordinates of the same points in the fiducial system of the tilted photograph. A pair of Eqs. (B-36) can be written for each control point, and since there are eight unknown parameters in Eqs. (B-36), four control points are needed to obtain a unique solution for the unknowns. More than four control points enables a least squares solution to be performed.

When equations of the type of Eqs. (B-36) have been written for all control points, a solution is obtained for the unknown parameters. These parameters are then used in Eqs. (B-36) to compute the rectified and ratioed coordinates of all other points in the tilted photo whose xy coordinates have been measured. Besides using these equations for rectification, they can be used to transform comparator coordinates into the photo coordinate system defined by the fiducial marks, and in this instance X and Y are calibrated coordinates of fiducial marks and x and y are their comparator coordinates.

REFERENCES

American Society of Photogrammetry: "Manual of Photogrammetry," 3d ed., Falls Church, Va., 1966, chap. 2.

———: "Manual of Photogrammetry," 4th ed., Falls Church, Va., 1980, chap. 2.

Baetsle, P. L.: Conformal Transformations in Three Dimensions, *Photogrammetric Engineering,* vol. 32, no. 5, p. 816, 1966.

Blais, J. A. R.: Three-Dimensional Similarity, *Canadian Surveyor,* vol. 26, no. 1, p. 71, 1972.

Erio, G.: Three-Dimensional Transformations for Independent Models, *Photogrammetric Engineering and Remote Sensing,* vol. 41, no. 9, p. 1117, 1975.

Light, D. L.: The Orientation Matrix, *Photogrammetric Engineering,* vol. 32, no. 3, p. 434, 1966.

Mikhail, E. M.: Simultaneous Three Dimensional Transformation of Higher Degree, *Photogrammetric Engineering,* vol. 30, no. 4, p. 588, 1964.

———: Discussion Paper: Simultaneous Three Dimensional Transformation, *Photogrammetric Engineering,* vol. 32, no. 2, p. 180, 1966.

Schmidt, E.: "Transformation of Rectangular Space Coordinates," Technical Bulletin No. 15, U.S. Coast and Geodetic Survey, Washington, D.C., 1960.

Schut, G. H.: Conformal Transformations and Polynomials, *Photogrammetric Engineering,* vol. 32, no. 5, p. 826, 1962.

Tewinkel, G. C.: A Trigonometric Derivation of the Formulas for the Three-Dimensional Rotation Matrix, *Photogrammetric Engineering,* vol. 30, no. 4, p. 635, 1964.

Umbach, M. J.: "Aerotriangulation: Tranformation of Surveying and Mapping Coordinate Systems," Technical Report No. 34, U.S. Coast and Geodetic Survey, Washington, D.C., 1967.

Vlcek, J.: Discussion Paper: Simultaneous Three Dimensional Transformation, *Photogrammetric Engineering,* vol. 32, no. 2, p. 178, 1962.

Yassa, G.: Orthogonal Transformations, *Photogrammetric Engineering,* vol. 40, no. 8, p. 961, 1974.

PROBLEMS

B-1 The following table contains arbitrary X and Y coordinates of a group of points determined from radial-line triangulation. The table also includes state plane coordinates for three of the points. Using a *two-dimensional conformal* coordinate transformation, calculate state plane coordinates for the other points.

Point	Arbitrary coordinates from radial-line triangulation		State plane coordinates	
	X, ft	Y, ft	X, ft	Y, ft
A	171.922	47.078	2,101,067.0	104,400.6
B	181.725	204.970	2,101,050.1	101,540.8
C	199.012	178.975	2,100,712.1	101,922.1
1	198.185	50.980		
2	367.871	15.046		
3	488.333	77.258		
4	597.505	171.608		
5	645.170	191.923		

B-2 The following table contains X and Y comparator coordinates for four side fiducial marks and six additional photo images. The table also contains calibrated coordinates for the four fiducials. Using a *two-dimensional affine* coordinate transformation, determine corrected photocoordinates for the six image points.

Point	Comparator coordinates		Calibrated Coordinates	
	X, mm	Y, mm	X, mm	Y, mm
1000	66.280	182.115	−113.000	0.000
2000	178.827	295.704	0.000	113.000
3000	292.111	182.928	113.000	0.000
4000	179.691	69.161	0.000	−113.000
1	239.609	127.251		
2	281.418	222.171		
3	270.191	253.286		
4	232.091	247.215		
5	135.063	256.258		
6	112.940	126.413		

B-3 For the data of Prob. B-2, use the two-dimensional projective transformation equations to determine corrected photo coordinates for the six image points.

B-4 Coordinates X_1, Y_1, and Z_1 for model I and X_2, Y_2, and Z_2 for model II of an independent model aerotriangulation are contained in the table below. Transform the model II coordinates into the model I coordinate system using a *three-dimensional conformal* coordinate transformation.

Point	Model I coordinates			Model II coordinates		
	X_1, mm	Y_1, mm	Z_1, mm	X_2, mm	Y_2, mm	Z_2, mm
R10	607.54	501.63	469.09	390.35	499.63	469.43
A	589.98	632.36	82.81	371.68	630.84	81.25
B	643.65	421.28	83.50	425.65	419.07	82.49
C	628.58	440.51	82.27	410.50	438.31	81.13
D	666.27	298.16	98.29	448.22	295.83	97.79
E	632.59	710.62	103.01	414.60	709.39	101.77
R11				611.37	498.98	470.45
F				637.49	323.67	85.67
G				573.32	401.51	84.48
H				647.00	373.97	83.76
I				533.51	285.01	87.13

DEVELOPMENT OF COLLINEARITY CONDITION EQUATIONS

C-1 INTRODUCTION

Collinearity, as illustrated in Fig. C-1, is the condition in which the exposure station of any photograph, an object point, and its photo image all lie on a straight line. The equations expressing this condition are called the *collinearity condition equations*. They are perhaps the most useful of all equations to the photogrammetrist.

In Fig. C-2 exposure station L of an aerial photo has coordinates X_L, Y_L, and Z_L with respect to the object (ground) coordinate system XYZ. Image a of object point A, shown in a *rotated* image plane, has image space coordinates x'_a, y'_a, and z'_a, where the rotated image space coordinate system $x'y'z'$ is parallel to object-space coordinate system XYZ.

C-2 ROTATION IN TERMS OF OMEGA, PHI, AND KAPPA

By means of the three-dimensional rotation formulas developed in Appendix B, an image point a having coordinates x_a, y_a, and z_a in a tilted photo such as that of Fig. C-1 may have its coordinates rotated into the $x'y'z'$ coordinate system (parallel to XYZ), as shown in Fig. C-3. The rotated image coordinates x'_a, y'_a, and z'_a may be expressed in terms of the measured photocoordinates x_a and y_a, the camera focal length f, and the three rotation angles omega, phi, and kappa. The rotation formulas are Eqs. (B-20), which were developed in Appendix B. For convenience they are repeated here:

$$x_a = m_{11}x'_a + m_{12}y'_a + m_{13}z'_a$$
$$y_a = m_{21}x'_a + m_{22}y'_a + m_{23}z'_a \qquad \text{(C-1)}$$
$$z_a = m_{31}x'_a + m_{32}y'_a + m_{33}z'_a$$

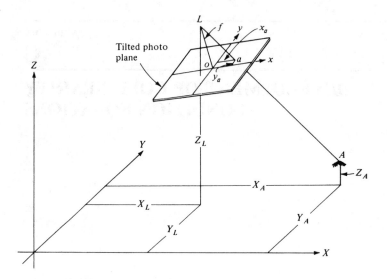

Figure C-1 The collinearity condition.

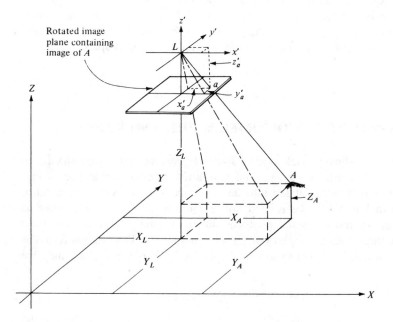

Figure C-2 Image coordinate system rotated such that it is parallel to the object space coordinate system.

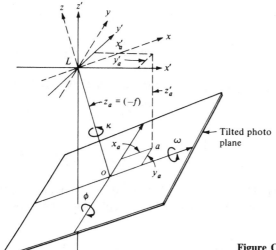

Figure C-3 Measurement x-y-z and rotated x'-y'-z' image coordinate systems.

In Eqs. (C-1) the m's are functions of the rotation angles omega, phi, and kappa. These functions are given as Eqs. (B-21) in Appendix B. Also note on Fig. C-3 that the value for z_a is equal to $(-f)$.

C-3 DEVELOPMENT OF THE COLLINEARITY CONDITION EQUATIONS

The collinearity condition equations are developed from similar triangles of Fig. C-2 as follows:

$$\frac{x'_a}{X_A - X_L} = \frac{y'_a}{Y_A - Y_L} = \frac{-z'_a}{Z_L - Z_A}$$

Reducing,

$$x'_a = \left(\frac{X_A - X_L}{Z_A - Z_L}\right) z'_a \tag{a}$$

$$y'_a = \left(\frac{Y_A - Y_L}{Z_A - Z_L}\right) z'_a \tag{b}$$

Also, by identity,

$$z'_a = \left(\frac{Z_A - Z_L}{Z_A - Z_L}\right) z'_a \tag{c}$$

Substituting (a), (b), and (c) into Eqs. (C-1),

$$x_a = m_{11}\left(\frac{X_A - X_L}{Z_A - Z_L}\right) z_a' + m_{12}\left(\frac{Y_A - Y_L}{Z_A - Z_L}\right) z_a' + m_{13}\left(\frac{Z_A - Z_L}{Z_A - Z_L}\right) z_a' \quad \text{(C-2)}$$

$$y_a = m_{21}\left(\frac{X_A - X_L}{Z_A - Z_L}\right) z_a' + m_{22}\left(\frac{Y_A - Y_L}{Z_A - Z_L}\right) z_a' + m_{23}\left(\frac{Z_A - Z_L}{Z_A - Z_L}\right) z_a' \quad \text{(C-3)}$$

$$z_a = m_{31}\left(\frac{X_A - X_L}{Z_A - Z_L}\right) z_a' + m_{32}\left(\frac{Y_A - Y_L}{Z_A - Z_L}\right) z_a' + m_{33}\left(\frac{Z_A - Z_L}{Z_A - Z_L}\right) z_a' \quad \text{(C-4)}$$

Factoring the term $(z_a'/Z_A - Z_L)$ from Eqs. (C-2) through (C-4), dividing (C-2) and (C-3) by (C-4), and substituting $(-f)$ for z_a, the following collinearity equations result:

$$x_a = -f\left[\frac{m_{11}(X_A - X_L) + m_{12}(Y_A - Y_L) + m_{13}(Z_A - Z_L)}{m_{31}(X_A - X_L) + m_{32}(Y_A - Y_L) + m_{33}(Z_A - Z_L)}\right] \quad \text{(C-5)}$$

$$y_a = -f\left[\frac{m_{21}(X_A - X_L) + m_{22}(Y_A - Y_L) + m_{23}(Z_A - Z_L)}{m_{31}(X_A - X_L) + m_{32}(Y_A - Y_L) + m_{33}(Z_A - Z_L)}\right] \quad \text{(C-6)}$$

C-4 LINEARIZATION OF THE COLLINEARITY EQUATIONS

Equations (C-5) and (C-6) are nonlinear and involve nine unknowns: the three rotation angles omega, phi, and kappa which are inherent in the m's; the three exposure station coordinates X_L, Y_L, and Z_L; and the three object point coordinates X_A, Y_A, and Z_A. The nonlinear equations are linearized using Taylor's theorem. In linearizing the collinearity equations, Eqs. (C-5) and (C-6) are rewritten as follows:

$$F = 0 = qx_a + rf \quad \text{(C-7)}$$

$$G = 0 = qy_a + sf \quad \text{(C-8)}$$

where

$$q = m_{31}(X_A - X_L) + m_{32}(Y_A - Y_L) + m_{33}(Z_A - Z_L)$$

$$r = m_{11}(X_A - X_L) + m_{12}(Y_A - Y_L) + m_{13}(Z_A - Z_L)$$

$$s = m_{21}(X_A - X_L) + m_{22}(Y_A - Y_L) + m_{23}(Z_A - Z_L)$$

According to Taylor's theorem, Eqs. (C-7) and (C-8) may be expressed in linearized form as:

$$0 = (F)_0 + \left(\frac{\partial F}{\partial x_a}\right)_0 dx_a + \left(\frac{\partial F}{\partial \omega}\right)_0 d\omega + \left(\frac{\partial F}{\partial \phi}\right)_0 d\phi + \left(\frac{\partial F}{\partial \kappa}\right)_0 d\kappa$$

$$+ \left(\frac{\partial F}{\partial X_L}\right)_0 dX_L + \left(\frac{\partial F}{\partial Y_L}\right)_0 dY_L + \left(\frac{\partial F}{\partial Z_L}\right)_0 dZ_L \tag{C-9}$$

$$+ \left(\frac{\partial F}{\partial X_A}\right)_0 dX_A + \left(\frac{\partial F}{\partial Y_A}\right)_0 dY_A + \left(\frac{\partial F}{\partial Z_A}\right)_0 dZ_A$$

$$0 = (G)_0 + \left(\frac{\partial G}{\partial y_a}\right)_0 dy_a + \left(\frac{\partial G}{\partial \omega}\right)_0 d\omega + \left(\frac{\partial G}{\partial \phi}\right)_0 d\phi + \left(\frac{\partial G}{\partial \kappa}\right)_0 d\kappa$$

$$+ \left(\frac{\partial G}{\partial X_L}\right)_0 dX_L + \left(\frac{\partial G}{\partial Y_L}\right)_0 dY_L + \left(\frac{\partial G}{\partial Z_L}\right)_0 dZ_L \tag{C-10}$$

$$+ \left(\frac{\partial G}{\partial X_A}\right)_0 dX_A + \left(\frac{\partial G}{\partial Y_A}\right)_0 dY_A + \left(\frac{\partial G}{\partial Z_A}\right)_0 dZ_A$$

In Eqs. (C-9) and (C-10), $(F)_0$ and $(G)_0$ are the functions F and G of Eqs. (C-7) and (C-8) evaluated at the initial approximations for the nine unknowns; the terms $(\partial F/\partial x_a)_0$, $(\partial F/\partial \omega)_0$, $(\partial F/\partial \phi)_0$, etc., are partial derivatives of the functions F and G with respect to the indicated unknowns evaluated at the initial approximations; and dx_a, $d\omega$, $d\phi$, etc., are unknown corrections to be applied to the initial approximations. The units of $d\omega$, $d\phi$, and $d\kappa$ are radians. Since dx_a and dy_a are corrections to measured photocoordinates x_a and y_a, they may be interpreted as residual errors in the measurements. These two terms may therefore be replaced by v_{x_a} and v_{y_a} which are the customary symbols adopted for residual errors. Note from Eqs. (C-7) and (C-8) that partial derivatives $\partial F/\partial x_a$ and $\partial G/\partial y_a$ are both equal to q. Substituting q for these terms in Eqs. (C-9) and (C-10), transposing $q dx_a$ and $q dy_a$ to the left-hand side of the equations, dividing each equation by q, and replacing dx_a and dy_a by v_{x_a} and v_{y_a}, respectively, the following simplified forms of the linearized collinearity equations are obtained:

$$v_{x_a} = b_{11}d\omega + b_{12}d\phi + b_{13}d\kappa - b_{14}dX_L - b_{15}dY_L - b_{16}dZ_L + b_{14}dX_A$$

$$+ b_{15}dY_A + b_{16}dZ_A + J \tag{C-11}$$

$$v_{y_a} = b_{21}d\omega + b_{22}d\phi + b_{23}d\kappa - b_{24}dX_L - b_{25}dY_L - b_{26}dZ_L + b_{24}dX_A$$

$$+ b_{25}dY_A + b_{26}dZ_A + K \tag{C-12}$$

In Eqs. (C-11) and (C-12), J and K are equal to $(F)_{o/q}$ and $(G)_{o/q}$, respectively. The b's are coefficients equal to the partial derivatives. For convenience these coefficients are given below. In these coefficients ΔX, ΔY, and ΔZ are equal to $(X_A -$

X_L), $(Y_A - Y_L)$, and $(Z_A - Z_L)$, respectively. Numerical values for these coefficient terms are obtained by using initial approximations for the unknowns:

$$b_{11} = \frac{x}{q}(-m_{33}\Delta Y + m_{32}\Delta Z) + \frac{f}{q}(-m_{13}\Delta Y + m_{12}\Delta Z)$$

$$b_{12} = \frac{x}{q}[\Delta X \cos \phi + \Delta Y(\sin \omega \sin \phi) + \Delta Z(-\sin \phi \cos \omega)]$$

$$+ \frac{f}{q}[\Delta X(-\sin \phi \cos \kappa) + \Delta Y(\sin \omega \cos \phi \cos \kappa)$$

$$+ \Delta Z(-\cos \omega \cos \phi \cos \kappa)]$$

$$b_{13} = \frac{f}{q}(m_{21}\Delta X + m_{22}\Delta Y + m_{23}\Delta Z)$$

$$b_{14} = \frac{x}{q}(m_{31}) + \frac{f}{q}(m_{11})$$

$$b_{15} = \frac{x}{q}(m_{32}) + \frac{f}{q}(m_{12})$$

$$b_{16} = \frac{x}{q}(m_{33}) + \frac{f}{q}(m_{13})$$

$$J = \frac{(qx + rf)}{q}$$

$$b_{21} = \frac{y}{q}(-m_{33}\Delta Y + m_{32}\Delta Z) + \frac{f}{q}(-m_{23}\Delta Y + m_{22}\Delta Z)$$

$$b_{22} = \frac{y}{q}[\Delta X \cos \phi + \Delta Y(\sin \omega \sin \phi) + \Delta Z(-\cos \omega \sin \phi)]$$

$$+ \frac{f}{q}[\Delta X(\sin \phi \sin \kappa) + \Delta Y(-\sin \omega \cos \phi \sin \kappa)$$

$$+ \Delta Z(\cos \omega \cos \phi \sin \kappa)]$$

$$b_{23} = \frac{f}{q}(-m_{11}\Delta X - m_{12}\Delta Y - m_{13}\Delta Z)$$

$$b_{24} = \frac{y}{q}(m_{31}) + \frac{f}{q}(m_{21})$$

$$b_{25} = \frac{y}{q}(m_{32}) + \frac{f}{q}(m_{22})$$

$$b_{26} = \frac{y}{q}(m_{33}) + \frac{f}{q}(m_{23})$$

$$K = \frac{(qy + sf)}{q}$$

C-5 APPLICATIONS OF COLLINEARITY

The collinearity equations are applicable to the analytical solution of almost every photogrammetry problem. As examples, Sec. 11-12 describes their use in *space resection*, in which the six elements of exterior orientation of a tilted photograph are computed, and Sec. 14-13 explains how collinearity is applied in analytic *relative orientation*, which is necessary in analytically extending control photogrammetrically. Other applications are described elsewhere in this book. Regardless of the particular problem, an *x* equation [Eq. (C-11)] and *y* equation [Eq. (C-12)] are written for each point whose image or images appear in the photo or photos involved in the problem. The equations will contain unknowns, the number of which will vary with the particular problem. If the number of equations is equal to or greater than the number of unknowns, a solution is possible.

Initial approximations are needed for all unknowns, and these are generally easily obtained by making certain assumptions, such as vertical photography. The initial approximations do not have to be extremely close, but the closer they are to the unknowns, the faster a satisfactory solution will be reached; and the result is a savings in computer time.

In solving a system of collinearity equations of the form of Eqs. (C-11) and (C-12) for any problem, the quantities which are determined are corrections to the initial approximations. After the first solution, the computed corrections are added to the initial approximations to obtain revised approximations. The solution is then repeated to find new corrections. This procedure is continued (iterated) until the magnitudes of the corrections become insignificant.

A system of collinearity equations of the form of Eqs. (C-11) and (C-12) may be expressed in matrix form as

$$_mV_1 = {_mA_n}\, {_nX_1} - {_mL_1} \qquad (\text{C-13})$$

In Eq. (C-13), *m* is the number of equations; *n* is the number of unknowns; *V* is the matrix of residual errors in the measured *x* and *y* photocoordinates; *A* is the matrix of *b*'s, the coefficients of the unknowns; *X* is the matrix of unknown corrections to the initial approximations; and *L* is the matrix of constant terms *J* and *K*. If the number of equations exceeds the number of unknowns, a least squares solution may be obtained for the most probable values for the unknowns by using matrix Eq. (A-12) or (A-13) of Appendix A. Precisions of the unknowns may be computed by applying matrix Eqs. (A-14) through (A-17) of Appendix A.

C-6 ROTATION IN TERMS OF AZIMUTH, TILT, AND SWING

Instead of using omega, phi, and kappa as the rotation angles for transforming the tilted photocoordinates into an $x'y'z'$ coordinate system parallel to the ground system, the rotation angles azimuth, tilt, and swing may be used. A tilted photo illustrating the azimuth (α), tilt (t), and swing (s) angles is shown in Fig. C-4. In the figure the photo principal plane intersects the datum plane along line $N_d P_d$. The rotation formulas are developed by initially assuming an $x'y'z'$ coordinate system parallel to XYZ, and then, by means of rotations, converting it into the xyz photo measurement system. The origins of both image coordinate systems are taken at the exposure station L.

The rotation equations are developed in a sequence of three separate two-dimensional rotations. The $x'y'z'$ coordinate system is first rotated about the z' axis through a clockwise angle α to create an $x^\alpha y^\alpha z^\alpha$ coordinate system. After rotation the y^α axis will be in the principal plane of the photo. With reference to Fig. C-5a the coordinates of any point in the $x^\alpha y^\alpha z^\alpha$ system are

$$x^\alpha = x' \cos \alpha - y' \sin \alpha$$

$$y^\alpha = x' \sin \alpha + y' \cos \alpha \qquad \text{(C-14)}$$

$$z^\alpha = z'$$

The second rotation is a counterclockwise rotation t about the x^α axis to create an $x^{\alpha t} y^{\alpha t} z^{\alpha t}$ coordinate system. After rotation, the $x^{\alpha t}$ and $y^{\alpha t}$ axes are in the plane of

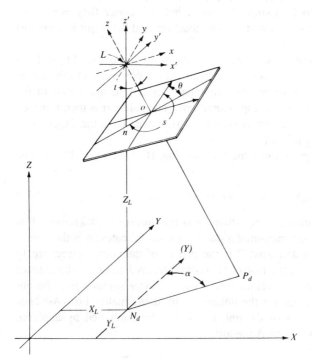

Figure C-4 Azimuth, tilt, and swing rotation angles.

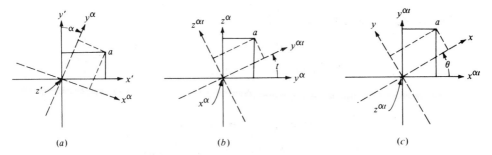

Figure C-5 Rotation in azimuth, tilt, and swing. (*a*) First rotation; (*b*) second rotation; (*c*) third rotation.

the tilted photograph. With reference to Fig. C-5*b* the coordinates of any point in the $x^{\alpha t}y^{\alpha t}z^{\alpha t}$ system are

$$x^{\alpha t} = x^{\alpha}$$
$$y^{\alpha t} = y^{\alpha} \cos t + z^{\alpha} \sin t \qquad \text{(C-15)}$$
$$z^{\alpha t} = -y^{\alpha} \sin t + z^{\alpha} \cos t$$

The third rotation is about the $z^{\alpha t}$ axis through the counterclockwise angle θ. Angle θ is defined as:

$$\theta = (s - 180°)$$

This third rotation creates an $x^{\alpha t\theta}y^{\alpha t\theta}z^{\alpha t\theta}$ coordinate system which coincides with the *xyz* tilted photo system. With reference to Fig. C-5*c* the coordinates of any point in the *xyz* system are

$$x = x^{\alpha t\theta} = x^{\alpha t} \cos \theta + y^{\alpha t} \sin \theta$$
$$y = y^{\alpha t\theta} = -x^{\alpha t} \sin \theta + y^{\alpha t} \cos \theta \qquad \text{(C-16)}$$
$$z = z^{\alpha t\theta} = z^{\alpha t}$$

Because $\sin \theta$ equals $- \sin s$, and $\cos \theta$ equals $- \cos s$, these substitutions may be made into Eqs. (C-16), from which

$$x = -x^{\alpha t} \cos s - y^{\alpha t} \sin s$$
$$y = x^{\alpha t} \sin s - y^{\alpha t} \cos s \qquad \text{(C-17)}$$
$$z = z^{\alpha t}$$

Substituting Eq. (C-14) into Eq. (C-15), in turn substituting into Eq. (C-17), and factoring, the following expressions result for the *x*, *y*, and *z* coordinates of any point:

$$x = m_{11}x' + m_{12}y' + m_{13}z'$$
$$y = m_{21}x' + m_{22}y' + m_{23}z' \qquad \text{(C-18)}$$
$$z = m_{31}x' + m_{32}y' + m_{33}z'$$

In Eqs. (C-18), the m's are

$$m_{11} = -\cos \alpha \cos s - \sin \alpha \cos t \sin s$$

$$m_{12} = \sin \alpha \cos s - \cos \alpha \cos t \sin s$$

$$m_{13} = -\sin t \sin s$$

$$m_{21} = \cos \alpha \sin s - \sin \alpha \cos t \cos s$$

$$m_{22} = -\sin \alpha \sin s - \cos \alpha \cos t \cos s \qquad \text{(C-19)}$$

$$m_{23} = -\sin t \cos s$$

$$m_{31} = -\sin \alpha \sin t$$

$$m_{32} = -\cos \alpha \sin t$$

$$m_{33} = \cos t$$

C-7 COLLINEARITY EQUATIONS USING AZIMUTH-TILT-SWING ROTATION

By simply substituting Eqs. (C-19) for the m's into Eqs. (C-5) and (C-6), collinearity equations are obtained which include azimuth, tilt, and swing as unknowns instead of omega, phi, and kappa. By applying Taylor's theorem, these azimuth, tilt, and swing equations can be linearized and used to solve photogrammetry problems analytically. More frequently, however, the omega-phi-kappa equations are used, and if the tilt, swing, and azimuth angles are desired, they are determined from omega, phi, and kappa as described in Sec. C-8.

C-8 CONVERTING FROM ONE ROTATION SYSTEM TO THE OTHER

Although the azimuth-tilt-swing expressions [Eqs. (C-19)] for the m's differ from their corresponding omega-phi-kappa expressions [Eqs. (B-21)], their numerical values are equal. This is true because the m's are actually direction cosines relating image and object coordinate systems as described in Sec. B-7 of Appendix B. Because of their equality, corresponding m's may be set equal to one another; for example, $m_{11} = \cos \phi \cos \kappa = -\cos \alpha \cos s - \sin \alpha \cos t \sin s$. This equality of m's enables converting back and forth between the omega-phi-kappa and azimuth-tilt-swing systems.

If omega, phi, and kappa for a particular photograph are known, numerical values for the m's can be calculated and tilt, swing, and azimuth determined from the following:

$$\cos t = m_{33} \qquad \text{(C-20)}$$

$$\tan s = \frac{-\sin t \sin s}{-\sin t \cos s} = \frac{m_{13}}{m_{23}} \tag{C-21}$$

$$\tan \alpha = \frac{-\sin \alpha \sin t}{-\cos \alpha \sin t} = \frac{m_{31}}{m_{32}} \tag{C-22}$$

The quadrant of s is established on the basis of the algebraic signs of m_{13} and m_{23}. Since t is always between $0°$ and $90°$, $\sin t$ is always positive. Therefore algebraic signs of m_{13} and m_{23} are derived from the sine and cosine of s, respectively. If both m_{13} and m_{23} are negative, then the sine and cosine of s are both positive and s is in quadrant I, having a value between $0°$ and $90°$. Using similar analyses, the following table was developed to provide the proper quadrant for both s and α:

Algebraic sign of m_{13} or m_{31}	Algebraic sign of m_{23} or m_{32}	Quadrant of s or α
−	−	I
−	+	II
+	+	III
+	−	IV

If the tilt, swing, and azimuth are known for a particular photo, conversion to omega, phi, and kappa is also readily made as follows:

$$\sin \phi = m_{31} \tag{C-23}$$

$$-\tan \omega = \frac{-\sin \omega \cos \phi}{\cos \omega \cos \phi} = \frac{m_{32}}{m_{33}} \tag{C-24}$$

$$-\tan \kappa = \frac{-\cos \phi \sin \kappa}{\cos \phi \cos \kappa} = \frac{m_{21}}{m_{11}} \tag{C-25}$$

In Sec. B-7 of Appendix B, a sign convention was adopted making omega, phi, and kappa positive if they were counterclockwise rotations when viewed from the positive end of their respective x, y, or z axes. Using this sign convention and recognizing that for aerial photography omega and phi cannot exceed $90°$, then phi is positive and counterclockwise if m_{31} is positive. Also, if m_{31} is negative, phi is negative and clockwise. Whether phi is positive or negative, since its value is less than $90°$, its cosine is always positive. Therefore if the quotient m_{32}/m_{33} is negative, omega is positive and counterclockwise, and if the quotient is positive, omega is negative and clockwise.

Kappa can take on any value between 0 and $360°$. If m_{11} is positive, then $\cos \kappa$ is positive because $\cos \phi$ is always positive. Also if the quotient m_{21}/m_{11} is negative, $\tan \kappa$ is positive. Now if both the cosine and tangent of kappa are positive, then kappa is in quadrant I and between 0 and $90°$. Using similar analyses, the following table was derived for establishing the proper quadrant of kappa:

Algebraic sign of m_{11}	Algebraic sign of m_{21}/m_{11}	Quadrant of κ
+	−	I
−	+	II
−	−	III
+	+	IV

REFERENCES

American Society of Photogrammetry: "Manual of Photogrammetry," 3d ed., Falls Church, Va., 1966, chap. 2.

——: "Manual of Photogrammetry," 4th ed., Falls Church, Va., 1980, chap. 2.

Keller, M., and G. C. Tewinkel: "Space Resection in Photogrammetry," ESSA Technical Report C&GS 32, U.S. Coast annd Geodetic Survey, Washington, D.C., 1966.

PROBLEMS

C-1 State the condition of collinearity in photogrammetry.

C-2 Explain why linearized collinearity equations must be iterated a number of times before a satisfactory solution is achieved.

C-3 For the following matrix M, what are the values of omega, phi, and kappa? What are azimuth, tilt, and swing?

$$M = \begin{bmatrix} -0.922345 & 0.386341 & -0.004423 \\ -0.386333 & -0.922054 & 0.023741 \\ 0.005094 & 0.023606 & 0.999708 \end{bmatrix}$$

C-4 Repeat Prob. C-3, except that the matrix M is composed of the following values:

$$M = \begin{bmatrix} -0.918650 & 0.392071 & -0.048606 \\ -0.392262 & -0.919834 & -0.005954 \\ -0.047044 & 0.013597 & 0.998800 \end{bmatrix}$$

INDEX